INEQUALITIES IN
ANALYSIS AND
PROBABILITY

Third Edition

INEQUALITIES IN
ANALYSIS AND
PROBABILITY

Third Edition

Odile Pons

National Institute for Agronomical Research, France

World Scientific

NEW JERSEY · LONDON · SINGAPORE · BEIJING · SHANGHAI · HONG KONG · TAIPEI · CHENNAI · TOKYO

Published by

World Scientific Publishing Co. Pte. Ltd.
5 Toh Tuck Link, Singapore 596224
USA office: 27 Warren Street, Suite 401-402, Hackensack, NJ 07601
UK office: 57 Shelton Street, Covent Garden, London WC2H 9HE

Library of Congress Cataloging-in-Publication Data
Names: Pons, Odile, author.
Title: Inequalities in analysis and probability / Odile Pons,
    National Institute for Agronomical Research, France.
Description: Third edition. | New Jersey : World Scientific Publishing Co. Pte. Ltd., [2022] |
    Includes bibliographical references and index.
Identifiers: LCCN 2021031175 | ISBN 9789811231346 (hardcover) |
    ISBN 9789811231353 (ebook)
Subjects: LCSH: Inequalities (Mathematics)
Classification: LCC QA295 .P66 2022 | DDC 512.9/7--dc23
LC record available at https://lccn.loc.gov/2021031175

British Library Cataloguing-in-Publication Data
A catalogue record for this book is available from the British Library.

For any available supplementary material, please visit
https://www.worldscientific.com/worldscibooks/10.1142/12135#t=suppl

Printed in Singapore

Preface

The third edition proves new inequalities for martingales and their supremum; thus improving the classical inequalities. These are presented in two different chapters, for discrete and time-continuous martingales. They are applied to the weak convergence of (local) martingales and semi-martingales and to Gaussian and point processes. The chapter on stochastic integration is completed and new results are established for local times.

February 2021

Preface of the Second Edition. The most important changes made in this edition are the insertion of two chapters: Chapter 5 on stochastic calculus with first-order differentiation and exponential (sub)-martingales, and Chapter 7 on time-continuous Markov processes, the renewal equations and the Laplace transform of the processes. I have added to Chapter 4 a section on the p-order variations of a process and modified the notations to extend the inequalities to martingales with a discontinuous process of quadratic variations, the notations of the previous edition were not modified for a continuous martingale and for a point process with a continuous predictable compensator. I have also corrected a few misprints and errors.

Examples of Poisson and Gaussian processes illustrate the text and their investigation leads to general results for processes with independent increments and for semi-martingales.

February 2016

Preface of the First Edition. The inequalities in vector spaces and functional Hilbert spaces are naturally transposed to random variables, martingales and time indexed stochastic processes with values in Banach spaces. The inequalities for transforms by convex functions are examples

v

of the diffusion of simple arithmetic results to a wide range of domains in mathematics. New inequalities have been developed independently in these fields. This book aims to give an account of inequalities in analysis and probability and to complete and extend them.

The introduction gives a survey of classical inequalities in several fields with the main ideas of their proofs and applications of the analytic inequalities to probability. This is not an exhaustive list. They are compared and sometimes improved with simple proofs. Further developments in the literature are mentioned. The book is organized according to the main concepts and it provides new inequalities for sums of random variables, their maximum, martingales, Brownian motions and diffusion processes, point processes and their suprema.

The emphasis on the inequalities is aimed at graduate students and researchers having the basic knowledge of courses in Analysis and Probability. The concepts of integration theory and of probabilities are supposed to be known, so the fundamental inequalities in these domains are acquired and references to other publications are added to complete the topic whenever possible. The book contains many proofs, in particular basic inequalities for martingales with discrete or continuous parameters in detail and the progress in several directions are easily accessible to the readers. They are illustrated by applications in probability.

I undertook this work in order to simplify the approach of uniform bounds for stochastic processes in functional classes. In the statistical applications, the assumptions for most results of this kind are specific to another distance than the uniform distance. Here, the results use inequalities of Chapter 4 between the moments of martingales and those of their predictable variations, then the conditions and the constants of the probabilistic bound differ from those of the other authors. During the preparation of the book, I added other inequalities while reading papers and books containing errors and unproved assertions; it should therefore fill some gaps. It does not cover the convex optimization problems and the properties of their solutions. It can be used as an introduction to more specific domains of the functional analysis or probability theory and as a reference for new applications to the asymptotic behaviour of non-standard empirical processes in statistics. Several applications to the tail behaviour of processes are developed in the following chapters.

Odile M.-T. Pons
April 2012

Contents

Chapter 1

Preliminaries

1.1 Introduction

The origin of the inequalities for convex functions is the inequalities in real vector spaces which have been extended to functional spaces by limits in Lebesgue integrals. They are generalized to inequalities for the tail distribution of sums of independent or dependent variables, under conditions for the convergence of their variance, and to inequalities for the distribution of martingales indexed by discrete or continuous sets. These inequalities are the decisive arguments for bounding series, integrals or moments of transformed variables and for proving other inequalities.

The convergence rate of sums of variables with mean zero is determined by probability inequalities which prove that a sum of variables normalized by the exact convergence rate satisfies a compactness property. If the normalization has a smaller order than its convergence rate, the upper bound of the inequality is one and it tends to zero if the order of the normalization is larger.

Many probability results are related to the Laplace transform, such as Chernoff's large deviations theorem, Bennett's inequalities and other exponential inequalities for sums of independent variables. This subject has been widely explored since the review papers of the Sixth Berkeley Symposium in Mathematical Statistics and Probability (1972) which covers many inequalities for martingales, Gaussian and Markov processes and the related passage problems and sojourn times. Some of them are revisited and extended after a brief review in this chapter. The upper bounds for the tail probability of the maximum of n variables depend on n, in the same way, the tail probability of the supremum of functional sums have upper bounds depending on the dimension of the functional classes.

1.2   Cauchy and Hölder inequalities

Inequalities for finite series were first written as inequalities in a vector space V provided with an Euclidean norm $\|x\|$, for x in V. The scalar product of x and y in V is defined from the norm by

$$< x, y > = \frac{1}{2}\{\|x + y\|^2 - \|x\|^2 - \|y\|^2\} \tag{1.1}$$

and, conversely, an Euclidean norm is a ℓ_2-norm related to the scalar product as

$$\|x\| = < x, x >^{\frac{1}{2}}.$$

From the definition (1.1), the norms of vectors x and y of an Euclidean vector space V satisfy the geometric equalities

$$\|x + y\|^2 + \|x - y\|^2 = 2(\|x\|^2 + \|y\|^2) \tag{1.2}$$
$$\|x + y\|^2 - \|x - y\|^2 = 4 < x, y >.$$

The space $\ell_2(V)$ is the space of series of V with a finite Euclidean norm. An orthonormal basis $(e_i)_{1 \leq i}$ of V is defined by the orthogonality property $< e_i, e_j > = 0$ for $i \neq j$ and by the normalization $\|e_i\| = 1$ for $i \geq 1$. Let V_n be a vector space of dimension n, for example $V_n = \mathbb{R}^n$, for an integer n and V_∞ be its limit as n tends to infinity. Every vector x of $\ell_2(V_n)$, $n \geq 1$, is the sum of its projections in the orthonormal basis

$$x = \sum_{i=1}^{n} < x, e_i > e_i,$$

its coordinates in the basis are $x_i = < x, e_i >$, $i = 1, \ldots, n$, and its norm is $\|x\|_2 = (\sum_{i=1}^{n} x_i^2)^{\frac{1}{2}}$. In $\ell_2(V_\infty)$, a vector x is the limit as n tends to infinity of $\sum_{i=1}^{n} < x, e_i > e_i$ and its norm is the finite limit of $(\sum_{i=1}^{n} x_i^2)^{\frac{1}{2}}$ as n tends to infinity. The space $\ell_p(V_n)$, $1 \leq p < \infty$, is defined with respect to the norm

$$\|x\|_p = \left(\sum_{i=1}^{n} |x_i|^p\right)^{\frac{1}{p}}$$

and the space $\ell_\infty(V_n)$ is the space of vector with a finite uniform norm $\|x\|_\infty = \max_{1 \leq i \leq n} |x_i|$. In $\ell_p(V_\infty)$ and $\ell_\infty(V_\infty)$, the norms are defined as the limits of the norms of $\ell_p(V_n)$ as n tends to infinity. The norms $(\|x\|_p)_{0 < p \leq \infty}$ are an increasing sequence for every x in a vector space V. The triangular inequality is $\|x + y\| \leq \|x\| + \|y\|$ with equality if and only if $< x, y > = 0$. Consequently, for all x and y in a vector space $|\|x\| - \|y\|| \leq \|x - y\|$.

The Cauchy inequality (1821) in the vector space $V_n = \mathbb{R}^n$, for an integer n, or V_∞ is

$$< x, y > \leq \|x\|_2 \|y\|_2$$

for all x and y in V_n, with equality if and only if x and y are proportional. It is established recursively from the triangular inequality in V_2.

All norms are equivalent in an n-dimensional vector space V_n: For every x in $\mathbb{R}^n$ and for $1 \leq p, q \leq \infty$, there exist constants $c_{p,q,n}$ and $C_{p,q,n}$ depending only on n, p and q, such that $c_{p,q,n}\|x\|_p \leq \|x\|_q \leq C_{p,q,n}\|x\|_p$

$$\|x\|_\infty \leq \|x\|_p \leq n^{\frac{1}{p}}\|x\|_\infty,$$
$$n^{-1}\|x\|_1 \leq \|x\|_\infty \leq \|x\|_1,$$
$$n^{-\frac{1}{p}}\|x\|_p \leq \|x\|_1 \leq n^{\frac{1}{p'}}\|x\|_p, \qquad (1.3)$$
$$n^{-\frac{1}{p'}}\|x\|_1 \leq \|x\|_p \leq n^{\frac{1}{p}}\|x\|_1,$$
$$n^{-\frac{1}{p'}-\frac{1}{q}}\|x\|_q \leq \|x\|_p \leq n^{\frac{1}{p}+\frac{1}{q'}}\|x\|_q,$$

with $p^{-1} + p'^{-1} = 1$ and $q^{-1} + q'^{-1} = 1$.

Extensions of Cauchy's inequality to bilinear series have been studied by Hilbert, who proved that for positive real series $(x_n)_{n \geq 1}$ in ℓ_p and $(y_m)_{m \geq 1}$ in ℓ_q

$$\sum_{n \geq 1} \sum_{m \geq 1} \frac{x_n y_m}{n+m} \leq \frac{\pi}{\sin(p^{-1}\pi)} \| \sum_{i \leq n} x_i \|_p \| \sum_{j \leq m} y_j \|_{p'}$$

where p and p' are conjugates such that $p^{-1} + p'^{-1} = 1$. Other examples are given by Hardy, Littlewood and Pólya (1952).

Cauchy's inequality is extended to an inequality for integrals with respect to the Lebesgue measure on $\mathbb{R}$. Let f and g be square-integrable functions in $L^2(\mathbb{R})$, the Cauchy-Schwarz inequality is

$$\left| \int_{\mathbb{R}} f(x)g(x)\,dx \right| \leq \left(\int_{\mathbb{R}} f^2(x)\,dx \right)^{\frac{1}{2}} \left(\int_{\mathbb{R}} g^2(x)\,dx \right)^{\frac{1}{2}},$$

with equality if and only if f and g are proportional. Let μ be a positive measure on $\mathbb{R}$ and w be a positive weighting function, the Cauchy-Schwarz weighted inequality is

$$\left| \int_{\mathbb{R}} wfg\,d\mu \right| \leq \left(\int_{\mathbb{R}} wf^2\,d\mu \right)^{\frac{1}{2}} \left(\int_{\mathbb{R}} wg^2\,d\mu \right)^{\frac{1}{2}},$$

with equality under the same condition. A simple proof for both inequalities relies on the inequality $\int_{\mathbb{R}} (tf(x) - g(x))^2 w(x)\,d\mu(x) \geq 0$ which develops as an equation of the second-order with respect to t, with a negative discriminant $(\int wfg\,d\mu)^2 - (\int wf^2\,d\mu)(\int wg^2\,d\mu)$.

It is extended to the Hölder inequalities for L^p-integrable real functions with a finite or infinite support $[a,b]$, where $-\infty \leq a < b \leq \infty$. For every real $p > 1$, let $L^{p,\mu}(a,b) = \{f : (\int_a^b |f|^p\,d\mu)^{\frac{1}{p}} < \infty\}$, if $1 \leq p < \infty$, and $L^\infty(a,b) = \{f : \sup_{(a,b)} |f| < \infty\}$. Let p and q be conjugates and let f in $L^{p,\mu}(a,b)$ and g in $L^{q,\mu}(a,b)$, Hölder inequality is a consequence of the inequality $uv \leq p^{-1}u^p + q^{-1}v^p$ for (u,v) in $\mathbb{R}_+^2$ such that $\|u\|_{L^p} = 1$ and $\|v\|_{L^q} = 1$

$$\int_{\mathbb{R}} |fg|\,d\mu \leq \left(\int_{\mathbb{R}} |f|^p\,d\mu\right)^{\frac{1}{p}} \left(\int_{\mathbb{R}} |g|^q\,d\mu\right)^{\frac{1}{q}}. \tag{1.4}$$

Let $r \geq 1$ and let p and q be conjugates such that $p^{-1} + q^{-1} = r^{-1}$ Hölder inequality generalizes as

$$\left(\int_{\mathbb{R}} |fg|^r\,d\mu\right)^{\frac{1}{r}} \leq \left(\int_{\mathbb{R}} |f|^p\,d\mu\right)^{\frac{1}{p}} \left(\int_{\mathbb{R}} |g|^q\,d\mu\right)^{\frac{1}{q}}. \tag{1.5}$$

The L^p norms are increasing. This implies Khintchine's inequality for $p \geq 1$: Let $(r_k)_{k\geq 0}$ be the Rademacher functions on $[0,1]$, for every measurable function $f = \sum_{k\geq 0} a_k r_k$ on $[0,1]$ satisfying $\sum_{k\geq 0} a_k^2 < \infty$, there exist constants $A_p > 0$ and B_p such that

$$A_p\left(\sum_{k\geq 0} a_k^2\right)^{\frac{1}{2}} \leq \|f\|_p \leq B_p \sum_{k\geq 0} (a_k^2)^{\frac{1}{2}}. \tag{1.6}$$

Since the constants do not depend on the dimension of the projection of the function onto the basis, (1.6) implies the equivalence between the norms of the function, $A_p\|f\|_2 \leq \|f\|_p \leq B_p\|f\|_2$.

Inequalities for a countable sequence $x = (x_i)_{i\geq 0}$ in the space $\ell_p(\mathbb{R}^\infty)$ are deduced from functional inequalities for piecewise constant integrable functions of a space $L^{p,\mu}$ or by setting $x_i = \int_{a_{i-1}}^{a_i} f\,d\mu$ for some function f of $L^{p,\mu}$ and $\cup_{i\geq 0}(a_{i-1}, a_i) = (a,b)$. Let $r < p$ and let q be the conjugate of p such that $r^{-1} = p^{-1} + q^{-1}$, then $\|xy\|_r \leq \|x\|_p \|y\|_q$.

The Cauchy and Hölder inequalities are recursively generalized to more than two series or functions. Let $k \geq 2$ and let $p_1, \ldots, p_k \geq 1$ such that $\sum_{i=1}^k p_k^{-1} = r^{-1}$ and let $f_1, \ldots, f_k$ be functions in $L^{p_1,\mu}(a,b), \ldots, L^{p_k,\mu}(a,b)$ respectively, the Hölder inequality for the functions $f_1, \ldots, f_k$ is

$$\left(\int_a^b |\prod_{i=1}^k f_i|^r\,d\mu\right)^{\frac{1}{r}} \leq \prod_{i=1}^k \|f_i\|_{p_i,\mu} \tag{1.7}$$

with equality if and only if the functions f_i are proportional.

The analogous Cauchy inequality for a set of k series $(x_{ij})_{1 \leq i \leq n, 1 \leq j \leq k}$ of L^{p_k}, respectively, is obtained with piecewise constant functions

$$\Big(\sum_{i=1}^{n} | \prod_{j=1}^{k} x_{ij} |^r \Big)^{\frac{1}{r}} \leq \prod_{j=1}^{k} \|x_j\|_{p_j}. \tag{1.8}$$

It is extended to infinite arrays $(x_{ij})_{1 \leq i < \infty, 1 \leq j < \infty}$.

The Hölder inequality is a consequence of the concavity of the function $\varphi_\alpha(u, v) = u^\alpha v^{1-\alpha}$ with $\alpha = p^{-1}$ (Neveu 1970) and the same argument applies for the other inequalities. More generally, let f be a real function of $L^1(a, b)$, then for every convex function φ from $\mathbb{R}$ to $\mathbb{R}$, Jensen's inequality for integrals is

$$\varphi \Big(\int_a^b f \, d\mu \Big) \leq \int_a^b \varphi(f) \, d\mu$$

and it extends the property $\varphi(\sum_{i=1}^n \alpha_i a_i) \leq \sum_{i=1}^n \alpha_i \varphi(a_i)$ for convex functions in vector spaces, with $\sum_{i=1}^n \alpha_i = 1$, and the inequality is an equality if and only if the a_i are proportional. The Minkowski inequality is the additivity property of the norms in the spaces $\ell_p(\mathbb{R}^n)$, $1 \leq n \leq \infty$

$$\|x + y\|_p \leq \|x\|_p + \|y\|_p, 1 \leq p \leq \infty$$

and it is extended to $L^p(\mathbb{R})$

$$\|f + g\|_p \leq \|f\|_p + \|g\|_p, 1 \leq p \leq \infty,$$

with equality if and only if x and y and, respectively, f and g are proportional. In $\ell_p(\mathbb{R}^n)$, it is a consequence of the concavity and the homogeneity of the function $\varphi_p(x) = \|x\|_p$ and it generalizes to $L^p(\mathbb{R})$.

In the complex space $\mathbb{C}$, the Euclidean norm of $x = x_1 + ix_2$ is

$$\|x\| = (x\bar{x})^{\frac{1}{2}} = (x_1^2 + x_2^2)^{\frac{1}{2}},$$

with the complex conjugate $\bar{x} = x_1 - ix_2$, and the scalar product $< x, y >$ of x and y is still defined by (1.1). In the product spaces $\mathbb{C}^n$ and $\mathbb{C}^\infty$, let $x = (x_i)_{i \in I}$ and $y = (y_i)_{i \in I}$ be vectors of finite or countable dimension, where I is a subset of integers $I = \{1, \ldots, n\}$, $I = \mathbb{N}$ or $I = \mathbb{Z}$, its Euclidean norm is $\|x\| = (\sum_{i \in I} x_i \bar{x}_i)^{\frac{1}{2}}$ and $< x, y >$ follows by (1.1). The inequalities of Cauchy and Minkowski are written in the same form as in $\mathbb{R}^2$.

The Fourier transform of a variable X with density function f_X is defined as $\widehat{f}_X(\omega) = Ee^{2\pi i \omega X} = \int_{-\infty}^{\infty} e^{2\pi i \omega x} f_X(x) \, dx$. Its norms $L^p([0, 1])$ satisfy the properties

$$\|\widehat{f}_X\|_p \leq \|f_X\|_p,$$

$$\|\widehat{f}_X\|_1 \leq \|\widehat{f}_X\|_p^p \|\widehat{f}_X\|_{p'}^{p'}, \ p^{-1} + p'^{-1} = 1.$$

It is generalized to functions of $L^p(\mathbb{R}^d)$. The Fourier transform of the sum of independent variables is the product of their transforms, hence by Hölder's inequality $\|\widehat{f}_{X+Y}\|_r \leq \|\widehat{f}_X\|_p \|\widehat{f}_Y\|_{p'}$, $r^{-1} = p^{-1} + p'^{-1}$.

1.3   Inequalities for transformed series and functions

The Cauchy inequality has been extended by convex transformations of the series. A convex real function f on $\mathbb{R}$ satisfies

$$f\left(\frac{\sum_{i=1}^{k} \alpha_i x_i}{\sum_{i=1}^{k} \alpha_i}\right) \leq \frac{\sum_{i=1}^{k} \alpha_i f(x_i)}{\sum_{i=1}^{k} \alpha_i}$$

for every linear combination $\sum_{i=1}^{k} \alpha_i x_i$ with all $\alpha_i > 0$, or equivalently $f(\sum_{i=1}^{k} \alpha_i x_i) \leq \sum_{i=1}^{k} \alpha_i f(x_i)$ for every linear combination such that $\alpha_i > 0$ for every i and $\sum_{i=1}^{k} \alpha_i = 1$. With $k = 2$ and $\alpha > 0$, consider $x > y$ and f increasing at y then

$$0 \leq f(y + \alpha(x - y)) - f(y) \leq \alpha\{f(x) - f(y)\} \to 0$$

as α tends to zero. Otherwise, let f be decreasing at x, $0 \leq f(x - (1 - \alpha)(x - y)) - f(x) \leq (1 - \alpha)\{f(y) - f(x)\}$ which tends to zero as α tends to 1, so every convex function is continuous.

Let f be increasing and belong to $C^2(\mathbb{R})$ and let y satisfy $f'(y) = 0$. Let α in $[0, 1]$ and $x > y$. As α tends to zero, a Taylor expansion of f in a neighbourhood of y is written $f(y + \alpha(x - y)) - f(y) = \frac{\alpha^2}{2}(x - y)^2 f''(y) + (\alpha^2)$ therefore $f''(y) > 0$. If f is a decreasing convex function of $C^2(\mathbb{R})$, its second derivative at y such that $f'(y) = 0$ is also strictly positive, so the second derivative of a convex function is strictly positive where its first derivative is zero and it is a minimum of the function. Conversely, a function f is concave if $-f$ is convex, its second derivative is strictly negative where its first derivative is zero and it is a maximum of the function. The polygons generate the space of convex functions.

The Hadamard inequality for a convex function f on a finite interval $[a, b]$ is

$$f\left(\frac{a + b}{2}\right) \leq \frac{1}{b - a} \int_a^b f(x)\, dx \leq \frac{f(a) + f(b)}{2}.$$

Cauchy (1821) proved other inequalities for convex functions of a countable number of variables, for example let $(x_i)_{i \geq 1}$ be a sequence of real numbers, then

$$\prod_{i \geq 1}(1 + x_i) < \exp\left\{\sum_{i \geq 1} x_i\right\}.$$

Let f be a positive and concave function on a subinterval $[a, b]$ of $\mathbb{R}_+$ and $(x_i)_{i=1,\dots,k}$ be k points in $[a, b]$ and let $\bar{x}_\alpha = \sum_{i=1}^{k} \alpha_i x_i$ be a weighted mean of these points, with $\sum_{i=1}^{k} \alpha_i = 1$. By concavity

$$\prod_{i=1}^{k} \{f(x_i)\}^{\alpha_i} \leq f(\bar{x}_\alpha), \tag{1.9}$$

with equality if $\alpha_1 x_1 = \cdots = \alpha_k x_k$. The exponential function is convex and the equality is fulfilled for all sequences $(x_i)_{i=1,\ldots,k}$ and $(\alpha_i)_{i=1,\ldots,k}$ then conditions for equality in (1.9) depend on the function f. The inequality (1.9) is also written $\sum_{i=1}^{k} \alpha_i \log f(x_i) \leq \log f(\bar{x}_\alpha)$ and it is satisfied by Jensen's inequality, due to the concavity of the logarithm function.

With $f(x) = 1 + x$, (1.9) implies

$$\prod_{i=1}^{k}(1 + x_i) \leq (1 + \bar{x}_k)^k,$$

with $\bar{x}_k = k^{-1} \sum_{i=1}^{k} x_i$. Applying this inequality with $x_i = 2i$ provides a bound for the $2k$-th moments $m_{2k} = 1.3\ldots(2k-1) = \{2^k(k)!\}^{-1}(2k)!$ of the normal distribution, $\prod_{i=0}^{k-1}(1 + 2i) \leq k^k$. This is a mean value bound but it is not very sharp. By the convexity of the exponential function, the converse inequality at points $(t_1, \ldots, t_k)$ yields $(1 + e^{\bar{t}_k})^k \leq \prod_{i=1}^{k}(1 + e^{t_i})$, which it is equivalent to

$$1 + \prod_{i=1}^{k} x_i^{\frac{1}{k}} \leq \prod_{i=1}^{k}(1 + x_i)^{\frac{1}{k}}.$$

Replacing x_i by $x_i y_i^{-1}$ and multiplying both sides by $\prod_{i=1}^{k} y_i^{\frac{1}{k}}$ implies

$$\prod_{i=1}^{k} x_i^{\frac{1}{k}} + \prod_{i=1}^{k} y_i^{\frac{1}{k}} \leq \prod_{i=1}^{k}(x_i + y_i)^{\frac{1}{k}}$$

for all sequences $(x_i)_{i=1,\ldots,k}$ and $(y_i)_{i=1,\ldots,k}$. More generally, let $(\alpha_1, \ldots, \alpha_k)$ be positive numbers such that $\sum_{i=1}^{k} \alpha_i = 1$, then

$$\prod_{i=1}^{k} x_i^{\alpha_i} + \prod_{i=1}^{k} y_i^{\alpha_i} \leq \prod_{i=1}^{k}(x_i + y_i)^{\alpha_i}.$$

For an array $(x_{ij})_{i=1,\ldots,k, j=1,\ldots,n}$, a recursive argument implies

$$n^{-1} \sum_{j=1}^{n} \prod_{i=1}^{k} x_{ij}^{\alpha_i} \leq \prod_{i=1}^{k} \bar{x}_i^{\alpha_i} \tag{1.10}$$

with the partial means $\bar{x}_i = n^{-1} \sum_{j=1}^{n} x_{ij}$. This inequality differs from the inequality (1.8) and from the Minkowski inequality for a sum of n series in $\mathbb{R}^k$, $\|\bar{x}\|_p \leq n^{-1} \sum_{j=1}^{n} \|x_j\|_p$, for $p \geq 1$.

With the convex function $f(t) = \log(1 + e^t)$, Equation (1.9) implies that for every $x_i > 0$, for $i = 1, \ldots, k$

$$\log\Big(1 + \prod_{i=1}^{k} x_i^{\alpha_i}\Big) \leq \prod_{i=1}^{k} \{\log(1 + x_i)\}^{\alpha_i}.$$

Replacing x_i by $x_i y_i^{-1}$, with $y_i > 0$ for $i = 1, \ldots, k$, and adding to the left member $\log \prod_{i=1}^{k} y_i^{\alpha_i} \leq \log \bar{y}_\alpha$ yields

$$\prod_{i=1}^{k} \log\{(x_i + y_i)^{\alpha_i}\} \leq \log \prod_{i=1}^{k} (x_i + y_i)^{\alpha_i} - \log \prod_{i=1}^{k} y_i^{\alpha_i} + \log \bar{y}_\alpha.$$

For every convex function φ from $\mathbb{R}$ to $\mathbb{R}$, Jensen's inequality for the integral of a function f with respect to the positive measure μ on $[a, b]$ is

$$\varphi\{\frac{1}{M(b) - M(a)} \int_a^b f \, d\mu\} \leq \frac{1}{M(b) - M(a)} \int_a^b \varphi \circ f \, d\mu,$$

with $M(a) = \mu(] - \infty, a])$ and $M(b) = \mu(] - \infty, b])$. As a consequence, for every real function f on $[a, b]$

$$\exp\left\{\frac{1}{M(b) - M(a)} \int_a^b \log f \, d\mu\right\} \leq \frac{1}{M(b) - M(a)} \int_a^b f \, d\mu.$$

Other integral inequalities for convex function can be found in Maurey (2004).

An integral equation similar to Equation (1.9) is obtained from Jensen's inequality for every concave function $f > 0$

$$\exp\left\{y^{-1} \int_0^y \log f \, d\mu\right\} \leq f\left\{y^{-1} \int_0^y x \, d\mu(x)\right\}, \quad y > 0. \tag{1.11}$$

The bound

$$y^{-1} \int_0^y f \, d\mu \leq f\left\{y^{-1} \int_0^y x \, d\mu(x)\right\}$$

is sharper and it is an equality for affine functions $f > 0$.

For example, with $f(x) = 1 + x$, Equation (1.11) is written

$$\int_0^y \log(1 + x) \, dx \leq y \log\left(1 + \frac{y}{2}\right)$$

for every $y > 0$. With $f(x) = x^\alpha$, with $x > 0$ and $0 < \alpha < 1$, (1.10) implies $\int_0^y \log x \, dx \leq y \log \frac{y}{2}$ and Jensen's inequality yields $y^\alpha \leq \alpha(\alpha + 1) \log \frac{y}{2}$.

Most inequalities for bounding finite series are intervals for the error of their approximation by a Taylor expansion or they are obtained by induction. For example, Cauchy (1833) defined the exponential function as

$$e^x = \lim_{n \to \infty, n\alpha \to x} (1 + \alpha)^n = \lim_{n \to \infty, n\alpha \to x} (1 - \alpha)^{-n}$$

which provided him the following interval

$$(1 + \alpha)^{\frac{x}{\alpha}} < e^x < (1 - \alpha)^{-\frac{x}{\alpha}},$$

for every real $\alpha > 0$. This definition entails immediately the main properties of the exponential. Reciprocally, the unique function ϕ which satisfies

$$\phi(x + y) = \phi(x)\phi(y),$$

for all x and y, is the exponential function $\phi(x) = \{\phi(1)\}^x \equiv e^x$.

Other classical bounds for functions have been established for the first n terms of their Taylor expansions. Alzer (1990b) studied a lower bound of the form $I_{n-1}(x)I_{n+1}(x) > c_n I_n^2(x)$ for the sum $I_n(x)$ of the first n terms in the Taylor expansion of the exponential function. More generally, Darboux's sums for a function f on an interval $[a, b]$ are defined using a partition $\pi_n = (a_k)_{k=0,\ldots,a_n}$ of increasing numbers of this interval such that $a = a_0$ and $a_n = b$

$$S(f, \pi_n) = \sum_{k=1}^{n} m_k(a_k - a_{k-1}),$$

$$T(f, \pi_n) = \sum_{k=1}^{n} M_k(a_k - a_{k-1}),$$

with $m_i = \inf_{a_{k-1} \leq x < a_k} f(x)$ and $M_i = \sup_{a_{k-1} \leq x < a_k} f(x)$ and the convergence of the sum $T(f, \pi_n)$ implies the convergence of the integral $I = \int_a^b f(x)\, dx$, conversely the convergence of I implies the convergence of the sum $S(f, \pi_n)$. The sum $1 + 3 + \cdots + (2k - 1)$ has then the bounds $k^2 - k - 1 \leq \sum_{i=0}^{k-1}(1 + 2i) \leq k^2 + k - 1$ and the product $m_{2k} = 1.3\ldots(2k-1) = \exp\{\sum_{i=0}^{k-1} \log(1 + 2i)\}$ satisfies

$$\prod_{i=0}^{k-1}(1 + 2i) \leq (1 + 2k)^{1+2k}e^{-2k} \leq e^{(2k)^2} = 2ke^{2k}, \qquad (1.12)$$

the inequality $m_{2k} \leq k^k$ obtained from (1.9) is therefore better than the upper bound of (1.12) for m_{2k}. Close approximations by Darboux's sums require a thin partition π_n.

1.4   Applications in probability

Let $(\Omega, \mathcal{F}, P)$ be a probability space and let X be a real variable defined on $(\Omega, \mathcal{F}, P)$, with distribution function F respectively. The norms $L^p(\mathbb{R})$ define norms for variables

$$\|X\|_p = (E|X|^p)^{\frac{1}{p}} = \left\{ \int_{\mathbb{R}} |x|^p dF(x) \right\}^{\frac{1}{p}},$$

$$\|X\|_\infty = \sup\{x : P(|X| > x) > 0\}.$$

Let X be a random vector of $\mathbb{R}_+^n$, by concavity of the norm, we have $E\|X\|_p \leq \|EX\|_p$ for every $p \geq 1$. The sequence $(\|X\|_p)_{0<p\leq\infty}$ is increasing and the additive and multiplicative properties of the norms are consequences of the Hölder and Minkowski inequalities. For dependent variables X and Y, the Cauchy-Schwarz inequality entails

$$E(XY) \leq \|X\|_p\|Y\|_q$$

for all conjugates $p, q \geq 2$. Let $(X_i)_{i\geq1}$ be a sequence of random variables in a vector space V, $\|\max_{i=1,...,n} X_i\| \leq \max_{i=1,...,n} \|X_i\|$ and the inequality extends as n tends to infinity.

The density of the sum $X+Y$ of independent variables with distribution functions F and G and densities f and g is the convolution

$$f * g(t) = \int_{\mathbb{R}} f(t-s)\,dG(s) = \int_{\mathbb{R}} g(t-s)\,dF(s).$$

The density of the ratio XY^{-1} of independent variables X and $Y > 0$ is the multiplicative convolution $h(t) = \int_{\mathbb{R}} f(s^{-1}t)\,dG(s)$, its norms has the bound $\|h\|_r \leq \|f\|_p\|g\|_q$, for all p, q, r such that $p^{-1} + q^{-1} = r^{-1}$, $1 \leq p \leq \infty$. If the variables have no densities, the convolution is $P(X + Y \leq t) = \int_{\mathbb{R}} G(t-s)\,dF(s) = \int_{\mathbb{R}} F(t-s)\,dG(s)$ and it is sufficient that F or G is continuous (respectively, has a density) to ensure the continuity (respectively, the derivability) of the distribution function of their sum. Their ratio has the distribution function $\int_{\mathbb{R}} F(s^{-1}t)\,dG(s)$.

The inequalities of Section 1.3 imply the following inequalities for the sequence of variables. Let $(X_i)_{i=1,...,n}$ be a sequence of random variables and let $\bar{X}_n = n^{-1}\sum_{i=1}^n X_i$ be their empirical mean, Cauchy's inequality implies that for every integer n

$$E\left\{n^{-1}\sum_{i=1}^n \log(1 + X_i)\right\} < E\{\bar{X}_n\}.$$

Let $(\alpha_i)_{i=1,...,n}$ be a real sequence such that $\sum_{i=1}^n \alpha_i = 1$. For every positive concave function f on $\mathbb{R}_+$, inequality (1.9) entails

$$E\left\{\sum_{i=1}^n \alpha_i \log f(X_i)\right\} \leq E \log f\left(\sum_{i=1}^k \alpha_i X_i\right) \quad ,$$

in particular $E\{n^{-1}\sum_{i=1}^n \log f(X_i)\} \leq E \log f(\bar{X}_n)$. By Jensen's inequality, for every convex function φ and for every variable X

$$\varphi(EX) \leq E\varphi(X).$$

Bienaymé-Chebychev's inequality has been formulated for a variable X of $L^2(\mathbb{R})$ as $P(|X| > a) \leq a^{-2}E|X|^2$, $a > 0$. For a variable X in $L^p(\mathbb{R})$, $1 \leq p < \infty$, and such that $EX = m$, it is

$$P(|X - m| > a) \leq a^{-2}\text{Var}X, \text{ for every } a > 0, \tag{1.13}$$

where $\text{Var}X = E(X - EX)^2$ is the variance of X. It extends to $p \geq 2$

$$P(X - m > a) \leq a^{-p}E|X - m|^p, \text{ for every } a > 0.$$

Another Chebychev's inequality for the mean of n variables with mean m and variance σ^2 is

$$P(|\bar{X} - m| > a) \leq \frac{1}{1 + \sigma^{-2}na^2}, \ a > 0.$$

Kolmogorov's inequalities for the sum $S_n = \sum_{i=1}^{n} X_i$ of n independent random variables $(X_1, \ldots, X_n)$ are

$$P\left(\sup_{1 \leq k \leq n} |S_k - E(S_k)| \geq a\right) \leq a^{-2}\text{Var}S_n, \text{ if } X_i \in L^2,$$

$$P\left(\sup_{1 \leq k \leq n} |S_k - E(S_k)| \geq a\right) \leq 1 - \frac{(a + k)^2}{E(S_n^2)}, \text{ if } \|X_i\|_\infty \leq k.$$

The Laplace transform of a variable X with distribution function F is

$$L_X(t) = Ee^{-tX} = Lf_X(t)$$

where f_X is the density of X and by convexity, $L_X(t) \geq e^{-tEX}$. Its moment generating function is defined as $G_X(t) = E(t^X)$ or $\varphi_X(t) = Ee^{tX}$. The moments $\mu_k = EX^k$ of the variable X are obtained from the value at zero of the derivatives of L_X or G_X, $\mu_k = (-1)^k L_X^{(k)}(0) = \varphi_X^{(k)}(0)$ and $G_X^{(k)}(0) = E\{X(X-1)\cdots(X-k+1)\}$, $k \geq 1$. The inversion formula is the same for φ_X and L_X. As a consequence of the expansion of the exponential, at every point of continuity x of F_X

$$F_X(x) = \lim_{\lambda \to \infty} \sum_{k \leq \lambda x} \frac{(-1)^k \lambda^k}{k!} \varphi_X^{(k)}(\lambda).$$

Let $(X_i)_{i=1,\ldots,n}$ be a sequence of independent random variables with means zero and respective variances σ_i^2 and let $S_n = \sum_{i=1}^{n} X_i$, $\bar{X}_n = n^{-1}S_n$ and $V_n = \sum_{i=1}^{n} \sigma_i^2$. If $\bar{\sigma}_n^2 = n^{-1}V_n$ converges to a finite limit $\sigma^2 > 0$, then for every $\beta > 0$ and $a > 0$, the inequality (1.13) implies

$$P(n^{\frac{\beta}{2}}|\bar{X}_n| > a) \leq n^\beta a^{-2}E\bar{X}_n^2 = n^{\beta-1}\bar{\sigma}_n^2.$$

As the upper bound tends to zero for $\beta < 1$, the inequality cannot be satisfied for every a. The upper bound being infinity for every $\beta > 1$, the convergence rate of the mean $\bar{X}_n = n^{-1}\sum_{i=1}^{n} X_i$ to zero is $\beta = 1$. For independent variables with respective means μ_i, the convergence rate of $n^{-\alpha}S_n$ to a limit μ is determined by the convergence rates of the mean $\mu_n = n^{-\alpha}\sum_{i=1}^{n}\mu_i$ to μ and of the mean variance $\sigma_n^2 = n^{-\beta}V_n$ to a limit $\sigma^2 > 0$. For every $a > 0$

$$P(n^{\frac{\beta}{2}}|n^{-\alpha}S_n - \mu| > a) \leq n^{2(\beta-\alpha)}a^{-2}n^{-\beta}V_n,$$

therefore the convergence rates α and β are equal.

Chow and Lai (1971) proved the following equivalences for a sequence of independent and identically distributed random variables $(X_i)_{i \geq 1}$ with mean $EX_1 = 0$, for α in $]0, 1[$

$$E \exp(t|X_1|^\alpha) < \infty, \, t > 0,$$

$$\lim_{n \to \infty} (\log n)^{\alpha^{-1}} X_n = 0, \text{ a.s.},$$

$$\lim_{n \to \infty} (\log n)^{\alpha^{-1}} \sum_{i=1}^{n} c_{n-i}X_i = 0, \text{ a.s.},$$

for a sequence of weighting constants such that $c_n = O(n^{-\nu})$, with $\nu > \frac{1}{2}$. Other equivalences between convergences proved by Chow and Lai (1971) are given in the appendix.

For every $a > 0$, Bienaymé-Chebychev's inequality for the maximum of n positive centered variables of L^2 with the same variance σ^2 is

$$\limsup_{n \to \infty} P\left(n^{-\frac{1}{2}} \max_{k=1,\ldots,n} |S_k| > a\right) \leq \frac{\sigma^2}{a^2}.$$

Billingsley (1968) proved many inequalities for partial sums and their moments and for the empirical process of independent random variables. In particular $E(|S_n - S_k|^2|S_k - S_j|^2) = (V_n - V_k)(V_k - V_j)$, $0 \leq j \leq k \leq n$, the variable

$$M'_n = \max_{0 \leq k \leq n} \min\{|S_k|, |S_n - S_k|\}$$

satisfies the inequalities

$$M'_n \leq M_n \leq M'_n + |S_n|, \quad M'_n \leq M_n \leq M'_n + 3 \max_{0 \leq k \leq n} |X_k|$$

and for every $\lambda > 0$, there exists a constant K such that

$$P(M'_n > \lambda) \leq K\lambda^{-4}V_m^4.$$

For dependent variables, the variance of S_n is bounded by a sum which also depends on the means $E|X_iX_j|$ for every $i \neq j$ and a mixing coefficient

η_{j-i} determines the convergence rate of this sum through the inequality $E|X_iX_j| \le \eta_{j-i}\sigma_i\sigma_j$. If there exist constants $\alpha > 0$ and $\beta > 0$ such that $\mu_n = n^{-\alpha}\sum_{i=1}^{n}\mu_i$ converges to μ and $n^{-\beta}\sum_{i=1,\dots,n}\sum_{j=1,\dots,n}\eta_{j-i}\sigma_i\sigma_j$ converges to a limit $\sigma^2 > 0$, then the convergence rate of $n^{-\alpha}S_n$ to μ is still $n^{-\beta}$.

The Berry-Essen inequality for independent variables concerns the convergence of the distribution of a normalized sum of variables $V_n^{-\frac{1}{2}}S_n$ to normal variable with distribution function Φ. If $E|X_i|^3$ is finite, there exists a constant K such that

$$\sup_{x\in\mathbb{R}}|P(V_n^{-\frac{1}{2}}(S_n - EX) \le x) - \Phi(x)| \le K\frac{\sum_{i=1}^{n}E|X_i - EX|^3}{V_n^{\frac{3}{2}}}.$$

For dependent variables with mixing coefficients such that $\sum_{k>0}k^2\eta_k$ is finite, the convergence rate of S_n is $n^{\frac{1}{2}}$ (Billingsley, 1968) and the Berry-Essen inequality is satisfied.

The Fourier transform of the correlation function $R_X(t) = E\{X_0X_t\}$ of a stationary process X is defined as its spectral density

$$\widehat{f}_X(\omega) = \int_{-\infty}^{\infty} R_X(t)e^{2\pi\omega t}\,dt.$$

Inverting the mean and the integrals, its norms satisfy the properties

$$\|\widehat{f}_X\|_p \le \|R_X\|_p,$$
$$\|\widehat{f}_X\|_1 \le \|\widehat{f}_X\|_p^p\|\widehat{f}_X\|_{p'}^{p'}, \quad \text{where } p^{-1} + p'^{-1} = 1.$$

1.5 Hardy's inequality

Hardy's inequalities have been developed in different forms presented in the monograph by Hardy, Littlewood and Pólya (1952). The functional inequality is a consequence of the Hölder inequality. Let $p > 1$ and let f be a positive function of $L^p(\mathbb{R}_+)$, with primitive F and such that $\lim_{x\to 0} x^{\frac{1}{p}-1}F(x) = 0$ and $\lim_{x\to\infty} x^{\frac{1}{p}-1}F(x) = 0$. The integral

$$I_p(y) := \int_0^y \left\{\frac{F(x)}{x}\right\}^p dx = \frac{p}{p-1}\int_0^y \left\{\frac{F(x)}{x}\right\}^{p-1} f(x)\,dx - \frac{1}{p-1}\frac{F^p(y)}{y^{p-1}}, \tag{1.14}$$

and $I_p = I_p(\infty)$ have the bounds

$$I_p \le \left(\frac{p}{p-1}\right)^p \int_0^\infty f^p(x)\,dx, \quad 1 < p < \infty, \tag{1.15}$$

and the inequality is strict except if f is a constant. With $p = 1$, for every function $f > 0$ on $\mathbb{R}_+$ having a primitive F such that $\lim_{x \to 0} f(x) = 0$, $\lim_{x \to \infty} x^{-1} F(x) = 0$, the integral cannot be calculated in the same way, however we have

$$I_1 = \int_0^\infty \frac{F(x)}{x^2} \, dx = \int_0^\infty \frac{f(x)}{x} \, dx.$$

If $0 < p < 1$, integrating by parts yields

$$I_p(y) = \frac{1}{1-p} \left[p \int_0^y \left\{ \frac{F(x)}{x} \right\}^{p-1} f(x) \, dx - y^{1-p} F^p(y) \right]$$

where

$$\int_0^y \left\{ \frac{F(x)}{x} \right\}^{p-1} f(x) \, dx \geq I_p^{1-\frac{1}{p}}(y) \|f\|_p.$$

Since $y^{1-p} F^p(y) > 0$ and tends to infinity as y tends to infinity, the inequality (1.15) cannot be inversed for $p < 1$.

For positive real series $(a_i)_{i \geq 1}$, let $A_n = \sum_{i=1}^n a_i$ and let $p > 1$ such that $(a_i)_{i \geq 1}$ belongs to ℓ_p. Hardy's inequality is

$$\sum_{n \geq 1} \left(\frac{A_n}{n} \right)^p < \left(\frac{p}{p-1} \right)^p \sum_{i \geq 1} a_i^p.$$

A special case is Euler's limit of the series $\sum_{k \geq 1} k^{-2} = \frac{\pi^2}{6}$ (Euler, 1735) which was obtained as the coefficient of x^2 in the expansion of the function $x^{-1} \sin x$ as a polynomial with roots $\pm k\pi$. Integrating the approximation of $(1-x)^{-1}$ by $(1-x^n)(1-x)^{-1}$, with normalization, it appears that

$$\int_0^1 \frac{1}{x} \left(\int_0^x \frac{1-t^n}{1-t} \, dt \right) dx = \frac{\pi^2}{6}$$

(Bradley, d'Antonio and Sandifer, 2007; Dunham, 1999; Sandifer, 2007). It follows that $\sum_{n \geq 1} n^{-2} (\sum_{k=1}^n k^{-1})^2 < \frac{2}{3} \pi^2$. Hardy and Littlewood (1930) extended the inequality to a function $f(z) = f(re^{i\theta})$ with maximum on the radius $F(\theta) = \sup_{r > 0} |f(z)|$, for $p > 1$

$$\int \frac{F^p(\theta)}{x^p} \, d\theta \leq \left(\frac{p}{p-1} \right)^p \int f^p(z) \, dz.$$

Bickel *et al.* (1993) mentioned a statistical application of the inequality to the derivative of the density of a randomly right-censored time variable. Let S be a positive real random variable with distribution function H and density h, for every square integrable function a such that $\int a \, dH = 0$

$$E \left(\frac{\int_S^\infty a \, dH}{H(S)} \right)^2 \leq 4 E a^2(S).$$

Kufner and Persson (2003) and others later proved weighted inequalities of the Hardy type. Pachpatte (2005) also provided a review of the classical inequalities of the same type and several generalizations of the inequality

$$\int_0^x \int_0^y \frac{f(s)g(t)}{s+t} \, ds \, dt \le pq \frac{\sqrt{xy}}{2} \left\{ \int_0^x (x-s){f'}^2(s) \, ds \int_0^y (y-t){g'}^2(t) \, dt \right\}^{\frac{1}{2}}.$$

Let f be a complex function defined on $\mathbb{R}^d$, with the Lebesgue measure λ_d and let $B_r(x)$ be the ball of $\mathbb{R}^d$, centred at x and with radius r. The maximum average value of f is

$$Mf(x) = \sup_{r>0} \frac{1}{\lambda_d(B_r(x))} \int_{B_r(x)} |f| \, d\lambda_d.$$

Hardy-Littelwood's weak inequality for Mf is

$$\lambda_d\{x \in \mathbb{R}^d : Mf(x) > a\} \le a^{-1} C_d \|f\|_{L^1(\mathbb{R}^d)}, \text{ for every } a > 0,$$

where C_d is a constant. A stronger inequality for Mf is due to Marcinkiewicz for every $p > 1$

$$\|Mf\|_{L^p(\mathbb{R}^d)} \le C_{d,p} \|f\|_{L^p(\mathbb{R}^d)}$$

where $C_{d,p}$ is a constant.

Another type of analytical inequalities for integrals is the Mean Value Theorem (Lagrange) and its extensions to higher order approximations. Hardy, Littlewood and Pólya (1952) proved a bound for the difference between the integral of the square of a continuously differentiable function f on $[0, 1]$ and the square of its integral

$$0 \le \int_0^1 f^2(x) \, dx - \left\{ \int_0^1 f(x) \, dx \right\}^2 \le \frac{1}{2} \int_0^1 x(1-x){f'}^2(x) \, dx.$$

Ostrowski's type inequalities (Sahoo and Riedel, 1998, Mitrinović, Pecarić and Fink 1957, Dragomir and Sofo 2000, Dragomir and Rassias 2002) have been written for the approximation of a continuously differentiable function f defined in an interval (a, b), having a derivative bounded by a constant M, in the form

$$\left| f(x) - \frac{1}{b-a} \int_a^b f(t) \, dt \right| \le (b-a) M \left[\frac{1}{4} + \frac{\{x - \frac{1}{2}(a+b)\}^2}{(b-a)^2} \right].$$

Gruss's inequalities are bounds for the difference between the integral of a product of bounded functions and the product of their integrals. Let $\varphi \le f(x) \le \Phi$ and $\gamma \le g(x) \le \Gamma$

$$\left| \frac{1}{b-a} \int_a^b f(x)g(x) \, dx - \frac{1}{b-a} \left\{ \int_a^b f(x) \, dx \right\} \frac{1}{b-a} \left\{ \int_a^b g(x) \, dx \right\} \right|$$

$$\le \frac{1}{4}(\Phi - \varphi)(\Gamma - \gamma).$$

These inequalities have been developed for several classes of functions and extended to inequalities of higher order approximations (Barnett and Dragomir, 2001, 2002).

1.6 Inequalities for discrete martingales

On a probability space $(\Omega, \mathcal{F}, P)$, let $(\mathcal{F}_n)_{n \geq 0}$ be a filtration of $\mathcal{F}$, *i.e.* a nondecreasing sequence of subsigma-algebras of $\mathcal{F}$, and let $(X_n)_{n \geq 0}$ be a sequence of $\mathcal{F}_n$-measurable real variables. It is a *martingale* with respect to $(\mathcal{F}_n)_n$ if $E(X_m | \mathcal{F}_n) = X_n$ for every $m > n \geq 0$. Then $EX_n = EX_0$ and $E\{(X_m - X_n)Y_n\} = 0$, for every $\mathcal{F}_n$-measurable variable Y_n. It follows that the conditional variance of a square integrable $(\mathcal{F}_n)_n$-martingale $(X_n)_{n \geq 0}$ is $E\{(X_m - X_n)^2 | \mathcal{F}_n\} = E(X_m^2 | \mathcal{F}_n) - X_n^2$.

The sequence $(X_n)_{n \geq 0}$ is a *submartingale* if $E(X_m | \mathcal{F}_n) \geq X_n$ for every $m > n \geq 0$ and it is a *supermartingale* if $E(X_m | \mathcal{F}_n) \leq X_n$. Let $(X_n)_{n \geq 0}$ be a real martingale and let φ be a function defined from $\mathbb{R}$ to $\mathbb{R}$. If φ is convex, $E\{\varphi(X_{n+1}) | \mathcal{F}_n\} \geq \varphi(E\{X_{n+1} | \mathcal{F}_n\}) = \varphi(X_n)$ then $(\varphi(X_n))_{n \geq 0}$ is a submartingale. If φ is concave, then $(\varphi(X_n))_{n \geq 0}$ is a supermartingale.

Proposition 1.1 (Neveu, 1972). *Let* $(X_n)_{n \geq 0}$ *be a positive supermartingale then for every* $c > 0$

$$P\left(\sup_{n \geq 0} X_n > c \mid \mathcal{F}_0\right) \leq \min(c^{-1}X_0, 1)$$

and $\sup_{n \geq 0} X_n$ *is a.s. finite on* $\{X_0 < \infty\}$.

Theorem 1.1. *Every positive supermartingale* $(X_n)_{n \geq 0}$ *converges a.s. to* $X_\infty = \lim_{n \to \infty} X_n$ *and* $E(X_\infty \mid \mathcal{F}_n) \leq X_n$. *Every positive submartingale* $(X_n)_{n \geq 0}$ *such that* $\sup_{n \in \mathbb{N}} E(X_n^+)$ *is finite converges a.s. to a limit* X_∞ *in* L^1 *such that* $E(X_\infty | \mathcal{F}_n) \geq X_n$ *for every integer* n.

The following examples of discrete martingales are classical.

(1) A random walk $X_n = \sum_{i=1}^{n} \zeta_i$ is defined by a sequence of independent and identically distributed random variables $(\zeta_i)_i = (X_i - X_{i-1})_{i=1,\ldots,n}$. Let $\mu = E\zeta_i$ be the mean increment of X_n, if $\mu = 0$ then X_n is a martingale, if $\mu > 0$ then X_n is a submartingale and if $\mu < 0$ then X_n is a supermartingale.

(2) For every increasing sequence of random variables $(A_n)_{n \geq 1}$ such that A_n is $\mathcal{F}_n$-measurable and $E(A_\infty | \mathcal{F}_0) < \infty$ a.s., $X_n = E(A_\infty | \mathcal{F}_n) - A_n$ is a supermartingale and $EM_n = (A_\infty | \mathcal{F}_n)$ is a martingale.

(3) Let

$$V_n(X) = X_0^2 + \sum_{i=1}^{n}(X_i - X_{i-1})^2$$

be the quadratic variations of a $L^2(P, \mathcal{F}_n)_n)$ martingale $(X_n)_{n \geq 1}$ with respect to a filtration $(\mathcal{F}_n)_{n \geq 1}$, this is an increasing sequence. For every $n \geq 0$, $EX_n^2 = \sum_{i=1}^n E(X_n^2 - X_{n-1}^2) + EX_0^2 = E\{\sum_{i=1}^n (X_i - X_{i-1})^2\} + EX_0^2 = EV_n(X)$. Let

$$E\{V_{n+1}(X)|\mathcal{F}_n\} = V_n(X) + E\{(X_{n+1} - X_n)^2|\mathcal{F}_n\} \geq V_n(X),$$

$(V_n(X))_{n \geq 1}$ is a submartingale and it converges a.s. to a limit $V(X)$ in $L^1(P)$.

Theorem 1.2. *On a filtered probability space* $(\Omega, \mathcal{F}, (\mathcal{F}_n)_{n \geq 0}, P)$, *let* $(X_n)_{n \geq 1}$ *be a sequence of* $\mathcal{F}_n$-*measurable variables with* $X_0 = 0$ *and let*

$$\widetilde{X}_n = \sum_{i=1} E(X_n - X_{n-1} \mid F_{n-1}),$$

then $(\widetilde{X}_n)_{n \geq 1}$ *is the unique predictable process such that* $(X_n - \widetilde{X}_n)_{n \geq 1}$ *is a* $(\mathcal{F}_n)_n$-*martingale.*

Theorem 1.3. *Let* $(X_n)_{n \geq 1}$ *be a square integrable martingale on a filtered probability space* $(\Omega, \mathcal{F}, (\mathcal{F}_n)_{n \geq 0}, P)$ *then* $(X_n^2 - V_n(X))_{n \geq 1}$ *and* $(\sum_{k=1}^n \{E(X_k^2 \mid \mathcal{F}_{k-1}) - X_{k-1}^2\})_{n \geq 1}$ *are* $(\mathcal{F}_n)_n$-*martingales.*

This is proved by Theorem 1.2 and the equality

$$E(X_n^2 - X_{n-1}^2|\mathcal{F}_{n-1}) = E\{(X_n - X_{n-1})^2|\mathcal{F}_{n-1}\}$$
$$= E(V_n - V_{n-1}|\mathcal{F}_{n-1}).$$

A transformed martingale on $(\Omega, \mathcal{F}, (\mathcal{F}_n)_{n \geq 0}, P)$ is defined by two sequences of $L^1(P)$ random variables $(X_n)_{n \geq 1}$ and $(A_n)_{n \geq 1}$ by $Y_0 = X_0$ and

$$Y_{n+1} = Y_n + A_n(X_{n+1} - X_n), n \geq 1, \tag{1.16}$$

where A_n is $\mathcal{F}_{n-1}$-measurable, for every integer n, and X_n is a $(\mathcal{F}_n)$-martingale, then $E\{Y_{n+1}|\mathcal{F}_n\} = Y_n + A_n E(X_{n+1} - X_n|\mathcal{F}_n)$, so Y_n is a $(\mathcal{F}_n)$-martingale. If X_n is a $\mathcal{F}_n$-submartingale (respectively supermartingale), then Y_n is a $\mathcal{F}_n$-submartingale (respectively supermartingale). The quadratic variations $V_n(Y)$ of $(Y_n)_{n \geq 1}$ satisfy

$$E\{V_{n+1}(Y)|\mathcal{F}_n\} - V_n(Y) = A_n^2 E\{(X_{n+1} - X_n)^2|\mathcal{F}_n\} \geq 0, \tag{1.17}$$

hence the process $V_n(Y) = X_0^2 + \sum_{i=1}^{n-1} A_i^2 \{V_{i+1}(X) - V_i(X)\}$ defines a convergent submartingale.

Kolmogorov's inequality for a $(\mathcal{F}_n)$-martingale $(X_n)_{n\geq1}$ of $L^p(\mathbb{R})$, for $1 \leq p < \infty$, is similar to Bienaymé-Chebychev's inequality for independent variables

$$P(|X_{n+1}| > a|\mathcal{F}_n) \leq a^{-p}|X_n|^p, \text{ for every } a > 0. \qquad (1.18)$$

A stopping time T of a uniformly integrable martingale sequence $(X_n)_{n\geq1}$ defined on a filtered probability space $(\Omega, \mathcal{F}, (\mathcal{F}_n)_{n\geq0}, P)$ satisfies the next measurability property: $\{T \leq n\}$ is $\mathcal{F}_n$-measurable, for every integer n. Doob established that $E(X_T \mid \mathcal{F}_S) = X_S$ for all stopping times S and $T > S$, let $X_n = E(X_\infty \mid \mathcal{F}_n)$

$$E(X_T \mid \mathcal{F}_S) = \sum_{i=S+1}^{\infty} E(X_i 1_{\{T=i\}} \mid \mathcal{F}_S)$$

$$= \sum_{i=S+1}^{\infty} E\{E(X_\infty \mid \mathcal{F}_i) 1_{\{T=i\}} \mid \mathcal{F}_S\}$$

$$= \sum_{i=S+1}^{\infty} E(X_\infty 1_{\{T=i\}} \mid \mathcal{F}_S) = E(X_\infty \mid \mathcal{F}_S),$$

therefore $E(X_T \mid \mathcal{F}_S) = X_{S\wedge T}$ and $EX_T = EX_0$, for all S and T.

Let $(X_n)_{n\geq1}$ be a sequence of $(\mathcal{F}_n)_n$-adapted random variables with values in $[0, 1]$, for every stopping time τ, the martingale $M_n = E(X_{n+1}|\mathcal{F}_n)$, for every n, satisfies the property (Freedman, 1973)

$$P\Big(\sum_{i=1}^{\tau} X_n \leq a, \sum_{i=1}^{\tau} M_n \geq b\Big) \leq \Big(\frac{b}{a}\Big)^a e^{a-b}$$

$$\leq \exp\Big\{-\frac{(a-b)^2}{2c}\Big\},$$

$$P\Big(\sum_{i=1}^{\tau} X_n \geq a, \sum_{i=1}^{\tau} M_n \leq b\Big) \leq \Big(\frac{b}{a}\Big)^a e^{a-b},$$

where $0 \leq a \leq b$ and $c = \max(a, b)$, and the bound reduced to 1 if $a = b = 1$. The inequalities for the maximum of variables extend to martingales. Let

$$X_n^* = \max_{1\leq i\leq n} |X_i|$$

be the maximal variable of $(X_1, \ldots, X_n)$ and $X_n^+ = \max(X_n, 0)$.

Theorem 1.4. *Let* $\tau_1 = 1$ *and for* $k > 1$, *let*

$$\tau_k = \min\{n > \tau_{k-1} : X_n^* = X_n\}$$

be stopping times for the martingale $(X_n)_n$. For every $k \geq 1$ such that τ_k is finite and for every $\lambda > 0$

$$\lambda P(X^*_{\tau_k} > \lambda) < E(X_{\tau_k} 1_{\{X^*_{\tau_k} > \lambda\}}) < EX^+_{\tau_k}$$

and the inequality holds only at the stopping times τ_k.

Proof. The variables τ_k are stopping times for the martingale $(X_n)_n$, $X^*_{\tau_k} = X_{\tau_k}$ for every $k \geq 1$ and $X^*_n > X_n$ for every n which does not belong to the sequence of stopping times. For every $\lambda > 0$

$$E(X_{\tau_k} 1_{\{X^*_{\tau_k} > \lambda\}}) = E(X_{\tau^*_k} 1_{\{X^*_{\tau_k} > \lambda\}}) > \lambda P(X^*_{\tau_k} > \lambda).$$

Otherwise $E(X_n 1_{\{X^*_n > \lambda \geq |X_n|\}}) \leq \lambda P(X^*_n > \lambda \geq |X_n|) \leq \lambda P(X^*_n > \lambda)$ and $E(|X_n| 1_{\{X^*_n > \lambda\}})$ is smaller than $E(X^*_n 1_{\{X^*_n > \lambda\}}) > \lambda P(X^*_n > \lambda)$. □

Note that the a.s. convergence of the sequence $n^{-\alpha} X^*_n$ to a limit, for some $\alpha > 0$, implies the a.s. convergence of the sequence $n^{-\alpha}|X_n|$. Let $X^* = \sup_{n \geq 1} X_n$.

Theorem 1.5 (Doob's inequality). Let $(X_n)_{n \geq 1}$ be a martingale such that $\sup_{n \geq 1} \|X_n\|_p$ is finite, for $p > 1$, and let $p^{-1} + q^{-1} = 1$, then

$$\sup_{n \geq 1} \|X_n\|_p \leq \|X^*\|_p \leq q \sup_{n \geq 1} \|X_n\|_p.$$

Proof. The proof of the upper bound relies on Theorem 1.4 with $\tau = \tau_1$, for $\lambda > 0$

$$\|X^*\|^p_p = p \int_0^\infty \lambda^{p-1} E(1_{\{X^* > \lambda\}}) \, d\lambda$$

$$\leq p \int_0^\infty \lambda^{p-2} E(X_\tau 1_{\{X^* > \lambda\}}) \, d\lambda$$

$$\leq \frac{p}{p-1} \sup_{n \geq 1} E(X_n X^{*p-1}) \leq q \sup_{n \geq 1} \|X_n\|_p \{E(X^{*p})\}^{\frac{1}{q}},$$

using Hölder's inequality. □

Burkholder's inequalities (1973) for martingales of L^p, $1 < p < \infty$ are

$$c_p E\{V^{\frac{p}{2}}(X)\} \leq \sup_{n \geq 1} E(X^p_n) \leq C_p E\{V^{\frac{p}{2}}(X)\} \tag{1.19}$$

with constants $c_p > 0$ and C_p depend only on p and not on the martingale $(X_n)_{n \geq 0}$, they have the orders $c_p^{-1} = O(p^{\frac{1}{2}} q)$ and $C_p = O(q^{\frac{1}{2}} p)$, where $p^{-1} + q^{-1} = 1$. The Burkholder-Davis-Gundy inequality is similar for the maximum variable $X^* = \sup_{n \geq 1} X_n$ and it extends to convex functions Φ from $\mathbb{R}_+$ to $\mathbb{R}_+$ such that $\Phi(2\lambda) \leq c\Phi(\lambda)$, $\lambda > 0$

$$cE\Phi(V^{\frac{1}{2}}(X)) \leq E\Phi(X^*) \leq CE\Phi(V^{\frac{1}{2}}(X)). \tag{1.20}$$

Mona (1994) proved that Burkholder's inequality is not true for discontinuous martingales and p in $]0,1[$. Meyer's proof (1969) of the Burkholder, Davis and Gundy inequality holds in the space $\mathcal{M}_0^p$ of L^p centered martingales, for every $p > 1$. Barlow and Yor (1982) improved the inequality (1.20) by several inequalities such as

$$\left\| \sup_{0 \leq s < t < \infty} \frac{|M_t - M_s|}{(< M >_t < M >_s)^{\frac{1}{2}-\varepsilon}} \right\|_p \leq C_{p,\varepsilon} \| < M >_\infty^\varepsilon \|_p$$

for every $p \geq 1$ and for ε in $]0, \frac{1}{2}]$, with a constant $C_{p,\varepsilon}$ depending only on p and ε.

As a consequence of the inequality (1.3) between norms $L^p(\Omega, \mathcal{F}, P)$ and $L^\infty(\Omega, \mathcal{F}, P)$ in vector spaces of dimension n, for every random variable $X = (X_1, \ldots, X_n)$

$$E(X_n^*)^p \leq E\|X\|_p^p \leq nE(X_n^*)^p,$$

which is equivalent to $n^{-1}E\|X\|_{n,p}^p \leq E(X_n^*)^p \leq E\|X\|_{n,p}^p$. The previous inequalities and the Kolmogorov inequality imply

$$P(X_n^* > a) \leq a^{-p}E\|X_n\|_{n,p}^p, \tag{1.21}$$

for every $a > 0$.

The inequality for the norms has been extended to transformed martingale by a convex function (Burkholder, Davis and Gundy, 1972).

Theorem 1.6. *Let ϕ be a non-negative convex function with $\phi(0) = 0$ and $\phi(2x) \leq 2\phi(x)$. There exist constants c_ϕ and C_ϕ such that for every martingale $(X_n)_n$ on a filtered probability space $(\Omega, \mathcal{F}, (\mathcal{F}_n)_{n \geq 0}, P)$*

$$c_\phi E\phi(X_n) \leq E\phi(X_n^*) \leq C_\phi E\phi(X_n). \tag{1.22}$$

Burkholder (1973) proved that for every $p > 1$ and for independent variables X_i, there exist constants such that for every random walk $S_n = \sum_{i=1,\ldots,n} X_i$

$$ES_n^p \leq A_p E(V_n(X)^{\frac{p}{2}}). \tag{1.23}$$

Some results have been extended to submartingales indexed by a multidimensional set of integers (Cairoli and Walsh, 1975). Due to the partial order of the sets of indexes, several notions of martingales are defined and the conditions for the filtration are generally stronger than previously. In $\mathbb{R}^d$, $d \geq 2$, the total order $m \leq n$ means $m_k \leq n_k$ for $k = 1, \ldots, d$ and an assumption of conditional independence of the marginal σ-algebras, given the

d-dimensional σ-algebra is usual. Christofides and Serfling (1990) proved that under the condition $E\{E(\cdot|\mathcal{F}_k)|\mathcal{F}_1\} = E(\cdot|\mathcal{F}_{k\wedge 1})$, $k = 2, \ldots, d$, a martingale $(X_k)_{k\in\mathbb{N}^d}$ with respect to a filtration $(\mathcal{F}_k)_{k\in\mathbb{N}^d}$ satisfies

$$P\left(\max_{k\leq n}|X_k| > a\right) \leq 4^{d-1}a^{-2}E(X_n^2), \ a > 0, \ n \in \mathbb{N}^d.$$

1.7 Martingales indexed by continuous parameters

A time-continuous martingale $X = (X_t)_{t\geq 0}$ on $(\Omega, \mathcal{F}, (\mathcal{F}_t)_{t\geq 0}, P)$ is defined with respect to a right-continuous and increasing filtration $(\mathcal{F}_t)_{t\geq 0}$ of $\mathcal{F}$ i.e. $\mathcal{F}_s \subset \mathcal{F}_t$ for every $s < t$, $\mathcal{F}_t = \cap_{s>t}\mathcal{F}_s$ and $\mathcal{F} = \mathcal{F}_\infty$. For the natural filtration $(\mathcal{F}_t)_{t\geq 0}$ of $(X_t)_{t\geq 0}$, $\mathcal{F}_t$ is the σ-algebra generated by $\{X_s; s \leq t\}$. A random variable τ is a stopping time with respect to $(\mathcal{F}_t)_{t\geq 0}$ if for every t, the set $\{\tau \leq t\}$ belongs to $\mathcal{F}_t$.

A martingale X is uniformly integrable if

$$\lim_{C\to\infty} \sup_{t\geq 0} E(|X_t|1_{\{|X_t|>C\}}) = 0.$$

Then $\lim_{t\to\infty} E|X_t - X_\infty| = 0$ and $X_t = E(X_\infty|\mathcal{F}_t)$ for every t, with the natural filtration $(\mathcal{F}_t)_{t\geq 0}$ of X. Doob's theorem (1975) for stopping times applies to continuously indexed martingales.

The Brownian motion $(B_t)_{t\geq 0}$ is a martingale with independent increments, defined by Gaussian marginals, a mean zero and the variance $B_t^2 = t$, hence $B_0 = 0$. It satisfies the following properties

(1) $E(B_sB_t) = s \wedge t$,
(2) $B_t - B_s$ and B_{t-s} have the same distribution, for every $0 < s < t$,
(3) $(B_t^2 - t)_{t\geq 0}$ is a martingale with respect to $(\mathcal{F}_t)_{t\geq 0}$,
(4) for every θ, $Y_\theta(t) = \exp\{\theta B_t - \frac{1}{2}\theta^2 t\}$ is an exponential martingale with respect to $(\mathcal{F}_t)_{t\geq 0}$, with expectation $EY_\theta(t) = EY_\theta(0) = 1$.

For every $t > 0$, the variable $t^{-\frac{1}{2}}B_t$ is a normal variable and the odd moments of the Brownian motion satisfy $\|B_t\|_{2k} \leq (kt)^{\frac{1}{2}}$, $k \geq 1$. Wiener's construction of the Brownian motion is reported by Itô and Mc Kean (1996) as the limit of the convergent series

$$X_t = \frac{t}{\sqrt{\pi}}g_0 + \sum_{k=2^{n-1}}^{2^n-1} \frac{\sqrt{2}}{\sqrt{\pi}}\frac{\sin(kt)}{k}g_k,$$

where the variables g_k are independent and have the normal distribution. Let $T_a = \inf\{s : B_s = a\}$, it is a stopping time for B and the martingale property of the process Y_θ implies that $E \exp\{\frac{1}{2}\theta^2 T_a\} = e^{a\theta}$, therefore the Laplace transform of T_a is $L_{T_a}(x) = e^{-a\sqrt{2x}}$. Its density function is

$$f_{T_a}(x) = \frac{a}{\sqrt{2\pi x^3}} \exp\left\{-\frac{a^2}{2x}\right\}.$$

Let a and b be strictly positive, the independence of the increments of the process B implies that the times T_a and $T_{a+b} - T_a$ are independent and the density function of T_{a+b} is the convolution of the densities of T_a and T_b, according to the property (2) of the Brownian motion. Let P_x be the probability distribution of $x + B$, for all a and b

$$P_x(T_a < T_b) = \frac{b-x}{b-a}, \quad P_x(T_b < T_a) = \frac{x-a}{b-a}, \quad \text{if } a < x < b,$$

and $E(T_a \wedge T_b) = (x-a)(b-x)$, this is a consequence of the expansion of the mean as $E_x B(T_a \wedge T_b) = a P_x(T_a < T_b) + b P_x(T_b < T_a) = x$. The rescaled Brownian motion B_{at} has the same distribution as $\sqrt{a} B_t$, for every real $a > 0$. Hence the variable T_a has the same distribution as $a^2 T_1$.

Let $\sigma_a = \inf\{t : B_t < t - a\}$, it is an a.s. finite stopping time for every finite a and $E e^{\frac{\sigma_a}{2}} = e^a$. Let $\sigma_{a,b} = \inf\{t : B_t < bt - a\}$, then $E e^{\frac{1}{2}b^2 \sigma_{a,b}} = e^{ab}$ (Revuz and Yor, 1991). Other bounds are proved by Freedman (1975).

Let $T = \arg\max_{t \in [0,1]} B_t$, it is a stopping time of the Brownian motion B and its distribution is determined as follows (Feller, 1966; Krishnapur, 2003; Durrett, 2010)

$$P(T \leq t) = P\left(\sup_{s \in [0,t]} B_s \geq \sup_{x \in [t,1]} B_x\right)$$

$$= P\left(\sup_{u \in [0,t]} B_{t-u} - B_t \geq \sup_{v \in [0,1-t]} B_{t+v} - B_t\right).$$

Let $X_t = \sup_{u \in [0,t]} B_{t-u} - B_t$ and $Y_t = \sup_{v \in [0,1-t]} B_{t+v} - B_t > 0$ a.s., they are independent and the variables $X = t^{-\frac{1}{2}} X_t$ and $Y = (1-t)^{-\frac{1}{2}} Y_t$ have the same distribution as $\sup_{t \in [0,1]} B_t - B_0$, then

$$P(T \leq t) = P(X_t \geq Y_t) = P\left(\frac{Y}{(X^2 + Y^2)^{\frac{1}{2}}} \leq t^{\frac{1}{2}}\right)$$

where the variable $Y(X^2 + Y^2)^{-\frac{1}{2}}$ is the sine of a uniform variable on $[0, \frac{\pi}{2}]$, hence

$$P(T \leq t) = \frac{2}{\pi} \arcsin(t^{\frac{1}{2}}). \tag{1.24}$$

The arcsine distribution has the density

$$f_{\text{arcsin}}(x) = \frac{1}{\pi} \frac{1}{x(1-x)}, \quad 0 < x < 1,$$

it is a symmetric function with respect to $\frac{1}{2}$ and it tends to infinity at zero or one. Geetor and Sharpe (1979) established that for every x, the time variable $S_{t,\varepsilon} = \int_{G_t}^{D_t} 1_{[0,\varepsilon]}(|B_s|)\,ds$ spent in $[-\varepsilon, \varepsilon]$ between two passages at zero of the Brownian motion has a Laplace transform satisfying $\lim_{\varepsilon \to 0} E \exp\{-\lambda \varepsilon^{-2} S_{t,\varepsilon}\} = (\cosh \sqrt{2\beta})^{-2}$.

The Brownian bridge $(B_t^0)_{t \in [0,1]}$ is the Gaussian process with mean zero and covariance $E\{B^0(s)B^0(t)\} = s \wedge t - ts$, B_t^0 has the same distribution as B_{1-t}^0 and as $B_t - t$, where B is the Brownian motion. For all $s < t$ in $[0,1]$, $B_t^0 - B_s^0$ has the same distribution as B_{t-s}^0. The Brownian bridge is the limit in distribution of the sum process $S_n(t) = n^{-\frac{1}{2}} \sum_{i=1}^n \{1_{\{X_i \le t\}} - P(X_i \le t)\}$ as a process with paths in the space $C([0,1])$ provided with the uniform norm and the Borel σ-algebra. Doob proved

$$P\left(\sup_{t \in [0,1]} |B^0(t)| \le x\right) = \sum_{i \in \mathbb{Z}} (-1)^j e^{-2j^2 x^2},$$

$$P\left(\sup_{t \in [0,1]} B^0(t) \le x\right) = e^{-2x^2}.$$

Dudley (1973) gave a survey of the sample properties of Gaussian processes, including processes satisfying Hölder conditions with various rates.

On a space $(\Omega, \mathcal{F}, P, \mathbb{F})$, let M be a square integrable $\mathbb{F}$-martingale, in $\mathcal{M}^2$, it is the sum of a continuous process M^c and a discrete process sum of the jumps $\Delta M_s = M_s - M_{s-}$ of M, $M_t^d = \sum_{0 < s \le t} \Delta M_s$. There exists an unique increasing and $\mathcal{F}$-predictable process $< M >$ such that the process

$$M^2 - < M >$$

is a martingale. The process $< M >$ is the process of the predictable quadratic variations of M

$$< M >_t = \widetilde{[M]}_t, \quad [M]_t = < M^c >_t + M_t^d.$$

It satisfies

$$E\{(M_t - M_s)^2 | \mathcal{F}_s\} = E(M_t^2 | \mathcal{F}_s) - EM_s^2 = E(< M >_t | \mathcal{F}_s) - < M >_s,$$

for $s < t$. It defines a scalar product for square integrable martingales M_1 and M_2 with mean zero

$$\begin{aligned} < M_1, M_2 > &= \frac{1}{2}(< M_1 + M_2, M_1 + M_2 > \\ &\quad - < M_1, M_1 > - < M_2, M_2 >), \end{aligned} \tag{1.25}$$

then $E < M_1, M_2 >_t = E M_{1t} M_{2t}$ for every $t > 0$. Two square integrable martingales M_1 and M_2 are orthogonal if and only if $< M_1, M_2 > = 0$ or, equivalently, if $M_1 M_2$ is a martingale. Let $\mathcal{M}_0^2$ be the space of the right-continuous square integrable martingales with value zero at zero, provided with the norm $\|M\|_2 = \sup_t (E M_t^2)^{\frac{1}{2}} = \sup_t (E < M >_t)^{\frac{1}{2}}$.

A process $(M_t)_{t\geq 0}$ is a square integrable local martingale if there exists an increasing sequence of stopping times $(S_n)_n$ such that $(M(t \wedge S_n))_t$ belongs to $\mathcal{M}^2$ and S_n tends to infinity. The space of L^2 local martingales is denoted $\mathcal{M}_{loc}^2$, or $\mathcal{M}_{0,loc}^2$ if they are centered. Let $(M_t)_{t\geq 0}$ be in $\mathcal{M}_{0,loc}^2$, it is written as the sum of a continuous part M^c and a discrete part $M_t^d = \sum_{0<s\leq t} \Delta M_s$, with jumps $\Delta M_s = M_s - M_{s-}$. The increasing process of its quadratic variations is $< M >_t = < M^d >_t + < M^c >_t$ where $M_t^{c2} - < M^c >_t$ and $\sum_{0<s\leq t} (\Delta M_s)^2 - < M^d >_t$ belong to $\mathcal{M}_{0,loc}$.

Let M be a local martingale of $\mathcal{M}_{0,loc}^2$ with bounded jumps, Lepingle (1978) proved that

$$Z_t(\lambda) = \exp\left\{ \lambda M_t - \frac{\lambda^2}{2} < M^c >_t - \int (e^{\lambda x} - 1 - \lambda x)\, dM^d(x) \right\} \quad (1.26)$$

is a positive local supermartingale. The process Z_t called the exponential supermartingale of M is a martingale for the Brownian motion. The exponential of a semi-martingale $X = M + A$ where A is a process with bounded quadratic variations is

$$Z_t = \exp\left\{ X_t - \frac{< X^c >_t}{2} \right\} \prod_{0\leq s\leq t} (1 + \Delta X_s) \exp(-\Delta X_s) \quad (1.27)$$

it is the unique solution of the equation

$$Z_t = 1 + \int_0^t Z_-\, dX.$$

Proposition 1.2. *Let $\lambda > 0$ be real, let M be a local martingale of $\mathcal{M}_{0,loc}^2$ and let T be a stopping time, then*

$$P\left(\sup_{t\in[0,T]} |M_t| > \lambda \right) \leq \lambda^{-2} E < M >_T.$$

For a local supermartingale M such that $E \sup_{t\geq 0} |M_t|$ is finite

$$P\left(\sup_{t\geq 0} M_t > \lambda \right) \leq \lambda^{-1} E |M_0|.$$

Let $\delta > 0$ and $\lambda = T^{\delta + \frac{1}{2}}$ and let M be a local martingale of $\mathcal{M}_{0,loc}^2$ with an increasing process $< M >$ satisfying a strong law of large numbers, then $\lim_{T \to \infty} P(T^{-\frac{1}{2}} \sup_{t \in [0,T]} |M_t| > T^\delta) = 0$. Let M be a bounded local supermartingale, then

$$\lim_{A \to \infty} P\left(\sup_{t \geq 0} M_t > A\right) = 0.$$

The Birnbaum and Marshal inequality is extended as an integral inequality for a right-continuous local submartingale $(S_t)_{t \geq 0}$, with a mean quadratic function $A(t) = E S_t^2$ and a increasing function q

$$P\left(\sup_{t \in [0,s]} q^{-1}(t) |S_t| \geq 1\right) \leq \int_0^s q^{-2}(t) \, dA(s).$$

Proposition 1.3 (Lenglart, 1977). *Let M be a local martingale of $\mathcal{M}_{0,loc}^2$ and let T be a stopping time, then for all $\lambda > 0$ and $\eta > 0$*

$$P\left(\sup_{t \in [0,T]} |M_t| \geq \lambda\right) \leq \frac{\eta}{\lambda^2} + P(< M >_T \geq \eta).$$

For every stopping time of a martingale M of $\mathcal{M}_0^2$, Doob's theorem entails $E(M_T^2) = E < M >_T$. By monotonicity

$$E\left(\sup_{t \in [0,T]} M_t^2\right) = E(< M >_T).$$

The Burkholder, Davis and Gundy inequality (1.23) has been extended to L^p local martingales indexed by $\mathbb{R}$, for every $p \geq 2$.

There exist several notions of martingales in the plane, according to the partial or total order of the two-dimensional indices. Walsh (1974) and Cairoli and Walsh (1975) presented an account of the results in $\mathbb{R}^2$. Cairoli (1970) established maximal inequalities for right-continuous martingales of L^p with parameter set $\mathbb{R}^2$. For $\lambda > 0$ and $p > 1$

$$E\left(\sup_{z \in \mathbb{R}^2} |M_z|^p\right) \leq \left(\frac{p}{p-1}\right)^{2p} \sup_{z \in \mathbb{R}^2} E(|M_z|^p).$$

A stochastic integral on an interval $[0,t]$ may be defined path-by-path as limit of sums on a real subdivision $\tau_n = (t_i)_{i=0,\ldots,n}$ of $[0,t]$ with $t_0 = 0$ and $t_n = t$ with $\delta_n = t_n - t_{n-1}$ tending to zero as n tends to infinity. Let $Y = (Y_t)_{t \geq 0}$ be a predictable process on a probability space $(\Omega, \mathcal{A}, P, \mathbb{F})$ provided with a filtration $\mathbb{F} = (\mathcal{F}_t)_{t \geq 0}$. For every $t > 0$, Y_t is the limit in probability of a sequence of step-wise constant processes with left-hand continuous processes with right-hand limits on the subdivision $\tau_n = (t_k)_{k=0,\ldots,n}$ of $[0,t]$

$$Y_n(t) = \sum_{k=1}^n Y_k 1_{]t_k, t_{k+1}]}$$

and the stochastic integral of Y with respect to a positive measure μ is the limit in probability of $\sum_{k=1}^{n} Y_{t_k} \mu(]t_{k-1}, t_k])$.

The stochastic integral of Y with respect to a process A with locally bounded variations, such that $A = A_+ - A_-$ is the difference of two increasing processes, is the limit in probability of $\sum_{i=1}^{n} Y_{t_i} \{A_+(]t_i, t_{i+1},]) - A_-(]t_i, t_{i+1},])\}$.

For a predictable process Y approximated in L^2 by a sequence of step-wise constant processes $(Y_n)_{n \geq 1}$ on partitions $\tau_n = (t_k)_{k=0,\ldots,n}$ of $[0, t]$, the stochastic integral of Y_n^2 with respect to the Lebesgue measure is a Cauchy sequence for every $t > 0$

$$\lim_{n \to \infty} E \int_0^t |Y_n(s) - Y_m(s)|^2 \, ds = 0$$

so there exists a limit which is the stochastic integral of Y^2 with respect to the Lebesgue measure. The stochastic integral $Y.B$ of Y with respect to B is defined as the L^2-limit of $Y_n.B = \sum_{k=1}^{n} Y_n(t_k)(B_{t_k} - B_{t_k-1})$ such that

$$\lim_{n \to \infty} E \left| \int_0^t Y_n(s) \, dB_s - \int_0^t Y_s \, dB_s \right|^2 \leq \lim_{n \to \infty} E \int_0^t |Y_s - Y_n(s)|^2 \, ds = 0.$$

The stochastic integral $Y.M_t$ of a predictable process Y with respect to a local martingale M of $\mathcal{M}_{0,loc}^2$, with a predictable increasing process of quadratic variations $< M >$, is uniquely defined as the L^2-limit of Y_n with respect to the increasing process $< M >$ and

$$\lim_{n \to \infty} E \left| \int_0^t Y_n(s) \, dM_s - \int_0^t Y_s \, dM_s \right|^2$$

$$\leq \lim_{n \to \infty} E \int_0^t |Y_s - Y_n(s)|^2 \, d < M >_s = 0.$$

Then $Y.M$ is a square integrable local martingale such that

$$E(Y.M_t)^2 = E \int_0^t Y_s^2 \, d < M >_s.$$

For M' of $\mathcal{M}_{0,loc}^2$ and for a predictable process X of $L^1(< M, M' >)$ we have

$$< X.M, M' >_t = \int_0^t X \, d < M, M' >.$$

1.8 Large deviations and exponential inequalities

Let $(X_i)_{i=1,\ldots,n}$ be a sequence of independent and identically distributed real random variables defined on a probability space $(\Omega, \mathcal{F}, P)$, with mean zero and moment generating function φ_X^n, and let $S_n = \sum_{i=1}^n X_i$. For all $a > 0$ and $t > 0$, the Bienaymé-Chebyshev inequality for the variable $\exp\{n^{-1} t S_n - at\}$ implies

$$P(S_n > a) \leq E\{\exp(t S_n - at)\} \leq e^{-at} \varphi_X^n(t). \tag{1.28}$$

The maximal variable satisfies $P(X_n^* > a) \geq 1 - \{1 - e^{-at} \varphi_X(t)\}^n$ and the minimum $X_{*n} = \min(X_1, \ldots, X_n)$ is such that $P(X_{*n} > a) \leq e^{-ant} \varphi_X^n(t)$.

Theorem 1.7 (Chernoff's theorem). *On a probability space* $(\Omega, \mathcal{F}, P)$, *let* $(X_i)_{i=1,\ldots,n}$ *be a sequence of independent and identically distributed real random variables with mean zero, having a finite moment generating function* φ_X, *and let* $S_n = \sum_{i=1}^n X_i$. *For all* $a > 0$ *and* $n > 0$

$$\log P(S_n > a) = \inf_{t>0} \{n \log \varphi_X(t) - at\}.$$

Proof. It is a direct consequence of (1.28) and of the concavity of the logarithm, which implies $\log E e^{S_n t} \geq t E X_1$, so that the inequality is an equality and $\log P(S_n > a) = \inf_{t>0} \{n \log \varphi_X(t) - at\}$. □

This equality entails

$$\lim_{n \to \infty} n^{-1} \log P(n^{-1} S_n > a) = \inf_{t>0} \{\log \varphi_X(t) - at\}.$$

The function

$$\psi_a(t) = \{\log \varphi_X(t) - at\} \tag{1.29}$$

is minimum at $t_a = \arg\min_{t>0} \psi_a(t)$, then for every integer n, $n^{-1} \log P(n^{-1} S_n > a) = \{\log \varphi_X(t_a) - a t_a\}$ and t_a is solution of the equation $\psi_a'(t) = 0$, i.e. $E(X e^{X t_a}) = a E(e^{X t_a})$.

With the norm L^1 of S_n, the equality in Chernoff's theorem is replaced by an upper bound. For all $a > 0$ and $n > 0$

$$\lim_{n \to \infty} n^{-1} \log P(n^{-1} |S_n| > a) \leq \inf_{t>0} \{\log L_{|X|}(t) - at\}.$$

A symmetric variable X has zero odd moments and its moment generating function has the bound

$$\phi_X(t) = E \sum_{k \geq 0} \frac{(tX)^{2k}}{(2k)!} \leq E \sum_{k \geq 0} \frac{1}{k!} \left\{ \frac{t^2 X^2}{2} \right\}^k = E e^{\frac{t^2 X^2}{2}}$$

then for every $a > 0$

$$\log P(X > a) \leq \inf_{t>0}\{\log Ee^{\frac{t^2 X^2}{2}} - at\}.$$

The moment generating function of a Gaussian variable X with mean zero and variance σ^2 is $\varphi_X(t) = e^{\frac{t^2\sigma^2}{2}}$ and $e^{-at}\varphi_X(t)$ is minimum at $t_a = \sigma^{-2}a$, then by (1.28), for every $a > 0$

$$P(X > a) = \exp\left\{-\frac{a^2}{2\sigma^2}\right\}$$

and by symmetry

$$P(|X| > a) = 2\exp\left\{-\frac{a^2}{2\sigma^2}\right\}.$$

Using the inequality for symmetric variables provides a larger bound for a Gaussian variable

$$P(X > a) \leq (1+a)^{\frac{1}{2}}\exp\left\{-\frac{a^{\frac{3}{2}}}{\sigma(1+a)^{\frac{1}{2}}}\right\}.$$

For the minimum of n independent Gaussian variables

$$P(n^{\frac{1}{2}}X_{*n} > a) = e^{-a^2(2\sigma^2)^{-1}}.$$

The standard Brownian motion on $[0,1]$ has the Laplace transform $L_{B_t}(x) = e^{\frac{x^2 t}{2}}$ and, for every $0 < s < t$, $L_{B_t - B_s}(x) = e^{\frac{1}{2}x^2(t-s)}$ and

$$P(B_t > a) = e^{-\frac{a^2}{2t}},$$

$$P(B_t - B_s > a) = e^{-\frac{a^2}{2(t-s)}},$$

$$P(B_t > a, B_t - B_s > b) = e^{-a^2\{(2t)^{-1}+(2(t-s))^{-1}\}}.$$

In particular, the probability of the event $\{B_t \leq a\sqrt{t}\}$ is bounded by $1 - e^{-\frac{a^2}{2}}$ and it tends to one as a tends to infinity.

The moment generating function at t of an exponential variable X with parameter λ is $\varphi_\lambda(t) = \lambda(\lambda - t)^{-1}$ for every $0 < t < \lambda$ and it is infinite if $\lambda < t$, therefore Chernoff's theorem does not apply. The probability of the event $(n^{-1}S_n > a)$ can however be bounded if a is sufficiently large

$$\log P(n^{-1}S_n > a) \leq \inf_{0<t<\lambda} n\{\log\varphi_X(t) - at\},$$

where the function ψ_a defined by (1.29) is minimum at $t = \lambda - \sqrt{a\lambda^{-1}}$ and the bound is negative if a is sufficiently large.

Chernoff's theorem applies to a sequence of independent and identically distributed random variables with values in $\mathbb{R}^d$. For a sequence of Gaussian vectors with mean zero and a nonsingular covariance matrix Σ, a diagonalization of the covariance allows us to write the quadratic form $X^t \Sigma^{-1} X$ as a sum of independent variables $X^t \Sigma^{-1} X = \sum_{k=1}^d \lambda_k Y_k^2$, where $\lambda_1, \ldots, \lambda_d$ are the strictly positive eigenvalues of Σ^{-1} and $Y_1, \ldots, Y_d$ are independent normal variables, then the moment generating function of the variable X at $t = (t_1, \ldots, t_d)$ equals

$$\varphi_X(t) = \exp\left\{- \sum_{i=1,\ldots,d} t_i^2 (4\lambda_i)^{-1}\right\}.$$

Let $a = (a_1, \ldots, a_d)$ and a^t be its transpose, the minimum in $\mathbb{R}^d$ of the function $\psi(t) = \log \varphi_X(t) - a^t t$ is reached at $t_i = 2a_i \lambda_i$ and $\lim_{n \to \infty} n^{-1} \log P(n^{-1} S_n > a) = -2 \sum_{i=1,\ldots,d} \lambda_i a_i^2$.

The isoperimetric inequalities in $\mathbb{R}^2$ concern the optimal inequality for the ratio of the squared perimeter and the area of a closed set $L^2 \geq 4\pi A$, with equality for the circle. In $\mathbb{R}^3$, the surface and the volume of a closed set satisfy the relationship $S^{\frac{3}{2}} \geq 6\sqrt{\pi} V$, with equality for the balls, and similar geometric inequalities are proved in higher dimensions with the minimal ratio for the hypersphere $\{x \in \mathbb{R}^n : \|x\|_n = r\}$. Talagrand's isoperimetric inequalities (1995) are probabilistic. Let A be a subset of a space $(\Omega, \mathcal{A}, P)$ and let

$$f(x, A) = \min_{y \in A} \sum_{i \leq d} 1_{\{x_i \neq y_i\}}$$

be a distance of x of Ω to A, for X having the probability distribution P

$$E_{P^*} e^{tf(X,A)} \leq P^{-1}(A) e^{\frac{t^2 d}{4}},$$

$$P^*\{x : f(x, A) \leq t\} \leq P^{-1}(A) e^{-\frac{t^2}{d}}, \ t > 0,$$

using an outerprobability P^* for non-measurable functions. More generally, in Talagrand's Proposition 2.2.1 (1995), the bound is replaced by an expression with an exponent $\alpha > 0$, for $t > 0$

$$E_{P^*} e^{tf(X,A)} \geq P^{-\alpha}(A) a^d(\alpha, t),$$

$$P^*\{x : f(x, A) \geq t\} \leq P^{-\alpha}(A) \exp\left\{-\frac{2t^2}{d} \frac{\alpha}{\alpha+1}\right\},$$

where $a(\alpha, t) = \sup_{u \in [0,1]}\{1 + u(e^t - 1)\}\{1 - u(1 - e^{-\frac{t}{\alpha}})\} \leq \exp\{\frac{t^2}{8}(1 + \alpha^{-1})\}$.

The results are generalized to functions f indexed by another function such as $f_h(x, A) = \inf_{y \in A} \sum_{i \leq d} h(x_i, y_i)$, with $h(x, x) = 0$ and $h > 0$ on

$\Omega^{\otimes 2d}$, and $f_h(x, A) := h(x, A) = \inf_{y \in A} h(x, y)$. The concentration measure of a Gaussian probability is obtained for $h(x, y) = K^{-1}(x - y)^2$ on $\mathbb{R}^2$, with a constant K.

Chernoff's theorem extends to sequences of independent and non-identically distributed random variables $(X_i)_{i=1,\ldots,n}$ having moment generating functions φ_{X_i} such that $n^{-1} \sum_{i=1}^{n} \log \varphi_{X_i}$ converges to a limit $\log \varphi_X$, it is written in the same form

$$\lim_{n \to \infty} n^{-1} \log P(S_n > a) = \lim_{n \to \infty} \inf_{t > 0} \left\{ \sum_{i=1}^{n} \log \varphi_{X_i}(t) - at \right\}$$

$$= \lim_{n \to \infty} \inf_{t > 0} \left\{ n \log \varphi_X(t) - at \right\}.$$

Bennett's inequality for independent random variables is proved as an application of Chernoff's theorem under a boundedness condition. It is an exponential inequality for $P(S_n \geq t)$ under a boundedness condition for the variables X_i.

Theorem 1.8. *Let $(X_i)_{i=1,\ldots,n}$ be a vector of independent random variables having moment generating functions and such that $EX_i = 0$, $EX_i^2 = \sigma_i^2$ and $|X_i| \leq \sigma_n^* M$ with $\sigma_n^* = \max_{i=1,\ldots,n} \sigma_i$ finite for every n. For every $t > 0$ and every integer n*

$$P(|S_n| \geq t) \leq 2 \exp\left\{ -n\phi\left(\frac{t}{n\sigma_n^* M} \right) \right\}$$

where $\phi(x) = (1 + x) \log(1 + x) - x$.

Proof. First let $(X_i)_{i=1,\ldots,n}$ be a vector of independent random and identically distributed variables such that $EX_i = 0$, $EX_i^2 = \sigma^2$ and X_i have a moment generating function φ_X. A bound for L_X is obtained from an expansion of the exponential function and using the bound $|X_i| \leq b = \sigma M$, a.s.

$$\varphi_X(\lambda) \leq 1 + \sum_{k=2}^{\infty} \frac{\lambda^k}{k!} (\sigma M)^k = 1 + \{ \exp(b\lambda) - 1 - b\lambda \}$$

$$\leq \exp\{ \exp(b\lambda) - 1 - b\lambda \}$$

with $1 + x \leq e^x$. From Chernoff's theorem, for every $t > 0$

$$P(S_n > t) \leq \inf_{\lambda > 0} \exp\{ -\lambda t + n(e^{b\lambda} - 1 - b\lambda) \},$$

where the bound is denoted $\inf_{\lambda > 0} \exp\{ \psi_t(\lambda) \}$. Its minimum is reached at $\lambda_t = b^{-1} \log\{ 1 + (nb)^{-1} t \}$ where the bound of the inequality is written as

$$\psi_t(\lambda_t) = -n\phi\left(\frac{t}{nb} \right).$$

With non-identically distributed random variables, the condition implies that the functions φ_{X_i} satisfy the condition of convergence of $n^{-1} \sum_{i=1}^{n} \log \varphi_{X_i}$ to a limit $\log \varphi_X$. The bound $M\sigma$ is replaced by $b_n = M \max_{i=1,\dots,n} \sigma_i := M\sigma_n^*$ and the upper bound of the limit $\varphi_X(t)$ has the same form as in the case of i.i.d. variables. □

Bennett's inequality applies to variable X_i satisfying the same condition and such that X_i has values in a bounded interval $[a_i, b_i]$, for every i. Weighted inequalities and other inequalities for independent variables are presented by Shorack and Wellner (1986). Inequalities are deduced from lower bounds of the function $\phi(x) = (1 + x) \log(1 + x) - x$. As $2\phi(x) \geq x \log(1 + x)$, it follows that

$$P(|S_n| \geq t) \leq 2 \exp\left\{ -\frac{t}{2M\sigma_n^*} \log\left(1 + \frac{t}{nM\sigma_n^*}\right) \right\}. \qquad (1.30)$$

Varadhan's Large Deviation Principle (1984) extends Chernoff's theorem in the following sense. A sequence of probabilities $(P_n)_n$ on a measurable space $(\mathbb{X}, \mathcal{X})$ follows the Large Deviation Principle with a rate left-continuous function I, with values in $\mathbb{R}_+$, if the sets $\{x : I(x) \leq \lambda\}$ are compact subsets of $\mathbb{X}$ and for every closed set C and for every open set G of $\mathbb{X}$

$$\limsup_{n \to \infty} n^{-1} \log P_n(C) \leq - \inf_{x \in C} I(x),$$

$$\liminf_{n \to \infty} n^{-1} \log P_n(G) \geq - \inf_{x \in G} I(x).$$

It follows that for every function φ of $C_b(\mathbb{X})$

$$\limsup_{n \to \infty} n^{-1} \log \int_{\mathbb{X}} \exp\{n\varphi(x)\} \, dP_n(x) = - \inf_{x \in C} \{\varphi(x) + I(x)\}.$$

These methods have been applied to random walks, Markov or Wiener processes by several authors, in particular Deuschel and Stroock (1984).

1.9 Functional inequalities

Let $(X_i)_{i \geq 1}$ be a sequence of independent random variables defined on a probability space $(\Omega, \mathcal{F}, P)$ and with values in a separable and complete metric space $(\mathcal{X}, \mathcal{B})$ provided with the uniform norm. Let $\mathcal{F}$ be a family of measurable functions defined from $(\mathcal{X}, \mathcal{B})$ to $\mathbb{R}$ and let $S_n(f) = \sum_{i=1}^{n} f(X_i)$. For a variable X having the same distribution probability P_X as the variables X_i, $P_X(f) = Ef(X) = n^{-1} ES_n(f)$ and

$$P_X(f^2) - P_X^2(f) = \operatorname{Var} f(X) = n^{-1} \{ ES_n^2(f) - E^2 S_n(f) \}.$$

The weak convergence of the functional empirical process

$$\nu_n(f) = n^{-\frac{1}{2}}\{S_n(f) - P_X(f)\}$$

to a Brownian bridge with sample-paths in the space $C(\mathcal{X})$ of the continuous functions on $(\mathcal{X}, \mathcal{B})$ has been expressed as a uniform convergence on $\mathcal{F}$ under integrability conditions and conditions for the dimension of $\mathcal{F}$ (Dudley, 1984; Massart, 1983; van der Vaart and Wellner, 1996, Theorem 2.5.2). For the intervals $[0, t]$ of $\mathbb{R}$, the empirical process reduces to $\nu_n(t) = n^{-\frac{1}{2}} \sum_{i=1}^{n} \{1_{\{X_i \leq t\}} - P(X_i \leq t)\}$.

The Kolmós-Major-Tusnady (1975a) representation theorem states the existence of a probability space where a triangular sequence of independent and identically distributed variables $(X_{in})_{i=1,\cdot,n,n \geq 1}$ and a sequence of independent Brownian bridges $(B_n)_{n \geq 1}$ are defined and such that for the empirical process ν_n of $(X_{in})_{i=1,\cdot,n}$, the process $D_n = \sup_{t \in \mathbb{R}} |\nu_n(t) - B_n(t)|$ satisfies

$$P(n^{\frac{1}{2}}|D_n| \geq x + a \log n) \leq be^{-cx}, \ x > 0$$

for positive constants a, b and c. A variant where $\log n$ is replaced by $\log d$, for an integer between 1 and n was given by Mason and van Zwet (1987). Major (1990) proved similar inequalities for the approximation near zero of the empirical process by a sequence of Poisson processes $(P_n(t))_n$ with parameter nt

$$P\Big(n^{\frac{1}{2}} \sup_{t \in [0, n^{-\frac{2}{3}}]} |\nu_n(t) - (P_n(t) - nt)| > C\Big) < K \exp\Big\{-\frac{\sqrt{n} \log n}{8}\Big\}$$

and replacing the Kolmós-Major-Tusnady representation by the tail probability for the supremum over $[t_n, \infty[$, where $nF_n(t_n) = k$ and conditionally on this equality, for all $k \geq 0$.

The class of the quadrants $\mathcal{C} = (\mathcal{C}_x)_{x \in \mathbb{R}^d}$, where $\mathcal{C}_x = \{y \in \mathbb{R}^d : y \leq x\}$ for every x in $\mathbb{R}^d$, has the Vapnik-Chervonenkis index $d+1$ and exponential inequalities for $\|\nu_n\|_{\mathcal{C}_d}$ have been considered. According to the Dvoretzky-Kiefer-Wolfowitz inequality, there exists a constant C such that for every x in $\mathbb{R}^d$, $P(\|\nu_n\|_{\mathcal{C}_d} > x) \leq Ce^{-2x^2}$. Inequalities for general classes have been established by Massart (1986).

The inequalities for the maximum of sums of variables over a class $\mathcal{F}$ such that $\|S_n\|_{\mathcal{F}}$ belongs to L^p are proved in the same way as in $\mathbb{R}$

$$P\Big(\max_{k=1,\dots,n} \|S_k\|_{\mathcal{F}} > \lambda\Big) \leq \frac{1}{\lambda^p} E\|S_n\|_{\mathcal{F}}^p.$$

Let $\sigma^2(\mathcal{F}) = \sup_{f \in \mathcal{F}} \{P_X(f^2) - P_X^2(f)\}$ be the maximum variance of the variables $f(X_i)$ over $\mathcal{F}$ and let $Z_n(\mathcal{F}) = \sup_{f \in \mathcal{F}} \nu_n(f)$ be a functional maximum empirical process. Conditions for the uniform weak convergence of $Z_n(\mathcal{F})$ were studied by Vapnik and Cervonenkis (1981), Dudley (1984), Giné and Zinn (1984). Necessary conditions are $\mathcal{F}$ is totaly bounded for the variance metric ρ and ν_n is asymptotically uniform equicontinuity on $\mathcal{F}$ i.e. for every $\varepsilon > 0$

$$\lim_{\delta \to 0} \limsup_{n \to \infty} P^* \left(\sup_{f,g \in \mathcal{F}, \rho(f-g) < \delta} |\nu_n(f - g)| > \varepsilon \right) = 0.$$

A uniform central limit theorem for set-indexed processes was established by Alexander and Pyke (1986) under moment inequalities and a convergent entropy integral of the family of sets. Exponential inequality for Z_n has been also written for countable classes $\mathcal{F}$ of real functions on $\mathcal{X}$ such that $\sup_{f \in \mathcal{F}} P_X(f)$ and $\sigma^2(\mathcal{F})$ are finite (Massart, 1983; Van der Vaart and Wellner, 1995).

1.10 Content of the book

The next chapters develop extensions and applications of the classical results presented in this introduction. Chapter 2 extends the Cauchy, Hölder and Hilbert inequalities for arithmetic and integral means in real spaces. Hardy's inequality is generalized, several extensions and applications are presented, in particular new versions of weighted inequalities in real analysis. The inequalities for convex transforms of a primitive F are written with a general integration measure and the conditions for the inequalities are discussed. Similar results are established for multilinear functions. The applications in probability concern moments of the maximal variables of independent and identically distributed variables and moments of transformed time variables.

Chapter 3 presents some of the more important analytic inequalities for the arithmetic and geometric means. These include functional means for the power and the logarithm functions. Carlson's inequality (1966) provided upper and lower bounds for the logarithm mean function on $\mathbb{R}_+$, it is improved and the same approach is applied to other functions. For the expansion as a partial sum $A_n(x) = \sum_{k=0}^{n} a_k$ of differentiable functions of $C^{n+2}(\mathbb{R})$, intervals for the ratio $A_{n-1} A_{n+1} A_n^{-2}$ are considered. For the exponential function, the best interval is provided. Inequalities for the arithmetic and the geometric means extend Cauchy's results. Inequalities for the median, the mode and the mean of density or distribution functions are also established. Functional equations, Young's integral inequality and several results about the entropy and the information are proved.

The six following chapters deal with random processes. Chapter 4 begins with inequalities for sums and maximum of n independent random variables. They are extended to discrete martingales, their increments and their maximum with inequalities of the same kind as Burkholder-Davis-Gundy inequalities. The Chernoff and Bennett theorems and other exponential inequalities are generalized to martingales under different conditions. Inequalities for p-order variations of martingales are developed.

Chapter 5 extends these inequalities to time-continuous martingales. They are generalized in several forms to martingales of $\mathcal{M}_0^2$, to their increments, their supremum and p-order conditional moments. Bennet's inequality extends to time-continuous local martingales using Lenglart's inequality under boundedness and integrability conditions, for every $T > 0$ and for every $\lambda > 0$ and $\eta > 0$

$$P\left(\sup_{t \in [0,T]} |M(t)| > \lambda \right) \leq \exp\{-\phi(\lambda \eta^{-\frac{1}{2}})\} + P(< M >_T > \eta).$$

The results are applied to stochastic integrals with respect to martingales and to tightness conditions for martingales and semi-martingales, to Brownian motions and Poisson processes which have deterministic squared variation processes, and to jump processes with random jump sizes. Other questions related to the Brownian motion are also considered, solutions of diffusion equations

$$dX(t) = \alpha(t, X_t)\, dt + \beta(t)\, dB(t)$$

are explicitly established in models for the function α. Martingales in the plane are also considered.

Chapter 6 studies stochastic integrals and exponential martingales or sub-martingales related to martingales and processes with independent increments, with necessary and sufficient conditions for a process to have independent increments, and applications to transformed Brownian motions and Poisson processes. Finally, it presents inequalities for the tail behaviour of random walks and Gaussian processes, and level crossing probabilities. The distribution of the local times for the Brownian motion and its stochastic integrals are determined. Properties of the local times of

Gaussian processes and martingales are studied and central limit theorem for local times are proved.

Chapter 7 concerns inequalities in functional spaces, for sums of real functions of random variables and their supremum on the class of functions defining the transformed variables. Uniform versions of the Burkholder-Davis-Gundy inequality and of the Chernoff, Hoeffding and Bennett theorems are established, they are extended to functionals of discrete or time-continuous martingales. Several applications to the weak convergence of nonstandard empirical processes are detailed.

Chapter 8 focuses on Markov processes with properties of their transition probabilities and applications of the ergodic theorem to recurrent processes, diffusion processes, branching and renewal processes. It presents the Chapman-Kolmogorov equation, properties of the Laplace transform for strong Markov processes X_t and operators of its conditional mean $E(e^{-\lambda X_t} \mid X_s)$, they are extended to time-space Markov processes.

Chapter 9 deals with inequalities for other stochastic processes. First, the inequalities for stationary Gaussian processes deduced from those for their covariance functions are proved, with new results. Then, we consider the distribution of the ruin time of the Sparre Anderson ruin model and in several more optimistic stochastic models with a diffusion term. Finally, we study spatial Poisson, Gaussian and stationary measures, their weak convergence, tail inequalities and properties of their stopping sets are obtained.

Chapter 10 focuses on complex spaces and on the Fourier transform. The classical theory is extended to higher dimensions in order to generalize the expansions of analytic functions of several variables in series. Expansions of functions in the orthonormal basis of the Hermite polynomials and their properties are studied, with the orders of the approximations and their Fourier transforms. The isometry between $\mathbb{R}^2$ and $\mathbb{C}$ is extended to an isometry between $\mathbb{R}^3$ and a complex space where the Fourier transform is also defined. The Cauchy conditions for the differentiability of complex functions and expansions of complex functions are established in this space. The same arguments apply to higher dimensions.

Chapter 2

Inequalities for Means and Integrals

2.1 Introduction

The inequalities presented in the introduction are upper bounds for norms of real vectors and functions. Results for L^p-norms of functions are related to those for vectors as limits of step functions. Some vectorial inequalities depend on the dimension of the vector space, like the equivalence of vectorial norms, and they cannot be immediately adapted to functional spaces. In the next section, the lower bounds for norms of real vectors are also specific to finite vectors.

New functional inequalities of the same kind as those of the first chapter are developed. Most inequalities presented in this chapter rely on the convexity inequalities and they are applied to the theory of integration and to probability inequalities. They are also adapted to bilinear maps by integrating with respect to product measures. Some of them use arithmetic inequalities of the next section. In particular, generalizations of the Hardy and Minkowski inequalities provide inequalities of moments for random variables and for the maximum of n independent and identically distributed random variable. They are also applied to functions of a right-censored time variable and to Laplace transforms of dependent variables.

2.2 Inequalities for means in real vector spaces

Cauchy proved that the arithmetic mean of n real numbers is always larger than their geometric mean defined as the n-root of the product of the corresponding terms. With two terms, Cauchy's inequality is written as

$$\frac{a+b}{2} \geq (ab)^{\frac{1}{2}}, \ a > 0, \ b > 0. \tag{2.1}$$

It is equivalent to $a^2 + b^2 \geq 2ab$ or $(a + b)^2 \geq 4ab$, which holds true for all real numbers a and b.

The concavity of the logarithm implies a generalization of higher powers, for every integer n

$$\left(\frac{a^n + b^n}{2}\right)^{\frac{1}{n}} \geq (ab)^{\frac{1}{2}},$$

for all $a > 0$ and $b > 0$, with equality if and only if $a = b$. For smaller exponents

$$\frac{a^{\frac{1}{n}} + b^{\frac{1}{n}}}{2} \geq (ab)^{\frac{1}{2n}}.$$

These inequalities extend to a real exponent $x > 0$, $a^x + b^x \geq (2\sqrt{ab})^x$, and to means and products of n terms.

Proposition 2.1. *For every positive real numbers* $(a_i)_{i=1,\ldots,n}$

$$n^{-1}\sum_{i=1}^{n} a_i \geq \left(\prod_{i=1}^{n} a_i\right)^{\frac{1}{n}} \qquad (2.2)$$

and for every real $x > 0$

$$\left(n^{-1}\sum_{i=1}^{n} a_i^x\right)^{\frac{1}{x}} \geq \left(\prod_{i=1}^{n} a_i\right)^{\frac{1}{n}}, \qquad (2.3)$$

with equality if and only if $a_1 = a_2 = \cdots = a_n$.

Proof. The first inequality is due to the concavity of the logarithm $x^{-1}\log(n^{-1}\sum_{i=1}^{n} a_i^x) \geq n^{-1}\sum_{i=1}^{n}\log a_i$, with equality if and only if all terms are equal. It is also written as

$$\log\left(n^{-1}\sum_{i=1}^{n} a_i^x\right) \geq xn^{-1}\sum_{i=1}^{n}\log a_i, \ x > 0$$

which yields (2.3). $\qquad\qquad\qquad\qquad\qquad\qquad\qquad\qquad\qquad\square$

Proposition 2.2. *Let* μ *be a positive measure on* $I \subset \mathbb{R}$, *for every real* $p > 0$ *and every positive real function* a *on* I

$$\left(\int_I a^p(t)\,d\mu(t)\right)^{\frac{1}{p}} \geq \exp\left\{\int_I \log a(t)\,d\mu(t)\right\},$$

with equality if and only if a *is constant. For every random variable* X *and for every positive real function* a *such that* $\|a(X)\|_p$ *is finite*

$$\exp\{E\log a(X)\} \leq \|a(X)\|_p.$$

Proof. Let $(x_{i,n})_{i \leq I_n, n \geq 1}$ be partition of the interval I such that $\mu([x_{i,n}, x_{i+1,n}]) = n^{-1}$. The inequality is an application of (2.2) to the real numbers $a_{i,n}$ defining a sequence of functions $a_n(t) = \sum_{i=1}^n a_{i,n} 1_{[x_{i,n}, x_{i+1,n}]}$ converging to a as n tends to infinity. The second equation is an equivalent formulation of the same inequality with a random variable having the distribution μ. $\qquad\square$

These inequalities apply directly to bounds of moments of random variables. On a probability space $(\Omega, \mathcal{F}, P)$, let X and Y be positive real random variables, their covariance is bounded by the square root of the product of their variances from the Cauchy-Schwarz inequality. The following inequalities are deduced from Proposition 2.1, for real variables X and Y

$$\mathrm{Cov}(X, Y) \leq \frac{1}{2}(\mathrm{Var}X + \mathrm{Var}Y)$$

and for every $p > 0$

$$E(XY) \leq E\left\{ \left(\frac{X^{2p} + Y^{2p}}{2} \right)^{\frac{1}{p}} \right\}.$$

If variables X and Y are colinear, the first inequality is reduced to the Cauchy inequality and these inequalities are strict unless $X = Y$. By convexity we have

$$E\{(X + Y)^p\} \leq 2^{p-1}\{E(X^p) + E(Y^p)\}.$$

Let $(X_1, \ldots, X_n)$ be a random vector and let $\bar{X}_n = n^{-1} \sum_{i=1}^n X_i$ be the empirical mean, by convexity

$$E\left\{ \left(\sum_{i=1}^n X_i \right)^p \right\} \leq n^{p-1} \left(\sum_{i=1}^n X_i^p \right),$$

and the higher moments of the sequence of variables satisfy

$$E\left\{ \left(n^{-1} \sum_{i=1}^n X_i^p \right)^{\frac{1}{p}} \right\} \geq E\left(\prod_{i=1}^n X_i^{\frac{1}{n}} \right),$$

for every integer n and for $p > 0$. From (2.3), moments of independent and identically distributed variables satisfy

$$E\left(\bar{X}_n^{\frac{1}{p}} \right) \geq \left(EX^{\frac{1}{np}} \right)^n,$$

$$E\left\{ \left(n^{-1} \sum_{i=1}^n X_i^p \right)^{\frac{1}{p}} \right\} \geq \left(EX^{\frac{1}{n}} \right)^n.$$

From (2.2), $E(X^p) \leq (EX^{\frac{p}{n}})^n$, for every integer n and for $p > 0$.

Proposition 2.3. *Let $X > 0$ be a real random variable with distribution function F_X and let ϕ be a convex function of $C(\mathbb{R})$ with a convex derivative $\phi^{(1)}$ belonging to $L^p(F_X)$, $p > 1$. The covariance of X and $\phi(X)$ is such that*

$$\text{Cov}(X, \phi(X)) \leq \|X\|_{L^p} \|\phi^{(1)}(X)\|_{L^{p'}}$$

with the conjugate p' of p.

Proof. Let $\mu_X = EX$, there exists θ in $[0,1]$ such that $\phi(X) = \phi(\mu_X) + (X - \mu_X)\phi^{(1)}(\mu_X + \theta(X - \mu_X))$ is lower than $\theta\phi^{(1)}(X) + (1 - \theta)\phi^{(1)}(\mu_X)$. By convexity, $E\phi(X) \geq \phi(\mu_X)$,

$$E\{X\phi(X)\} \leq \mu_X\phi(\mu_X) + E\{(X - \mu_X)\phi^{(1)}(\mu_X + \theta(X - \mu_X))\},$$
$$\text{Cov}\{X, \phi(X)\} \leq \|X\|_{L^p}\{\theta\|\phi^{(1)}(X)\|_{L^{p'}} + (1 - \theta)|\phi^{(1)}(\mu_X)|$$

and $|\phi^{(1)}(\mu_X)| = \|\phi^{(1)}(\mu_X)\|_{L^{p'}} \leq \|\phi^{(1)}(X)\|_{L^{p'}}$. $\square$

Let a, b and c be real numbers, Proposition 2.1 states that

$$(a^3 + b^3 + c^3) \geq 3abc \quad \text{and} \quad \frac{a + b + c}{3} \geq (abc)^{\frac{1}{3}},$$

by a change of variables. Closer inequalities for sums of products rather than their powers are obtained by multiplying both sides of inequalities similar to $a^2 + b^2 \geq 2ab$ for each pair of (a, b, c) by another term and adding them, or by reparametrization of another inequality. The first inequality of the next proposition comes from Lohwater (1982) and it is generalized in several equivalent forms.

Proposition 2.4. *The following inequalities apply all positive real numbers a, b and c, and they are all equivalent*

$$a^2 + b^2 + c^2 \geq ab + ac + bc,$$
$$a^2b^2 + a^2c^2 + b^2c^2 \geq abc(a + b + c),$$
$$c^{-1}ab + b^{-1}ac + a^{-1}bc \geq a + b + c,$$
$$a^3 + b^3 + c^3 \geq (ab)^{\frac{3}{2}} + (bc)^{\frac{3}{2}} + (ac)^{\frac{3}{2}},$$
$$a + b + c \geq (ab)^{\frac{1}{2}} + (bc)^{\frac{1}{2}} + (ac)^{\frac{1}{2}},$$

with equalities if and only if $a = b = c$.

The results of Proposition 2.4 are generalized to p terms, by the same method.

Proposition 2.5. *Let p be an integer and let $(a_i)_{i=1,\ldots,p}$ be a vector of positive real numbers, then the following inequality*

$$\sum_{i=1}^{p} a_i \geq \sum_{i=1}^{p} \sum_{j \neq i, j=1}^{p} (a_i a_j)^{\frac{1}{2}},$$

is equivalent to

$$\sum_{i=1}^{p} a_i^2 \geq \sum_{i=1}^{p} \sum_{j \neq i, j=1}^{p} a_i a_j,$$

$$\sum_{i=1}^{p} a_i^3 \geq \sum_{i=1}^{p} \sum_{j \neq i, j=1}^{p} (a_i a_j)^{\frac{3}{2}},$$

$$\sum_{i=1}^{p} \sum_{j \neq i, j=1}^{p} a_i^2 a_j^2 \geq \Big(\prod_{i=1}^{p} a_i\Big)\Big(\sum_{j=1}^{p} a_j\Big),$$

$$\sum_{i=1}^{p} \sum_{j \neq i, j=1}^{p} \sum_{k \neq i,j; k=1}^{p} a_i a_j a_k^{-1} \geq \sum_{i=1}^{p} a_i,$$

and the equality holds if and only if $a_1 = a_2 = \cdots = a_p$.

Replacing the constants a_i by real functions $a_i(t)$ and integrating with respect to a positive measure μ yields functional inequalities similarly to Proposition 2.2. They are proved by convexity. For all positive real functions and for every random variable X, Proposition 2.5 implies

$$\sum_{i=1}^{p} E\{a_i(X)\} \geq \sum_{i=1}^{p} \sum_{j \neq i, j=1}^{p} E[\{a_i(X)a_j(X)\}^{\frac{1}{2}}],$$

$$\sum_{i=1}^{p} E[\{a_i(X)\}^2] \geq \sum_{i=1}^{p} \sum_{j \neq i, j=1}^{p} E\{a_i(X)a_j(X)\},$$

$$\sum_{i=1}^{p} E[\{a_i(X)\}^3] \geq \sum_{i=1}^{p} \sum_{j \neq i, j=1}^{p} E[\{a_i(X)a_j(X)\}^{\frac{3}{2}}],$$

$$\sum_{i=1}^{p} \sum_{j \neq i, j=1}^{p} E[\{a_i(X)\}^2 \{a_j(X)\}^2] \geq \sum_{i=1}^{p} E\{a_i(X) \prod_{j=1}^{p} a_j(X)\},$$

$$\sum_{i=1}^{p} \sum_{j \neq i, j=1}^{p} \sum_{k \neq i,j; k=1}^{p} E\frac{a_i(X)a_j(X)}{a_k(X)} \geq \sum_{i=1}^{p} E\{a_i(X)\}$$

with equality if and only if a is constant. Let ϕ be a convex function, for every integer p and every vector $(a_i)_{i=1,\ldots,p}$ of $[0,1]^p$

$$E\Big\{\phi\Big(\sum_{i=1}^{p} a_i X_i\Big)\Big\}^k \leq E\Big\{\sum_{i=1}^{p} a_i \phi(X_i)\Big\}^k \leq p^k \sum_{i=1}^{p} E\{a_i^k \phi^k(X_i)\}.$$

2.3   Hölder and Hilbert inequalities

Extensions of Cauchy's inequality to positive bilinear series have been introduced by Hilbert. The next bound for the series $\sum_{n\geq 1}\sum_{m\geq 1}(m+n)^{-1}x_n\,y_m$ is obtained by using the inequality $m+n \geq 2(mn)^{\frac{1}{2}}$ and Hölder's inequality.

Proposition 2.6. *Let $\alpha > 2$ and let $p > 1$ and $q > 1$, then for all positive series $(x_n)_{n\geq 1}$ and $(y_m)_{m\geq 1}$*

$$\sum_{n\geq 1}\sum_{m\geq 1}\frac{x_n y_m}{(n+m)^\alpha} \leq \frac{1}{2^\alpha}\Big(\sum_{n\geq 1}n^{-\frac{\alpha}{2}}x_n\Big)\Big(\sum_{m\geq 1}m^{-\frac{\alpha}{2}}y_m\Big)$$

$$\leq c_{p,q}\|(x_n)_n\|_p\,\|(y_n)_n\|_q,$$

with the constant

$$c_{p,q} = \Big\{\sum_{n\geq 1}n^{-\frac{\alpha p}{2(p-1)}}\Big\}^{1-\frac{1}{p}}\Big\{\sum_{n\geq 1}n^{-\frac{\alpha q}{2(q-1)}}\Big\}^{1-\frac{1}{q}}.$$

Proposition 2.7. *For $p > 1$ and $q > 1$, let α be a real number strictly larger than the minimum of $(p-1)p^{-1}$ and $(q-1)q^{-1}\}$ and let f and g be real functions on intervals $[a,b)$ and respectively $[c,d)$, with $a > 0$ and $c > 0$, then there exists a constant $C_{p,q}$ such that*

$$\left|\int_a^b\int_c^d\frac{f(x)g(y)}{(x+y)^{2\alpha}}\,dxdy\right| \leq \frac{1}{2^{2\alpha}}\left|\int_a^b\frac{f(x)}{x^\alpha}\,dx\right|\left|\int_c^d\frac{g(y)}{y^\alpha}\,dy\right|$$

$$\leq C_{p,q}\|f\|_p\,\|g\|_q.$$

This is a consequence of the Hölder inequality with the constant

$$C_{p,q} = \frac{1}{2^{2\alpha}}\frac{p-1}{p}\left(\frac{1}{b^{(\alpha p-1)(p-1)^{-1}}}-\frac{1}{a^{(\alpha p-1)(p-1)^{-1}}}\right)^{\frac{p-1}{p}}$$

$$\times\frac{q-1}{q}\left(\frac{1}{b^{(\alpha q-1)(q-1)^{-1}}}-\frac{1}{a^{(\alpha q-1)(q-1)^{-1}}}\right)^{\frac{q-1}{q}}.$$

The integration bounds b and d may be finite or not.

Let X and Y be independent random variables with real distribution functions F and G and densities f and g. Proposition 2.7 implies that for every $\beta > 0$

$$|E(X+Y)^{-2\alpha}| \leq 2^{-2\alpha}|EX^{-\alpha}|\,|EY^{-\alpha}|$$

$$\leq 2^{-2\alpha}\|X^{-\beta}\|_p\,\|X^{\beta-\alpha}\|_{p'}\,\|Y^{-\beta}\|_q\,\|Y^{\beta-\alpha}\|_{q'}.$$

With the distribution functions, the first inequality of Proposition 2.7 is also written as

$$\left| \int_a^b \int_c^d \frac{1}{(x+y)^{2\alpha}} \, dF(x) dG(y) \right| \le \frac{1}{2^{2\alpha}} \left| \int_a^b \frac{1}{x^\alpha} \, dF(x) \right| \left| \int_a^b \frac{1}{x^\alpha} \, dG(x) \right|.$$

Hilbert's inequality has been extended by Schur (1912) and by Hardy, Littlewood, Pólya (1952) to bound a scalar product of f and g defined with respect to a measure on $\mathbb{R}^2$. Here is another version of this inequality.

Proposition 2.8. *Let p and p' be conjugates and let K be a positive homogeneous weighting function on $\mathbb{R}_+^2$ such that*

$$k_p = \min \left\{ \int_0^\infty x^{-\frac{1}{p}} K(1,x) \, dx, \int_0^\infty x^{-\frac{1}{p'}} K(x,1) \, dx \right\}$$

is finite. For all functions f in $L^p(\mathbb{R}_+)$ and g in $L^{p'}(\mathbb{R}_+)$

$$\left| \int_0^\infty \int_0^\infty K(x,y) f(x) g(y) \, dx \, dy \right| \le k_p \|f\|_p \|g\|_{p'}.$$

With $K(x,y) = (x+y)^{-1}$, $k_p = \min\{p^{-1}, p'^{-1}\}$ is always smaller than the bound $\pi \sin^{-1}(p^{-1}\pi)$ of the Hilbert inequality. Let $\lambda > 0$ be real and let $K(x,y) = (x+y)^{-\lambda}$ be a kernel function, then the constant is $k_p = \lambda \min\{p^{-1}, p'^{-1}\}$.

Let (X,Y) be a random variable defined from a probability space $(\Omega, \mathcal{F}, P)$ to $\mathbb{R}^2$. Hölder's inequality can be expressed as an inequality for (X,Y) and functions of (X,Y) as X and Y are dependent.

Proposition 2.9. *Let φ and ψ be real functions on $\mathbb{R}$ and let $p > 1$, then there exists a constant k_p such that for every random variable (X,Y) with values in $\mathbb{R}^2$ and such that $\varphi(X)$ belongs to L^p and $\psi(Y)$ to $L^{p'}$*

$$|E\{\varphi(X)\psi(Y)\}| \le k_p E\|\varphi(X)\|_p E\|\psi(Y)\|_{p'},$$

where the best constant is 1 if X and Y are independent.

More generally, a similar inequality applies to multidimensional random variables. Let (X,Y) be a random variable defined from Ω to $\mathbb{R}^{2n}$ and such that $\varphi(X)$ belongs to L^p and $\psi(Y)$ to $L^{p'}$ and let $X = (X_i)_{i=1,\dots,n}$ and $Y = (Y_i)_{i=1,\dots,n}$, then

$$E| < \varphi(X), \psi(Y) > | = E| \sum_{i=1}^n \varphi(X_i)\psi(Y_i)|$$

$$\le k_p E\|\varphi(X)\|_p E\|\psi(Y)\|_{p'},$$

with the $L^p(\mathbb{R}^n)$ norm $\|x\|_p = (\sum_{i=1}^n |x_i|^p)^{\frac{1}{p}}$. The converse of the Hölder inequality for vectors can also be written in the L^p spaces. Let $p > 1$, a necessary condition for the existence of a constant k such that $\int_0^\infty |f(x)g(x)|\, dx \le k\|f\|_p$, for all functions f in L^p, is the existence of a finite norm $\|g\|_{p'}$, for every g, where p' is the conjugate of p.

2.4 Generalizations of Hardy's inequality

Let $p > 1$ with conjugate p', such that $p^{-1} + p'^{-1} = 1$, and let f and g be positive integrable real functions on $\mathbb{R}_+$ such that f belongs to L^p and g to $L^{p'}$. Let F and G be their primitive functions such that

$$\lim_{x \to 0} x^{\frac{1}{p}-1}F(x) = 0, \ \lim_{x \to \infty} x^{\frac{1}{p}-1}F(x) = 0, \tag{2.4}$$

$$\lim_{x \to 0} x^{-\frac{1}{p'}}G(x) = 0 \ \lim_{x \to \infty} x^{-\frac{1}{p'}}G(x) = 0.$$

Hardy's inequality yields the next inequality

$$I_p = \int_0^\infty \left\{\frac{F(x)}{x}\right\}^p dx \le \left(\frac{p}{p-1}\right)^p \|f\|_p^p, \tag{2.5}$$

$$I_{p'} = \int_0^\infty \left\{\frac{G(x)}{x}\right\}^{\frac{p-1}{p}} dx \le p^{\frac{p}{p-1}} \|g\|_{p'}^{p'}.$$

The Hardy's inequality (2.5) is proved by integrating by parts the Hölder inequality

$$I_p = \frac{p}{p-1} \int_0^\infty \left\{\frac{\bar{F}(x)}{x}\right\}^{p-1} f(x)\, dx,$$

$$\int_0^\infty \left\{\frac{\bar{F}(x)}{x}\right\}^{p-1} f(x)\, dx \le \bar{I}_p^{1-\frac{1}{p}} \|f\|_p.$$

By convexity, the integral

$$N_p(F) = \left[\int_0^\infty \left\{\frac{F(x)}{x}\right\}^p dx\right]^{\frac{1}{p}}$$

defines a norm in the space of the primitives of functions of $L^p(\mathbb{R}_+)$. It is related to a scalar product by (1.1), with the norm N_2, hence it is a weighted L^2 scalar product of the primitive functions

$$< F, G > = \int_0^\infty x^{-2}F(x)G(x)dx.$$

By the geometric equalities (1.2)

$$< F, G > = \frac{1}{4}\{N_2(F + G) - N_2(F - G)\}.$$

The integral $\int_0^\infty x^{-2}F(x)G(x)\,dx$ is bounded using successively Hölder's inequality and Hardy's inequality.

For all conjugates $p > 1$ and p'

$$\int_0^\infty \frac{F(x)G(x)}{x^2}\,dx \le \frac{p^2}{p-1}\Big\{\int_0^\infty f^p(x)\,dx\Big\}^{\frac{1}{p}}\Big\{\int_0^\infty g^{\frac{p}{p-1}}(x)\,dx\Big\}^{1-\frac{1}{p}}. \quad (2.6)$$

For a probabilistic application, let S be a positive real random variable with distribution function H, and let $p > 1$ and let p' be its conjugate, such that $p^{-1} + p'^{-1} = 1$. Let a be a function in $L^p(\mathbb{R}_+)$ and b be a function in $L^{p'}(\mathbb{R}_+)$, such that $\int a\,dH = 0$ and $\int b\,dH = 0$, the inequalities (1.15) and (2.6) are expressed in terms of mean functions of S as

$$E\Big(\frac{\int_0^S a\,dH}{H(S)}\Big)^p \le \Big(\frac{p}{p-1}\Big)^p Ea^p(S), \quad (2.7)$$

$$E\frac{\int_0^S a\,dH}{H(S)}\frac{\int_0^S b\,dH}{H(S)} \le \frac{p^2}{p-1}\{Ea^p(S)\}^{\frac{1}{p}}\{Eb^{\frac{p}{p-1}}(S)\}^{1-\frac{1}{p}}.$$

This is a reparametrization of inequalities (2.6). The variable $X = H(S)$ has a uniform distribution in $[0,1]$ and, with the notations $H(s) = x$, $F(s) = \int_0^s a\,dH$ and $G(s) = \int_0^s b\,dH$, we get

$$\frac{\int_0^S a\,dH}{H(S)} = \frac{F \circ H^{-1}(X)}{X},$$

$$Ea^p(S) = \int_0^\infty a^p(s)\,dH(s) = \int_0^\infty \{a \circ H^{-1}(x)\}^p\,dx,$$

moreover

$$E\Big(\frac{\int_0^S a\,dH}{H(S)}\Big)^p = \int_0^\infty \Big(\frac{F}{H}\Big)^p\,dH = \int_0^1 \Big(\frac{F \circ H^{-1}(x)}{x}\Big)^p\,dx,$$

where the derivative of $F \circ H^{-1}(x)$ is $a \circ H^{-1}(x)$. The first inequality is equivalent to the Hardy inequality since $\int a\,dH = 0$. The integrals of the function b have the same reparametrization and the second inequality is deduced from the Hölder inequality. The inequalities (2.7) are also proved directly by the same arguments as (2.5), with integrals in $\mathbb{R}_+$.

There are no inequalities of the same kind as (1.14), (1.15) and (2.6) for the decreasing survival function $\bar{F}(x) = \int_x^\infty f(t)\,dt$ on $\mathbb{R}_+$ since $\bar{F}(0) = 1$. Let f be an integrable positive function on $\mathbb{R}_+$ with a survival function $\bar{F}$. Let $a > 0$ and $p > 1$, an integration by parts implies

$$\bar{I}_p(a) := \int_a^\infty \Big\{\frac{\bar{F}(x)}{x}\Big\}^p\,dx = \frac{1}{p-1}\Big[\frac{\bar{F}^p(a)}{a^{p-1}} - p\int_a^\infty \Big\{\frac{\bar{F}(x)}{x}\Big\}^{p-1}f(x)\,dx\Big],$$

and by the Hölder inequality, we obtain

$$\int_a^\infty \left\{ \frac{\bar{F}(x)}{x} \right\}^{p-1} f(x)\, dx \leq \bar{I}_p^{1-\frac{1}{p}}(a) \left(\int_a^\infty f^p \right)^{\frac{1}{p}},$$

$$\bar{I}_p(a) \geq \frac{1}{(p-1)^p} \left\{ \bar{I}_p^{(\frac{1}{p}-1)}(a) \frac{\bar{F}^p(a)}{a^{p-1}} - p \left(\int_a^\infty f^p \right)^{\frac{1}{p}} \right\}^p.$$

These inequalities are strict except if f is a constant. Let f be the density of a positive variable X, $F(x) = P(X \leq x)$ and $\bar{F}(x) = P(X \geq x)$, therefore F and $\bar{F}$ are bounded by one, $F(0) = 0$ and $\bar{F}(0) = 1$, $\lim_{x \to \infty} F(x) = 1$ and $\lim_{x \to \infty} \bar{F}(x) = 0$. The limiting condition (2.4) for F at infinity is satisfied only if $p > 1$.

Proposition 2.10. *Let $p > 1$ and let $\bar{F}$ be defined in $\mathbb{R}$ and satisfy the condition $\lim_{x \to \pm\infty} x^{\frac{1}{p}-1}\bar{F}(x) = 0$, then*

$$\int_{-\infty}^\infty \left\{ \frac{\bar{F}(x)}{x} \right\}^p dx \geq \left(\frac{p}{1-p} \right)^p \|f\|_p^p.$$

Proof. It is proved as above, integrating by parts

$$\bar{I}_p = \int_{-\infty}^\infty \left\{ \frac{\bar{F}(x)}{x} \right\}^p dx = \frac{p}{1-p} \int_{-\infty}^\infty \left\{ \frac{\bar{F}(x)}{x} \right\}^{p-1} f(x)\, dx$$

then, by the Hölder inequality

$$\int_{-\infty}^\infty \left\{ \frac{\bar{F}(x)}{x} \right\}^{p-1} dx \leq \bar{I}_p^{1-\frac{1}{p}} \|f\|_p.$$

$\square$

Since $\bar{F}$ is decreasing, the integral $\bar{I}_p$ is negative and its lower bound is necessarily negative. These results are generalized by replacing the normalization of F and $\bar{F}$ by any distribution function and the bounds are not modified. The proof of this new result is similar to the previous ones.

Proposition 2.11. *Let $p > 1$ and let F and G be distribution functions in $\mathbb{R}_+$ such that $H^{-1}F$ belongs to $L^p(H)$, F has a density f and*

$$\lim_{x \to \infty} \frac{F^p}{H^{p-1}}(x) = 0 = \lim_{x \to 0} \frac{F^p}{H^{p-1}}(x),$$

then

$$\int_0^\infty \left\{ \frac{\bar{F}(x)}{H(x)} \right\}^p dH(x) \leq \left(\frac{p}{1-p} \right)^p \|f\|_p^p. \tag{2.8}$$

Proposition 2.12. *Let F and G be distribution functions in $\mathbb{R}$ such that $H^{-1}F$ belongs to $L^p(H)$, F has a density f and*

$$\lim_{x \to \pm\infty} \frac{F^p}{H^{p-1}}(x) = 0,$$

then

$$\int_0^\infty \left\{ \frac{\bar{F}(x)}{H(x)} \right\}^p dH(x) \geq \left(\frac{p}{1-p} \right)^p \|f\|_p^p. \tag{2.9}$$

The norms

$$N_p(F) = \left[\int_0^\infty \left\{ \frac{F(x)}{H(x)} \right\}^p dH(x) \right]^{\frac{1}{p}}, \ p \geq 1,$$

and the scalar products

$$< F, G > = \int_0^\infty H^{-2}(x) F(x) G(x) dx$$

are deduced from Proposition 2.11.

Inequalities (1.14)–(1.15) and Proposition 2.11 extend to fractional and real convex power functions and to other convex functions.

Proposition 2.13. *Let $p > 1$ and $1 \leq q < p$ and let $r = pq^{-1}$. Let f be a positive function of $L^r(\mathbb{R}_+)$ such that $\lim_{x \to 0} x^{\frac{1}{r}-1}F(x) = 0$ and $\lim_{x \to \infty} x^{\frac{1}{r}-1}F(x) = 0$, then*

$$I_r = \int_0^\infty \left(\frac{F(x)}{x} \right)^r dx \leq \left(\frac{p}{p-q} \right)^r \int_0^\infty f^r(x) dx, \tag{2.10}$$

and there is equality if and only if f is a constant.

Proof. Integrating by parts implies

$$I_r = \frac{p}{q} \left\{ I_r - \int_0^\infty \left(\frac{F(x)}{x} \right)^{r-1} f(x) dx \right\},$$

$$I_r = \frac{p}{p-q} \int_0^\infty \left(\frac{F(x)}{x} \right)^{r-1} f(x) dx.$$

Let $p = p(p-q)^{-1} > 1$, then the Hölder inequality yields

$$\int_0^\infty \left(\frac{F(x)}{x} \right)^{r-1} f(x) dx \leq I_r^{\frac{1}{p}} \left(\int_0^\infty f^\beta \right)^{\frac{1}{\beta}}$$

with $\beta^{-1} + p^{-1} = 1$, hence $\beta = r$. It follows that

$$I_r \leq \frac{p}{p-q} I_r^{(p-q)p^{-1}} \left(\int_0^\infty f^r \right)^{\frac{1}{r}}.$$

$\square$

The same inequality holds for the integral $I_r(y)$, as in Equation (1.14). For $\bar{I}_r$, the integration by parts leads to an opposite inequality with an additive constant, as previously.

Theorem 2.1. *Let $\lambda > 1$ be a real number and let f be a positive function in $L^\lambda(\mathbb{R}_+)$ with primitive F such that $\lim_{x \to 0} x^{\lambda^{-1}-1} F(x)$ and $\lim_{x \to \infty} x^{\lambda^{-1}-1} F(x)$ are zero, then*

$$I_\lambda = \int_0^\infty \left(\frac{F(x)}{x}\right)^\lambda dx \leq \left(\frac{\lambda}{\lambda-1}\right)^\lambda \int_0^\infty f^\lambda(x)\, dx, \qquad (2.11)$$

with equality if and only if f is constant.

The inequalities extend to weighted inequalities more general than the inequalities by Kufner and Persson (2003), with similar proofs as Theorem 2.1.

Theorem 2.2. *Let $\lambda > 1$ and $\alpha > 0$ be real numbers and let f be a positive function such that $\int_0^\infty f^\lambda(x)x^\alpha\, dx < \infty$. If $\lambda > \alpha+1$, $\lim_{x \to 0} x^{\frac{\alpha+1}{\lambda-1}} F(x) = 0$ and $\lim_{x \to 0} x^{\frac{\alpha+1}{\lambda-1}} F(x) = 0$, then*

$$I_{\lambda,\alpha} := \int_0^\infty \left(\frac{F(x)}{x}\right)^\lambda x^\alpha\, dx = \frac{\lambda}{\lambda-\alpha-1} \int_0^\infty \left(\frac{F(x)}{x}\right)^{\lambda-1} f(x)x^\alpha\, dx$$

$$\leq \left(\frac{\lambda}{\lambda-\alpha-1}\right)^\lambda \int_0^\infty f^\lambda(x)x^\alpha\, dx,$$

with equality if and only if f is constant.

Proof. Integrating by parts implies and using the Hölder inequality yields

$$I_{\lambda,\alpha} = \frac{\lambda}{\alpha+1}\left\{I_{\lambda,\alpha} - \int_0^\infty \left(\frac{F(x)}{x}\right)^{\lambda-1} x^{\frac{\alpha(\lambda-1)}{\lambda}} f(x)x^{\frac{\alpha}{\lambda}}\, dx\right\},$$

$$I_{\lambda,\alpha} \leq \frac{\lambda}{\lambda-\alpha-1} I_{\lambda,\alpha}^{\frac{\lambda-1}{\lambda}} \left\{\int_0^\infty f^\lambda(x)x^\alpha\, dx\right\}^{\frac{1}{\lambda}}.$$

$\square$

Corollary 2.1. *Let $X_1, \ldots, X_n$ be a sequence of independent and identically distributed random variables with a density f on $\mathbb{R}_+$. For every $1 \leq p \leq n-1$, its maximum X_n^* satisfies*

$$E\{(X_n^*)^{-p}\} = n\left(\frac{n}{p}\right)^{n-1} \int_0^\infty f^n(x)x^{n-p-1}\, dx.$$

Proof. First, let $p = n - 1$. Let F be the distribution function of the variables X_i, the distribution of their maximum X_n^* is F^n and

$$E\{(X_n^*)^{-(n-1)}\} = n \int_0^\infty \left(\frac{F(x)}{x}\right)^{n-1} f(x) \, dx \leq n I_n^{1-\frac{1}{n}} \|f\|_n,$$

by Hölder's inequality. Applying Hardy's inequality to I_n leads to

$$I_n^{1-\frac{1}{n}} \leq \left(\frac{n}{n-1}\right)^{n-1} \left(\int_0^\infty f^n(x) \, dx\right)^{1-\frac{1}{n}}$$

and the result for $p = n - 1$ follows. For $1 \leq p < n - 1$, the inequality is proved using Theorem 2.2, with $\lambda = n$ and $\alpha = n - p - 1$, and the inequality

$$E\{(X_n^*)^{-(p)}\} = n \int_0^\infty \left(\frac{F(x)}{x}\right)^{n-1} x^{n-p-1} f(x) \, dx$$

$$\leq n I_{n,n-p-1}^{1-\frac{1}{n}} \left(\int_0^\infty f^n(x) x^{n-p-1} \, dx\right)^{\frac{1}{n}},$$

$$I_{n,n-p-1}^{1-\frac{1}{n}} \leq \left(\frac{n}{p}\right)^{n-1} \left(\int_0^\infty f^n x^{n-p-1} \, dx\right)^{1-\frac{1}{n}}.$$

$\square$

This result extends to real moments $E\{(X_n^*)^{-\alpha}\}$, $\alpha > 0$. Theorem 2.2 does not apply to $E\{(X_n^*)^p\}$, with $p > 0$, since the condition $\lambda > \alpha + 1$ is not fulfilled with $\lambda = n$ and $\alpha = n + p - 1$. Integrations similar to the calculus of $E\{(X_n^*)^{-p}\}$ cannot be bounded for positive moments $E\{(X_n^*)^p\}$.

The previous inequalities are specific to power functions. For a general convex function Φ, the more general inequality proved by Kufner and Persson (2003) is the next theorem, with the best constant 1. For power functions Φ on $\mathbb{R}_+$, it does not reduce to Hardy's inequality (1.14).

Theorem 2.3 (Hardy-Knopp's inequality). *Let Φ be a convex and increasing function, for every positive function f with primitive F and such that $x^{-1}\Phi \circ f(x)$ belongs to $L^1(\mathbb{R}_+)$ and $\Phi \circ f(0) = 0$*

$$\int_0^\infty \Phi\left\{\frac{F(x)}{x}\right\} \frac{dx}{x} \leq \int_0^\infty \Phi(f(x)) \frac{dx}{x}. \tag{2.12}$$

Proof. This is a consequence of the convex inequality

$$\Phi\left\{\frac{1}{x} \int_0^x f(t) \, dt\right\} \leq \frac{1}{x} \int_0^x \Phi \circ f(t) \, dt,$$

integrating by parts the integrals of both sides of this inequality implies

$$\int_0^\infty \Phi\left\{\frac{F(x)}{x}\right\} \frac{dx}{x} \leq \int_0^\infty \Phi(f(x)) \frac{dx}{x}.$$

$\square$

Weighted extensions of Hardy-Knopp's inequality are straightforward.

Theorem 2.4. *Let Φ be a convex positive monotone function and let $\alpha > 0$ be a real number. For every positive function f with primitive F and such that the function $x^{-\alpha}\Phi \circ f(x)$ belongs to $L^1(\mathbb{R}_+)$*

$$\int_0^\infty \Phi\left\{\frac{F(x)}{x}\right\}\frac{dx}{x^\alpha} \leq \frac{1}{\alpha}\int_0^\infty \Phi(f(x))\frac{dx}{x^\alpha}. \qquad (2.13)$$

Proof. For every $x > 0$, the convexity of Φ is expressed as

$$\Phi\left(\frac{F(x)}{x}\right) \leq \frac{\int_0^x \Phi(f)(t)\,dt}{x}.$$

Denoting $g = \Phi(f)$, it is still written as

$$\Phi\left\{x^{-1}\int_0^x \Phi^{-1}(g)(t)\,dt\right\} \leq x^{-1}G(x)$$

with the primitive G of g, and by Fubini integration lemma, it implies

$$\int_0^\infty \Phi\left\{\frac{F(x)}{x}\right\}\frac{dx}{x^\alpha} = \int_0^\infty \Phi\left\{\frac{\int_0^x \Phi^{-1}(g)(t)\,dt}{x}\right\}\frac{dx}{x^\alpha}$$

$$\leq \int_0^\infty \left(\int_t^\infty \frac{dx}{x^{\alpha+1}}\right)g(t)\,dt$$

$$= \frac{1}{\alpha}\int_0^\infty \Phi(f(t))\frac{dt}{t^\alpha}.$$

$\square$

Example 2.1. With the exponential function Φ, for every function f such that $x^{-1}e^{f(x)}$ belongs to $L^1(\mathbb{R}_+)$, inequality (2.13) is written

$$\int_0^\infty e^{x^{-1}F(x)}x^{-1}\,dx \leq \int_0^\infty e^{f(x)}x^{-1}\,dx.$$

If $x^{-1}f(x)$ is integrable, it becomes

$$\int_0^\infty x^{-1}\exp\left\{x^{-1}\int_0^x \log\{f(t)\}\,dt\right\}dx \leq \int_0^\infty x^{-1}f(x)\,dx$$

and, for an integrable function f

$$\int_0^\infty x^{-1}\exp\left\{x^{-1}\int_0^x \log\{tf(t)\}\,dt\right\}dx \leq \int_0^\infty f(x)\,dx.$$

Example 2.2. Let $f(x) = -x^{-1}$, the inequality (2.12) implies

$$\int_0^\infty x^{-(1+x^{-1})}\,dx \leq \int_0^\infty e^{-x^{-1}}x^{-1}\,dx < \infty.$$

By convexity, Theorem 2.4 extends to a Minkowski type inequality for every convex and monotone real function $\Phi > 0$

$$\int_0^\infty \Phi\Big\{\frac{(F+G)(x)}{x}\Big\}\,\frac{dx}{x^\alpha} \leq \frac{1}{\alpha}\int_0^\infty \{\Phi(f(x)) + \Phi(g(x))\}\,\frac{dx}{x^\alpha}.$$

The general expression of Theorem 2.4 is obtained by integrating the convex transform of F by other measures.

Theorem 2.5. *Let Φ be a convex positive monotone function and let u be a strictly positive function on $\mathbb{R}_+$ such that $U(x) = \int_x^\infty t^{-1}u(t)\,dt$ is finite for every $t > 0$. For every positive function f with primitive F*

$$\int_0^\infty \Phi\Big\{\frac{F(x)}{x}\Big\}u(x)\,dx \leq \int_0^\infty \Phi(f(x))\,U(x)\,dx. \tag{2.14}$$

Let $X > 0$ be a random variable with density u and let $U(x) = P(X > x)$. Theorem 2.5 expresses that for every positive function f with primitive F, $E\Phi\{X^{-1}F(X)\} \leq \int_0^\infty \Phi(f(x))\,U(x)\,dx$.

Let k be a function defined from $\mathbb{R}_+^2$ to $\mathbb{R}_+$ and let $K(x) = \int_0^x k(x,t)\,dt$. Let $u \geq 0$ be a real function defined from $\mathbb{R}_+$ and

$$v(x) = x\int_0^x \frac{k(s,x)}{K(s)}\frac{u(s)}{s}\,ds.$$

Kaijser, Nikolova, Perrson and Wedestig (2005) proposed further generalizations of the Hardy-Knopp inequalities, in particular for a weighted mean.

Theorem 2.6. *For every convex function φ on $\mathbb{R}_+$*

$$\int_0^\infty \phi\Big(\frac{1}{K(x)}\int_0^x k(x,t)f(t)\,dt\Big)\frac{u(x)}{x}\,dx \leq \int_0^\infty \phi\circ f(x)\frac{v(x)}{x}\,dx.$$

The constant $\{p(p-1)^{-1}\}^p$ in Hardy's inequality is larger than one and it is specific to the power function $\Phi(x) = x^p$, with a weight $u(x) = x^{-1}$. It does not appear in Theorem 2.5 where it is included in the integral $U(x)$. With the function $u = id$, the weighting function $U(x)$ is not bounded for any x, so the integral $\int_0^\infty \Phi\{x^{-1}F(x)\}\,dx$ cannot be bounded as a consequence of Theorem 2.5.

Let $\alpha > 1$ be a real number, let u and v be strictly positive functions on $\mathbb{R}_+$ and let

$$U(x) = \int_x^\infty t^{-1}u(t)\,dt, \quad V(y) = \int_y^\infty t^{-1}v(t)\,dt. \tag{2.15}$$

Theorem 2.5 and the Hölder inequality entail a bilinear inequality

$$\int_0^\infty \int_0^\infty \left\{ \frac{F(x)G(y)}{(x+y)^2} \right\}^\alpha u(x)v(y)\, dx\, dy \leq \frac{1}{2^{2\alpha}} \left\{ \int_0^\infty f^\alpha(x)U(x)\, dx \right\}$$
$$\times \left\{ \int_0^\infty g^\alpha(y)V(y)\, dy \right\}.$$

Special cases and inequalities of the same kind are obtained with other convex and monotone functions Φ.

Example 2.3. Let Φ be the exponential function

$$\int_0^\infty \int_0^\infty \exp\left\{ \frac{F(x)+G(y)}{x+y} \right\} u(x)v(y)\, dx\, dy$$
$$\leq \int_0^\infty \exp\left\{ \frac{F(x)}{x} \right\} u(x)\, dx \int_0^\infty \exp\left\{ \frac{G(y)}{y} \right\} v(y)\, dy$$
$$\leq \int_0^\infty \exp\{f(x)\} U(x)\, dx \int_0^\infty \exp\{g(y)\} V(y)\, dy.$$

Weighted probability inequalities are deduced from the above results, using the same reparametrization as inequality (2.7). Let S be a positive real random variable with distribution function H. For every function a such that $Ea(S) = 0$, Equation (2.12) is equivalent to

$$E\left\{ H^{-1}(S)\Phi\left(\frac{\int_0^S a\, dH}{H(S)} \right) \right\} \leq E\{ H^{-1}(S)\Phi \circ a(S) \} \qquad (2.16)$$

and, for $\alpha > 0$, (2.13) is equivalent to

$$E\left\{ H^{-\alpha}(S)\Phi\left(\frac{\int_0^S a\, dH}{H(S)} \right) \right\} \leq E\{ H^{-\alpha}(S)\Phi \circ a(S) \}.$$

The general form of the weighted inequalities is written

$$E\left\{ \Phi\left(\frac{\int_0^S a\, dH}{H(S)} \right) u \circ H(S) \right\} \leq E\{ U \circ H(S)\, \Phi \circ a(S) \},$$

where u is a strictly positive function on $\mathbb{R}_+$ and U is defined by (2.15).

Let S and T be positive real random variables with respective distribution functions H_S and H_T, and let a and b be functions such that $\int a\, dH_S = 0$ and $\int b\, dH_T = 0$. Let u and v be strictly positive functions on $\mathbb{R}_+$ and let U and V be defined by (2.15). Example 2.3 is equivalent to

$$E\left[\exp\left\{ \frac{\int_0^S a\, dH_S + \int_0^T b\, dH_T}{X_S + X_T} \right\} u(X_S)v(X_T) \right]$$
$$\leq E[\exp\{a(X_S) + b(X_T)\} U(X_S)\, V(X_T)]$$

with the uniform variables $X_S = H_S(S)$ and $X_T = H_T(T)$.

Let $\alpha > 1$ be a real number, the inequality with the power transforms becomes

$$E\left[\left\{\frac{\int_0^S a\, dH_S \int_0^T b\, dH_T}{(X_S + X_S)^2}\right\}^\alpha u(X_S)v(X_T)\right]$$

$$\leq \frac{1}{2^{2\alpha}} E[\{a(X_S)b(X_T)\}^\alpha U(X_S)\, V(X_T)].$$

2.5 Carleman's inequality and generalizations

Carleman's inequality was established for maps of series of $\ell_1(\mathbb{R}_+)$

$$\sum_{n \geq 1} \exp\left(n^{-1} \sum_{k=1}^n \log x_k\right) \leq \sum_{n \geq 1} x_k$$

and it has been extended to an integral form. Let f be a positive function of $L^1(\mathbb{R}_+)$, this inequality can be expressed as

$$\int_0^\infty \exp\left(\frac{\int_0^x \log f(t)\, dt}{x}\right) dx \leq \int_0^\infty f(x)\, dx.$$

It is an application of Jensen's inequality for the exponential function on $\mathbb{R}_+$ in the form $\exp(x^{-1}\int_0^x g(t)\, dt) \leq \exp \circ g(x)$ for every positive function g of $L^1(\mathbb{R}_+)$. It generalizes by integrating a transformed primitive with respect to a positive measure μ on $\mathbb{R}_+$.

Theorem 2.7. *Let Φ be a convex and increasing function on $\mathbb{R}_+$ and F be the primitive of a positive function f on $\mathbb{R}_+$ such that $\Phi \circ f$ belongs to $L^1(\mu)$, then*

$$\int_0^\infty \Phi\left(\frac{F(x)}{x}\right) d\mu(x) \leq \int_0^\infty \Phi \circ f(x)\, d\mu(x). \qquad (2.17)$$

Example 2.4. Let $f : \mathbb{R} \mapsto \mathbb{R}_+$ be a density with derivative f', by Theorem 2.7, for every convex and increasing function Φ on $\mathbb{R}$

$$E\Phi\left(\frac{f'}{f}(X)\right) \geq E\Phi\left(\frac{\log f(X)}{X}\right).$$

In particular, the Fisher information of a parametric family of densities

$$\mathcal{F}_\Theta = \{f_\theta, \theta \in \Theta; f \in C(\Theta)\},$$

with derivative f_θ' with respect to the parameter, satisfies

$$I_\theta(f) = E\left\{\frac{f_\theta'}{f_\theta}(X)\right\}^2 \geq E\left\{\left(\frac{\log f_\theta(X)}{X}\right)^2 1_{\{X \neq 0\}}\right\}, \theta \in \Theta.$$

Carleman's inequality and Equation (2.17) are extended to an inequality involving a function Φ and the primitive of $\Phi^{-1}(f)$, in a form similar to the initial inequality with the exponential function Φ. Let Φ be a convex, positive and monotone function on $\mathbb{R}_+$ and let f be a positive function of $L^1(\mu)$, then

$$\int_0^\infty \Phi\left(\frac{\int_0^x \Phi^{-1} \circ f(t)\, dt}{x}\right) d\mu(x) \leq \int_0^\infty f(x)\, d\mu(x). \qquad (2.18)$$

By the same change of variables as in the inequality (2.7), Theorem 2.7 is rewritten in the following forms.

Proposition 2.14. *Let S be a positive real random variable with distribution function H. For every function a such that $\Phi \circ a(S)$ belongs to $L^1(P)$*

$$E\left\{\Phi\left(\frac{\int_0^S a\, dH}{H(S)}\right)\right\} \leq E\{\Phi \circ a(S)\},$$

if $Ea(S) = 0$

$$E\left\{\Phi\left(\frac{\int_0^S \Phi^{-1} \circ a\, dH}{H(S)}\right)\right\} \leq E\{a(S)\}.$$

2.6   Minkowski's inequality and generalizations

Let $p > 1$, Minkowski's inequality states the additivity property of the L^p norm, $\|f + g\|_p \leq \|f\|_p + \|g\|_p$. Mulholland (1950) searched to determine classes of functions $\varphi : I \subset \mathbb{R} \mapsto \mathbb{R}$ such that for all functions f and $g : J \subset \mathbb{R}^n \mapsto I$ of $L^1(\mu)$ the following inequality should be satisfied

$$\varphi^{-1}\left(\int \varphi(f+g)\, d\mu\right) \leq \varphi^{-1}\left\{\int \varphi(f)\, d\mu\right\} + \varphi^{-1}\left\{\int \varphi(f)\, d\mu\right\}, \qquad (2.19)$$

with equality if and only if f and g are proportional. On a probability space $(\Omega, \mathcal{A}, P)$, this inequality is expressed in terms of means of random variables. The question is to determine classes of functions $\varphi : I \subset \mathbb{R} \mapsto \mathbb{R}$ such that

$$\varphi^{-1}\{E\varphi(X+Y)\} \leq \varphi^{-1}\{E\varphi(X)\} + \varphi^{-1}\{E\varphi(Y)\}, \qquad (2.20)$$

for all random variables X and Y on $(\Omega, \mathcal{A}, P)$ and with values in I, such that $\varphi(X)$ and $\varphi(Y)$ are in $L^1(P)$, with equality if and only if X and Y are proportional.

Assume that the inequality (2.20) is satisfied. Let f and g be real functions defined on $\mathbb{R}$, and let $\widetilde{X}$ and $\widetilde{Y}$ be real random variables with joint distribution functions F and marginals $F_{\widetilde{X}}$ and $F_{\widetilde{Y}}$. With the variables $X = f(\widetilde{X})$ and $Y = g(\widetilde{Y})$, Equation (2.20) is written

$$\varphi^{-1}\left\{\int_{\mathbb{R}} \varphi(f+g)\,dF\right\} \leq \varphi^{-1}\left\{\int_{\mathbb{R}} \varphi(f)\,dF_{\widetilde{X}}\right\} + \varphi^{-1}\left\{\int_{\mathbb{R}} \varphi(g)\,dF_{\widetilde{Y}}\right\}.$$

Inequality (2.19) is obtained with variables $\widetilde{X}$ and $\widetilde{Y}$ having the same distribution μ.

The next proposition presents a weaker inequality due to monotonicity and convexity. It is extended as $E(X+Y) \leq \varphi^{-1}\{E\varphi(X)\} + \varphi^{-1}\{E\varphi(Y)\}$ for all random variables X and Y such that $\varphi(X)$ and $\varphi(Y)$ are $L^1(P)$, with equality if and only if X and Y are proportional.

Proposition 2.15. *For every convex and strictly increasing function φ and for all real sequences $(x_i)_{i\geq 1}$ and $(y_i)_{i\geq 1}$ such that $\{\varphi(x_i)\}_{i\geq 1}$ and $\{\varphi(y_i)\}_{i\geq 1}$ belong to $\ell_1(\mathbb{R})$*

$$n^{-1}\sum_{i\geq 1}(x_i + y_i) \leq \varphi^{-1}\left\{n^{-1}\sum_{i\geq 1}\varphi(x_i)\right\} + \varphi^{-1}\left\{n^{-1}\sum_{i\geq 1}\varphi(y_i)\right\},$$

with equality if and only if x_i and y_i are proportional, for every integer i.

Proof. Under the assumptions, every sequence $(x_i)_{i\geq 1}$ is such that $\varphi(n^{-1}\sum_{i\geq 1}x_i) \leq n^{-1}\sum_{i\geq 1}\varphi(x_i)$, by convexity. This implies

$$n^{-1}\sum_{i\geq 1}x_i \leq \varphi^{-1}\left\{n^{-1}\sum_{i\geq 1}\varphi(x_i)\right\}, \ n^{-1}\sum_{i\geq 1}y_i \leq \varphi^{-1}\left\{n^{-1}\sum_{i\geq 1}\varphi(y_i)\right\}$$

and the result follows by their sum. $\qquad\square$

Let $\mathcal{F}_1(I)$ be the set of strictly increasing real functions φ defined on a subset I of $\mathbb{R}$ and such that the function

$$\psi_\varphi(u,v) = \varphi(\varphi^{-1}(u) + \varphi^{-1}(v))$$

is concave on $\varphi(I) \times \varphi(I)$ and let $\mathcal{F}_2(I)$ be the set of strictly increasing real functions φ defined on a subset I of $\mathbb{R}$ and such that the function ψ_φ is convex on $\varphi(I) \times \varphi(I)$.

Theorem 2.8. *Let I be a subset of $\mathbb{R}$. For every function φ of $\mathcal{F}_1(I)$ and for all random variables X and Y on $(\Omega, \mathcal{A}, P)$, with values in I and such that $\varphi(X)$ and $\varphi(Y)$ are $L^1(P)$*

$$\varphi^{-1}\{E\varphi(X+Y)\} \leq \varphi^{-1}\{E\varphi(X)\} + \varphi^{-1}\{E\varphi(Y)\}. \tag{2.21}$$

For every function φ of $\mathcal{F}_2(I)$ and for all random variables X and Y on $(\Omega, \mathcal{A}, P)$, with values in I and such that $\varphi(X)$ and $\varphi(Y)$ are $L^1(P)$

$$\varphi^{-1}\{E\varphi(X+Y)\} \geq \varphi^{-1}\{E\varphi(X)\} + \varphi^{-1}\{E\varphi(Y)\}. \tag{2.22}$$

Proof. Jensen's inequality for a concave function ψ implies that for every variable (U, V), $E\psi(U, V) \leq \psi(EU, EV)$. By the monotone change of variables $U = \varphi(X)$ and $V = \varphi(Y)$, this is equivalent to

$$E\varphi(X + Y) \leq \varphi[\varphi^{-1}\{E\varphi(X)\} + \varphi^{-1}\{E\varphi(Y)\}]$$

for an increasing function φ and the result follows by the monotonicity of the function φ^{-1}. These inequalities are reversed under the assumptions of an increasing function φ and a convex function ψ.   □

If φ is a decreasing function, the inequalities are inverted. In Minkowski's inequality, the function $\psi(u, v) = (u^{\frac{1}{p}} + v^{\frac{1}{p}})^p$ is increasing and concave as proved by Neveu (1970).

The first two derivatives of the function ψ are

$$\psi_u^{(1)}(u, v) = \frac{\varphi^{(1)}(\varphi^{-1}(u) + \varphi^{-1}(v))}{\varphi^{(1)} \circ \varphi^{-1}(u)},$$

$$\psi_{u,u}^{(2)}(u, v) = \frac{\varphi^{(2)}(\varphi^{-1}(u) + \varphi^{-1}(v))}{\varphi^{(1)2} \circ \varphi^{-1}(u)}$$

$$- \varphi^{(1)}(\varphi^{-1}(u) + \varphi^{-1}(v))\frac{\varphi_u^{(2)}}{\varphi^{(1)3}}(\varphi^{-1}(u)),$$

$$\psi_{u,v}^{(2)}(u, v) = \frac{\varphi^{(2)}(\varphi^{-1}(u) + \varphi^{-1}(v))}{\varphi^{(1)} \circ \varphi^{-1}(u)\varphi^{(1)} \circ \varphi^{-1}(v)}$$

and the signs of $\psi_{u,v}^{(2)}(u, v)$ and $\psi_{u,u}^{(2)}(u, v)$ may be different. They cannot be determined from general inequalities between the derivatives of the function φ at a single point.

Example 2.5. The function $\psi(u, v) = \log(e^u + e^v)$ related to the logarithm function φ has the derivatives

$$\psi_u^{(1)}(u, v) = \frac{e^u}{e^u + e^v}, \quad \psi_{u,v}^{(2)}(u, v) = -\frac{e^{u+v}}{(e^u + e^v)^2}, \quad u \neq v,$$

it is concave for real numbers $u \neq v$. Theorem 2.8 implies

$$\exp\{E\log(X + Y)\} \leq \exp(E\log X) + \exp(E\log Y).$$

By the change of variables $X = e^U$ and $Y = e^V$, this is equivalent to

$$E\{\log(e^U + e^V)\} \leq \log(e^{EU} + e^{EV})$$

and this inequality is also a direct consequence of the concavity of the function $\psi(u, v)$.

Example 2.6. With discrete variables X and Y, Theorem 2.8 is written in the form

$$\varphi^{-1}\left\{\sum_i \varphi(x_i + y_i)\mu_i\right\} \leq \varphi^{-1}\left\{\sum_i \varphi(x_i)\mu_{X,i}\right\} + \varphi^{-1}\left\{\sum_i \varphi(y_i)\mu_{Y,i}\right\}$$

for every function φ of $\mathcal{F}_1$, where μ_i is the probability that $X = x_i$ and $Y = y_i$, $\mu_{X,i}$ and $\mu_{Y,i}$ are the probability that $X = x_i$ and, respectively, $Y = y_i$.

Let $X_1, \ldots, X_n$ be strictly positive and identically distributed random variables such that $\log X_1$ is integrable and let $\varphi(x) = \log x$. By concavity of the logarithm,

$$E\{\log(n^{-1}S_n)\} \geq n^{-1}\sum_{i=1}^n E(\log X_i) = E(\log X_1)$$

which entails

$$\exp\{E(\log S_n)\} \geq n\exp\{E(\log X_1)\}.$$

This inequality and the extension of the inequality of Example 2.5 to a sum of positive and identically distributed variables imply

$$\exp\{E(\log S_n)\} = n\exp\{E(\log X_1)\}.$$

Theorem 2.8 extends to sums of n variables.

Theorem 2.9. *Let $X_1, \ldots, X_n$ be random variables on a probability space $(\Omega, \mathcal{A}, P)$. For every real function φ of $\mathcal{F}_1(\mathbb{R})$ and such that the variables $\varphi(X_i)$ belong to L^1 for $i = 1, \ldots, n$*

$$\varphi^{-1}\left\{E\varphi(\sum_{i=1}^n X_i)\right\} \leq \sum_{i=1}^n \varphi^{-1}\{E\varphi(X_i)\}.$$

For every real function φ of $\mathcal{F}_2(\mathbb{R})$ and such that the variables $\varphi(X_i)$ belong to L^1 for $i = 1, \ldots, n$

$$\varphi^{-1}\left\{E\varphi(\sum_{i=1}^n X_i)\right\} \geq \sum_{i=1}^n \varphi^{-1}\{E\varphi(X_i)\}.$$

For independent and identically distributed variables $X_1, \ldots, X_n$ and a convex function φ, the inequality of Theorem 2.9 is written as

$$\varphi^{-1}\{n^{-1}E\varphi(S_n)\} \leq \varphi^{-1}\{E\varphi(X_1)\}. \tag{2.23}$$

Example 2.7. Let X be a uniform variable on $[0,1]$, then

$$E|\log X| = -\int_0^1 \log x \, dx = 2.$$

Let a be in $[0,1]$ and let $X_1, \ldots, X_n$ be independent and identically distributed variables with the same distribution as X. Since $\log(n^{-1}S_n) < 0$

$$E|\log(n^{-1}S_n)| = -E\log(n^{-1}S_n) \leq 2$$

and inequality (1.13) is written

$$
\begin{aligned}
P(n^{-1}S_n > a) &= P(|\log(n^{-1}S_n)| < |\log a|) \\
&\geq 1 - (|\log a|)^{-1}E|\log(n^{-1}S_n)| \\
&\geq 1 - 2(|\log a|)^{-1}, \\
P(n^{-1}S_n < a) &= P(|\log(n^{-1}S_n)| > |\log a|) \leq 2(|\log a|)^{-1}.
\end{aligned}
$$

2.7 Inequalities for the Laplace transform

The Laplace transform $L_X(t) = Ee^{-tX}$ of a positive variable X defined on $\mathbb{R}_+$ is also the Laplace transform of the probability density of X. Let H be a real distribution on $\mathbb{R}_+$ and let U be a uniform variable on $[0,1]$. The Laplace transform of the variable $X = H^{-1}(U)$, with distribution function H, is

$$L_X(\lambda) = \int_0^1 e^{-\lambda H^{-1}(u)} \, du = \int_{\mathbb{R}_+} e^{-\lambda x} \, dH(x).$$

For an integer-valued variable, $L_X(\lambda) = \sum_{n \geq 0} e^{-\lambda n} P(X = n)$ is the generating function of X at $e^{-\lambda}$. More generally, the Laplace transforms of a variable X with distribution function $F = F^c + F^d$ is the sum of the generating function at $e^{-\lambda}$ of X^d, with the discrete distribution function F^d, and the Laplace transforms of X^c with the continuous distribution function F^c.

Let X be a symmetric variable, the odd moments EX^{2n+1} of X are zero, hence $L_X(t) = L_X(-t)$ and the odd derivatives of the Laplace transforms of X are zero. For every positive variable X, the derivatives of the Laplace transform of a variable X satisfy

$$(-1)^k L_X^{(k)}(t) = \int_0^\infty x^k e^{-tx} \, dF_X(x)$$

and the k-th moment of X is $\mu_k = (-1)^k L_X^{(k)}(0)$. For every positive variable X having a distribution function such that $\lim_{x \to \infty} xF(x) = 0$, the derivative of $L_X(t)$ satisfies $L'_X(t) = -E(Xe^{-tX})$. The Laplace transform of the sum of variables with densities f and g is the transform of their convolution $L(f * g)(t) = Lf(t) \, Lg(t)$ and every integer n

$$Lx^n(t) = \int_0^\infty x^n e^{-tx} \, dx = \frac{1}{n! \, t^{n+1}}.$$

A mixture density $f = \int_\Theta f_\eta \, dG_\theta$ has the Laplace transform

$$L_f(t) = \int_\Theta L_{f_\theta}(t) \, dG_\theta$$

in particular for an exponential mixture where $f_\theta(x) = \theta e^{-\theta x}$, we get

$$L_f(x) = \int_\Theta \frac{\theta}{t + \theta} \, dG_\theta.$$

By convexity of the exponential function, L_X is a convex function and $L_X(t) \geq e^{-tEX}$ therefore $L_X(t) \geq 1$ for every variable X. By the same argument, the moment generating function of a positive convex function g is

$$\varphi_{g(X)}(t) = E e^{tg(X)} \geq e^{tEg(X)}.$$

Let X be a random variable on a space $\mathcal{X}$, for a function $g > 0$ defined from $\mathcal{X}$ to $\mathbb{R}^*_+$ and for every $\lambda > 0$, we have

$$E e^{\lambda g(X)} \geq e^{\lambda g(EX)}, \text{ if } g \text{ is convex},$$
$$E e^{-\lambda g(X)} \geq e^{-\lambda g(EX)}, \text{ if } g \text{ is concave}.$$

For every variable X, the logarithm of $l_X(t)$ of $L_X(t)$ has the second derivative

$$l_X^{(2)}(t) = \frac{L_X^{(2)}(t) L_X(t) - L_X^{(1)2}(t)}{L_X^2(t)}$$

and by the Bienaymé-Chebychev inequality (1.13)

$$l_X^{(2)} \geq 0$$

so $\log L_X$ and $\log \varphi_X$ are convex functions.

The following bounds are consequences of the convexity of the functions exponential and φ.

Proposition 2.16. *Let X be a random variable with values in a bounded interval $[a, b]$*

$$\frac{E e^{tX} - e^{ta}}{EX - a} \leq \frac{e^{tb} - e^{ta}}{b - a} \leq \frac{e^{tb} - E e^{tX}}{b - EX},$$
$$\{e^{-ta} - L_X(t)\}(EX - a) \vee \{L_X(t) - e^{-tb}\}(b - EX) \leq (e^{tb} - e^{ta})(b - a).$$

Proposition 2.17. *Let X be a random variable with values in a bounded interval $[a, b]$ and φ be a convex function defined on $[a, b]$, then*

$$Ee^{\varphi(X)} \leq \frac{EX - a}{b - a}e^{\varphi(b)} + \frac{b - EX}{b - a}e^{\varphi(a)}$$

and the moment generating function of X has the bound $Ee^{tX} \leq e^{\frac{1}{8}t^2(b-a)^2}$.

Proposition 2.18. *The Laplace transform of a variable X satisfies*

$$t^{-1}L_X(t)|\log L_X(t)| \leq |L'_X(t)| \leq \{L''_X(0)L_X(2t)\}^{\frac{1}{2}}.$$

Proof. Let $Y = e^{-tX}$, the first inequality is due to the convexity of the function $y \log y$, for $y > 0$, and to Jensen's inequality.   □

The Laplace transform of a variable is generally characterized by a differential equation. For a centered Gaussian variable X with expectation μ and variance σ^2, it is the unique solutions of the equation

$$\frac{L''_X(t)}{L_X(t)} - \left\{\frac{L'_X(t)}{L_X(t)}\right\}^2 = \sigma^2$$

with the initial values $L_X(0) = 1$ and $L'_X(0) = -\mu$.

The Laplace transform of $(X - \mu)^2$ is $L_{X^2}(t) = (1 + 2\sigma^2 t)^{-\frac{1}{2}}$, it is the unique solution of the equation $L'_{X^2}(t) = -\sigma^2 L^3_{X^2}(t)$ with the initial value $L_{X^2}(0) = 1$.

The independence of the components of a vector $(X_1, \ldots, X_n)$ is equivalent to the factorization of their moment generating function and their Laplace transform, their sum is the product $L_{\sum_{i=1}^n X_i}(t) = \prod_{i=1}^n L_{X_i}(t)$, for any real t. The moment generating functions φ_{X_i} have the same properties. Theorem 2.8 allows us to write a reciprocal inequality for a sum of variables.

Proposition 2.19. *Let (X, Y) be a real random variable on a probability space $(\Omega, \mathcal{A}, P)$, their moment generating functions satisfy*

$$\varphi_{X+Y}(x) \geq \varphi_X(x)\varphi_Y(x), \ x \geq 0,$$

with equality if and only if X and Y are independent.

Proof. The function $\varphi_\lambda(x) = e^{\lambda x}$ has the inverse $\varphi_\lambda^{-1}(x) = \lambda^{-1}\log u$, then the second derivative of the function $\psi_\alpha(u, v) = uv$ of Theorem 2.8 are $\psi_{u,v}^{(2)}(u, v) = 1$ and $\psi_{u,u}^{(2)}(u, v) = 0$ for every $0 < u \neq v$, so that ψ belongs to $\mathcal{F}_2(\mathbb{R}_+)$ and the result is the inequality (2.22).   □

Example 2.8. Let X and Y be exponential variables $\mathcal{E}_\theta$, $\theta > 0$, then for every $x \geq 0$

$$\frac{\theta}{\theta - 2t} = \varphi_{2X}(x) \geq \varphi_X(x)\varphi_Y(x) = \frac{\theta^2}{(\theta - t)^2}.$$

A linear combination $Y = \sum_{i=1}^{n} a_i X_i$ of the components of a vector $(X_1, \ldots, X_n)$, such that $a_i \geq 0$, $i = 1, \ldots, n$, and $\sum_{i=1}^{n} a_i = 1$, has a Laplace transform such that for every x

$$L_Y(x) \leq \sum_{i=1}^{n} a_i L_{X_i}(x),$$

for $Y = n^{-1} \sum_{i=1}^{n} X_i$, $L_Y(x) \leq n^{-1} \sum_{i=1}^{n} a_i L_{X_i}(x)$.

For positive variables U and V, let $X = \prod_{i=1}^{n} \varphi_\alpha^{-1}(U) = \alpha^{-1} \log U$ and $Y = \alpha^{-1} \log V$, the inequality of Proposition 2.20 is equivalent to

$$E(UV) \geq (EU)(EV) \tag{2.24}$$

which completes the Hölder inequality. It implies that the covariance of positive variables is always positive. For example, let $U = \cos\theta$ and $V = \sin\theta$, with a uniform variable θ on the interval $[0, \frac{\pi}{2}]$, then $E(UV) = \frac{1}{2} E \sin(2\theta) = \frac{2}{\pi}$ while $EU = \frac{2}{\pi}$ and $EV = 0$.

Let X be a random variable on $[0, 1]$, Chernoff's theorem implies that for every integer $n \geq 1$ and for every $a > 0$

$$\inf_{t>0}\{\log \varphi_X(t) - at\} \geq 1 - (|\log a|)^{-1} E(|\log X|).$$

Proposition 2.20. *Let X be a random variable such that $E \log X$ is finite and strictly positive. For every $a > 1$*

$$\inf_{t>0}\{\log \varphi_X(t) - at\} \leq \frac{E(\log X)}{\log a}.$$

Proof. Equation (2.23) implies that for a sum of n independent variables having the same distribution as X, $0 < E \log(n^{-1} S_n) \leq \log EX$ and the result is obtained from the large deviations theorem for every integer $n \geq 1$

$$P(n^{-1} S_n > a) = \inf_{t>0}\{\log \varphi_X(t) - at\}$$
$$= P(\log(n^{-1} S_n) > \log a)$$
$$\leq \frac{E\{\log(n^{-1} S_n)\}}{\log a}.$$

$\square$

2.8   Inequalities for multivariate functions

Let F and G be the primitives of real and positive functions f and g on $\mathbb{R}_+$. The product inequality (2.6) is generalized by transforms. Let Φ and Ψ be convex functions on $\mathbb{R}_+$, let u be a positive density on $\mathbb{R}_+^2$ and let U be defined on $\mathbb{R}_+^2$ like (2.15), as

$$U(x,y) = \int_x^\infty \int_y^\infty s^{-1} t^{-1} u(s,t)\, ds\, dt,$$

then

$$\int_0^\infty \int_0^\infty \Phi\Big(\frac{F(x)}{x}\Big) \Psi\Big(\frac{G(y)}{y}\Big) u(x,y)\, dx\, dy$$

$$\leq \int_0^\infty \int_0^\infty \Big\{ \int_0^x \int_0^y \Phi(f(s)) \Psi(g(t))\, ds\, dt \Big\} U(dx, dy)$$

$$\leq \int_0^\infty \int_0^\infty \Phi(f(x)) \Psi(g(y))\, U(x,y)\, dx\, dy.$$

Extending Proposition 2.7, there exists a constant $k_u = \|U\|_2$ such that for every $p > 1$

$$\Big| \int_0^\infty \int_0^\infty \Phi\Big(\frac{F(x)}{x}\Big) \Psi\Big(\frac{G(y)}{y}\Big) u(x,y)\, dx\, dy \Big| \leq k_u \|\Phi(f)\|_p \|\Psi(g)\|_{p'},$$

where $p^{-1} + p'^{-1} = \frac{1}{2}$. With exponential functions Φ and Ψ, this inequality becomes

$$\int_0^\infty \int_0^\infty \exp\Big\{ \frac{\int_0^x \log f}{x} + \frac{\int_0^x \log g}{y} \Big\} u(x,y)\, dx\, dy \leq k_u \|f\|_p \|g\|_{p'}$$

for all functions f of L^p and g of $L^{p'}$ such that $p^{-1} + p'^{-1} = \frac{1}{2}$. This result is directly extended to a convex functions Φ on $\mathbb{R}_+^2$

$$\int_0^\infty \int_0^\infty \Phi\Big(\frac{F(x)}{x}, \frac{G(y)}{y}\Big) u(x,y)\, dx\, dy$$

$$\leq \int_0^\infty \int_0^\infty \Big\{ \int_0^x \int_0^y \Phi(f(s), g(t))\, ds\, dt \Big\} U(dx, dy)$$

$$\leq \int_0^\infty \int_0^\infty \Phi(f(x), g(y))\, U(x,y)\, dx\, dy. \tag{2.25}$$

If the bounds in the integrals are modified, they are normalized by the surface of the domain of integration. Let f and g be defined on $[a, \infty[$ and, respectively, $[b, \infty[$ and let $U_{ab}(x, y) = \int_x^\infty \int_y^\infty (s-a)^{-1}(t-b)^{-1}u(s, t)\, ds\, dt$, for $x > a$ and $y > b$, then

$$\int_0^\infty \int_0^\infty \Phi\left(\frac{\int_a^x f}{x-a}\right)\Psi\left(\frac{\int_b^y g}{y-b}\right) u(x, y)\, dx\, dy$$

$$\leq \int_0^\infty \int_0^\infty \left\{\int_a^x \int_b^y \Phi(f(s))\Psi(g(t))\, ds\, dt\right\} U_{ab}(dx, dy)$$

$$\leq \int_0^\infty \int_0^\infty \Phi(f(x))\Psi(g(y))\, U_{ab}(x, y)\, dx\, dy.$$

In particular

$$\left|\int_0^\infty \int_0^\infty \Phi\left(\frac{\int_a^x f}{x-a}\right)\Psi\left(\frac{\int_b^y g}{y-b}\right) u(x, y)\, dx\, dy\right| \leq \|U_{ab}\|_2 \|\Phi(f)\|_p \|\Psi(g)\|_{p'},$$

for all p and p' such that $p^{-1} + p'^{-1} = \frac{1}{2}$, and

$$\left|\int_0^\infty \int_0^\infty \Phi\left(\frac{\int_a^x f}{x-a}, \frac{\int_b^y g}{y-b}\right) u(x, y)\, dx\, dy\right| \leq \|U_{ab}\|_p \|\Phi(f, g)\|_{p'}$$

for all conjugates p and p'.

As an application, let S and T be positive real random variables with respective distribution functions H_S and H_T, and let $X_S = H_S(S)$ and $X_T = H_T(T)$ be the uniform transformed variables. Let a and b be functions such that $\int a\, dH_S = 0$ and $\int b\, dH_T = 0$. The weighted mean inequality (2.25) for the integral

$$E\left\{\Phi\left(\frac{\int_0^S a\, dH_S}{X_S}, \frac{\int_0^T b\, dH_T}{X_T}\right) u(X_S, X_T)\right\}$$

$$= E\left\{\Phi\left(\frac{\int_S^\infty a\, dH_S}{X_S}, \frac{\int_T^\infty b\, dH_T}{X_T}\right) u(X_S, X_T)\right\}$$

is also written as

$$E\left\{\Phi\left(\frac{\int_0^S a\, dH_S}{X_S}, \frac{\int_0^T b\, dH_T}{X_T}\right) u(X_S, X_T)\right\}$$

$$\leq E\{\Phi(a(X_S), b(X_T))\, U(X_S, X_T)\}.$$

Theorem 2.7 gives an inequality without weighting function, it is extended to a convex and increasing function $\Phi : \mathbb{R}_+^2 \mapsto \mathbb{R}$

$$\int_0^\infty \Phi\left(\frac{F(x)}{x}, \frac{G(x)}{x}\right) d\mu(x) \leq \int_0^\infty \Phi \circ (f, g)(x)\, d\mu(x),$$

more generally

$$\int_0^\infty \int_0^\infty \Phi\left(\frac{F(x)}{x}, \frac{G(y)}{y}\right) \mu(dx, dy)$$
$$\leq \int_0^\infty \int_0^\infty \Phi(f(x), g(y)) \mu(dx, dy), \qquad (2.26)$$

and for functions f defined on $[a, \infty[$ and g defined on $[b, \infty[$, this inequality becomes

$$\int_a^\infty \int_b^\infty \Phi\left(\frac{\int_a^x f}{x-a}, \frac{\int_b^y g}{y-b}\right) \mu(dx, dy) \leq \int_a^\infty \int_b^\infty \Phi \circ (f(x), g(y)) \mu(dx, dy).$$

These inequalities differ from the generalized Hardy inequality (2.25) by the integration measures. They are extended to functions $f : [a, \infty[\mapsto \mathbb{R}^n$, where a belongs to $\mathbb{R}$ or $\mathbb{R}^n$, for every integer n. Let (X, Y) be a random vector with a distribution function μ on $\mathbb{R}^2$, Equation (2.26) is also written as

$$E\left\{\Phi\left(\frac{F(X)}{X}, \frac{G(Y)}{Y}\right)\right\} \leq E\{\Phi(f(X), g(Y))\}$$

for every convex function Φ.

Carleman's inequality is also extended to an inequality for a multivariate function in a similar form as (2.18) but without weighting functions. Let $\Phi : \mathbb{R}^2 \mapsto \mathbb{R}^2$ be a convex and monotone function and let f and g be positive real functions defined on $\mathbb{R}$

$$\int_a^\infty \int_b^\infty \Phi\left\{\frac{\int_a^x \int_b^y \Phi^{-1}(f(s), g(t)) \, ds \, dt}{(x-a)(y-b)}\right\} dx \, dy$$
$$\leq \int_a^\infty \int_b^\infty (f(x), g(y)) \, dx \, dy. \qquad (2.27)$$

Let $(X_i, Y_i)_{i=1,\dots,n}$ be a sequence of independent random vectors on $\mathbb{R}^2$, with respective distribution functions F_{X_i, Y_i} and denote their mean distribution function $F_n = n^{-1} \sum_{i=1}^n F_{X_i, Y_i}$. Inequality (2.26) implies

$$E\Phi(\bar{X}_n, \bar{Y}_n) \leq E \int_{\mathbb{R}^2} \Phi(x, y) \, F_n(dx, dy).$$

With identically distributed variables, $E\Phi(\bar{X}_n, \bar{Y}_n) \leq E\Phi(X_1, Y_1)$. As a special case

$$E\{(\bar{X}_n - E\bar{X}_n)^2 + (\bar{Y}_n - E\bar{Y}_n)^2\}^{\frac{1}{2}}$$
$$\leq n^{-1} \sum_{i=1}^n E\{(X_i - EX_i)^2 + (Y_i - EY_i)^2\}^{\frac{1}{2}}.$$

In Section 1.5, a mean integral was defined on balls of $\mathbb{R}^d$, a similar result holds on rectangles or other convex connex domains of $\mathbb{R}^d$. By convexity, for every x in $\mathbb{R}^d$ and for every function $f : \mathbb{R}^d \mapsto \mathbb{R}$

$$I_{p,2} = \int_{[0,\infty[} \left(\frac{\left| \int_{[0,x]} f(t)\,dt \right|}{\prod_{i=1}^d x_i} \right)^p \frac{dx}{\prod_{i=1}^d x_i}$$

$$\leq \int_{[0,\infty[} \int_{[0,x]} |f(t)|^p\,dt \frac{\prod_{i=1}^d dx_i}{\prod_{i=1}^d x_i^2}$$

$$\leq \int_{[0,\infty[} |f(t)|^p \frac{\prod_{i=1}^d dt_i}{\prod_{i=1}^d t_i}.$$

For a positive function f defined on $\mathbb{R}^d$, Hardy's inequality and an integration by parts allows us to write the integral of the function $(\prod_{i=1}^d x_i)^{-1} \int_{[0,x]} f(t)^p\,dt$ with respect to the Lebesgue measure on $\mathbb{R}^d$ as

$$I_{p,1} = \frac{p}{p-1} \int_{[0,\infty[} \left(\frac{\int_{[0,x]} f(t)\,dt}{\prod_{i=1}^d x_i} \right)^{p-1} dx$$

$$I_{p,1} \leq \left(\frac{p}{p-1} \right)^p \int_{[0,\infty[} f^p(x)\,dx.$$

These inequalities are generalized to integrals with respect to a positive measure μ on $\mathbb{R}^d$

$$I_{p,\mu} = \int_{[0,\infty[} \left(\frac{\left| \int_{[0,x]} f(t)\,d\mu(t) \right|}{\prod_{i=1}^d x_i} \right)^p d\mu(x)$$

$$\leq \int_{[0,\infty[} \int_{[0,x]} |f(t)|^p\,dt \frac{d\mu(x)}{\prod_{i=1}^d x_i}$$

$$\leq \int_{[0,\infty[} |f(t)|^p \int_{[t,\infty[} \frac{d\mu(x)}{\prod_{i=1}^d x_i}\,d\mu(t).$$

Let $X = (X_1, \ldots, X_d)$ be a variable with distribution μ on $\mathbb{R}_+^d$, this inequality is equivalently written

$$E\left\{ \left(\frac{F(X)}{\prod_{i=1}^d X_i} \right)^p \right\} \leq E\left\{ |f(X)|^p \int_{[X,\infty[} \frac{d\mu(x)}{\prod_{i=1}^d x_i} \right\}.$$

Theorem 2.6 extends to a kernel k defined from $\mathbb{R}_+^{2d}$ to $\mathbb{R}_+$ which allows to restrict the integral on a subset of $\mathbb{R}_+^{2d}$. Let $K(x) = \int_{[0,x]} k(x,t)\,dt$ for x in $\mathbb{R}_+^d$. Let $u \geq 0$ and $v \geq 0$ be real functions defined from $\mathbb{R}_+^d$, with

$$v(x) = x \int_{[x,\infty[} \frac{k(s,x)\,u(s)}{K(s)}\,\frac{ds}{s}.$$

Theorem 2.10. *For every convex and real function φ on $\mathbb{R}_+^d$ and for every $p > 1$*

$$\int_{[0,\infty[^d} \phi\left(\frac{1}{K(x)} \int_{[0,x]} k(x,t)f(t)\,dt\right) \frac{u(x)}{x}\,dx \leq \int_{[0,\infty[^d} \phi \circ f(x) \frac{v(x)}{x}\,dx.$$

Replacing ϕ by its L^p norm in this inequality entails

$$\left\{\int_{[0,\infty[^d} \phi^p\left(\frac{1}{K(x)} \int_{[0,x]} k(x,t)f(t)\,dt\right) \frac{u(x)}{x}\,dx\right\}^{\frac{1}{p}}$$

$$\leq \left\{\int_{[0,\infty[^d} \phi^p \circ f(x) \frac{v(x)}{x}\,dx\right\}^{\frac{1}{p}}.$$

For integrals on the ball $B_r(x)$ centered at x in $\mathbb{R}_+^d$, the Lebesgue measure of $B_r(x)$ is a constant proportional to r^d which does not depend on x and

$$I_p(r) = \int_{[r,\infty[^d} \left(\frac{1}{\lambda_d(B_r)} \int_{B_r(x)} |f(t)|\,dt\right)^p dx \leq \int_{[r,\infty[^d} |f(t)|^p\,dt,$$

since $\int_{[r,\infty[^d} 1_{B_r(x)}(t)\,dx = \lambda_d(B_r(t)) = \lambda_d(B_r)$. The same result is true for the integration on every convex and connex set having a Lebesgue measure independent of x.

Carleman's inequality on balls becomes

$$\int_{[r,\infty[^d} \exp\left(\frac{1}{\lambda_d(B_r)} \int_{B_r(x)} \log f(t)\,dt\right) dx \leq \int_{[r,\infty[^d} f(x)\,dx$$

and on rectangular sets $[0,x]$ it is

$$\int_{[0,\infty[} \exp\left(\frac{\int_{[0,x]} \log f(t)\,dt}{\prod_{i=1}^d x_i}\right) d\mu(x) \leq \int_{[0,\infty[} f(t)\,dt \int_{[t,\infty[} \left(\prod_{i=1}^d x_i\right)^{-1} d\mu(x).$$

For every increasing and convex function ϕ, we deduce the multivariate inequalities on the balls

$$\int_{[r,\infty[^d} \phi\left(\frac{1}{\lambda_d(B_r)} \int_{B_r(x)} \phi^{-1} \circ f(t)\,dt\right) dx \leq \int_{[r,\infty[^d} f(x)\,dx,$$

and on the rectangles

$$\int_{[0,\infty[} \phi\left(\frac{\int_{[0,x]} \phi^{-1} \circ f(t)\,dt}{\prod_{i=1}^d x_i}\right) d\mu(x) \leq \int_{[0,\infty[} f(t)\,dt \int_{[t,\infty[} \left(\prod_{i=1}^d x_i\right)^{-1} d\mu(x).$$

Chapter 3

Analytic Inequalities

3.1 Introduction

Many inequalities rely on Taylor expansions of functions in series such as the trigonometric series

$$\cos(x) = 1 + \sum_{k=1}^{\infty}(-1)^k \frac{x^{2k}}{(2k)!}, \quad \sin(x) = \sum_{k=0}^{\infty}(-1)^k \frac{x^{2k+1}}{(2k+1)!},$$

the expansions of the exponential and the logarithm functions, with

$$\log x = 2\arg\operatorname{th}\left(\frac{x-1}{x+1}\right) = 2\int_{-1}^{\frac{x-1}{x+1}} \frac{dy}{1-y^2}$$

$$= 2\sum_{k=0}^{\infty}\frac{1}{2k+1}\left\{\left(\frac{x-1}{x+1}\right)^{2k+1} + 1\right\}, \; x > 0.$$

Numerical tables of these functions and many other functions have been published during the 17th and 18th centuries (Hutton 1811). Expansions in a series provide a simple method for calculating approximations of constants or functions by their partial sums. For example $e = \sum_{k=0}^{\infty}(n!)^{-1}$ and it is also expressed by de Moivre-Stirling's formula $(n^{-1}e)^n \sim \sqrt{2\pi}(n!)^{-1}$ as n tends to infinity. The number π is expanded as

$$\pi = 4\arctan 1 = 4\sum_{k=0}^{\infty}\frac{(-1)^k}{2k+1}.$$

Approximations of the hyperbolic and trigonometric functions are easily obtained. Inequalities for partial sums in the expansions have been considered for the evaluation of the approximation errors and they can generally be proved by induction. Adler and Taylor (2007) presented expansions in a series for the probability $P(\sup_{t \in A} f(t) \geq u)$ for a general parameter space

A and where P is a Gaussian probability distribution and f a function of $C^n(A)$.

For every integer $n \geq 1$ and for every $0 < x < y$, the ratio of $x^n - y^n$ and $x - y$ is expanded as the finite sum $\sum_{i=0}^{n-1} y^k x^{n-k-1}$ which provides the bounds

$$\frac{1}{ny^{n-1}} \leq \frac{x-y}{x^n - y^n} \leq \frac{1}{nx^{n-1}}$$

and for vectors $x = (x_1, \ldots, x_n)$ and $y = (y_1, \ldots, y_n)$ such that $0 < x_k < y_k$ for $k = 1, \ldots, n$

$$n^{-n} \prod_{k=1}^{n} y_k^{1-n} \leq \frac{\prod_{k=1}^{n}(x_k - y_k)}{\prod_{k=1}^{n}(x_k^n - y_k^n)} \leq n^{-n} \prod_{k=1}^{n} x_k^{1-n}.$$

Inequalities for the partial sum A_n of the first $n + 1$ terms of the Taylor expansion of a function have been considered. For the exponential function, Alzer (1990b) stated that

$$(n + 1)(n + 2)^{-1} A_n^2(x) < A_{n-1}(x)A_{n+1}(x) < A_n^2(x),$$

this inequality is improved in Section 3.2. The existence of constants such that

$$c_n \left(\frac{A_n(x)}{A_{n+1}(x)}\right)^2 < \frac{A_{n-1}(x)}{A_{n+1}(x)} < C_n \left(\frac{A_n(x)}{A_{n+1}(x)}\right)^2$$

is established in Section 3.2 for the partial sums of other functions, sufficient conditions for $C_n = 1$ are given in Proposition 3.1.

Cauchy's inequalities for two real sequences $(x_i)_{i=1,\ldots,n}$ and $(a_i)_{i=1,\ldots,n}$ are illustrated by several cases in Section 3.3 and they are applied to the comparison of the geometric and arithmetic means, generalizing other inequalities proved by Alzer (1990a). Section 3.4 provides inequalities for the mean, the modes and median of random variables. Section 3.5 deals with other specific points of curves related to the Mean Value Theorem and solutions of implicit functional equations are established. Carlson's inequality is an inequality of the same kind for the logarithmic mean function and it is due to a convexity argument. In Section 3.6 it is generalized to concave or convex functions. Inequalities for the functional means of power functions are then proved.

3.2 Bounds for series

Proposition 3.1. *Let* $(a_n)_{n\geq 1}$ *be strictly positive and strictly decreasing sequence of real functions and let* $A_n = \sum_{k=0}^n a_k$, *then for every integer* n

$$A_n^2 > A_{n-1}A_{n+1}. \tag{3.1}$$

The same inequality holds for every strictly negative and decreasing sequence of functions $(a_n)_{n\geq 1}$.

Proof. For every integer n

$$A_n^2 - A_{n-1}A_{n+1} = a_n A_n - a_{n+1}A_{n-1}$$
$$> a_{n+1}(A_n - A_{n-1}) > 0.$$

$\square$

Example 3.1. The power function $f_n(x) = (1+x)^n$, $x > 0$ and $n > 0$, develops as a series $A_n(x) = \sum_{k=0}^n a_k(x)$ with $a_k(x) = C_n^k x^k$, hence the sequence $(a_k(x))_{\frac{n}{2}\leq k\leq n}$ is decreasing on $]0,1[$. Therefore, the inequality (3.1) is fulfilled on $]0,1[$, for $A_n - A_{\frac{n}{2}}$ if n is even and for $A_n - A_{\frac{n-1}{2}}$ if n is odd.

Example 3.2. The logarithm function $\log(1-x)$ on $]0,1[$ has an expansion with functions $a_n(x) = n^{-1}x^n$ satisfying the conditions of the proposition. The logarithm function $f(x) = \log(1+x)$, $x > -1$, has an expansion $A_n(x) = \sum_{k=1}^n (-1)^{k+1}k^{-1}x^k = A_{1n}(x) - A_{2n}(x)$, where

$$A_{1n}(x) = \sum_{k=0}^{[\frac{n}{2}]} \frac{x^{2k+1}}{2k+1}, \quad A_{2n}(x) = \sum_{k=1}^{[\frac{n-1}{2}]} \frac{x^{2k}}{2k}.$$

For x in $]0,1[$, $A_{1n}(x)$ and $A_{2n}(x)$ are strictly decreasing and the inequality (3.1) applies but not to A_n.

The inequality (3.1) also holds for non decreasing sequences of functions, therefore it does not characterize a class of functions.

Example 3.3. Let $A_n(x) = \sum_{k=0}^n x^k = (1 - x^{n+1})(1-x)^{-1}$, for $x > 0$, $x \neq 1$, and for every integer $n \geq 1$. The sequence $(x^n)_{n\geq 1}$ is increasing for $x > 1$ and decreasing for $0 < x < 1$. Obviously, $n < A_n(x) < nx^n$ if $x > 1$ and $nx^n < A_n(x) < n$ if belongs to $]0,1[$. For every $x > 0$, $x \neq 1$, $A_n^2(x) - A_{n-1}(x)A_{n+1}(x) = x^{n-1}(x-1)^2$ is strictly positive. Reversely, $A_{n-1}A_{n+1} > c_n A_n^2$ for some function $c_n(x)$ with values in $]0,1[$ if

$$c_n(x) < 1 - x\frac{(1-x)^2}{(1-x^{n+1})}.$$

If $x > 1$, $(n+1)(x-1) < x^{n+1} - 1 < (n+1)x^n (x-1)$ and

$$1 - \frac{x}{(n+1)^2} < c_n(x) < 1 - \frac{1}{x^{2n+1}(n+1)^2},$$

if $0 < x < 1$, $(n+1)x^{n+1}(x-1) < x^{n+1} - 1 < (n+1)(x-1)$ and

$$1 - \frac{1}{x^{2n+1}(n+1)^2} < c_n(x) < 1 - \frac{x}{(n+1)^2}.$$

The exponential function develops as a series $e^x = A_n(x) + R_n(x)$ with

$$A_n(x) = \sum_{k=0}^{n} \frac{x^k}{k!} = A_{n-1}(x) + a_n(x)$$

and the sequence $(a_n(x))_{n \geq 1}$ is not decreasing for every $x > 0$.

Theorem 3.1. *The expansion of the exponential function satisfies*

$$\frac{A_n^2(x)}{2(n+1)} < A_{n-1}(x)A_{n+1}(x) < A_n^2(x)$$

for every real $x > 0$ and for every integer $n \geq 1$.

Proof. The upper bound is true for every x and for $n = 1, 2$. Let $n > 2$ be an integer, for every $x > 0$

$$(A_{n+1}^2 - A_n A_{n+2})(x) = (a_{n+1}A_{n+1} - a_{n+2}A_n)(x)$$

$$= a_{n+1}(x)\left\{ \sum_{k=0}^{n+1} \frac{x^k}{k!} - \frac{1}{n+2} \sum_{k=1}^{n+1} \frac{x^k}{(k-1)!} \right\}$$

$$= a_{n+1}(x)\left\{ 1 + \sum_{k=1}^{n+1} \frac{x^k}{(k-1)!} \left(\frac{1}{k} - \frac{1}{n+2} \right) \right\} > 0.$$

For the lower bound, let $c_n = (n+1)^{-1}k_n > 0$ with $k_n < 1$

$$(A_{n-1}A_{n+1} - c_n A_n^2)(x) = \{(1-c_n)A_n A_{n-1} + a_{n+1}A_{n-1} - c_n a_n A_n\}(x),$$

$$(a_{n+1}A_{n-1} - c_n a_n A_n)(x) = a_n(x)\left\{ \sum_{k=1}^{n} \frac{x^k}{(k-1)!} \left(\frac{1}{n+1} - \frac{c_n}{k} \right) - c_n \right\},$$

therefore $A_{n-1}A_{n+1} - c_n A_n^2 > 0$ as $c_n < (n+1)^{-1}$ if $(1-c_n)A_n A_{n-1} - a_n c_n \geq 0$. Let $c_n = \{2(n+1)\}^{-1}$, then $A_n A_{n-1} > a_n$ and this condition is fulfilled. $\qquad\square$

The bounds of Theorem 3.1 for the expansion of the exponential function have the best constant uniformly in n.

The variations between $A_{n-1}A_{n+1} - A_n^2$ have the bounds

$$\frac{A_n^2(x)}{2(n+1)} - A_n^2(y) \leq A_{n-1}(x)A_{n+1}(x) - A_{n-1}(y)A_{n+1}(y)$$

$$\leq A_n^2(x) - \frac{A_n^2(y)}{2(n+1)},$$

and the difference

$$\Delta_n(x,y) = A_{n-1}(x)A_{n+1}(x) - A_{n-1}(y)A_{n+1}(y) - A_n^2(x) - A_n^2(y)$$

satisfies

$$-A_n^2(x)\frac{2n+1}{2(n+1)} \leq \Delta_n(x,y) \leq A_n^2(y)\frac{2n+1}{2(n+1)}.$$

Let $A_n(x)$ be the partial sum of order n in the Taylor expansion of a real function f of $C^{n+1}(\mathbb{R})$ in a neighbourhood of x_0

$$A_n(x) = \sum_{k=0}^{n} \frac{x^k f^{(k)}(x)}{k!},$$

$$A_n^2(x) = f^2(x_0) + \sum_{k=1}^{\frac{n}{2}} (x - x_0)^{2k} \frac{f^{(k)2}(x_0)}{(k!)^2}$$

$$+ 2 \sum_{1 \leq j < m \leq n} (x - x_0)^{j+m} \frac{f^{(j)}(x_0) f^{(m)}(x_0)}{j! m!}$$

and a similar expansion can be written for $A_{n-1}A_{n+1}$, which allows to bound the difference $A_{n-1}A_{n+1} - A_n^2$.

For every sum $A_n(x) = \sum_{k=0}^{n} a_k(x)$

$$\frac{A_{n-1}A_{n+1}}{A_n^2} = 1 - \frac{a_n^2}{A_n^2},$$

$$1 - \frac{a_n^2}{n^2 \min_{0 \leq k \leq n} a_k^2} \leq \frac{A_{n-1}A_{n+1}}{A_n^2} \leq 1 - \frac{a_n^2}{2\sum_{k=1}^{n} a_k^2},$$

these inequalities entail the following proposition.

Proposition 3.2. *If the series* $(a_n)_{n \geq n_0}$ *is decreasing to zero, for* $n \geq n_0$

$$1 - \frac{1}{n^2} \leq \frac{A_{n_0}^2 + (n - n_0 - 1)^2 a_n^2}{A_{n_0}^2 + (n - n_0)^2 a_n^2} \leq \frac{A_{n-1}A_{n+1}}{A_n^2}$$

$$\leq 1 - \frac{a_n^2}{A_{n_0}^2 + (n - n_0)^2 a_{n_0}^2}. \tag{3.2}$$

By the Stirling formula, the exponential function has a Taylor expansion with terms $a_n(x) = x^n(n!)^{-1} < 1$ for $x \leq e^{-1}n$, and a_n is decreasing if $0 < x < n+1$, under the last condition the inequality (3.1), hence (3.2) are true for the exponential function.

3.3 Cauchy's inequalities and convex mappings

Cauchy (1821) established the following results for arithmetic means and transformed series.

Proposition 3.3 (Cauchy). *Let* $(x_i)_{i=1,...,n}$ *be a real sequence, then for every sequence of non zero real numbers* $(a_i)_{i=1,...,n}$

$$\min_{1 \leq k \leq n} \frac{x_k}{a_k} \leq \frac{\sum_{i=1}^{n} x_i}{\sum_{i=1}^{n} a_i} \leq \max_{1 \leq k \leq n} \frac{x_k}{a_k},$$

$$\min_{1 \leq k \leq n} \frac{x_k}{a_k} \leq \frac{(\sum_{i=1}^{n} x_i^2)^{\frac{1}{2}}}{(\sum_{i=1}^{n} a_i^2)^{\frac{1}{2}}} \leq \max_{1 \leq k \leq n} \frac{x_k}{a_k}$$

and the equality

$$\frac{x_1}{a_1} = \frac{x_2}{a_2} = \cdots = \frac{x_n}{a_n}$$

implies

$$\frac{1}{n} \sum_{k=1}^{n} \frac{x_k}{a_k} = \frac{\sum_{i=1}^{n} x_i}{\sum_{i=1}^{n} a_i^2} = \frac{(\sum_{i=1}^{n} x_i^2)^{\frac{1}{2}}}{(\sum_{i=1}^{n} a_i^2)^{\frac{1}{2}}}. \tag{3.3}$$

Consider two sequences of non zero real numbers $(x_i)_{i=1,...,n}$ and $(a_i)_{i=1,...,n}$, all a_i having the same sign and being different from their mean $\bar{a}_n = n^{-1} \sum_{j=1}^{n} a_j$. The first equality of Proposition 3.3 is also written $\bar{x}_n \bar{a}_n = n^{-1} \sum_{i=1}^{n} a_i x_i$ and it is obviously false without further conditions. It is equivalent to

$$\zeta_n(a, x) = \sum_{k=1}^{n} \left(\frac{\bar{a}_n}{a_k} - 1 \right) x_k = 0$$

and generally it is not satisfied under the above conditions. It must then be an inequality.

Example 3.4. Let $a_k = k!$ and $x_k = x^k$, then $\lim_{n \to \infty} \zeta_n(a, x) > 0$ is equivalent to $x^{-1}(1 - x)(e^x - 1) > \lim_{n \to \infty} n(\sum_{k=1}^{n} k!)^{-1} = 0$ if $0 < x < 1$, and it is equivalent to $x^{-n}(e^x - 1) > 0$ if $x > 1$.

Example 3.5. Let $x_k = x^k$ for x in $[0, 1]$ and a decreasing sequence $a_k = k^{-1}$, the reverse inequality $\lim_{n \to \infty} \zeta_n(a, x) < 0$ is equivalent to $\log(1 - x) < 0$.

Example 3.6. Let $x_k = x^{2k}$ and $a_k = 2^k$, the inequality $\lim_{n \to \infty} \zeta_n(a, x) > 0$ is fulfilled with equality if $x > 2$ and $(1 - x^2)(2 - x)^{-1} > n2^{-n}$ if $0 < x < 1$. With a faster rate of convergence to zero for the sequence $a_k = 2^{-k}$, we have $\zeta_n(a, x) > 0$ if $x \geq \frac{1}{2}$ and $\zeta_n(a, x) < 0$ if $x \leq \frac{1}{2}$.

In conclusion, the condition of a constant sequence $x_k a_k^{-1}$ is not necessary for the equality in (3.3). For sequences that do not satisfy the condition of Proposition 3.3, the equalities in Equation (3.3) may be replaced by an upper or lower inequality, according to the sequence. With a convex transformation of the series, a general inequality holds for the transformed sequence of Proposition 3.3.

Proposition 3.4. *Let $(x_i)_{i=1,\ldots,n}$ be a real sequence and φ be a convex function, then for every sequence of non zero real numbers $(a_i)_{i=1,\ldots,n}$ such that $a_k^{-1}\varphi(x_k)$ is constant*

$$\frac{1}{n} \sum_{k=1}^{n} \frac{\varphi(x_k)}{a_k} \geq \varphi\left(\frac{\sum_{i=1}^{n} x_i}{\sum_{i=1}^{n} a_i}\right).$$

However, several inequalities similar to Cauchy's Theorem 3.3 are satisfied with power, logarithm and n-roots transforms. In particular, it applies with the logarithm and in a multiplicative form such as

$$\min_{1 \leq k \leq n} x_k^{a_k^{-1}} \leq \left(\prod_{i=1}^{n} x_i\right)^{(\sum_{i=1}^{n} a_i)^{-1}} \leq \max_{1 \leq k \leq n} x_k^{a_k^{-1}}.$$

Let $(x_i)_{i=1,\ldots,n}$ and $(a_i)_{i=1,\ldots,n}$ be non zero real sequences such that $x_i > 0$ for every i. Under the condition of a constant sequence $a_k^{-1} \log x_k$

$$\prod_{i=1}^{n} x_i^{\frac{1}{a_n}} = \prod_{i=1}^{n} x_i^{\frac{1}{a_k}}.$$

Let $x = \log y$ on $]0, \infty[$. By an exponential mapping and for all non negative sequences $(y_i)_{i=1,\ldots,n}$ and $(a_i)_{i=1,\ldots,n}$

$$\exp\left(\frac{\sum_{i=1}^{n} a_i x_i}{\sum_{i=1}^{n} a_i}\right) = \prod_{i=1}^{n} y_i^{a_i(\sum_{k=1}^{n} a_k)^{-1}} \leq \frac{\sum_{i=1}^{n} a_i y_i}{\sum_{i=1}^{n} a_i},$$

with equality if and only if all x_k are equal. Assuming $a_i = 1$ for every i, this inequality becomes $\prod_{i=1}^{n} y_i^{\frac{1}{n}} \leq \bar{y}_n$, therefore the arithmetic mean of a real sequence is larger than its geometric mean, as Cauchy had already proved it. The comparison of the arithmetic mean and the geometric mean is extended as follows.

Proposition 3.5. *Let $(x_i)_{i=1,\ldots,n}$ be a real sequence in $[0,1[$*

$$\prod_{i=1}^{n} \left(\frac{x_i}{1-x_i} \right)^{\frac{1}{n}} < \frac{\bar{x}_n}{1-\bar{x}_n}, \quad \text{if } 0 < x < \frac{1}{2},$$

$$\prod_{i=1}^{n} \left(\frac{x_i}{1-x_i} \right)^{\frac{1}{n}} > \frac{\bar{x}_n}{1-\bar{x}_n}, \quad \text{if } \frac{1}{2} < x < 1,$$

with equality if and only if $x = \frac{1}{2}$.
 For a real sequence $(x_i)_{i=1,\ldots,n}$ in $]1,\infty[$

$$\prod_{i=1}^{n} \left(\frac{x_i}{x_i - 1} \right)^{\frac{1}{n}} > \frac{\bar{x}_n}{\bar{x}_n - 1}.$$

Proof. Let $a_i = 1 - x_i$ and let $\varphi(x) = \log\{x(1-x)^{-1}\}$ for x in $[0,1]$. The function φ is increasing on $[0,1]$. On the subinterval $[\frac{1}{2}, 1[$, it is convex with values in $[0, \infty[$ and

$$\prod_{i=1}^{n} \left(\frac{x_i}{1-x_i} \right)^{\frac{1}{n}} = \exp\left\{ n^{-1} \sum_{i=1}^{n} \varphi(x_i) \right\} \geq \exp\{\varphi(\bar{x}_n)\} = \frac{\bar{x}_n}{1-\bar{x}_n}.$$

On $[0, \frac{1}{2}]$, the function φ is concave with values in $]-\infty, 0]$ and the inequality is reversed. On $]1, \infty[$, φ is replaced by the decreasing and convex function $\psi(x) = \log\{x(x-1)^{-1}\}$ and the convexity argument yields the inequality as in the first case. $\qquad\square$

Alzer (1990a) proved stronger inequalities between the arithmetic and the geometric means on $[0, \frac{1}{2}]$, $A_n(x) = \bar{x}_n$ and $G_n(x) = \prod_{i=1}^{n} x_i^{\frac{1}{n}}$, by introducing the means $A_n'(x) = 1 - \bar{x}_n$ and $G_n'(x) = \prod_{i=1}^{n} (1-x_i)^{\frac{1}{n}}$

$$\frac{G_n}{G_n'} \leq \frac{A_n}{A_n'}, \quad \frac{1}{G_n'} - \frac{1}{G_n} \leq \frac{1}{A_n'} - \frac{1}{A_n}.$$

The first inequality is proved in Proposition 3.5 and the second one is immediately deduced from the first one and from the inequality $A_n \geq G_n$, writing

$$\frac{1}{G_n'} - \frac{1}{G_n} = \frac{1}{G_n}\left\{ \frac{G_n}{G_n'} - 1 \right\} \leq \frac{1}{A_n}\left\{ \frac{A_n}{A_n'} - 1 \right\} = \frac{1}{A_n'} - \frac{1}{A_n}.$$

Proposition 3.5 extends on $\mathbb{R}_+$ by the same arguments.

Proposition 3.6. *Let* $x_n = (x_i)_{i=1,\ldots,n}$ *be a real sequence*

$$\frac{1}{G'_n}(x_n) - \frac{1}{G_n}(x_n) < \frac{1}{A'_n}(x_n) - \frac{1}{A_n}(x_n), \quad \text{if } 0 < x_n < \frac{1}{2},$$

$$\frac{1}{G'_n}(x_n) - \frac{1}{G_n}(x_n) > \frac{1}{A'_n}(x_n) - \frac{1}{A_n}(x_n), \quad \text{if } \frac{1}{2} < x_n < 1,$$

$$\prod_{i=1}^{n}(x_i - 1)^{-\frac{1}{n}} - \prod_{i=1}^{n} x_i^{-\frac{1}{n}} > \frac{1}{\bar{x}_n - 1} - \frac{1}{\bar{x}_n}, \quad \text{if } x_n > 1.$$

If the sequences $\bar{x}_n$ and $n^{-1}\sum_{i=1}^{n} \log x_i$ converge respectively to finite limits μ and λ as n tends to infinity, these inequalities still hold for the limits. More generally, let X be a random variable with distribution function F on a real interval I_X and such that the means $\mu = EX$ and $\lambda = \int_{I_X} \log x \, dF(x) = E(\log X)$ are finite, then G_n converges to $e^{E(\log X)}$ and G'_n converges to $e^{E(\log(1-X))}$. By the concavity of the logarithm, $e^{E(\log X)} \leq EX$, therefore Proposition 3.6 implies the following inequalities.

$$e^{-E(\log(1-X))} - e^{-E(\log X)} < \frac{1}{1 - EX} - \frac{1}{EX}, \quad \text{if } I_X =]0, \frac{1}{2}[,$$

$$e^{-E(\log(1-X))} - e^{-E(\log X)} > \frac{1}{1 - EX} - \frac{1}{EX}, \quad \text{if } I_X =]\frac{1}{2}, 1[,$$

$$e^{-E(\log(X-1))} - e^{-E(\log X)} > \frac{1}{EX(EX - 1)}, \quad \text{if } I_X =]1, \infty[.$$

Similar inequalities cannot be established on other intervals.

The Cauchy distribution for a variable X is defined by its density and distribution functions on $\mathbb{R}_+$

$$f_X(x) = \frac{2}{\pi(1 + x^2)}, \quad F_X(x) = \frac{2}{\pi} \arctan x. \tag{3.4}$$

Thus Cauchy's distribution is identical to the distribution of the tangent of a uniform variable U over $[0, \frac{\pi}{2}]$ and it cannot be extended by periodicity on $[0, \pi]$ or $[-\frac{\pi}{2}, \frac{\pi}{2}]$ where the tangent has negative values. The variable X has no mean since the integral $\int_0^x s f_X(s) \, ds = \frac{1}{\pi} \log(1 + x^2)$ tends to infinity with x.

It is known that the variable X^{-1} has the Cauchy distribution on $\mathbb{R}_{*+}$. The function $G(x) = \frac{1}{2}(x - x^{-1})$ is increasing from $\mathbb{R}_{*+}$ to $\mathbb{R}$, with inverse function $G^{-1}(y) = y + (1 + y^2)^{\frac{1}{2}}$, then the variable $Y = G(X)$ has also the Cauchy distribution. Let H be the function $H(x) = (1 - x)^{-1}(x + 1)$ defined on $[0, 1[\cup]1, \infty[$, it is increasing in each subinterval and it is a bijection between $[0, 1[$ and $]1, \infty[$ and between $]1, \infty[$ and $] -\infty, -1[$. Let X' be a variable having the Cauchy distribution restricted to the interval $]1, \infty[$, namely $2F_X$. The variable $Y = H(X')$ has the same restricted Cauchy distribution as X'

$$P\left(\frac{X'+1}{X'-1} \leq y\right) = P(X' \leq H^{-1}(y)) = \frac{4}{\pi} \arctan \frac{y-1}{y+1},$$

with values in $[0, 1]$.

3.4   Inequalities for the mode and the median

The mode of a real function f is its maximum when it is finite, it is then reached at

$$M_f = \inf\{y \in I; f(y) \geq x \text{ for every } x \in I\}.$$

In the neighbourhood of its mode, a function is locally concave and it is locally convex in a neighbourhood of the mode of $-f$ where it reaches its minimum. Therefore the derivatives of f are such that $f^{(1)}(M_f) = 0$ and $f^{(2)}(M_f) \leq 0$. For some distributions, the mode can be explicitly determined in terms of the parameters of the distribution, such as the expectation μ for a Gaussian distribution $\mathcal{N}(\mu, \sigma^2)$ or the Laplace distribution with density $f_{\mu,b}(x) = (2b)^{-1} \exp\{-b^{-1}|x - \mu|\}$. Sometimes only an interval can be given, for example the mode of the Poisson distribution with parameter λ belongs to the interval $[\lambda - 1, \lambda]$. The power functions, the logarithm or the exponential have no mode. The Weibull distribution $F_{\lambda,k}(x) = 1 - \exp\{-\lambda^{-k}x^k\}$ has the mode $M_f = \lambda(1 - k^{-1})^{\frac{1}{k}}$, for $\lambda > 0$ and $k > 0$.

On a probability space $(\Omega, \mathcal{F}, P)$, let X be a random variable with real values and with a right-continuous distribution function $F(x) = P(X \leq x)$. The median m_X of a variable X is defined as

$$m_X = \arg\min_{y \in \mathbb{R}} E_F |X - y|$$

with the expectation $E_F |X - y| = \int_{\mathbb{R}} |x - y| \, dF(x)$. The function

$$g(y) = E_F |X - y| = y\{2F(y) - 1\} + E_F(X 1_{\{X \geq y\}}) - E_F(X 1_{\{X < y\}})$$

reaches its minimum at the median value, which is equivalent to $2F(m_X) - 1 = 0$ if the variable X has a density function. The median is equivalently defined by

$$m_F = \inf\left\{x \in \mathbb{R} : F(x) \geq \frac{1}{2}\right\} = \inf\{x \in \mathbb{R} : F(x) = 1 - F(x)\}.$$

At its minimum, $E_F|X - m_F| \geq 0$ and, by definition, it equals

$$E_F|X - m_F| = 2E_F(X1_{\{X \geq m_F\}}) - 1,$$

and it follows that

$$E_F(X1_{\{X \geq m_F\}}) \geq \frac{1}{2}, \ E_F(X1_{\{X < m_F\}}) \leq \frac{1}{2}.$$

Proposition 3.7. *For every random variable X of $L^1(\mathbb{R})$*

$$|EX - m_X| = E|X - m_X| \tag{3.5}$$

and $E|X - m_X| = 0$ if and only if the expectation of X equals its median.

Proof. For every real y, $E_F X - m_F = E_F|X - y| + y - m_F$ and the difference $|E_F X - m_F| \leq E_F|X - y| + |y - m_F|$ is minimum at m_F where the expectation of the variable X satisfies Equation (3.5). For a variable X having a expectation different from its median, $E_F|X - m_F| > 0$. $\square$

Equation (3.5) implies that two variables X_1 and X_2 having equal medians also have equal expectations if and only if $E|X_1 - m_{X_1}| = E|X_2 - m_{X_2}|$. The equality of the expectations of variables do not imply the equality of their medians, except if $E|X_1 - m_{X_1}| = E|X_2 - m_{X_2}|$. The distribution function of X has a left heavy tail if $EX > m_X$ and it has a right heavy tail if $EX < m_X$. For a symmetric random variable with a density function, the expectation, the median and the location of the mode of its density are equal.

The median of a transformed variable by a monotone function can be explicitly calculated by mapping the median of the variable. Let X be a random variable with a continuous distribution function on $\mathbb{R}$ and with median m_X, then $F(m_X) = \frac{1}{2}$ and

$$P(1 - X < m_{1-X}) = 1 - F(1 - m_{1-X}) = \frac{1}{2}$$

which entails $m_{1-X} = 1 - m_X$. This argument extends to every monotone and continuous real function φ. The median of $\varphi(X)$ is

$$m_{\varphi(X)} = \varphi(m_X).$$

In particular, $m_{\alpha X} = \alpha m_X$ for every real α. Similar to Proposition 3.6, if $m_X < EX$ we obtain

$$\frac{m_X}{m_{1-X}} < \frac{EX}{1 - EX},$$

$$\frac{1}{m_{1-X}} - \frac{1}{m_X} < \frac{1}{1 - EX} - \frac{1}{EX},$$

and if $EX < m_X$, the inequalities are reversed.

Ordered variables X_1 and X_2 have ordered distribution functions

$$X_1 < X_2 \Longrightarrow F_{X_2} < F_{X_1}$$

they have ordered expectations and medians. Let F_1 and F_2 be real distribution functions with respective medians m_{F_1} and m_{F_2} and such that there exists A satisfying $F_1 \leq F_2$ on $]-\infty, A]$ and $F_2 \leq F_1$ on $[A, \infty[$, then obviously, $1 = 2F_1(m_{F_1}) \leq 2F_2(m_{F_1})$ if $m_{F_1} < A$, hence

$$m_{F_1} \leq m_{F_2} \text{ if } m_{F_2} \leq A,$$

$$m_{F_2} \leq m_{F_1} \text{ if } m_{F_1} \geq A.$$

One cannot deduce an order for the expectations $E_{F_1}(X)$ and $E_{F_2}(X)$ of variables under F_1 and F_2, but only for the expectations of variables restricted to the subintervals defined by A.

Mixtures of k unimodal densities may be unimodal or multimodal according to the distance between the modes of the densities of the mixture and the number of local maxima of the mixture density is between one and k. Let us consider the mixture of two symmetric densities with proportions p and $1 - p$, and with location parameters μ and $-\mu$

$$f_{p,\mu} = pf_\mu + (1 - p)f_{-\mu}$$

and let X be a random variable having the mixture density $f_{p,\mu}$. The expectation of X is $EX = (2p - 1)\mu$, it is zero only if $p = .5$, when $f_{p,\mu}$ is symmetric. The first derivative of a mixture with Gaussian components equals $f_{p,\mu}^{(1)}(x) = -f_{-\mu}(x)\{(x - \mu)pe^{2x\mu} + (1 - p)(x + \mu)\}$ and it is zero as

$$e^{2x\mu} = \frac{(1 - p)(x + \mu)}{p(\mu - x)} > 0$$

therefore the mode of a unimodal Gaussian mixture density $f_{p,\mu}$ satisfies $-\mu < M_f < \mu$. If p belongs to the interval $]0, .5[$, $M_f > 0$, and for p in $].5, 1[$, $M_f < 0$.

Let f_1 be the density function of a variable with distribution $\mathcal{N}(\mu_1, \sigma_1^2)$, f_2 be the density function of $\mathcal{N}(\mu_2, \sigma_2^2)$ and let $\theta = \sigma_2^2 \sigma_1^{-2}$. Replacing $x - \mu$ by $y_1 = (x - \mu)\sigma_1^{-1}$ and $x + \mu$ by $y_2 = (x + \mu)\sigma_2^{-1}$, the first derivative of $f_p = pf_1 + (1 - p)f_2$ is

$$f_{p,\mu,\sigma}^{(1)}(x) = \frac{1}{\sigma_2^2\sqrt{2\pi}}e^{-y_2^2}\{p\theta y_1 e^{y_2^2 - y_1^2} + (1 - p)y_2\},$$

therefore the same inequality holds for the mode of the mixture of Gaussian densities with inequal variances.

The sub-densities pf_μ and $(1-p)f_{-\mu}$ have an intersection at x_p where $pf_\mu(x_p) = (1-p)f_{-\mu}(x_p)$, denoted y_p, therefore $f_{p,\mu}(x_p) = 2pf_\mu(x_p)$ and, by symmetry, $f_\mu(x_p) = f_{-\mu}(x_p)$ if and only if $p = .5$. A mixture density $f_{p,\mu}$ is bimodal if and only if it has a local minimum at x_p.

Proposition 3.8. *The Gaussian mixture density $f_{.5,\mu}$ is bimodal if $\mu > 1$ and it is unimodal if $\mu < 1$. For every $p \neq .5$ in $]0,1[$, the Gaussian mixture density $f_{p,\mu}$ is bimodal if $\mu > (1-x_p^2)^{\frac{1}{2}}$ and it is unimodal if $\mu < (1-x_p^2)^{\frac{1}{2}}$. A mixture density $f_{p,\mu_1,\mu_2,\sigma^2}$, with Gaussian components f_{μ_1,σ^2} and f_{μ_2,σ^2}, is bimodal if $|\mu_1 - \mu_2| > 2(1-x_p^2)^{\frac{1}{2}}$ and it is unimodal if $|\mu_1 - \mu_2| < 2(1-x_p^2)^{\frac{1}{2}}$.*

Proof. The second derivative of the Gaussian mixture density $f_{p,\mu}$ at x_p is $f_{p,\mu}^{(2)}(x_p) = [\{(x_p - \mu)^2 - 1\} + \{(x_p + \mu)^2 - 1\}]y_p = 2(x_p^2 + \mu^2 - 1)y_p$ and the sign of $x_p^2 + \mu^2 - 1$ determines whether $f_{p,\mu}$ has a maximum or a minimum at x_p. $\square$

With respective variances σ_1^2 and σ_2^2, the sign of the second derivative $f_{p,\mu,\sigma}^{(2)}$ is defined by the sign of $x_p + \mu$ with respect to a constant depending on the variances. These properties extend to mixtures of two symmetric densities with respective expectations μ_1 and μ_2. If their difference $|\mu_1 - \mu_2|$ is sufficiently large, a mixture density f_{p,μ_1,μ_2} is bimodal and it is unimodal if $|\mu_1 - \mu_2|$ is smaller than a threshold μ_p depending on the mixture proportion p. Proposition 3.8 can be generalized to mixtures with n symmetric components by cumulating the tails of the n components and it is necessary to increase the distance between two consecutive modes with the number of components to have a multimodal mixture density with n modes. The number of modes of mixtures with a countable number of components is necessarily smaller than the number of components.

Continuous mixture distributions describe the distribution of variable partially observed in models with a latent (unobserved) variable. Let (X, W) is a vector of dependent variables on $\mathcal{X} \times \mathcal{W}$ where W is a latent variable. The distribution function of X is

$$F_X(x) = \int_{\mathbb{R}} F_{X|W}(x, w) \, dF_W(w),$$

where F_W is the distribution function of a variable W and for every w, $F_{X|W}(\cdot, w)$ is the conditional distribution of a variable X given $W = w$. If $F_{X|W}(\cdot, w)$ is absolutely continuous with density $f_{X|W}(\cdot, w)$ for every real w, then F_X is absolutely continuous with density

$$f_X(x) = \int_{\mathbb{R}} f_{X|W}(x, w) \, dF_W(w)$$

and the Laplace transform of X is

$$L_X(x) = \int_{\mathbb{R}} L_{X|W}(x, w) \, dF_W(w) = EL_{X|W}(x, W).$$

In particular, for a scale mixture of a distribution $X = NV^{\frac{1}{2}}$, with independent random variables X and N, N being a standard normal variable and $V = X^2 Y^{-2}$ being a random variance, the Laplace transform of X is $L_X(t) = \int_0^\infty e^{-\frac{x^2}{2v}} dF_V(v)$.

Keilson and Steutel (1974) studied several classes of continuous mixtures. They defined a scale mixture variable $X = YW$ with independent variables X and Y having absolutely continuous densities and with $W = XY^{-1}$. The mixture density of X is

$$f_X(x) = \int_0^\infty f_Y\left(\frac{x}{w}\right) w^{-1} \, dF_W(w).$$

The norms of variables X and Y of $L^p(\mathbb{R}_+)$ are such that

$$\frac{\|X\|_{L^p}}{\|X\|_{L^1}} = \frac{\|Y\|_{L^p}}{\|Y\|_{L^1}} \frac{\|W\|_{L^p}}{\|W\|_{L^1}}.$$

They proved that the variances of the variables X and Y satisfy

$$\frac{\sigma_X^2}{\|X\|_{L^1}^2} \geq \frac{\sigma_Y^2}{\|Y\|_{L^1}^2},$$

with equality if and only if the mixing distribution is degenerate, and

$$\frac{\sigma_X^2}{\|X\|_{L^1}^2} - \frac{\sigma_Y^2}{\|Y\|_{L^1}^2} = \frac{\|X\|_{L^2}}{\|X\|_{L^1}^2} \frac{\sigma_W^2}{\|W\|_{L^1}^2}.$$

3.5 Mean residual time

Let X be a positive real random variable with a distribution function F and a survival function $\bar{F}$ such that $\lim_{x \to \infty} x\bar{F}(x) = 0$. The mean residual time of X at t is

$$e_F(t) = E_F(X - t|X > t) = \frac{1}{\bar{F}(t)} \int_t^\infty (x - t) \, dF(x)$$

$$= \frac{1}{\bar{F}(t)} \left\{ EX - t - \int_0^t (x - t) \, dF(x) \right\} = \int_t^\infty \frac{\bar{F}(x)}{\bar{F}(t)} \, dx,$$

with the convention $\frac{0}{0} = 0$, and $e_F(0) = EX$. The cumulated failure rate of F is defined as $R_F(x) = \int_0^x \bar{F}^{-1}(t) \, dF(t)$ and it defines uniquely the distribution function by $\bar{F}(x) = \exp\{-R_F(x)\}$. An exponential distribution with parameter λ has a constant failure rate $r(x) \equiv \lambda$ and a function $R_\lambda(x) = \lambda x$.

Proposition 3.9. *A distribution function F is determined from its mean residual time by*

$$\bar{F}(x) = \frac{e_F(0)}{e_F(x)} \exp\left\{-\int_0^x \frac{dt}{e_F(t)}\right\}.$$

Proof. By definition of the mean residual time, $e_F(t)\bar{F}(t) = \int_t^\infty \bar{F}(x)\,dx$. Deriving this expression yields

$$\frac{d\bar{F}(t)}{\bar{F}(t)} = -\frac{de_F(t)}{e_F(t)} - \frac{dt}{e_F(t)}$$

and the expression of $\bar{F}$ follows by integration. $\qquad\square$

An exponential distribution is therefore characterized by a constant function e_F. Let G be a distribution function of a positive real random variable satisfying the same condition as F at infinity and such that $R_F(t) \leq R_G(t)$ for every t in an interval $[a, \infty[$.

Proposition 3.10. *Let t be in the interval $[a, \infty[$ and let $x > t$, the inequality $\int_t^x dR_G < \int_t^x dR_F$ for non exponential distributions F and G implies $e_F(t) < e_G(t)$.*

Proof. Let $t > a$, the difference

$$\delta_{F,G}(t, x) = \int_t^x \frac{\bar{F}(s)}{\bar{F}(t)}\,ds - \int_t^x \frac{\bar{G}(s)}{\bar{G}(t)}\,ds$$

has the derivative

$$\delta'_{F,G}(x, t) = \frac{\bar{G}(x)}{\bar{G}(t)} - \frac{\bar{F}(x)}{\bar{F}(t)}\,dx = \exp\left\{-\int_t^x dR_G(s)\right\} - \exp\left\{-\int_t^x dR_F(s)\right\},$$

for $x > t$, and $\delta'_{F,G}(x, t)$ is strictly positive if

$$\int_t^x dR_G < \int_t^x dR_F, \tag{3.6}$$

for every $x > t$. By the mean value theorem for $\delta_{F,G}(t, x)$, the inequality (3.6) implies $e_F(t) > e_G(t)$. $\qquad\square$

If F has a density f, the function R has a derivative $r(t) = f(t)\bar{F}^{-1}(t)$.

Proposition 3.11. *The mean residual time of a variable X having a density is strictly increasing at t if and only if $r_F(t)e_F(t) > 1$.*

Proof. The derivative of $e_F(t)$ is

$$e_F'(t) = -1 + \frac{f(t)}{\bar{F}^2(t)} \int_t^\infty \bar{F}(x)\, dx = r_F(t) e_F(t) - 1.$$

$\square$

The variance of the residual time of X at t is

$$v_F(t) = E_F\{(X-t)^2 | X > t\} - e_F^2(t) = E_F[\{X - t - e_F(t)\}^2 | X > t].$$

Let X have an exponential distribution $\mathcal{E}_\lambda$, its failure rate is $r_\lambda(t) \equiv \lambda$ and $e_\lambda(t) \equiv \lambda^{-1}$, satisfying $r_\lambda(t) e_\lambda(t) = 1$. Its variance $v_F(t) \equiv \lambda^{-2}$ equals the variance of the variable X. In the same way, for all $t > 0$ and $k \geq 1$

$$E_\lambda\{(X-t)^k | X > t\} = E_\lambda(X^k),$$

that is the lack of memory of the exponential variable. The k-th moment of its residual time after any time t is identical to the k-th moment of the exponential variable and the generating function of the residual time after t is identical to the generating function of the exponential variable. The distribution of $X - t$ conditionally on $\{X > t\}$ is therefore the exponential distribution $\mathcal{E}_\lambda$.

Proposition 3.12. *Let X be a variable X having a density, then the function v_F is strictly increasing at t if and only if $v_F(t) < r_F^{-1}(t) e_F(t) - e_F^2(t)$.*

Proof. From the expression of e_F', the derivative of $v_F(t)$ is

$$v_F'(t) = 2e_F(t) - 2r_F(t)\{v_F(t) + e_F^2(t)\}$$

and v_F is increasing at t if and only if $v_F'(t) > 0$ then the condition is fulfilled. $\square$

3.6 Functional equations

Let $F : \mathbb{R} \to \mathbb{R}$ be a continuous monotone function and

$$\eta(x, y) = F^{-1}(F(x) + F(y)).$$

When F is increasing (respectively decreasing), η is increasing (respectively decreasing) with respect to (x, y) in $\mathbb{R}^2$. Thus, the function $F(x) = e^{-\lambda x}$, $\lambda > 0$ and x in $\mathbb{R}$, defines in $\mathbb{R}^2$ the function

$$\eta(x, y) = y - \lambda^{-1} \log\{1 + e^{-\lambda(x-y)}\} = x - \lambda^{-1} \log\{1 + e^{-\lambda(y-x)}\}.$$

Let $\eta(x, y) = x + y$, equivalently $F(x+y) = F(x) + F(y)$ hence $F(0) = 0$ and F is an odd function.

Let α and β in $\mathbb{R}$ and $\eta(x,y) = x^\alpha + y^\beta$, (x,y) in $\mathbb{R}^2$. With $y = -x$, it follows that $F(0) = 0$ and $F(x) = F(x^\alpha)$ for every x, which is true only by $\alpha = 1$. Similarly, $\beta = 1$ therefore $\eta(x,y) = x + y$ is the unique solution of the functional equation $F(x^\alpha + y^\beta) = F(x) + F(y)$.

Proposition 3.13. *For every real function ψ of $C^2(\mathbb{R}^2)$, a function f such that*

$$\frac{f(x) - 2f(\frac{x+y}{2}) + f(y)}{x - y} = \frac{\psi(x,y)}{2}, \quad x \neq y$$

is defined for $x > 0$ by

$$f(x) = f(0) + xf'(0) - \frac{1}{2}\int_0^x \psi(u,0)\, du,$$

$$f(-x) = f(0) - xf'(0) + \frac{1}{2}\int_{-x}^0 \psi(-u,0)\, du$$

under the necessary conditions that the partial derivatives of ψ satisfy $\psi'_x(x,x) = \psi'_y(x,x)$, and $\psi(x,0) = \psi(0,x)$.

Proof. The functions f and ψ satisfy the following properties

$$\psi(x, -x) = \frac{f(x) + f(-x) - 2f(0)}{2x} = -\psi(-x, x), \tag{3.7}$$

$$0 = \lim_{|x-y| \to 0} \frac{1}{2}\{f'(x) - f'(y)\}$$

$$= \lim_{|x-y| \to 0} \frac{1}{x - y}\left[\left\{f(x) - f\left(\frac{x+y}{2}\right)\right\} - \left\{f(y) - f\left(\frac{x+y}{2}\right)\right\}\right]$$

$$= \psi(x, x).$$

The derivatives of f are

$$f'(x) = \lim_{|x-y| \to 0} \frac{1}{2}\{f'(x) + f'(y)\}$$

$$= \lim_{|x-y| \to 0} \frac{1}{x - y}\left[\left\{f(x) - f\left(\frac{x+y}{2}\right)\right\} + \left\{f(y) - f\left(\frac{x+y}{2}\right)\right\}\right]$$

$$f''(x) = \lim_{|x-y| \to 0} \frac{1}{x - y}\left[\left\{f'(x) - f'\left(\frac{x+y}{2}\right)\right\} + \left\{f'(y) - f'\left(\frac{x+y}{2}\right)\right\}\right]$$

$$= \lim_{|x-y| \to 0} \frac{\psi(x,y) - \psi(x,x)}{2(x - y)} = \frac{1}{2}\psi'_x(x,x) = \frac{1}{2}\psi'_y(x,x).$$

Integrating the derivatives, for every $x > 0$

$$f'(x) = f'(0) - \frac{\psi(x,0)}{2} = f'(0) - \frac{\psi(0,x)}{2},$$

$$f(x) = f(0) + xf'(0) - \frac{1}{2}\int_0^x \psi(u,0)\, du,$$

and in $\mathbb{R}^-$

$$f'(-x) = f'(0) + \frac{\psi(-x,0)}{2},$$

$$f(-x) = f(0) - xf'(0) + \frac{1}{2}\int_{-x}^{0} \psi(-u,0)\, du.$$

$\square$

For a function ψ defined by an odd function g as $\psi(x,y) = g(x) - g(y)$, the necessary conditions $\psi(x,-x) = -\psi(-x,x)$ and $\psi(x,x) = 0$ are satisfied and there are constants a and b such that for every $x > 0$

$$f(x) = ax + b - \frac{1}{2}\int_0^x g(u)\, du,$$

$$f(-x) = -ax + b - \frac{1}{2}\int_{-x}^{0} g(u)\, du.$$

As a consequence of Proposition 3.13, a function f satisfies the equality in Hadamard's inequality (Section 1.3)

$$f\left(\frac{x+y}{2}\right) = \frac{f(x) + f(y)}{2}$$

if the function $\psi = 0$, therefore f is an affine function.

Sahoo and Riedel (1998) presented several functional "open problems". The first one is to find all functions $f :]0,1[\to \mathbb{R}$ such that

$$f(xy) + f(x(1-y)) + f((1-x)y) + f((1-x)(1-y)) = 0, \ 0 < x, y < 1. \quad (3.8)$$

Putting $y = .5$, the equation becomes

$$f\left(\frac{x}{2}\right) + f\left(\frac{1-x}{2}\right) = 0.$$

Hence the function f is skew symmetric with center at $x = \frac{1}{2}$ and $f(\frac{1}{4}) = 0$. Let x tend to zero and $y = .5$ implies $f(0^+) + f(\frac{1}{2}) = 0$. Equation (3.8) also implies that f is skew symmetric around $x = \frac{1}{2}$. Letting $x = y$ tend to zero in (3.8), then

$$0 = f(x^2) + 2f(x(1-x)) + f((1-x)^2) \to 3f(0^+) + f(1^-).$$

So Equation (3.8) admits a linear solution $f(x) = a(x - \frac{1}{4})$, $a \neq 0$, hence $f(0) = -\frac{1}{4}a$ and $f(1) = \frac{3}{4}a$ and there is no other polynomial solution.

Let $f(x) = a\{g(x) - \frac{1}{4}\}$, where the straight line $g(x) = x$ is replaced by any curve that respects the symmetry of f around the line. For example, a sinusoidal function g with an even number of periods is written as

$$g_k(x) = \pm \sin\left(\frac{2k\pi x}{\tau}\right), \quad \tau = \frac{(a^2+1)^{\frac{1}{2}}}{2\sqrt{2}},$$

with an even integer $k \geq 2$, and $f_k(x) = a\{g_k(x) - \frac{1}{4}\}$ is the solution of (3.8). Obviously, the sinusoidal function can be replaced by any continuous and twice differentiable function with first two derivatives having the same signs as g_k up to a translation of its maxima and minima. Every function g is continuous or discontinuous and skew symmetric around $x = \frac{1}{4}$ and such that $g(x + \frac{1}{2}) = g(x)$ provides a solution of the equation.

Equation (3.8) on the closed interval $[0,1]$ implies $f(x) + f(1-x) = 0$, hence $f(\frac{1}{2}) = 0$ instead of $f(\frac{1}{2}) = \frac{3}{4}a > 0$ for the same equation on $]0,1[$, therefore the condition on the boundary modifies the solution which reduces to zero.

The second problem is to find all functions $f :]0,1[\to \mathbb{R}$ such that

$$f(xy) + f((1-x)(1-y)) = f(x(1-y)) + f((1-x)y), \ 0 < x, y < 1. \ (3.9)$$

As y tends to zero, the equation implies $f((1-x)^-) = f(x^-)$ for every x in $]0,1[$. Let the function f be a polynomial of degree 2, $f(x) = ax^2 + bx + c$, then $a = b$ and c varies freely. With polynomials of degree ≥ 3, (3.9) reduces to the roots of higher order equation and it does not determine a function solution of the equation. Equation (3.9) is also fulfilled for Neper's logarithm and for $\log_a x$ for every $a > 0$, it has therefore an infinity of solutions.

Another problem is to find functions f, g, h and $k : \mathbb{R} \to \mathbb{R}$ solutions of the equation

$$\{f(x) - f(y)\} k(x + y) = \{g(x) - g(y)\} h(x + y), \ (x, y) \in \mathbb{R}^2.$$

First, f is even if and only if g is even and, under the condition that f is odd, $f(x) = h(0)\{g(x) - g(-x)\} \{2k(0)\}^{-2}$. For functions f and g of $C(\mathbb{R})$, dividing both terms of the equation by $x - y$ and considering the limit as $|x - y|$ tends to zero implies

$$f(x) - f(0) = \int_0^x g'(u) \frac{h(2u)}{k(2u)} \, du, \ x > 0.$$

Constraints on the functions in the Mean Value Theorem determine simpler classes of functions without the assumption of derivability.

Proposition 3.14. *The functions f and $g : \mathbb{R} \to \mathbb{R}$ satisfy the equation*

$$\frac{f(x) - f(y)}{x - y} = \frac{g(x) + g(y)}{2} \tag{3.10}$$

if and only if g is constant c and $f(x) = f(0) + cx^2$.

Proof. Equation (3.10) with $y = 0$ implies $f(x) = ax + b + \frac{1}{2}xg(x)$ where $a = \frac{1}{2}g(0)$ and $b = f(0)$ and

$$f(x) - f(y) = a(x - y) + \frac{xg(x) - yg(y)}{2} = (x - y)\frac{g(x) + g(y)}{2}$$

which is equivalent to $yg(x) - xg(y) = 0$ for all x and y, therefore $g(0) = 0$ and there exists a constant c such that $g(x) = 2cx$ for every x.   □

Proposition 3.15. *Functions f, g and $h : \mathbb{R} \to \mathbb{R}$ satisfy the equation*

$$\frac{f(x) - g(y)}{x - y} = h(x + y)$$

if and only if $f = g$, h is affine and f is a second-order polynomial.

Proof. Interchanging x and y in the equation implies that for every (x, y) in $\mathbb{R}^2$, $f(x) - g(y) = g(x) - f(y)$, hence $f(x) = g(x) + g(0) - f(0)$ at $y = 0$, and $f(x) - f(y) = g(x) - g(y)$ in $\mathbb{R}^2$. Combining both equalities entails $f \equiv g$. With $y = 0$, $f(x) = f(0) + xh(x)$ hence $h(0) = f'(0)$

$$f(x) - f(y) = xh(x) - yh(y) = (x - y)h(x + y)$$

and $f'(x) = h(2x)$, which is a necessary condition for the existence of a solution of the equation. Moreover

$$h(x + y) = \frac{xh(x) - yh(y)}{x - y} = h(x) + y\frac{h(x) - h(y)}{x - y}$$

as y tend to x, $h(2x) = h(x) + xh'(x)$ for every $x \neq 0$, which implies that $f'(x) = h(x) + xh'(x)$, therefore $h(x) = ax + b$ with real constants a and b. Solutions of the equation are quadratic polynomials.   □

Proposition 3.16. *Functions f, g and $h : \mathbb{R} \to \mathbb{R}$ satisfy the equation*

$$f(x + y) - g(x - y) = h(x)$$

if and only if $f + g$ is constant, $h = f - g$ and the function $H = h - h(0)$ satisfies $H(ax) = aH(x)$ for every real a.

Proof. Writing the functional equation of f, g and h with $y = x$, $y = -x$ then $y = 0$, yields

$$f(x) = h(\frac{x}{2}) + g(0),$$
$$g(x) = -h(\frac{x}{2}) + f(0),$$
$$f(x) + g(x) = a := f(0) + g(0),$$
$$f(x) - g(x) = h(x), \text{ with } y = 0.$$

Let $b = f(0) - g(0) = h(0)$, $F(x) = f(x) - f(0)$, $G(x) = g(x) - g(0)$ and $H(x) = h(x) - h(0)$, the above equalities imply $F(x) + G(x) = 0$, $F(x) - G(x) = H(x)$, therefore

$$F(x) = \frac{1}{2}H(x) = -G(x).$$

Introducing these functions in the functional equation yields

$$H(x) = \frac{1}{2}\{H(x+y) + H(x-y)\}, \tag{3.11}$$

for every (x, y) in $\mathbb{R}$. Moreover

$$F(x) = H\left(\frac{x}{2}\right) = \frac{1}{2}H(x) = \frac{1}{2}F(2x),$$

and this equality extends to positive rational numbers by changing the values of x and y in the previous equality, and therefore to every real number. □

As a consequence of the homogeneity of the function H, the functions f, g and h solutions of Proposition 3.16 satisfy $f(ax) = af(x) + (1-a)f(0)$, $g(ax) = ag(x) + (1-a)g(0)$ and $h(ax) = ah(x) + (1-a)h(0)$ for all real a and x.

Changing g in $-g$ defines another equation and its solutions are characterized by $g(x) = \frac{1}{2}\{h(x) + f(0) - g(0)\}$ and $f(x) = \frac{1}{2}\{h(x) - f(0) + g(0)\}$, moreover $F(x) + G(x) = H(x)$, $F(x) = G(x) = H(\frac{1}{2}x) = \frac{1}{2}H(x)$ and the function H still satisfies Equation (3.11).

Corollary 3.1. *Functions f, g and $h : \mathbb{R} \to \mathbb{R}$ satisfy the equation*

$$f(x+y) + g(x-y) = h(x)$$

if and only if $f - g$ is constant, $h = f + g$ and $f(ax) = af(x) + (1-a)f(0)$, $g(ax) = ag(x) + (1-a)g(0)$ and $h(ax) = ah(x) + (1-a)h(0)$ for every real $a > 0$.

3.7 Carlson's inequality

Carlson's inequality provides upper and lower bounds for the function

$$L(x, y) = \frac{x - y}{\log x - \log y}, \quad 0 < x \neq y \tag{3.12}$$

and it is extended by continuity to $L(x, x) = x$. It is a symmetric, homogeneous and continuous function on $\mathbb{R}_+^2$. Sahoo and Riedel (1998) gave another interesting form of the function L of Equation (3.12)

$$L(x, y) = y \int_0^1 \left(\frac{x}{y}\right)^t dt \tag{3.13}$$

and it is extended to vectors of $\mathbb{R}_+^{*n}$

$$L(x_1, \ldots, x_n) = x_n^{n-1} \int_{[0,1]^n} \prod_{k=1}^{n-1} \left(\frac{x_k}{x_n} \right)^{t_k} \prod_{j=1}^{n-1} dt_j.$$

Carlson (1972) established the lower bound $(xy)^{\frac{1}{2}} \leq L(x, y)$. Sharper upper and lower bounds are given here.

Proposition 3.17. *Let $\varphi : I \subset \mathbb{R} \mapsto \mathbb{R}$ be a function with a convex and decreasing first derivative and let*

$$L_\varphi(x, y) = \frac{x - y}{\varphi(x) - \varphi(y)}.$$

For every $x \neq y$ in $\mathbb{R}_+^2$

$$\frac{2}{\varphi'(x) + \varphi'(y)} \leq L_\varphi(x, y) \leq \frac{1}{\varphi'(x + y)}.$$

If φ has a concave and increasing first derivative, for every $x \neq y$ in $\mathbb{R}_+^2$

$$\frac{1}{\varphi'(x + y)} \leq L_\varphi(x, y) \leq \frac{2}{\varphi'(x) + \varphi'(y)}.$$

Proof. The lower bound of the first inequality is proved by convexity of the function φ'

$$L^{-1}(x, y) = \int_0^1 \varphi'(ux + (1 - u)y) \, du$$

$$\leq \int_0^1 \{u\varphi'(x) + (1 - u)\varphi'(y)\} \, du \leq \frac{1}{2}\{\varphi'(x) + \varphi'(y)\}.$$

The upper bound of $L_\varphi(x, y)$ is a consequence of the inequality $ux + (1 - u)y \leq x + y$ for every u in $[0, 1]$, hence $\varphi'(ux + (1 - u)y) \geq \varphi'(x + y)$ for a decreasing function φ'. The same arguments for an increasing and concave derivative provide the second inequality. $\square$

This general result applies to the function L defined for the logarithm and it is refined by smaller intervals.

Proposition 3.18. *For every $x \neq y$ in $\mathbb{R}_+^2$*

$$\frac{2xy}{x + y} \leq (xy)^{\frac{1}{2}} \leq \frac{(xy)^{\frac{1}{4}}}{2}(x^{\frac{1}{2}} + y^{\frac{1}{2}}) \leq L(x, y),$$

$$L(x, y) \leq \left(\frac{x^{\frac{1}{4}} + y^{\frac{1}{4}}}{2} \right)^2 (x^{\frac{1}{2}} + y^{\frac{1}{2}}) \leq \frac{(x^{\frac{1}{2}} + y^{\frac{1}{2}})^2}{2} \leq x + y.$$

Proof. The inequality is proved by iterative applications of the inequality

$$\frac{2xy}{x+y} \leq L(x,y) \leq x+y$$

which is proved by convexity of the inverse function $x \mapsto x^{-1}$. Applying this inequality to $(x^{\frac{1}{2}}, y^{\frac{1}{2}})$ implies

$$2\frac{(xy)^{\frac{1}{2}}}{x^{\frac{1}{2}}+y^{\frac{1}{2}}} \leq \frac{x^{\frac{1}{2}}-y^{\frac{1}{2}}}{\log x^{\frac{1}{2}} - \log y^{\frac{1}{2}}} \leq x^{\frac{1}{2}}+y^{\frac{1}{2}}$$

multiplying all terms by $\frac{1}{2}(x^{\frac{1}{2}}+y^{\frac{1}{2}})$ yields

$$(xy)^{\frac{1}{2}} \leq L(x,y) \leq \frac{(x^{\frac{1}{2}}+y^{\frac{1}{2}})^2}{2}$$

and the inequalities

$$\frac{2xy}{x+y} \leq (xy)^{\frac{1}{2}} \leq L(x,y) \leq \frac{(x^{\frac{1}{2}}+y^{\frac{1}{2}})^2}{2} \leq x+y.$$

With $(x^{\frac{1}{4}}, y^{\frac{1}{4}})$, we obtain

$$2\frac{(xy)^{\frac{1}{4}}}{x^{\frac{1}{4}}+y^{\frac{1}{4}}} \leq \frac{x^{\frac{1}{4}}-y^{\frac{1}{4}}}{\log x^{\frac{1}{4}} - \log y^{\frac{1}{4}}} \leq x^{\frac{1}{4}}+y^{\frac{1}{4}},$$

then multiplying all terms by $\frac{1}{4}(x^{\frac{1}{4}}+y^{\frac{1}{4}})(x^{\frac{1}{2}}+y^{\frac{1}{2}})$ yields

$$\frac{(xy)^{\frac{1}{4}}}{2}(x^{\frac{1}{2}}+y^{\frac{1}{2}}) \leq L(x,y) \leq \left(\frac{x^{\frac{1}{4}}+y^{\frac{1}{4}}}{2}\right)^2 (x^{\frac{1}{2}}+y^{\frac{1}{2}})$$

and these bounds are included in the previous ones. $\qquad\square$

More iterations like in the proof of Proposition 3.17 allow to improve the bounds of Carlson's function L. With exponential values, it follows that for all distinct $y > 0$ and $x > 0$,

$$e^{-\frac{x+y}{2}} \leq e^{-\frac{x+y}{4}} \frac{e^{-\frac{x}{2}}+e^{-\frac{y}{2}}}{2} \leq \frac{e^{-x}-e^{-y}}{y-x}$$

$$\leq \left(\frac{e^{-\frac{x}{4}}+e^{-\frac{y}{4}}}{2}\right)^2 (e^{-\frac{x}{2}}+e^{-\frac{y}{2}}) \leq \frac{(e^{-\frac{x}{2}}+e^{-\frac{y}{2}})^2}{2} \leq (e^{-x}+e^{-y}).$$

Equivalently

$$e^{-\frac{3(x+y)}{2}} \leq e^{-\frac{5(x+y)}{4}} \frac{e^{-\frac{x}{2}}+e^{-\frac{y}{2}}}{2} \leq \frac{e^y - e^x}{y-x}$$

$$\leq e^{-(x+y)} \left(\frac{e^{-\frac{x}{4}}+e^{-\frac{y}{4}}}{2}\right)^2 (e^{-\frac{x}{2}}+e^{-\frac{y}{2}}) \leq e^{-(x+y)} \frac{(e^{-\frac{x}{2}}+e^{-\frac{y}{2}})^2}{2}$$

$$\leq e^{-2(x+y)}(e^x + e^y).$$

On $[0, \frac{\pi}{2}]$, the cosine function is decreasing and concave and the sine function is increasing and concave.

Proposition 3.19. *For every $x \neq y$ in $[0, \frac{\pi}{2}]$*

$$1 < \frac{x - y}{\sin x - \sin y} < \min \left\{ \frac{1}{\cos(x + y)}, \frac{2}{\cos x + \cos y} \right\},$$

$$\frac{1}{\sin(x + y)} < \frac{x - y}{\cos y - \cos x} < \frac{2}{\sin x + \sin y}.$$

The exponential on $\mathbb{R}_+$ is an increasing and convex function, and Proposition 3.17 only provides lower bounds of the function $L_{\exp}$.

Proposition 3.20. *Let $\alpha > 0$, for every $x \neq y$ in $\mathbb{R}_+^2$*

$$\frac{\alpha(x - y)}{e^{\alpha x} - e^{\alpha y}} \geq \min \left\{ \frac{2}{e^{\alpha x} + e^{\alpha y}}, e^{-\alpha(x+y)} \right\}.$$

Proposition 3.21. *Let $0 < \alpha < 1$ and let $L_\alpha(x, y) = (x - y)(x^\alpha - y^\alpha)^{-1}$ be defined on $\mathbb{R}_+^2$. For every $x \neq y$*

$$\frac{(x + y)^{1-\alpha}}{\alpha} \leq L_\alpha(x, y) \leq \frac{2}{\alpha} \frac{(xy)^{1-\alpha}}{x^{1-\alpha} + y^{1-\alpha}}.$$

The function $\varphi(x) = x^\alpha$, $0 < \alpha < 1$, has an increasing and concave derivative and Proposition 3.17 applies, which provides the bounds. Applying this inequality to $(x^{\frac{1}{2}}, y^{\frac{1}{2}})$ and multiplying by $(x^{\frac{1}{2}} + y^{\frac{1}{2}})(x^{\frac{\alpha}{2}} + y^{\frac{\alpha}{2}})^{-1}$ implies

$$L_\alpha(x, y) \leq \frac{2}{\alpha}(xy)^{\frac{1-\alpha}{2}} \frac{x^{\frac{1}{2}} + y^{\frac{1}{2}}}{\{x^{\frac{1-\alpha}{2}} + y^{\frac{1-\alpha}{2}}\}\{x^{\frac{\alpha}{2}} + y^{\frac{\alpha}{2}}\}}$$

$$\leq \frac{2}{\alpha}(xy)^{\frac{1-\alpha}{2}}$$

but $(xy)^{\frac{1}{2}(1-\alpha)} \leq x^{1-\alpha} + y^{1-\alpha}$ and this upper bounds is larger than the upper bound of Proposition 3.21. A lower bound of L_α is

$$L_\alpha(x, y) \geq \frac{1}{\alpha} \frac{(x^{\frac{1}{2}} + y^{\frac{1}{2}})^{2-\alpha}}{x^{\frac{\alpha}{2}} + y^{\frac{\alpha}{2}}}.$$

The consecutive bounds of L_α are not decreasing intervals for every (x, y) in $\mathbb{R}_+^2$, as they are for the logarithm. The lower bounds would be increasing if $(x+y)^{1-\alpha}(x^{\frac{\alpha}{2}} + y^{\frac{\alpha}{2}}) \leq (x^{\frac{1}{2}} + y^{\frac{1}{2}})^{2-\alpha}$. By concavity, $x^{\frac{\alpha}{2}} + y^{\frac{\alpha}{2}} < 2(x+y)^{\frac{\alpha}{2}}$ and the previous inequality would be satisfied if

$$(x + y)^{1-\frac{\alpha}{2}} < \frac{1}{2}(x^{\frac{1}{2}} + y^{\frac{1}{2}})^{2-\alpha}$$

but the right-hand term of this inequality is lower than $2^{1-\alpha}(x+y)^{1-\frac{\alpha}{2}}$ which is smaller than the left-hand term. The upper bounds would be decreasing if

$$\frac{(x^{\frac{1}{2}}+y^{\frac{1}{2}})(x^{1-\alpha}+y^{1-\alpha})}{\{x^{\frac{1}{2}(1-\alpha)}+y^{\frac{1-\alpha}{2}}\}\{x^{\frac{\alpha}{2}}+y^{\frac{\alpha}{2}}\}} \leq (xy)^{\frac{1-\alpha}{2}}$$

but it does not seem to be generally true. Since $(x^{\frac{1}{2}}+y^{\frac{1}{2}})^{\alpha}$ is lower than $x^{\frac{\alpha}{2}}+y^{\frac{\alpha}{2}}$, the left-hand term of this inequality is smaller than

$$\frac{(x^{\frac{1}{2}}+y^{\frac{1}{2}})^{1-\alpha}(x^{1-\alpha}+y^{1-\alpha})}{x^{\frac{1-\alpha}{2}}+y^{\frac{1-\alpha}{2}}} < (x^{\frac{1}{2}}+y^{\frac{1}{2}})^{1-\alpha}(x^{1-\alpha}+y^{1-\alpha})^{\frac{1}{2}}$$

which is not always lower than $(xy)^{\frac{1}{2}(1-\alpha)}$. So no order can be established and the method of Proposition 3.18 cannot be generalized to all functions. It remains to determine a class of function for which such decreasing intervals can be defined by the same arguments as the logarithm.

3.8 Functional means

Functional means have been introduced by Stolarski (1975), Mays (1983), Sahoo and Riedel (1998). For a real function f in $\mathbb{R}$, let

$$M_f(x,y) = f'^{-1}\Big\{\frac{f(y)-f(x)}{y-x}\Big\}, \; x \neq y.$$

With $f(x) = x^{-1}$, $M_f(x,y) = (xy)^{\frac{1}{2}}$ is the geometric mean, with $f(x) = x^2$, $M_f(x,y)$ is the arithmetic mean of x and y. With other power functions $f_\gamma(x) = x^\gamma$, γ real, we obtain the following expressions called Cauchy's means

$$M_{f_\gamma}(x,y) = \Big\{\frac{x^\gamma-y^\gamma}{\gamma(x-y)}\Big\}^{\frac{1}{\gamma-1}}, \; 1 < \gamma \in \mathbb{R},$$

$$M_{f_k}(x,y) = \Big\{k^{-1}\sum_{j=0}^{k-1}y^j x^{k-j-1}\Big\}^{\frac{1}{k-1}}, \; k \in \mathbb{N},$$

$$M_{f_{\frac{1}{\gamma}}}(x,y) = \Big\{\frac{x^{\frac{1}{\gamma}}-y^{\frac{1}{\gamma}}}{\gamma(x-y)}\Big\}^{-\frac{\gamma}{\gamma-1}}, \; 1 < \gamma \in \mathbb{R},$$

$$M_{f_{\frac{1}{2}}}(x,y) = \Big\{\frac{x^{\frac{1}{2}}+y^{\frac{1}{2}}}{2}\Big\}^2,$$

and the logarithm function f yields the logarithm mean L. The functional means have been generalized to $\mathbb{R}^n$ and to functions with two parameters in Sahoo and Riedel (1998). For all real numbers x and y

$$M_{f_2}(x,y) \leq M_{f_3}(x,y)$$

but no order can be established between $M_{f_{\frac{1}{2}}}(x, y)$ and $M_{f_{\frac{1}{3}}}(x, y)$ for example and, more generally, between two functional means M_{f_γ} and M_{f_α} with $\alpha \neq \gamma$. The interval established for the function $L_\alpha(x, y) = \alpha^{-1} M_{f_\alpha}^{1-\alpha}(x, y)$ in Proposition 3.21 also applies to M_{f_α}, $0 < \alpha < 1$.

Proposition 3.22. *For all* $0 < \alpha < 1$ *and* (x, y) *in* $\mathbb{R}_+^{*2}$

$$M_{f_2}(x, y) \leq \frac{M_{f_\alpha}(x, y)}{2} \leq \frac{xy}{(x^{1-\alpha} + y^{1-\alpha})^{\frac{1}{1-\alpha}}}.$$

Proposition 3.23. $L(x, y) \leq \frac{1}{2}(x + y)$ *for every* (x, y) *in* $\mathbb{R}_+^{*2}$, *with equality if and only if* $x = y$.

Proof. Consider the difference between the logarithm mean and the arithmetic mean

$$\delta(x, y) = \frac{x - y}{\log x - \log y} - \frac{x + y}{2}, \quad x, y > 0;$$

Since the respective orders of x and y do not change the sign of the logarithm mean, one can assume that $x > y$ and write $y = \theta x$ with θ in $]0, 1[$. For $\theta = 1$, $\delta(x, x) = 0$ and for every $x > 0$ and θ in $]0, 1[$

$$\delta(x, y) = xh(\theta),$$
$$h(\theta) = \frac{1 - \theta}{\log(\theta^{-1})} - \frac{1 + \theta}{2}$$

and $h(1) = 0$. The sign of $h(\theta)$ is the same as the sign of

$$g(\theta) = 2\frac{1 - \theta}{1 + \theta} + \log \theta,$$

increasing on $]0, 1[$, therefore $g(\theta) \leq g(1) = 0$. □

The logarithm mean is extended to $\mathbb{R}_+^{*n}$ as

$$L_n(x) = \prod_{k=1}^{n-1} \frac{x_k - x_n}{\log x_k - \log x_n}, \quad x = (x_1, \dots, x_n).$$

For x in $\mathbb{R}_+^{*n}$, the difference $L_n(x) - n^{-1} \sum_{k=1}^n x_k$ is unchanged as the sign of $x_k - x_n$ changes, therefore x_k can be supposed lower than x_n and $x_k = \theta_k x_n$, $0 < \theta_k < 1$. Let

$$\delta_n(x) = x_n^{n-1} \prod_{k=1}^{n-1} \frac{\theta_k - 1}{\log \theta_k} - x_n \frac{\sum_{k=1}^{n-1} \theta_k + 1}{n},$$

since the sign of $\delta_n(x)$ depends on whether x_n is larger or smaller than an expression depending on the $(\theta_1, \dots, \theta_{n-1})$, there is no order between the logarithmic and the arithmetic means on $\mathbb{R}_+^{*n}$, $n \geq 3$.

For α in $]0,1[$, let $\eta_\alpha = M_{f_\alpha}$ be defined for strictly positive $x \neq y$ as the functional mean for the power function with exponent α, and let $L(x,y)$ be the limit of η_α as α tends to zero. Sahoo and Riedel (1998) mentioned Alzer's conjecture as follows. For every $\alpha \neq 0$

$$L(x,y) < \frac{\eta_\alpha(x,y) + \eta_{-\alpha}(x,y)}{2} < \frac{x+y}{2} = M_{f_2}(a,b). \qquad (3.14)$$

Using the reparametrization $y = \theta x$ as in the proof of Proposition 3.23

$$h_\alpha(x,y) = \eta_\alpha(x,y) + \eta_{-\alpha}(x,y) = x\left[\left\{\frac{1-\theta^\alpha}{\alpha(1-\theta)}\right\}^{\frac{1}{\alpha-1}} + \left\{\frac{\theta^{-\alpha}-1}{\alpha(1-\theta)}\right\}^{\frac{-1}{\alpha+1}}\right].$$

For every α in $]0,1[$, the inequality

$$L \leq M_{f_2} \leq \frac{M_{f_\alpha}}{2} \leq \frac{\eta_\alpha(x,y) + \eta_{-\alpha}(x,y)}{2}$$

is a consequence of Propositions 3.22 and 3.23. However, it is not proved that the upper bound of (3.14) is true and (3.14) is still a conjecture for $\alpha > 1$.

Proposition 3.24. *For every $\alpha > 1$, $L(x,y) \leq M_{f_\alpha}(x,y)$, (x,y) in $\mathbb{R}_+^{*2}$, with equality if and only if $x = y$.*

Proof. The sign of $L - M_{f_\alpha}$ is the same as the sign of

$$g(\theta) = \frac{\alpha(1-\theta)^2}{1-\theta^\alpha} + \log\theta$$

with a negative first derivative therefore g is decreasing. Since $\lim_{\theta \to 1} = 0$, $g(\theta) > 0$ on $]0,1[$ and it is zero only if $\theta = 1$. $\qquad \square$

Proposition 3.25. *On $\mathbb{R}_+^{*2}$*

$$M_{f_{-2}}(x,y) \leq L(x,y) \leq M_{f_2}(x,y),$$

with equality if and only if $x = y$.

Proof. The function $L(x,y) - M_{f_{-2}}(x,y)$ is written

$$x\left[\frac{1-\theta}{-\log\theta} - \left\{\frac{\theta^{-2}-1}{2(1-\theta)}\right\}^{\frac{-1}{3}}\right]$$

with $y = \theta x$. Its sign is the same as the sign of

$$g_{-2}(\theta) = (1-\theta)(1+\theta)^{\frac{1}{3}}(2\theta^2)^{-\frac{1}{3}} + \log\theta.$$

When θ tends to zero, $(2\theta^2)^{\frac{1}{3}}g_{-2}(\theta)$ is equivalent to $1 + (2\theta^2)^{\frac{1}{3}}\log\theta$ which tends to one and $g_{-2}(\theta)$ tends to infinity. The function g_{-2} is decreasing on $]0,1[$, moreover it tends to zero as θ tends to one and the lower bound for L follows. The upper bound is due to Proposition 3.24. $\qquad \square$

Proposition 3.25 is not generalized to every $\alpha > 1$ as in Proposition 3.24. The interval for the logarithmic mean given in Proposition 3.18

$$\frac{(xy)^{\frac{1}{4}}}{2}(x^{\frac{1}{2}} + y^{\frac{1}{2}}) \leq L(x,y) \leq \left(\frac{x^{\frac{1}{4}} + y^{\frac{1}{4}}}{2}\right)^2 (x^{\frac{1}{2}} + y^{\frac{1}{2}})$$

is smaller than the interval of Proposition 3.25. Intervals for the means M_{f_2} and $M_{f_{-2}}$ are deduced form those of Proposition 3.18

$$M_{f_2}(x,y) \leq \frac{(x^{\frac{1}{2}} + y^{\frac{1}{2}})^2}{2},$$

$$\frac{2xy}{x+y} \leq M_{f_{-2}}(x,y) \leq (xy)^{\frac{1}{2}}.$$

3.9   Young's inequalities

Young's inequality for power functions states that for all real x and y

$$xy \leq \frac{x^p}{p} + \frac{y^{p'}}{p'},$$

with equality if and only if $y = x^{p-1}$, where $p > 1$ and p' are conjugates. It implies that for every convex function φ

$$\varphi(xy) \leq \frac{\varphi(x^p)}{p} + \frac{\varphi(y^{p'})}{p'}.$$

Applying Young's inequality to the moment generating function of real variables X and Y, it follows that for every $\lambda > 0$

$$\varphi_{X+Y}(\lambda) \leq \frac{\varphi_X(\lambda p)}{p} + \frac{\varphi_Y(\lambda p')}{p'}.$$

Young's inequality generalizes in several forms.

Theorem 3.2. *Let $f : I \subset \mathbb{R} \mapsto \mathbb{R}$ be a function of $C(I)$ with a strictly monotone derivative and such that $f(0) = 0$. There exists a unique function $g : f(I) \mapsto \mathbb{R}$ such that $g(0) = 0$*

$$g(y) = \sup_{x \in I}\{xy - f(x)\},$$
$$f(x) = \sup_{y \in I}\{xy - g(y)\},$$
$$xf'(x) = f(x) + g(f'(x)).$$

Proof. Let $g(y) = \sup_{x \in I}\{xy - f(x)\}$, for every y in $f(I)$ there exists a value x in $\bar{I}$ such that $y = f'(x)$ and $x = h(y)$, where h is the inverse of f'. By definition of g,
$$g(y) = yh(y) - f(h(y))$$
and its derivative is $g'(y) = h(y)$, the definition of g is then equivalent to
$$f(x) = \sup_{y \in I}\{xy - g(y)\}.$$
Considering $y_x = f'(x)$ as a function, the derivative of xy_x equals $f'(x) + y'_x g'(y_x)$, which is equivalent to $xy = f(x) + g(y)$ under the conditions $f(0) = 0$ and $g(0) = 0$. $\qquad\qquad\square$

Removing the condition $f(0) = 0$, the equality is replaced by an inequality for positive convex functions f and g of $C(\mathbb{R}_+)$, such that $f'(x)$ is the inverse of $g'(y)$
$$xy_x \leq f(x) + g(y_x).$$
For a variable X with moment generating function φ_X, the Cramer transform
$$\psi(y) = \sup_{x \in I}\{xy - \log \varphi_X(x)\}$$
is a positive convex function by convexity of the function $\log \varphi_X$, and it satisfies Young's inequality
$$\log \varphi_X(x) = \sup_{y \in I}\{xy - \psi(y)\}.$$
For every x, there exists y_x such that
$$\psi(y_x) = x\frac{\varphi'_X}{\varphi_X}(x) - \log \varphi_X(x),$$
$$y_x = \frac{\varphi'_X}{\varphi_X}(x). \tag{3.15}$$
The moment generating function is $\log \varphi_X(x) = y_x \psi'(y_x) - \psi(y_x)$, where y_x is defined by $\psi'(y_x) = x$, and the inequality $\log \varphi_X(x) \geq -\psi(0)$ holds. At $x = 0$, $y_0 = EX$ and by (3.15)
$$\psi(y_0) = \sup_{x \in I}\{xEX - \log \varphi_X(x)\} = 0.$$

Theorem 3.3. *Let $a > 0$ and let $f : [0, a] \mapsto [0, f(a)]$ be a continuous and strictly increasing function. For every b in $[0, f(a)]$, the inverse function of f satisfies*
$$ab \leq \int_0^a f(x)\, dx + \int_0^b f^{-1}(x)\, dx,$$
$$af(a) \geq \int_0^a f(x)\, dx + \int_0^b f^{-1}(x)\, dx,$$
with equality if and only if $b = f(a)$.

Proof. The third equation of Theorem 3.2 applied to primitive functions is equivalent to

$$af(a) = \int_0^a f(x)\,dx + \int_0^{f(a)} f^{-1}(x)\,dx.$$

Since f^{-1} is a strictly increasing function

$$\int_b^{f(a)} f^{-1}(x)\,dx \geq \{f(a) - b\}f^{-1}(b)$$

which is positive if $b \leq f(a)$, it implies the inequality

$$af(a) \geq \int_0^a f(x)\,dx + \int_0^b f^{-1}(x)\,dx,$$

with equality if and only if $b = f(a)$.   □

Corollary 3.2. *Under the conditions of Theorem 3.2, for every b in $[0, f(a)]$*

$$ab \leq af(a) + bf^{-1}(b).$$

Let $F(a) = \int_0^a f(x)\,dx$ and $G(b) = \int_0^b f^{-1}(x)\,dx$, then for positive variables X and Y

$$E(XY) \leq E\{F(X)\} + E\{G(Y)\} \leq E\{Xf(X)\} + E\{Yf^{-1}(Y)\}.$$

3.10   Entropy and information

Let $(\mathcal{X}, \mathcal{B})$ be a separable Banach space, with a measure μ, and let $(\mathcal{F}, \|\cdot\|)$ be a vector space of strictly positive functions of $L^1(\mathcal{X})$. The entropy is defined on $(\mathcal{F}, \|\cdot\|)$ by

$$Ent_\mu(f) = \int_{\mathcal{X}} (\log f) f\,d\mu - \left(\int_{\mathcal{X}} f\,d\mu\right) \log\left(\int_{\mathcal{X}} f\,d\mu\right), \quad f > 0 \in \mathcal{F}. \quad (3.16)$$

By a normalization of f as a density $g_f = (\int_{\mathcal{X}} f\,d\mu)^{-1}f$, the entropy of a density g is written as $Ent_\mu(g) = \int_{\mathcal{X}} (\log g)\,g\,d\mu = E\{\log g(X)\}$, where X is a random variable with values in $\mathcal{X}$ with density g, and

$$Ent_\mu(f) = \left(\int_{\mathcal{X}} f\,d\mu\right) Ent_\mu\left\{\left(\int_{\mathcal{X}} f\,d\mu\right)^{-1} f\right\}.$$

For every density function g in $\mathcal{F}$

$$Ent_\mu(g) = \int_{\mathcal{X}} (\log g)\,g\,d\mu.$$

Let $\mathcal{F}$ be a family of densities such that for $F = \sup\{f \in \mathcal{F}\}$, $F \log F$ is $L^1(\mathcal{X})$, then the entropy of $\mathcal{F}$ is finite and equals to

$$Ent_\mu(\mathcal{F}) = \sup_{g \in \mathcal{F}} \int_{\mathcal{X}} (\log g) \, g \, d\mu = \int_{\mathcal{X}} (\log F) \, F \, d\mu.$$

This definition is extended to probability measures P and Q on $(\mathcal{X}, \mathcal{B})$ such that P is absolutely continuous with respect to Q

$$Ent_Q(P) = \int_{\mathcal{X}} \left(\log \frac{dP}{dQ}\right) dP.$$

For a density

$$Ent_\mu(g) = \int_{\mathcal{X}} (\log g) \, g \, d\mu \le \int_{\mathcal{X}} (\log g) \, g 1_{\{g>1\}} \, d\mu$$

and $Ent_\mu(g) \le 0$ for every density $g \le 1$, for example for the uniform densities on intervals larger than 1, for the Gaussian densities with variances larger than $(2\pi)^{-\frac{1}{2}}$ and for the exponential densities $g(x) = ae^{-ax}$ on $\mathcal{R}_+$, with constants $0 < a \le 1$.

Another upper bound for the entropy is deduced from the concavity of the logarithm.

$$Ent_\mu(f) \le \left\{\int_{\mathcal{X}} \left(\log f - \int_{\mathcal{X}} (\log f) \, d\mu\right)^2 d\mu\right\}^{\frac{1}{2}} \left\{\int_{\mathcal{X}} f^2 \, d\mu\right\}^{\frac{1}{2}}.$$

The entropy of a product of densities is

$$Ent_{\mu^{\otimes n}} \left(\prod_{k=1}^n f_k\right) = \sum_{k=1}^n Ent_\mu(f_k)$$

and more generally, for a product of functions

$$Ent_{\mu^{\otimes n}} \left(\prod_{k=1}^n f_k\right) = \sum_{k=1}^n Ent_\mu(f_k) \left\{\prod_{k \ne j=1}^n \int_{\mathcal{X}} f_j \, d\mu\right\}.$$

Let $\lambda > 0$, the entropy of λf is

$$Ent_\mu(\lambda f) = Ent_\mu(f) + \left(\int_{\mathcal{X}} f \, d\mu\right) \left\{(1 - \lambda) \log\left(\int_{\mathcal{X}} f \, d\mu\right) - \lambda \log \lambda\right\} + \log \lambda$$

and for a density g, $Ent_\mu(\lambda g) = Ent_\mu(g) + (1 - \lambda) \log \lambda$.

The Kullback-Leibler information for two densities f and g of a space of densities $(\mathcal{F}, \|\cdot\|)$ on $(\mathcal{X}, \mathcal{B}, \mu)$ is

$$I_f(g) = \int_{\mathcal{X}} \left(\log \frac{g}{f}\right) f \, d\mu. \tag{3.17}$$

Proposition 3.26. *The Kullback-Leibler information for two densities f and g with respect to a measure μ such that $\log f$ and $\log g$ are integrable with respect to the measure $\int_{-\infty}^{\cdot} f \, d\mu$ satisfies $I_f(g) \le 0$, with equality if and only if $f = g$, μ a.s.*

Proof. Since $\int_{\mathcal{X}}(g - f)\,d\mu = 0$ and $\log(1 + x) \le x$ for every $x > 0$

$$I_f(g) = \int_{\mathcal{X}}\left\{\log\frac{g}{f} - \frac{g}{f} + 1\right\}f\,d\mu$$

and by concavity of the function $\phi(u) = \log u + 1 - u$ with value zero for $u = 1$, we have $I_f(g) = E_P\phi(f^{-1}g) \le \phi(E_P(f^{-1}g)) = 0$. □

For probability measures P and Q on $(\mathcal{X}, \mathcal{B})$ such that P is absolutely continuous with respect to Q, the information is

$$I_P(Q) = \int_{\mathcal{X}}\left(\log\frac{dQ}{dP}\right)dP.$$

Proposition 3.27. *For absolutely continuous probability measures P and Q, we have $I_P(Q) = -Ent_Q(P)$ and $I_P(Q) \le 0$ with equality if and only if $P = Q$ a.s.*

Proof. Let h be the density of P with respect to Q, then the information f P with respect to Q is $I_P(Q) = -E_Q\{h(\omega)\log h(\omega)\}$ and the result follows from the convexity of the function $\varphi(x) = x\log x$ on $\mathbb{R}_+^*$. □

For example, the information of two Gaussian variables with means μ_1 and μ_2, and variances σ_1^2 and σ_2^2, is

$$I_{f_1}(f_2) = \left(\log\frac{\sigma_1}{\sigma_2} - \frac{\sigma_1^2}{\sigma_2^2} + 1\right) - \frac{1}{2}(\mu_1 - \mu_2)^2 \le -\frac{1}{2}(\mu_1 - \mu_2)^2.$$

Chernoff's theorem applied to the empirical distribution of independent variables with density f, $\widehat{F}_n(t) = \sum_{i=1}^n 1_{\{X_i \le t\}}$, has a limiting function $-I(f)$, where $I(f)$ is the Fisher information of the distribution of the variables with respect to the Lebesgue measure.

Let $(P_\theta)_{\theta \in \Theta}$ be a family of probabilities on a bounded parametric set Θ, with densities $(f_\theta)_{\theta \in \Theta}$ in $C^2(\mathbb{R})$ uniformly in a neighborhood of $f_0 = f_{\theta_0}$ and such that $\int f_0^{-1}f_0'^2$ is finite. By a Taylor expansion of f_θ in a neighborhood of f_0, its information develops as

$$I_{f_0}(g_{\theta_n}) = \int_{\mathbb{R}}\log\{1 + \lambda(f_{\theta_n} - f_0)f_0^{-1})\}f_0\,d\mu$$

$$= \lambda\int_{\mathbb{R}}(\theta_n - \theta_0)'\left(\int f_0^{-1}f_0'^2\right)(\theta_n - \theta_0) + o(\|\theta_n - \theta_0\|^2)$$

and $(P_\theta)_\Theta$ is locally asymptotically quadratic at θ_0. A set of probabilities $(P_\theta)_{\theta \in \Theta}$ on a bounded parametric set Θ is locally asymptotically quadratic at θ if the logarithm of the Radon-Nikodym derivative of the probabilities at $\theta_n = \theta + a_n t_n$ and θ has a quadratic expansion with a rate a_n tending to zero

$$\log \frac{dP_{\theta_n}}{dP_\theta} = a_n t_n U_n - \frac{a_n^2}{2} t_n' V_n t_n + o_{P_\theta}(\|a_n t_n\|^2)$$

where $a_n U_n$ is a centered random variable or a process converging weakly to a limit U and $a_n^2 V_n$ is a symmetric matrix converging in probability to a matrix V, under P_θ. Then

$$I_{P_\theta}(P_{\theta_n}) = -\frac{a_n^2}{2} t_n' E_{P_\theta} V_n t_n + o(\|a_n t_n\|^2)$$

and it converges to zero n as tends to infinity.

By concavity, the information of a linear combination of densities $\sum_{k=1}^{n} \alpha_k f_k$, for positive constants α_k such that $\sum_{k=1}^{n} \alpha_k = 1$, has the lower bound

$$I_{f_0}\left(\sum_{k=1}^{n} \alpha_k f_k\right) \geq \sum_{k=1}^{n} \alpha_k I_{f_0}(f_k).$$

The information of a product of marginal densities is the sum of their informations, if $f_0 = \prod_{k=1}^{n} f_{0k}$

$$I_{f_0}\left(\prod_{k=1}^{n} f_k\right) = \sum_{k=1}^{n} I_{f_{0k}}(f_k).$$

For vector of dependent variables (X, Y), information under P with respect to P_0 of the density $f_{X|Y} = f_Y^{-1} f_{X,Y}$ of X, conditionally on Y, is

$$I_{f_{0,X|Y}}(f_{X|Y}) = I_{f_{0,X,Y}}(f_{X,Y}) - I_{f_{0,Y}}(f_Y) \geq I_{f_{0,X,Y}}(f_{X,Y}).$$

The Hellinger distance $h(P, Q)$ between probabilities P and Q is

$$h^2(P, Q) = \frac{1}{2} \int (\sqrt{dP} - \sqrt{dQ})^2 = 1 - \int \left(\frac{dQ}{dP}\right)^{\frac{1}{2}} dP.$$

The L^1 norm, the affinity $\rho(P, Q) = \int \sqrt{dQ\, dP}$ and the Hellinger distance satisfy the inequalities

$$\rho(P, Q) \leq \rho\left(P, \frac{P+Q}{2}\right) = \int \left(\frac{1}{2} + \frac{dQ}{2\, dP}\right)^{\frac{1}{2}} dP$$

$$\leq \left(\frac{1}{2} + \int \frac{dQ}{2}\right)^{\frac{1}{2}} = 1,$$

therefore $h^2(P,Q) \geq 0$ and

$$1 - \rho(P,Q) = h^2(P,Q) \leq \frac{1}{2}\|P - Q\|_1 \leq \{1 - \rho^2(P,Q)\}^{\frac{1}{2}}.$$

The variational distance of probabilities P and Q absolutely continuous with respect to a dominating measure μ On $(\Omega, \mathcal{A})$ is

$$d(P,Q) = 2\sup\{|P(A) - Q(A)|, A \in \mathcal{A}\}$$
$$= 2\int_{\Omega} \left\| \frac{dP}{d\mu} - \frac{dQ}{d\mu} \right\| d\mu$$
$$= 2\Big\{1 - \int_{\Omega} (dP \wedge dQ)\, d\mu\Big\},$$
$$h^2(P,Q) \leq d(P,Q) \leq h(P,Q)\{1 - h^2(P,Q)\}^{\frac{1}{2}}$$

and the Cauchy-Schwarz inequality implies

$$\int \sqrt{dQ\, dP} = \int \sqrt{dP \wedge dQ} \sqrt{dP \vee dQ}$$
$$\leq \Big\{\int (dP \wedge dQ)\Big\}^{\frac{1}{2}} \Big\{\int (dP \vee dQ)\Big\}^{\frac{1}{2}}$$
$$\leq 1 \wedge \Big\{2\int (dP \wedge dQ)\Big\}^{\frac{1}{2}}.$$

The affinity of a product of independent probabilities is their product and their Hellinger distance is

$$h^2\Big(\prod_{i=1}^{n} P_i, \prod_{i=1}^{n} Q_i\Big) = 1 - \prod_{i=1}^{n}\{1 - h^2(P_i, Q_i)\}$$
$$\leq \sum_{i=1}^{n} h^2(P_i, Q_i).$$

Proposition 3.28. *For probability measures P and Q on $(\mathcal{X}, \mathcal{B})$ such that P is absolutely continuous with respect to Q, $I_P(Q) \leq h^2(P,Q)$.*

Proof. By concavity of the function $\phi(u) = \log u - u + \sqrt{u}$ for $u > 0$, the information and the Hellinger distance satisfy the inequality

$$I_P(Q) - h^2(P,Q) = E_P\phi\Big(\frac{dQ}{dP}\Big) \leq \phi\Big(E_P\frac{dQ}{dP}\Big) = \phi(1) = 0,$$

therefore $I_P(Q) \leq h^2(P,Q)$. □

Proposition 3.29. *For a real variable X such that $E_P(e^{tX})$ is finite for every t, $h(a) = \sup_{t \geq 0}\{at - E_P(e^{tX})\} = \inf_{Q:E_Q(X)=a} K(Q,P)$.*

Applying Proposition 3.29 to the bound of Chernoff's Theorem 1.7, with $h(a) = \sup_{t \geq 0} \{at - E_P(e^{tS_n})\}$, it becomes

$$P(S_n > a) = \exp\{-h(a)\} = \sup_{Q: E_Q(S_n) = a} \exp\{-K(Q, P)\}.$$

Let $(P_n)_{n \geq 0}$ and μ be probability measures such that $P_n << \mu$ with a real density $f_n = f_{\theta_n}$ on $\mathbb{R} \times \Theta$ where Θ is a complete subset of $\mathbb{R}^d$ and $\theta \mapsto f_\theta(x)$ belongs to $C_b^2(\Theta)$ uniformly on x and θ_n converges to θ_0 as n tends to infinity. The density f_{θ_n} has a second order Taylor expansion

$$f_{\theta_n} = f_0 + (\theta_n - \theta_0)^T \dot{f}_0 + \frac{1}{2}(\theta_n - \theta_0)^T \ddot{f}_0 (\theta_n - \theta_0) + o(\|\theta_n - \theta_0\|^2)$$

where $f_0 = f_{\theta_0}$, $\dot{f}_0$ and $\ddot{f}_0$ are the first two derivatives of the function $\theta \mapsto f_\theta$, it implies

$$\|f_0^{-\frac{1}{2}}(f_n - f_0)\|_{L^2}^2 = \int f_0^{-1}(f_{\theta_n} - f_0)^2 \, d\mu$$

$$= (\theta_n - \theta_0)^T \Big\{ \int f_0^{-1} \dot{f}_0^2 \, d\mu \Big\}(\theta_n - \theta_0) + o(\|\theta_n - \theta_0\|^2).$$

By the Taylor expansion $f_n^{\frac{1}{2}} - f_0^{\frac{1}{2}} = (\theta_n - \theta_0)^T \frac{\dot{f}_0}{2 f_0^{\frac{1}{2}}} + o(\|\theta_n - \theta_0\|)$, the Hellinger distance has the expansion

$$h^2(P_n, P_0) = \frac{1}{2} \int \Big(f_n^{\frac{1}{2}} - f_0^{\frac{1}{2}}\Big)^2 \, dP_0$$

$$= \frac{1}{8}(\theta_n - \theta_0)^T \Big\{ \int f_0^{-1} \dot{f}_0^2 \, d\mu \Big\}(\theta_n - \theta_0) + o(\|\theta_n - \theta_0\|^2).$$

Let P_1 and P_2 be the probability distributions of Poisson processes N_1 with parameter λ_1 and N_2 with parameter λ_2, the affinity of the probabilities P_{1t} of N_{1t} and P_{2t} of N_{2t} is

$$\rho(P_{1t}, P_{2t}) = e^{-\frac{1}{2}(\lambda_1 t + \lambda_2 t)} \sum_{k \geq 0} \frac{(\lambda_1 \lambda_2)^{\frac{k}{2}} t^k}{k!}$$

$$= \exp\Big\{-\frac{1}{2}(\lambda_1 t + \lambda_2 t) + (\lambda_1 \lambda_2 t^2)^{\frac{1}{2}}\Big\}$$

$$= \exp\Big[-\frac{1}{2}\{(\lambda_1 t)^{\frac{1}{2}} - (\lambda_2 t)^{\frac{1}{2}}\}^2\Big].$$

The information $I_{P_{1t}}(P_{2t})$ is

$$I_{P_{1t}}(P_{2t}) = e^{-\lambda_1 t} \sum_{k \geq 0} \frac{\lambda_1^k t^k}{k!} \log\Big\{e^{(\lambda_1 t - \lambda_2 t)} \frac{\lambda_2^k}{\lambda_1^k}\Big\}$$

$$= -(\lambda_1 - \lambda_2)t + e^{-\lambda_1 t} \log \frac{\lambda_2}{\lambda_1} \sum_{k \geq 1} \frac{\lambda_1^k t^k}{(k-1)!}$$

$$= (\lambda_2 - \lambda_1)t + \lambda_1 t \log \frac{\lambda_1}{\lambda_2}.$$

Let N_t, $t \geq 0$, be a point process having a predictable compensator $\widetilde{N}_{0,t} = \int_0^t Y_s \, d\Lambda_{0,s}$ under a probability measure P_0 and $\widetilde{N}_t = \int_0^t Y_s \, d\Lambda_s$ under a probability measure P absolutely continuous with respect to P_0, then Λ_s is absolutely continuous with respect to $\Lambda_{0,s}$ with derivative μ_s and $\widetilde{N}_t = \int_0^t Y_s \mu_s \, d\Lambda_{0,s}$. If $\Lambda_{0,t}$ is a continuous function, the likelihood ratio of N_t is

$$L_t = E\left(\frac{dP}{dP_0} \mid \mathcal{F}_t\right) = L_0 \prod_{s \leq t} \mu_s^{N_t} \exp\left\{-\int_0^t (\mu_s - 1) \, d\Lambda_{0,s}\right\}$$

$$= L_0 \exp\left\{\int_0^t \log \mu_s \, dN_s - \int_0^t (\mu_s - 1) \, d\Lambda_{0,s}\right\}$$

where $EL_t = EL_0 = 1$ and the information of the distribution of N_t is

$$I_{P_0}(P) = E_0\left\{\log(L_0) + \int_0^t \log \mu_s \, dN_s - \int_0^t (\mu_s - 1) Y_s \, d\Lambda_{0,s}\right\}$$

$$= E_0\left\{\log(L_0) + \int_0^t (\log \mu_s - \mu_s + 1) Y_s \, d\Lambda_{0,s}\right\}.$$

The Hellinger distance of $P_{0,t}$ and P_t is $h(P_{0,t}, P_t) = E_0(L_t^{\frac{1}{2}})$.

Chapter 4

Inequalities for Discrete Martingales

4.1 Introduction

On a probability space $(\Omega, \mathcal{F}, P)$, let $(X_n)_{n \geq 0}$ be a sequence of real variables. Let $X_{(n)} = (X_1, \ldots, X_n)$ be the vector of its first n components and let $X_n^* = \sup_{1 \leq i \leq n} X_i$ be its maximal variable. The equivalence of the norms in vector spaces (1.3) is transposed to random vectors in $\mathbb{R}^n$. Let $\|X_{(n)}\|_{n,p} = E\{\sum_{i=1}^n X_i^p\}^{\frac{1}{p}}$, thus $n^{-1}\|X_{(n)}\|_{n,1} \leq |X_n^*| \leq \|X_{(n)}\|_{n1}$ with the equality $\|X_{(n)}\|_{n,1} = n|X_n^*|$ if and only if all components of $X_{(n)}$ are equal. Let $X = (X_1, X_2, \ldots)$ belong to $\mathbb{R}^\infty$ then for every $p \geq 1$, $|X_\infty^*| \leq \|X\|_{\infty,p}$.

Proposition 4.1. *For all real numbers* $1 \leq p < q < \infty$

$$n^{\frac{1}{p}}\|X_{(n)}\|_{n,p} \leq |X_n^*| \leq \|X_{(n)}\|_{n,p} \leq n^{\frac{1}{p}}|X_n^*|,$$

$$n^{-\frac{1}{p}}\|X_{(n)}\|_{n,p} \leq \|X_{(n)}\|_{n,1} \leq n^{\frac{1}{p'}}\|X_{(n)}\|_{n,p},$$

$$n^{-\frac{1}{p'}}\|X_{(n)}\|_{n,1} \leq \|X_{(n)}\|_{n,p} \leq n^{\frac{1}{p}}\|X_{(n)}\|_{n,1},$$

$$n^{-\frac{1}{p'}-\frac{1}{q}}\|X_{(n)}\|_{n,q} \leq \|X_{(n)}\|_{n,p} \leq \|X_{(n)}\|_{n,q},$$

with the conjugates p' *and* q' *such that* $p^{-1} + p'^{-1} = 1$ *and* $q^{-1} + q'^{-1} = 1$.

Most bounds depend on the dimension of the vector and the inequalities are generalized as n tends to infinity with a normalization of $\|X_{(n)}\|_{n,p}$ by $n^{-\frac{1}{p}}$. Several inequalities between a sum $S_n = \sum_{i=1}^n X_i$ and the sum of the squared variables $V_n = \sum_{i=1}^n X_i^2$ are first established, for a sequence of independent centered random variables $(X_n)_{n \geq 0}$. They are based on the equivalence of the norms in $\mathbb{R}^n$. The same notation is used for independent variables or for dependent variables such as the increments of a discrete martingale and the inequalities are extended to martingales indexed by $\mathbb{R}$.

Other inequalities concern the distances between subsets of n-dimensional vector spaces. For martingales, they have been improved by Burkholder's inequalities.

4.2   Inequalities for sums of independent random variables

The Bienaymé-Chebychev inequality is an upper bound, a similar inequality is established for a lower bound of the probability $P(X > a)$, for a random variable X.

Proposition 4.2. *Let X be a random variable of $L^2(\mathbb{R})$ and let b such that $E|X| \geq b$. For every a in $]0, b[$*

$$P(|X| > a) \geq \frac{(b-a)^2}{EX^2}.$$

Proof. Splitting $|X|$ on the sets $\{|X| > a\}$ and $\{|X| \leq a\}$, we obtain
$$E|X| \leq \{EX^2 P(|X| > a)\}^{\frac{1}{2}} + E\{|X|1_{\{|X| \leq a\}}\},$$
moreover $E|X| - E\{|X|1_{\{|X| \leq a\}} \geq b - a$ and the result follows.   □

In particular, for every a in $]0, E|X|[$
$$P(|X| > a) \geq \frac{(E|X| - a)^2}{EX^2}.$$
The next bound is lower than the bound of the Bienaymé-Chebychev inequality.

Proposition 4.3. *Let X be a random variable of $L^2(\mathbb{R})$ such that $EX = m$ and $\mathrm{Var}X > 0$. For every $a > 0$ such that $a^2 \leq \mathrm{Var}X$*

$$P(|X - m| > a) \geq \frac{a^2}{\mathrm{Var}X}.$$

Proof. By the Cauchy-Schwarz inequality
$$a < E\{|X - m|1_{\{|X-m|>a\}}\} \leq \{(\mathrm{Var}X)P(|X - m| > a)\}^{\frac{1}{2}}.$$

□

Let $X_1, \ldots, X_n$ be n independent and identically distributed variables, if X_i belongs to L^p for a real $p \geq 2$, then X_i^2 belongs to $L^{\frac{p}{2}}$. Let $c_{p,n} = n^{-\frac{1}{p} - \frac{1}{2}}$ and $C_{p,n} = n^{\frac{p-1}{p} + \frac{1}{2}}$, Proposition 4.1 implies the following inequalities.

Proposition 4.4. *On $(\Omega, \mathcal{F}, P)$, let $(X_n)_{n \geq 0}$ be a sequence of random variables of L^p for a real $p \geq 2$*

$$c_{p,n}E(V_n^{\frac{p}{2}}) \leq E(|S_n|^p) \leq C_{p,n}E(V_n^{\frac{p}{2}}). \tag{4.1}$$

For positive variables $E(V_n^{\frac{p}{2}}) \leq E(|S_n|^p)$.

The constants in Proposition 4.4 depend on the number of variables in the sums S_n and V_n. With independent and centered random variables, we have $EX_iX_j = 0$, $i \neq j$, and the inequality becomes an equality with $EV_n = ES_n^2$. The inequalities are strict for $p > 2$, however, for every odd integer p, $ES_n^p = 0$. Therefore Proposition 4.4 applies only to $E(|S_n|^p)$.

Proposition 4.5. *For every integer $p \geq 1$, for every sequence of independent and symmetric random variables $(X_n)_{n \geq 0}$ of L^{2p} and for every integer $n \geq 1$*

$$\|V_n\|_p \leq \|S_n^2\|_p \leq 2\|V_n^p\|_p.$$

Proof. The inequalities are equalities for $p = 1$. For $p > 1$, by the independence of the symmetric variables and the equality $S_n^2 = V_n + \sum_{j \neq i} X_i X_j$, we have $E(S_n^{2j}V_n^{p-j}) = E(V_n^p)$ for every $j \leq p$, it follows that

$$E\{(S_n^2 - V_n)^p\} = E(S_n^{2p}) + \sum_{j=1}^{p}(-1)^j \binom{p}{j} E\left\{S_n^{2(p-j)}V_n^j\right\}$$

$$= E(S_n^{2p}) + E(V_n^p)\left\{\sum_{j=1}^{p}(-1)^j \binom{p}{j}\right\}$$

$$= E(S_n^{2p}) - E(V_n^p),$$

and by convexity $E\{(S_n^2 - V_n)^p\} \geq \{E(S_n^2 - V_n)\}^p = 0$, this proves the lower bound. The upper bound is deduced from the inequality $\sum_{j \neq i} X_i X_j \leq V_n$, writing

$$E\left\{\left(\sum_{j \neq i} X_i X_j\right)^p\right\} \leq E(V_n^p),$$

$$E(S_n^{2p}) = \left\{E\left(V_n + \sum_{j \neq i} X_i X_j\right)^p\right\} \leq 2^p E(V_n^p).$$

$\square$

A sequence of independent variables $(X_i)_{i=1,\ldots,n}$ is symmetrized by an independent sequence of mutually independent and uniform random variables $(\varepsilon_i)_{i=1,\ldots,n}$ with values in $\{1, -1\}$, as $Y_i = \varepsilon_i X_i$.

Lemma 4.1. *Let $(X_n)_{n \geq 0}$ be a sequence of independent random variables of L^{2p} for an integer $p \geq 1$. Then*

$$E\left(\sum_{i=1}^{n} \varepsilon_i X_i\right)^{2p} = E\left(\sum_{i=1}^{n} X_i^2\right)^p.$$

Proof. By independence of the centered variables X_i, the property is true for $p = 1$, with $E(\sum_{i=1}^n \varepsilon_i X_i)^2 = E(\sum_{i=1}^n X_i^2)$, and the same property is valid for $E\|\sum_{i=1}^n \varepsilon_i X_i\|_{L^{2p}}$ since it develops in sums containing only products of variables X_i with an even exponent.   □

The moment generating function of the symmetrized variables $(Y_i)_{i=1,\ldots,n}$ has the bound

$$\phi_Y(t) \leq Ee^{\frac{t^2 X^2}{2}}$$

Chernoff's Theorem 1.7 entails that for all $a > 0$ and $n > 0$

$$\log P\Big(\sum_{i=1}^n \varepsilon_i X_i > a\Big) = \inf_{t>0}\{n \log Ee^{\frac{t^2 X^2}{2}} - at\}.$$

For independent variables, the upper bound of Proposition 4.5 is always true. If the variables are independent, we have $E(S_n^2) = E(V_n)$ and the inequalities are equalities for $p = 1$, moreover $E(S_n^3) \leq E(V_n)$ for every n. Let $\mu = \lim_{n\to\infty} n^{-1} S_n$ and let $v = \lim_{n\to\infty} n^{-1} V_n$, if $\mu \neq 0$ then $S_n^{2p} = O_P(n^{2p})$ and $V_n^p = O_P(n^p)$, as n tends to infinity.

Proposition 4.6. *For every real $p \geq 1$, for every integer $n \geq 1$ and for every sequence of independent centered random variables $(X_n)_{n\geq 0}$ of L^{2p}*

$$\frac{p-1}{np}\|V_n\|_p \leq \|X_n^{*2}\|_p \leq \|V_n\|_p.$$

Proof. The upper bound is due to the inequality $X_n^{*2} \leq V_n$. For the lower bound, we first prove that $\lambda P(V_n > \lambda) \leq E(X_n^{*2} 1_{\{V_n>\lambda\}})$. We have

$$\lambda P(V_n > \lambda) \leq E(V_n 1_{\{V_n>\lambda\}})$$
$$\leq nE(X_n^{*2} 1_{\{V_n>\lambda\}}),$$

it follows that

$$E(V_n^p) = \int_0^\infty \lambda^p \, dP(V_n \leq \lambda)$$
$$\leq pE \int_0^\infty \lambda^{p-1} 1_{\{V_n>\lambda\}} \, d\lambda$$
$$\leq npE\Big\{X_n^{*2} \int_0^\infty \lambda^{p-2} 1_{\{V_n>\lambda\}} \, d\lambda\Big\}$$
$$= \frac{np}{p-1} E(V_n^{p-1} X_n^{*2})$$

and by Hölder's inequality $\|V_n^p\|_p \leq \frac{np}{p-1}\|X_n^{*2}\|_p$.   □

Lemma 1.2 of Lenglart, Lépingle and Pratelli (1980) allows to extend Propositions 4.4 and 4.6 to convex functions of the variables X_i.

Let $a = (a_i)_{i=1,\ldots,n}$ and $X = (X_i)_{i=1,\ldots,n}$ be vectors of $\mathbb{R}^n$ such that X_i belongs to L^{2p}, for an integer $p \geq 1$ and for $i = 1, \ldots, n$. The sum of the weighted variables $S_{n,a} = \sum_{i=1}^n a_i X_i$ and $V_{n,a} = \sum_{i=1}^n a_i^2 X_i^2$ satisfy the inequalities

$$E(|S_{n,a}|^{2p}) \leq \|a\|_2^{2p} E(\|S_n\|_2^{2p}),$$
$$E(V_{n,a}^{2p}) \leq \|a\|_4^{2p} E(\|X\|_4^{2p})$$

and they extend to the variable $(aX)_n^*$ by Proposition 4.4.

For a real p in $]0,1[$, the inequalities have another form, due to the concavity of the norms indexed in $]0,1[$.

Proposition 4.7. *For every p in $]0, \frac{1}{2}[$ and for every sequence of independent and centered random variables $(X_n)_{n\geq 0}$ of L^{2p}, $\|S_n^2\|_p \leq 2\|V_n^p\|_p$.*

The proof is similar to the proof of Proposition 4.5.

Proposition 4.8. *For every p in $]0,1[$, for every integer $n \geq 1$ and for every sequence of independent centered random variables $(X_n)_{n\geq 0}$ of L^{2p}*

$$n^{-1}\|V_n\|_p \leq \|X_n^{*2}\|_p \leq \|V_n\|_p.$$

Proof. The upper bound is due to the inequality $X_n^{*2} \leq V_n$. For the lower bound, $V_n \leq nX_n^{*2}$. □

Proposition 4.9. *Let $(X_n)_{n\geq 0}$ be a sequence of independent centered random variables of $L^2(\mathbb{R})$ then*

$$\|S_n^2 - V_n\|_{L^{\frac{p}{2}}} < E(V_n), \quad p \in]0,2[,$$

$$E|S_n^2 - V_n|^{\frac{p}{2}} > E\left\|\sum_{\substack{i\neq j=1}}^n X_i X_j 1_{\{\sum_{i\neq j=1}^n X_i X_j > 0\}}\right.$$

$$\left. - |\sum_{\substack{i\neq j=1}}^n X_i X_j 1_{\{\sum_{i\neq j=1}^n X_i X_j < 0\}}\right\| \geq 0, \quad p > 2.$$

Proof. Writing $(\sum_{i=1}^n X_i)^2 - \sum_{i=1}^n X_i^2 = \sum_{i\neq j=1}^n X_i X_j$, we have $E|S_n^2 - V_n|^{\frac{p}{2}} < (E|\sum_{i\neq j=1}^n X_i X_j|)^{\frac{p}{2}}$ and by monotonicity $\|S_n^2 - V_n\|_{L^{\frac{p}{2}}}$ is lower than

$$\sum_{i\neq j=1}^{n} E|X_i X_j| < \sum_{i\neq j=1}^{n} (EX_i^2 EX_j^2)^{\frac{1}{2}}$$

$$< \sum_{i=1}^{n} EX_i^2 = EV_n.$$

The second inequality is the consequence of the convexity of the power function and of the triangular inequality. $\qquad\square$

Proposition 4.4 is generalized as an inequality for products of sums of independent random variables, with integers $2 \le k < m < n$. For all real numbers $p \ge 2$ and $\beta \ge 2$

$$c_{p,n-m}c_{\beta,m-k}E\{(V_n - V_m)^{\frac{p}{2}}(V_m - V_k)^{\frac{\beta}{2}}\} \le E(|S_n - S_m|^p |S_m - S_k|^\beta)$$

$$\le C_{p,n-m}C_{\beta,m-k}E\{(V_n - V_m)^{\frac{p}{2}}(V_m - V_k)^{\frac{\beta}{2}}\}.$$

Let $j_n < k_n$ be increasing indices such that $j_n k_n^{-1}$ converges to p in $]0,1[$ as j_n tends to infinity. Let S_{j_n} and S_{k_n} be the sums of independent variables with zero means and variances σ_i^2, up to j_n and k_n respectively. The normalized sums $Y_{1n} = j_n^{-\frac{1}{2}} S_{j_n}$ and $Y_{2n} = k_n^{-\frac{1}{2}} S_{k_n}$ converge weakly to a random variable $Y \sim \mathcal{N}(0,\sigma^2)$ if $\bar{\sigma}_{j_n}^2 = j_n^{-1}\sum_{i=1}^{j_n}\sigma_i^2$ and $\bar{\sigma}_{k_n}^2 = k_n^{-1}\sum_{i=1}^{k_n}\sigma_i^2$ converge to a limit σ^2 as j_n tends to infinity. Then $Y_{3n} = (k_n - j_n)^{-\frac{1}{2}}(S_{k_n} - S_{j_n})$ is independent of Y_{2n} and it converges weakly to the variable $Y_3 = (1 - \alpha)^{-\frac{1}{2}}(Y_2 - \alpha^{\frac{1}{2}}Y_1)$ where $Y_1 = \lim j_n^{-\frac{1}{2}} S_{j_n}$ and Y_3 are independent, Y_1, Y_2 and Y_3 have the same distribution. The variable

$$\frac{Y_{2n}}{Y_{1n}} = \frac{S_{k_n}}{k_n^{\frac{1}{2}}}\left(\frac{S_{j_n}}{j_n^{\frac{1}{2}}}\right)^{-1} = \left(\frac{j_n}{k_n}\right)^{\frac{1}{2}} + \frac{S_{k_n} - S_{j_n}}{(k_n - j_n)^{\frac{1}{2}}}\left(\frac{k_n - j_n}{j_n}\right)^{\frac{1}{2}}\left(\frac{S_{j_n}}{j_n^{\frac{1}{2}}}\right)^{-1}$$

converges weakly to $\alpha^{\frac{1}{2}} + (1 - \alpha)^{\frac{1}{2}}\alpha^{-\frac{1}{2}}Y_1^{-1}Y_3$ and

$$E\left(\frac{Y_{2n}}{Y_{1n}} - \alpha^{\frac{1}{2}}\right)^2 = \frac{1-\alpha}{\alpha}\frac{EY_3^2}{EY_1^2} = \frac{1-\alpha}{\alpha}.$$

The Laplace transforms of Y_{1n}, Y_{2n} and Y_{3n} converge to the Laplace transform of X.

If the variables X_i have a expectation μ_i such that $\bar{\mu}_n$ converges to a limit μ, the variables Y_{kn} are centered

$$Y_{3n} = (k_n - j_n)^{\frac{1}{2}}\{(k_n - j_n)^{-1}(S_{k_n} - S_{j_n}) - (\bar{\mu}_{k_n} - \bar{\mu}_{j_n})\}$$

$$= (k_n - j_n)^{-\frac{1}{2}}(S_{k_n} - S_{j_n}) - (k_n - j_n)^{\frac{1}{2}}(\bar{\mu}_{k_n} - \bar{\mu}_{j_n})$$

and it converges weakly to Y only if $n^{-1}\bar{\mu}_n$ converges to a finite limit, hence $\beta_n = (k_n - j_n)^{\frac{1}{2}}(\bar{\mu}_{k_n} - \bar{\mu}_{k_n})$ converges to zero as n tends to infinity.

With identically distributed variables X_i, the Laplace transforms of S_{j_n} and S_{k_n} are asymptotically equivalent to $L_X^{j_n}$ and $L_X^{k_n}$, respectively, and the Laplace transform of $S_{k_n} - S_{j_n}$ is asymptotically equivalent to $L_X^{k_n - j_n}$.

Applying the Bienaymé-Chebychev inequality (1.13) to the partial sums entails that for every $a > 0$

$$P\{(k_n - j_n)^{-\frac{1}{2}}|S_{k_n} - S_{j_n}| > a\} \leq \frac{\sum_{i=j_n+1}^{k_n} \sigma_i^2}{a^2(k_n - j_n)} \qquad (4.2)$$

and it tends to zero as a tends to infinity. If the variables are not centered, Equation (4.2) applies to the variable $|S_{k_n} - S_{j_n} - \beta_n|$. With $p = 2$ and $k_n - j_n = bn$, the next bound is asymptotically independent on n

$$\sup_{n \geq 1} P\{(k_n - j_n)^{-1}|S_{k_n} - S_{j_n}| > a\} \leq \frac{n\sum_{i=j_n+1}^{k_n} \sigma_i^2}{a^2(k_n - j_n)^2} = \frac{\sum_{i=j_n+1}^{k_n} \sigma_i^2}{a^2 b(k_n - j_n)}.$$

From Proposition 4.2, if $0 < a < (k_n - j_n)^{-1}E|\sum_{i=j_n+1}^{k_n} X_i|$

$$P\{(k_n - j_n)^{-1}|S_{k_n} - S_{j_n}| \leq a\} \geq \frac{(E|S_{k_n} - S_{j_n}| - (k_n - j_n)a)^2}{E|S_{k_n} - S_{j_n}|^2}.$$

Chernoff's equality for $S_{k_n} - S_{j_n}$ is also modified by the number of variables of the sum, for all $a > 0$ and $t > 0$

$$P\{S_{k_n} - S_{j_n} > a\} = E\exp\{t(S_{k_n} - S_{j_n}) - at\} = e^{-at}L_X^{k_n - j_n}(t),$$

then $\lim_{n \to \infty} \log P\{S_{k_n} - S_{j_n} > a\} = \inf_{t>0}\{(k_n - j_n)\log L_X(t) - at\}$.

Under the conditions $X_i = 0$ and $|X_i| \leq \sigma_i M$ for every i, Bennett's inequality is written as

$$P((k_n - j_n)^{-1}(S_{k_n} - S_{j_n}) > t) \leq \exp\left\{-(k_n - j_n)\phi\left(\frac{t}{M\sum_{i=j_n+1}^{k_n} \sigma_i^2}\right)\right\}$$

where ϕ is defined in Theorem 1.8. Removing the boundedness condition, the right member of this inequality is a sum of two terms. For all $M > 0$ and $t > 0$

$$P\left(\frac{S_{k_n} - S_{j_n}}{k_n - j_n} > t\right) \leq \exp\left\{-(k_n - j_n)\phi\left(\frac{t}{\sum_{i=j_n+1}^{k_n} \sigma_i^2 M}\right)\right\}$$
$$+ \prod_{i=j_n}^{k_n} P(|X_i| > \sigma_i M).$$

By independence of $S_{k_n} - S_{j_n}$ and S_{j_n}, the inequalities are extended to product inequalities. Let $|X_i| \leq \sigma_i M$ for every i

$$\lim_{n \to \infty} P\{|S_{j_n}| > a, |S_{k_n} - S_{j_n}| > b\} \leq \frac{(\sum_{i=1}^{j_n} \sigma_i^2)(\sum_{i=j_n+1}^{k_n} \sigma_i^2)}{(ab)^2},$$

$$\lim_{n \to \infty} P(S_{j_n} > t_1, S_{k_n} - S_{j_n} > t_2) \leq \exp\left\{-j_n \phi\left(\frac{t_1}{M \sum_{i=1}^{j_n} \sigma_i^2}\right)\right\}$$

$$\exp\left\{-(k_n - j_n)\phi\left(\frac{t_2}{M \sum_{i=j_n+1}^{k_n} \sigma_i^2}\right)\right\}.$$

The moment generating function of the combination of independent partial sums $Y_{n,\alpha} = \alpha S_{j_n} + (1 - \alpha)(S_{k_n} - S_{j_n})$, $0 < \alpha < 1$, is

$$L_{n,\alpha}(t) = E[\exp\{t\alpha S_{j_n}\} \exp\{t(1 - \alpha)(S_{k_n} - S_{j_n})\}]$$

$$= \prod_{i=1}^{j_n} L_{X_i}(t\alpha) \prod_{i=j_n+1}^{k_n} L_{X_i}(t(1 - \alpha)).$$

With partial sums of Gaussian variables $\mathcal{N}(0, \sigma^2)$, Chernoff's inequality for $Y_{n,\alpha}$ implies

$$P\{Y_{n,\alpha} > t\} = \exp\left[-\frac{t^2}{\{(k_n - j_n)(1 - \alpha)^2 + j_n \alpha^2\}\sigma^2}\right].$$

Let S_n be the sum of independent and identically distributed vectors of $\mathbb{R}^d$ and such that the variance of the k-th component of $n^{-\frac{1}{2}} S_n$ is σ_k^2 and let $\sigma^2 = n^{-1} \sum_{k=1}^d \sigma_k^2$. From Proposition 4.4, an inequality of the same kind as (1.13) can be proved for the L^2-norm of S_n.

Theorem 4.1. *The sum of independent and identically distributed variables with values in $\mathbb{R}^d$ and mean zero satisfies*

$$P(\|S_n\|_{d,2} > a) \leq \frac{2\sigma\sqrt{d}}{a}, \ t > 0$$

where Let σ^2 is the sum of variances of the variables.

Proof. The $L^2(\mathbb{R}^d)$-norm of S_n has the variance

$$E\|n^{-1}S_n\|_{d,2}^2 = \sum_{k=1}^d E|S_{nk}|^2 = \sum_{k=1}^d EV_{nk} \leq d\sigma^2$$

and the result is obtained from (1.13). □

In $\mathbb{R}^d$, Chernoff's limit is replaced by an exponential inequality

$$P(n^{-1}\|S_n\|_1 > a) \leq \sum_{k=1}^{d} P(n^{-1}|S_{nk}| > a) \leq 2\sum_{k=1}^{d} \inf_{t_k>0} L_{X_k}^n(t_k)e^{-at_kn}$$

where $t = (t_1, \ldots, t_d)$ is a vector of $\mathbb{R}^d$. When n tends to infinity, $n^{-\frac{1}{2}}S_n$ converges to a centered Gaussian variable of $\mathbb{R}^d$ with variance matrix Σ and $\inf_{t_k>0}\exp[n\{\log L_{X_k}(t_k) - at_k\}]$ is equivalent to $\exp\{-n\frac{a^2}{2\sigma_k^2}\}$, therefore, as n tends to infinity

$$P(n^{-1}\|S_n\|_{d,2} > a) \leq 2\sum_{k=1}^{d} e^{-na^2(2\sigma_k^2)^{-1}}.$$

Since $En^{-\frac{1}{2}}\|S_n\|_{d,2}$ converges to $\|\Sigma\|_{d,2}$, the inequalities are also written with sharper bounds as n tends to infinity, then $P(\|n^{-\frac{1}{2}}S_n\|_{d,2} > a)$ is equivalent to $e^{-na^2(2\|\Sigma\|_2^2)^{-1}}$.

The dependence between σ-algebras $\mathcal{A}$ and $\mathcal{B}$ are measured by the coefficients

$$\varphi(\mathcal{A}, \mathcal{B}) = \sup_{A\in\mathcal{A}, B\in\mathcal{B}:P(B)>0} |P(A|B) - P(A)|,$$

$$\alpha(\mathcal{A}, \mathcal{B}) = \sup_{A\in\mathcal{A}, B\in\mathcal{B}:P(B)>0} |P(A\cap B) - P(A)P(B)|.$$

The convergence of dependent variables is classically established under mixing condition between the σ-algebras they generate. Let $\mathcal{M}_{1,n}$ be the σ-algebra generated by $(X_j)_{j=1,\ldots,k}$ and $\mathcal{M}_{n+k,\infty}$ be the σ-algebra generated by $(X_j)_{j\geq n+k}$. The φ-mixing coefficients for $(S_n)_{n\geq 1}$ and its strong mixing coefficients are defined for every integer $k \geq 1$ as

$$\varphi_k = \varphi(\mathcal{M}_{1,n}, \mathcal{M}_{n+k,\infty}),$$

$$\alpha_k = \alpha(\mathcal{M}_{1,n}, \mathcal{M}_{n+k,\infty}).$$

The variables $(X_n)_{n\geq 1}$ is φ-mixing or α-mixing if the coefficients φ_k or α_k tend to zero as k tends to infinity.

Lemma 4.2 (Serfling 1968). *Let X be a random variable and let $1 \leq p \leq \infty$, for all Borel set $\mathcal{F}$ and $r > p$*

$$\|E(X|\mathcal{F}) - EX\|_p \leq 2\{\varphi(\mathcal{F}, \mathcal{A})\}^{1-\frac{1}{r}}\|X\|_r,$$

$$\|E(X|\mathcal{F}) - EX\|_p \leq 2(2^{\frac{1}{p}} + 1)\{\alpha(\mathcal{F}, \mathcal{A})\}^{\frac{1}{p}-\frac{1}{r}}\|X\|_r.$$

The moments of Proposition 4.4 is replaced by the next ones. Let X_i be in $\mathcal{M}_{1,n}$ and X_j be in $\mathcal{M}_{n+k,\infty}$ and let $p > 1$. Under the condition of φ-mixing, $|E(X_iX_j) - EX_iEX_j| \leq \varphi_k^{\frac{1}{2}}\|X_i\|_p\|X_j\|_q$, where q is the conjugate of p, and the variance of $n^{-\frac{1}{2}}S_n$ is finite under the condition $\sum_{k\geq 0} k^2\varphi_k^{\frac{1}{2}} < \infty$. According to Billingsley (1968), the moments of S_n satisfy

$$E|S_n^p| \leq \sum_{k=0}^{\frac{p}{2}}(k+1)^{p-2}\varphi_k^{\frac{1}{2}}\{\max_{i=1,\ldots,n} E(X_i^2)\}^k, \; p \geq 2.$$

Under the α-mixing condition and if

$$\sum_{n=0}^{\infty}(n+1)^{p-2}\alpha_n^{\frac{1}{2}} < \infty, \tag{4.3}$$

then

$$|E(X_iX_j) - EX_iEX_j| \leq 12\alpha_k^{\frac{1}{2}}\|X_i\|_p\|X_j\|_q$$

where $p^{-1} + q^{-1} = \frac{1}{2}$, and there exists a constant $k(\alpha, p)$ such that

$$E|S_n^p| \leq k(\alpha, p)\sum_{k=0}^{\frac{p}{2}}\{\max_{i=1,\ldots,n} E(X_i^2)\}^k.$$

Under the conditions about the convergence rate of the mixing coefficients, the normalized sum $n^{-\frac{1}{2}}(S_n - ES_n)$ converges weakly to a Gaussian variable with mean zero and a finite variance.

For dependent variables $(X_i)_{i=1,\ldots,n}$, the moment generating function of the sum S_n is not a product of n identical terms L_X but it is sufficient that there exists a constant $\alpha > 0$ such that $n^{-\alpha}S_n$ converges a.s. to a limit to ensure a large deviations inequality.

Theorem 4.2. *Let* $(X_i)_{i=1,\ldots,n}$ *be a sequence of random variables on a probability space* $(\Omega, \mathcal{F}, P)$ *and let* $S_n = \sum_{i=1}^{n} X_i$. *If there exists a constant* $\alpha > 0$ *such that* $n^{-\alpha}S_n$ *converges a.s. to a limit* S_0 *having a finite moment generating function* $\varphi_{S_0}(t)$ *for every* $t > 0$, *then*

$$\lim_{n\to\infty} n^{-\alpha}\log P(n^{-\alpha}S_n > a) = \inf_{t>0}\{\log\varphi_{S_0}(t) - at\}$$

and it is finite.

Proof. By concavity, $\log Ee^{n^{-\alpha}S_nt} \geq n^{-\alpha}t\,ES_n$ which converges to tES_0. The a.s. convergence of $n^{-\alpha}S_n$ implies

$$\varphi_{S_n}(t) = Ee^{\{S_0+o(1)\}t} = \{\varphi_{S_0}^{n^{\alpha}}(n^{-\alpha}t)\}\{1 + o(1)\}$$

with a.s. $o(1)$. The proof ends like in Chernoff's theorem. $\qquad\square$

U-statistics are defined as

$$U_n = \frac{1}{\sqrt{n(n-1)}} \sum_{i \neq j=1}^{n} \{h(X_i, X_j) - E\{h(X_i, X_j)\}$$

for a sequence $(X_i)_{i=1,\dots,n}$ of independent and identically distributed real variables and a symmetric function $h : \mathbb{R}^2 \mapsto \mathbb{R}$ such that the variance σ^2 of the variables $h(X_i, X_j)$ is finite for $i \neq j$. Let $m = E\{h(X_i, X_j)$ and for every i, let $m(X_i)$ be the expectation of $h(X_i, X_j)$ conditionally on X_i and let σ_i^2 be its conditional variance.

Proposition 4.10. *The variable U_n converges weakly to a centered Gaussian variables with variance $\sigma^2 = E\sigma_i^2 + \mathrm{Var}\{m(X_i)\}$.*

Proof. Conditionally on X_i, the variables $h(X_i, X_j) - m$ and

$$H_n(X_i) = \frac{1}{n-1} \sum_{j \neq i, j=1}^{n} \{h(X_i, X_j) - m(X_i)\}$$

have the characteristic functions

$$\phi_{X_i}(t) = E[e^{it\{h(X_i, X_j) - m\}} \mid X_i],$$
$$\phi_{H_n(X_i)}(t) = e^{-it\{m(X_i) - m\}} \phi_{X_i}^{n-1}((n-1)^{-1} t)$$

and the logarithm of $\phi_{X_i}^{n-1}((n-1)^{-1} t)$ has a second order expansion

$$(n-1) \log \phi_{X_i}((n-1)^{-1} t) = -\frac{t^2}{2(n-1)} \sigma_i^2 \{1 + o(1)\}.$$

Conditionally on X_i, the convergence to zero as n tends to infinity of $\log \phi_{n, X_i}(t)$ for every t implies the convergence of $\phi_{n, X_i}(t)$ to 1 and $H_n(X_i)$ converges in probability to zero, it follows that $n^{-1} \sum_{i=1}^{n} H_n(X_i)$ converges in probability to zero. The variable $Y_n = \{n(n-1)\}^{-1} \sum_{i \neq j=1}^{n} h(X_i, X_j)$ is written as

$$Y_n = \frac{1}{n} \sum_{i=1}^{n} H_n(X_i) + \frac{1}{n} \sum_{i=1}^{n} m(X_i)$$

it converges therefore to $m = E\{m(X_i)\}$. The variable

$$U_{ni} = (n-1)^{-\frac{1}{2}} \sum_{j \neq i, j=1}^{n} \{h(X_i, X_j) - m(X_i)\}$$

has the conditional characteristic function

$$\phi_{U_{ni}}(t) = E(e^{itU_{ni}} \mid X_i) = \phi^{n-1}((n-1)^{-\frac{1}{2}} t, X_i)$$

by a second order expansion of its logarithm we have

$$\log \phi^{n-1}((n-1)^{-\frac{1}{2}}t, X_i) = -\frac{1}{2}t^2\sigma_i^2 + o(1)$$

and $\phi^{n-1}((n-1)^{-\frac{1}{2}}t, X_i)$ converges to the characteristic function of a Gaussian variable with random variance σ_i^2, conditionally on X_i. The variable $U_n = n^{-\frac{1}{2}}\sum_{i=1}^{n}U_{ni}$ has the variance σ^2 and it is the normalized sum of asymptotically independent centered Gaussian variables with respective variances σ_i^2 therefore it converges weakly to a centered Gaussian variables with variance σ^2. $\qquad\square$

The inequalities for the sums of variables S_n and V_n apply to U-statistics $S_n = \sum_{j\neq i,j=1}^{n}h(X_i, X_j)$ and the sum of the squared variables $V_n = \sum_{j\neq i,j=1}^{n}h^2(X_i, X_j)$. Let $p \geq 1$ and variables X_i in L^p, by convexity of the power function x^p and the Cauchy-Schwarz inequality we have

$$E(S_n^{2p}) \geq \Big[E\Big\{S_n^2 + \sum_{j_1\neq i_1}\sum_{j_2\neq i_2,(i_1,j_1)\neq(i_2,j_2)} h(X_{i_1}, X_{j_1})h(X_{i_2}, X_{j_2})\Big\}\Big]^p$$

$$\geq 2^p\{E(S_n^2)\}^p.$$

Proposition 4.11. *For all $a > 0$, $M > 0$ and $n \geq 1$ we have*

$$P(U_n > a) \leq \exp\Big\{-n(n-1)\phi\Big(\frac{a}{M\{n(n-1)\}^{\frac{1}{2}}}\Big)\Big\}$$
$$+ P(|h(X_1, X_2) - m(X_1)| > M),$$

where $\phi(x) = (1+x)\log(1+x) - x$.

Proof. With the notations used to prove Proposition 4.10, let

$$\varphi_{X_i}(t) = E\Big[e^{t\{h(X_i,X_j)-m(X_i)\}} \mid X_i\Big]$$

be the moment generating function of $h(X_i, X_j) - m(X_i)$ conditionally on X_i, the conditional moment generating function $\varphi_{U_{ni}}(t) = \varphi_{X_i}^{n-1}((n-1)^{-\frac{1}{2}}t)$ of the variables $U_{ni} = (n-1)^{-\frac{1}{2}}\sum_{j\neq i,j=1}^{n}\{h(X_i, X_j) - m(X_i)\}$ yields the moment generating function $\varphi_{U_n}(t) = \prod_{i=1}^{n} E\varphi_{U_{ni}}(n^{-\frac{1}{2}}t)$ of the variable $U_n = n^{-\frac{1}{2}}\sum_{i=1}^{n}U_{ni}$, by the independence of the variables $X_1, \ldots, X_n$. Let $b_n = n(n-1)$, applying Chernoff's theorem to the variable U_n entails

$$\log P(U_n > a) = \inf_{t>0}\Big[b_n \log E\Big\{\varphi_{X_i}(b_n^{-\frac{1}{2}}t)\Big\} - ta\Big].$$

If $\|h(X_i, X_j) - m(X_i)\|_k \mid X_i\} \leq M$ for all $i \neq j$, expanding the exponential in the function $E\varphi_{X_i}(s)$, its logarithm is bounded by $\exp(Ms) - 1 - Ms$ and $\log P(U_n > a) = \inf_{t>0} h_{n,a}(t)$ with the function

$$h_{n,a}(t) = b_n\{\exp(Mb_n^{-\frac{1}{2}}t) - 1 - Mb_n^{-\frac{1}{2}}t\} - at.$$

The minimum of $h_{n,a}$ is reached at $t_{n,a}$ such that $\exp(Mb_n^{-\frac{1}{2}}t_{n,a}) - 1 = aM^{-1}b_n^{-\frac{1}{2}}$ and $h_{n,a}(t_{n,a}) = b_n\phi(aM^{-1}b_n^{-\frac{1}{2}})$. □

As n tends to infinity $h_{n,a}(t_{n,a})$ is asymptotically equivalent to a^2M^{-2} and $P(U_n > a)$ has the asymptotic bound

$$\lim_{n\to\infty} P(U_n > a) \leq \exp(-a^2M^{-2}) + P(|h(X_1, X_2) - m(X_1)| > M).$$

4.3   Inequalities for discrete martingales

On a filtered probability space $(\Omega, \mathcal{F}, (\mathcal{F}_n)_{n\geq 0}, P)$, let $(X_n)_{n\geq 0}$ be a real martingale of $\mathcal{M}^2$ and let $(V_n(X))_{n\geq 0}$ be the process of its predictable quadratic variations. Let $U_i = X_i - X_{i-1}$, then X_n is written as the sum $X_n = X_0 + \sum_{i=1}^n U_i$, where $(U_n)_{n\geq 0}$ is a sequence of dependent random variables with respective variances

$$\sigma_i^2 = E(X_i - X_{i-1})^2 = E\{V_i(X) - V_{i-1}(X)\}$$

and their covariance are zero by the martingale property. The conditions about the moments of the variables X_n are different from those of mixing variables. Bounds of the even moments of discrete martingales are proved extending Propositions 4.4 and 4.6 to the martingale differences $U_n = X_n - X_{n-1}$.

The sum $S_n = \sum_{i=1}^n X_i$ of independent centered variables $X_1, \ldots, X_n$ is a discrete martingale with respect to the filtration $\mathcal{F}$ defined as $\mathcal{F}_n = \sigma(X_1, \ldots, X_n)$ so the inequalities of this section apply to $(S_n)_{n\geq 1}$. They are also valid for sums S_n of dependent centered variables $X_1, \ldots, X_n$ such that $E(X_n \mid \mathcal{F}_{n-1}) = 0$.

Proposition 4.12. *Let $(X_n)_{n\geq 1}$ be a martingale of $\mathcal{M}^p$, $p \geq 2$, on $(\Omega, \mathcal{F}, (\mathcal{F}_n)_{n\geq 0}, P)$ and let $(V_n(X))_{n\geq 1}$ be the process of its quadratic variations. For every stopping time N with respect to $(\mathcal{F}_n)_{n\geq 0}$*

$$E\left(c_{p,N}V_N^{\frac{p}{2}}\right) \leq E(|X_N|^p) \leq E\left(C_{p,N}V_N^{\frac{p}{2}}\right). \tag{4.4}$$

Proof. Writing X_n as the sum of conditionally independent and centered variables $X_n = \sum_{i=1}^n U_i$ and $V_n(X) = \sum_{i=1}^n U_i^2 = (X_i - X_{i-1})^2$, the inequalities are deduced from Proposition 4.4. The upper bounds are still valid when the integer n is replaced by a random stopping time N of the martingale, by Doob's theorem. □

In Burkholder's inequality (1.19), the constant of the lower bound is strictly smaller than one and decreases to zero as p tends to infinity, the constant $C_p > 1$ of the upper bound tends to infinity with p. The next proposition improves its lower bound of Burkholder's inequality for small values of p. Let $(X_n)_{n \geq 1}$ be a martingale of $\mathcal{M}^2$, by Theorem 1.3 there exists a unique predictable process A_n such that $M_n = X_n^2 - A_n$ is a martingale and $A_n = E(V_n \mid \mathcal{F}_{n-1})$.

Proposition 4.13. *For every real $p \geq 1$, there exists a constant $C_p > 1$ such that for every martingale $(X_n)_{n \geq 1}$ of $\mathcal{M}^{2p}$ and for every $n \geq 2$*

$$E(A_n^p) \leq E(X_n^{2p}) \leq C_p E(A_n^p).$$

Proof. The lower bound is established by convexity of the function x^p, for every real $p \geq 1$, it implies

$$E(X_n^{2p} \mid \mathcal{F}_{n-1}) \geq \{E(X_n^2 \mid \mathcal{F}_{n-1})\}^p = A_n^p$$

therefore $E(X_n^{2p}) \geq E(A_n^p)$. The upper bound is due to Burkholder's inequality for every $p \geq 1$. $\square$

Proposition 4.13 generalizes to real convex functions on $\mathbb{R}_+$ by Lemma 1.2 of Lenglart, Lépingle and Pratelli (1980). As a consequence of Proposition 4.13, $(A_n)_{n \geq 1}$ belongs to $\mathcal{M}^p$ if and only if $(X_n)_{n \geq 1}$ belongs to $\mathcal{M}^{2p}$, for every real $p \geq 1$ and there exists a constant $C_p > 1$ such that

$$P(A_n > x) \leq x^{-p} E(X_n^{2p})$$
$$P(X_n > x) \leq x^{-2p} C_p E\{A_n^p(X)\}.$$

Proposition 4.13 extends to the variations of martingales.

Proposition 4.14. *For every real $p \geq 1$, for every martingale $(X_n)_{n \geq 1}$ of $\mathcal{M}^{2p}$, and for all $n > m \geq 2$*

$$E[\{E(A_n \mid \mathcal{F}_m) - A_m\}^p] \leq E\{(X_n - X_m)^{2p}\}$$
$$\leq \left(\frac{2p}{p-1}\right)^p E[\{E(A_n \mid \mathcal{F}_m) - A_m\}^p].$$

Proof. The martingale property implies that for every $n > m$

$$E\{(X_n - X_m)^2 \mid \mathcal{F}_m\} = E\left[\left\{\sum_{k=m+1}^{n} (X_k - X_{k-1})\right\}^2 \mid \mathcal{F}_m\right]$$

$$= E\left\{\sum_{k=m+1}^{n} (X_k - X_{k-1})^2 \mid \mathcal{F}_m\right\}$$

$$= E(V_n - V_m \mid \mathcal{F}_m),$$

and by convexity of the power function x^p, for every real $p \geq 1$ we obtain

$$E\{(X_n - X_m)^{2p} \mid \mathcal{F}_m) \geq [E\{(X_n - X_m)^2 \mid \mathcal{F}_m\}]^p$$
$$= \{E(V_n - V_m \mid \mathcal{F}_m)\}^p$$
$$= \{E(A_n \mid \mathcal{F}_m) - V_m\}^p$$

therefore $E\{(X_n - X_m)^{2p}\} \geq E[\{E(A_n \mid \mathcal{F}_m) - A_m\}^p]$.

For the upper bound, for every $\lambda > 0$ let $B_{nm}(\lambda) = \{(X_n - X_m)^2 > \lambda\}$, we have

$$\lambda P((X_n - X_m)^2 > \lambda) \leq E\{(X_n - X_m)^2 1_{B_{nm}(\lambda)}\}$$
$$= E\{(V_n - V_m)1_{B_{nm}(\lambda)}\}$$
$$+ E\Big\{1_{B_{nm}(\lambda)} \sum_{j \neq k = m+1}^{n} (X_j - X_{j-1})(X_k - X_{k-1})\Big\}$$
$$\leq 2E\{(V_n - V_m)1_{B_{nm}(\lambda)}\}$$

by the Cauchy-Schwarz inequality. Arguing like in the proof of Doob's inequality, this entails that for all $n > m > 0$

$$\|(X_n - X_m)^2\|_p \leq K_p \|V_n - V_m\|_p = K_p \|E(A_n \mid \mathcal{F}_m) - V_m\|_p$$

with the constant $K_p = 2p(p-1)^{-1}$ depending only on p. $\qquad \square$

Proposition 4.14 generalizes to real convex functions on $\mathbb{R}_+$. Propositions 4.15 and 4.14 are modified for p in $]0, 1[$.

Proposition 4.15. *For every p in $]0, 1[$, there exists a constant $c_p < 1$ such that for every martingale $(X_n)_{n \geq 1}$ of $\mathcal{M}^{2p}$ and for every $n \geq 2$*

$$c_p E(A_n^p) \leq E(X_n^{2p}) \leq E(A_n^p).$$

Proof. If $0 < p < 1$, the function x^p is concave and the moment inequality of Proposition 4.13 is reversed

$$E(X_n^{2p} \mid \mathcal{F}_{n-1}) \leq \{E(X_n^2 \mid \mathcal{F}_{n-1})\}^p = A_n^p$$

therefore $E(X_n^{2p}) \leq E(A_n^p)$. The lower bound is due to Burkholder's inequality. $\qquad \square$

Proposition 4.16. *For every p in $]0, 1[$, there exists a constant K_p such that for every martingale $(X_n)_{n \geq 1}$ of $\mathcal{M}^{2p}$, and for all $n > m \geq 2$*

$$2^{1-p} E[\{E(A_n \mid \mathcal{F}_m) - A_m\}^p] \leq E\{(X_n - X_m)^{2p}\}$$
$$\leq E[\{E(A_n \mid \mathcal{F}_m) - A_m\}^p].$$

Proof. By the concavity of the power function x^p, for p in $]0,1[$, and by the martingale property $E\{(X_n - X_m)^2 \mid \mathcal{F}_m\} = \{E(V_n - V_m \mid \mathcal{F}_m)\}$, for every $n > m$, we obtain

$$E\{(X_n - X_m)^{2p} \mid \mathcal{F}_m) \leq [E\{(X_n - X_m)^2 \mid \mathcal{F}_m\}]^p$$
$$= \{E(V_n - V_m \mid \mathcal{F}_m)\}^p$$
$$= \{E(V_n \mid \mathcal{F}_m) - V_m\}^p$$

it follows that $E\{(X_n - X_m)^{2p}\} \leq E[\{E(A_n \mid \mathcal{F}_m) - A_m\}^p]$.

For the lower bound, the variable $g_{nm} = \sum_{k=m+1}^n X_{k-1}(X_k - X_{k-1})$ has the expectation zero and

$$g_{nm} - X_n X_{n-1} = \sum_{k=m+1}^{n-1} X_{k-1}(X_k - X_{k-1}) - X_{n-1}^2,$$

$$V_{n-1} - V_m = 2X_{n-1}X_n - X_{n-1}^2 - X_m^2 - 2g_{nm}$$

where the expectation of the last term is $E(X_{n-1}^2 - X_m^2) = E(X_{n-1} - X_m)^2$ by the martingale property. The convexity of the power function implies

$$E[\{2X_{n-1}X_n - X_{n-1}^2 - X_m^2 - (X_{n-1} - X_m)^2\}^p]$$
$$\leq [E\{2X_{n-1}X_n - X_{n-1}^2 - X_m^2 - (X_{n-1} - X_m)^2\}]^p,$$
$$E\{(-2g_{nm})^p\} \leq \{E(-2g_{nm})\}^p = 0$$

it follows that

$$E\{(V_{n-1} - V_m)^p\} \leq 2^{p-1} E\{(X_{n-1} - X_m)^{2p}\},$$

furthermore $E\{(V_{n-1} - V_m)^p \mid \mathcal{F}_m\} = E[\{E(A_n \mid \mathcal{F}_m) - A_m\}^p]$. $\square$

Propositions 4.14 and 4.16 prove that the convergence of the series $(X_n)_{n\geq 1}$ in L^{2p} is equivalent to the convergence of the series $(V_n)_{n\geq 1}$ in L^p, for every real $p > 0$, according to Cauchy's convergence property.

Proposition 4.17. *Let $X = (X_n)_{n\geq 0}$ be a martingale of $\mathcal{M}^p$, $p \leq 1$, then for every stopping time N of X*

$$\|X_N^2 - A_N\|_{L^{\frac{p}{2}}} \leq \|A_N\|_{L^{\frac{p}{2}}}.$$

Proof. Let $X_N = \sum_{i=1}^n (X_i - X_{i-1})$, by concavity of the power function we have $E(|X_n^2 - A_n|^{\frac{p}{2}}) \leq \{E| \sum_{i\neq j=1}^n (X_i - X_{i-1})(X_j - X_{j-1})|^{\frac{p}{2}}\}$, and by the Cauchy-Schwarz inequality in l_2, it is lower than

$$\left[E\left\{ \sum_{i=1}^n (X_i - X_{i-1})^2 \right\}^{\frac{p}{2}} \right] = E(A_n^{\frac{p}{2}}).$$

$\square$

Proposition 4.13 does not extend to products, for $0 < k < m < n$ the following expectations of products cannot be factorized, except for martingales with independent increments. Let $\alpha \geq 2$ and $\beta \geq 2$

$$E[E\{(A_n - A_m)^{\frac{\alpha}{2}}|\mathcal{F}_m\}(A_m - A_k)^{\frac{\beta}{2}}]$$
$$\leq E\{E(|S_n - S_m|^{\alpha}|\mathcal{F}_m)|S_m - S_k|^{\beta}\}$$
$$\leq C_{\alpha,n-m}C_{\beta,m-k}E[E\{(A_n - A_m)^{\frac{\alpha}{2}}|\mathcal{F}_m\}(A_m - A_k)^{\frac{\beta}{2}}].$$

A discrete version of the Birnbaum and Marshal inequality for a weighted martingale is the following. Let $X = (X_n)_n$ be a martingale and $(a_n)_n$ be a real sequence. Due to the Kolmogorov inequality, for every $\lambda > 0$, the maximum of the martingales

$$Y_n = \sum_{k=1}^{n} a_k(X_k - X_{k-1}), \tag{4.5}$$

and

$$Z_n = \sum_{k=1}^{n} a_k^{-1}(X_k - X_{k-1}) \tag{4.6}$$

have the bounds

$$P(\max_{k=1,\ldots,n} |a_k^{-1}Y_k| > \lambda) \leq \lambda^{-2} \sum_{k=1}^{n} \left(\frac{a_k}{a_n}\right)^2 (A_k - A_{k-1})(X), \tag{4.7}$$

$$P(\max_{k=1,\ldots,n} |a_k Z_k| > \lambda) \leq \lambda^{-2} \sum_{k=1}^{n} \left(\frac{a_n}{a_k}\right)^2 (A_k - A_{k-1})(X)$$

therefore $P(\max_{k=1,\ldots,n} |a_k^{-1}Y_k| > \lambda) \leq \lambda^{-2} A_n(X)$ if $(a_n)_n$ is increasing and $P(\max_{k=1,\ldots,n} |a_k Z_k| > \lambda) \leq \lambda^{-2} A_n(X)$ if $(a_n)_n$ is decreasing.

Proposition 4.18. *Let $X = (X_n)_n$ be in $\mathcal{M}^2$, let $(A_n)_{n\geq 1}$ be a predictable process. Let $Y_n = \sum_{k=1}^{n} A_k(X_k - X_{k-1})$, then for every $\lambda > 0$ and for every stopping time N of X*

$$P(Y_N > \lambda) \leq \lambda^{-p} C_p E\left\{V_N^{\frac{p}{2}}(Y)\right\},$$

$$P(Y_N^* > \lambda) \leq \lambda^{-p} C_p E\left\{\sum_{n=1}^{N} V_n^{\frac{p}{2}}(Y)\right\}.$$

The first inequality is an immediate consequence of Equations (1.17) and (4.4). The inequality $P(Y_N^* > \lambda) \leq E\{\sum_{i=1}^{N} P(Y_n > \lambda)\}$ implies the second bound of Proposition 4.18.

Proposition 4.19. *Let φ be a real convex function on $\mathbb{R}$, let $X = (X_n)_n$ be in $\mathcal{M}^2$ and let $Y_{n+1} = Y_n + \varphi(X_{n+1} - X_n)$, then Y and its quadratic variations are submartingales. For every $\lambda > 0$*

$$P\left(\sup_{n\geq 1} |Y_n| > \lambda\right) \leq 2\lambda^{-2} \sum_{n=1}^{\infty} E\varphi^2(X_n - X_{n-1}).$$

Proof. For every $m > n \geq 0$

$$E\{Y_m|\mathcal{F}_n\} = Y_n + E\left[E \sum_{i=n}^{m-1} E\{\varphi(X_{i+1} - X_i)|\mathcal{F}_i\}|\mathcal{F}_n\right]$$

$$\geq \varphi(Y_n) + E\left[E \sum_{i=n}^{m-1} \varphi(E\{X_{i+1}|\mathcal{F}_i\}) - \varphi(X_i)|\mathcal{F}_n\right] = Y_n.$$

The quadratic variations $V_n(Y) = \sum_{1\leq i\leq n} (Y_i - Y_{i-1})^2$ have conditional expectations

$$E\{V_{n+1}(Y)|\mathcal{F}_n\} = V_n(Y) + E(Y_{n+1}^2|\mathcal{F}_n) - Y_n^2$$

$$\geq V_n(Y) + \varphi^2(E(X_{n+1}|\mathcal{F}_n)) - \varphi^2(X_n) = V_n(Y).$$

$\square$

By Jensen's inequality

$$E\{e^{t(X_{i+1}-X_i)}e^{t(X_i-X_{i-1})}|\mathcal{F}_i\} = e^{t(X_i-X_{i-1})}E\{e^{t(X_{i+1}-X_i)}|\mathcal{F}_i\}$$

$$\geq e^{t(X_i-X_{i-1})},$$

it follows that the moment generating function of X_n is

$$L_{X_n}(t) = E\prod_{i=1}^{n} e^{t(X_i-X_{i-1})} = E\prod_{i=1}^{n} E\{e^{t(X_i-X_{i-1})}|\mathcal{F}_{i-1}\} \geq 1$$

and $\log L_{X_n}(t) \geq E\sum_{i=1}^{n} \log E\{e^{t(X_i-X_{i-1})}|\mathcal{F}_{i-1}\}$, which is the expectation of a sum of the random variables $\log L_i(t) = \log E\{e^{t(X_i-X_{i-1})}|\mathcal{F}_{i-1}\}$.

Theorem 4.3. *Let $(X_n)_{n\geq 0}$ be a real martingale with a moment generating function φ_{X_n} such that $n^{-1}\log \varphi_{X_n}(t)$ converges to a limit $\log \varphi_X(t)$ as n tends to infinity. Then*

$$\lim_{n\to\infty} n^{-1}\log P(X_n > a) = \inf_{t>0}\{n\log \varphi_X(t) - at\}.$$

Bennett's inequality for independent random variables can be adapted to martingales. Let $\sigma_i^2 = E\{V_i(X) - V_{i-1}(X)\}$, the variance of X_n is $n\bar{\sigma}_n^2$, where the expectation variance $\bar{\sigma}_n^2 = n^{-1}\sum_{i=1}^{n} \sigma_i^2$ is supposed to converge to a limit σ^2 as n tends to infinity.

Theorem 4.4. *Let $(X_n)_{n \geq 0}$ be a real martingale with mean zero and such that there exists a constant M for which the variations of the martingale satisfy $\sigma_n^{-1}|X_n - X_{n-1}| \leq M$ a.s., for every integer n. For every $t > 0$*

$$P(X_n > t) \leq \exp\left\{-\phi\left(\frac{t}{n\bar{\sigma}_n M}\right)\right\},$$

where $\phi(x) = (1 + x)\log(1 + x) - x$.

Proof. The moment generating function φ_{X_n} has the same expansion as φ_X in Bennett's Theorem 1.8, under the boundedness condition for $|X_n - X_{n-1}|$. This condition implies $\sum_{i=1}^{n} EX_n^k \leq M^k \sum_{i=1}^{n} \sigma_i^k \leq (Mn\bar{\sigma}_n)^k$, where the bound is denoted by b_n^k, and the expansion of L_{X_n} differs from the expansion of the moment generating function of n independent variables

$$E\{e^{\lambda X_n}\} \leq 1 + \sum_{k=2}^{\infty} \frac{\lambda^k}{k!} b_n^k = 1 + \{\exp(b_n\lambda) - 1 - b_n\lambda\}$$
$$\leq \exp\{\exp(b_n\lambda) - 1 - b_n\lambda\}.$$

From Chernoff's theorem, for every $t > 0$

$$\log P(X_n > t) = \inf_{\lambda > 0}\{\exp(b_n\lambda) - 1 - b_n\lambda - \lambda t\}.$$

The first derivative with respect to λ of $h_t(\lambda) = \exp(b_n\lambda) - 1 - b_n\lambda - \lambda t$ is $h_t'(\lambda) = b_n\exp(b_n\lambda) - b_n - t$, hence the function h_t is minimum at the value $\lambda_{n,t} = b_n^{-1}\log\{1 + b_n^{-1}t\}$ where $\exp(b_n\lambda_{n,t}) = 1 + b_n^{-1}t$ and

$$\exp\{h_t(\lambda_{n,t})\} = \exp\{-\phi(b_n^{-1}t)\} = \exp\{-\phi((Mn\bar{\sigma}_n)^{-1}t)\}.$$

$\square$

The bound (1.30) of Bennet's Theorem 1.8 is improved by the inequalities $\phi \geq \phi_1 \geq \phi_2$ for

$$\phi_1(x) = x\log\left(1 + \frac{x}{3}\right), \quad \phi_2(x) = \frac{x}{2}\log(1 + x), \quad x > 0.$$

Theorem 4.5. *Under the conditions of Theorem 4.4, for every $t > 0$*

$$P(X_n > t) \leq \exp\left\{-\frac{t}{n\bar{\sigma}_n M}\log\left(1 + \frac{t}{3n\bar{\sigma}_n M}\right)\right\}$$
$$\leq \exp\left\{-\frac{t}{2n\bar{\sigma}_n M}\log\left(1 + \frac{t}{n\bar{\sigma}_n M}\right)\right\}.$$

Proof. This is an application to Theorem 4.4 of the inequalities

$$\exp\{-\phi(x)\} \leq \exp\left\{-x\log\left(1 + \frac{x}{3}\right)\right\} \leq \exp\left\{-\frac{x}{2}\log(1 + x)\right\}.$$

$\square$

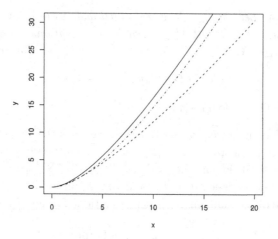

Fig. 4.1 Graph of ϕ, ϕ_1 and ϕ_2.

Under a weaker condition, another version of Theorem 4.4 is deduced from its application to the restriction of the process X to the set $\{V_n \leq \eta\}$.

Theorem 4.6. *Let $(X_n)_{n\geq 0}$ be a real martingale with mean zero and such that there exists a constant M for which the variations of the martingale satisfy $|X_n - X_{n-1}| \leq MV_n$ a.s., for every integer n. For every $t > 0$*

$$P(X_n > t) \leq \exp\left\{-\phi\left(\frac{t}{\sqrt{\eta}M}\right)\right\} + P(V_n > \eta).$$

Under the conditions of Theorem 4.6, Theorem 4.5 entails that for every $t > 0$

$$P(X_n > t) \leq \exp\left\{-\frac{t}{\sqrt{\eta}M} \log\left(1 + \frac{t}{3\sqrt{\eta}M}\right)\right\} + P(V_n > \eta),$$

$$\leq exp\left\{-\frac{t}{2\sqrt{\eta}M} \log\left(1 + \frac{t}{\sqrt{\eta}M}\right)\right\} + P(V_n > \eta).$$

For every monotone function H, Theorem 4.4 implies

$$P(H(X_n) > t) \leq \exp\left\{-\phi\left(\frac{H^{-1}(t)}{n\bar{\sigma}_n M}\right)\right\}, \quad t > 0.$$

Let $X = (X_n)_n$ be a martingale, let $(A_n)_n$ be a sequence of predictable random variables, and let Y_n be the transformed martingale

$$Y_n = \sum_{k=1}^{n} A_k(X_k - X_{k-1}). \tag{4.8}$$

The variance of Y_n is

$$\bar{\sigma}_n^2(Y) = \sum_{i=1}^n E\{A_i^2(V_i - V_{i-1})(X)\} \leq \|(A_i)_{1 \leq i \leq n}\|_4^2 \|(\sigma_i)_{1 \leq i \leq n}\|_4^2$$

and $\bar{\sigma}_n^2(Y)$ converges to a finite limit σ_Y^2 if $(A_n)_n$ belongs to $L^4(\mathbb{R}^n)$ and if $\bar{\sigma}_n^2(X)$ converges to a limit σ^2.

Under the boundedness condition

$$|Y_i - Y_{i-1}| \leq M\{A_i^2(V_i - V_{i-1})(X)\}^{\frac{1}{2}},$$

we have $EY_n^k = \sum_{i=1}^n E(Y_i - Y_{i-1})^k$ where each term is bounded by

$$E|Y_i - Y_{i-1}|^k \leq M^k E[\{A_i^2(V_i - V_{i-1})(X)\}^{\frac{k}{2}}] \leq \{Mn\bar{\sigma}_n(Y)\}^k.$$

Theorem 4.4 applies to the martingale Y_n using the bound $b_n = n\bar{\sigma}_n(Y)M$ in the expansion of the moment generating function of Y_n. If $(X_n)_{n \geq 0}$ is a supermartingale and satisfies the boundedness condition of Theorem 4.4, the same bound holds.

Let φ be a real concave function on $\mathbb{R}$ let $X = (X_n)_n$ be in $\mathcal{M}^2$ and let $Y_{n+1} = Y_n + \varphi(X_{n+1} - X_n)$. The mean quadratic variations of the supermartingale Y is

$$E\{V_{n+1}(Y)|\mathcal{F}_n\} = V_n(Y) + E(Y_{n+1}^2|\mathcal{F}_n) - Y_n^2$$
$$\geq V_n(Y),$$

hence $var Y_n \leq \sum_{i=1}^n EV_i(Y) \leq nE(X_1 - X_0)^2$ which is denoted $n\sigma^2$. Under the condition $|Y_i - Y_{i-1}| \leq M\{E(Y_i - Y_{i-1})^2\}^{\frac{1}{2}}$, Theorem 4.4 applies to the supermartingale Y_n with the bound $b_n = \bar{\sigma}_n M = M \sum_{i=1}^n \{E(Y_i - Y_{i-1})^2\}^{\frac{1}{2}}$. Assuming that φ is a real Lipschitz function on $\mathbb{R}$ implies that $(Y_n)_n$ is a supermartingle sequence and there exists a constant $c_\varphi > 0$ such that $E\varphi^2(X_{n+1} - X_n) \leq c_\varphi^2 E\{(X_{n+1} - X_n)^2|\mathcal{F}_n\}$ and the same inequality holds for the variances

$$E\{V_{n+1}(Y) - V_n(Y)|\mathcal{F}_n\} \leq c_\varphi^2 E\{V_{n+1}(X) - V_n(X)|\mathcal{F}_n\}.$$

Under the conditions of Theorem 4.4, for every Lipschitz function φ

$$P(X_n > t) \leq \exp\left\{-\phi\left(\frac{t}{c_\varphi n\bar{\sigma}_n M}\right)\right\}, \ t > 0.$$

4.4 Inequalities for first passage and maximum

Let $(M_n)_{n \geq 0}$ be a martingale on a probability space $(\Omega, \mathcal{F}, (\mathcal{F}_n)_{n \geq 0}, P)$ and let $\tau = \inf\{n : |M_n| \geq \lambda\}$, $\lambda > 0$, then τ is a stopping time with respect to the filtration $\mathbb{F} = (\mathcal{F}_n)_{n \geq 0}$.

Proposition 4.20. *On a filtered probability space* $(\Omega, \mathcal{F}, (\mathcal{F}_n)_{n \geq 0}, P)$, *let* $(X_n)_{n \geq 0}$ *be a real supermartingale then for every stopping time* τ, $Y_n = E^{\mathcal{F}_n}(X_{\tau \vee n})$ *is a supermartingale a.s. bounded by* X_n.

Proof. If $\tau = k$, then

$$Y_n = E^{\mathcal{F}_n}(X_{k \vee n}) = E^{\mathcal{F}_n}(X_k)1_{\{k > n\}} + X_n 1_{\{k \leq n\}}$$

and $Y_n \leq X_n$. For all integers $m > n$, the events $\{\tau \leq n\}$ and $\{\tau > n\}$ belong to $\mathcal{F}_n$ and

$$E^{\mathcal{F}_n}(Y_m) = E^{\mathcal{F}_n}(X_{\tau \vee m})1_{\{\tau \leq n\}} + E^{\mathcal{F}_n}(X_{\tau \vee m})1_{\{\tau > n\}}$$
$$\leq X_n 1_{\{\tau \leq n\}} + E^{\mathcal{F}_n}(X_{\tau \vee m})1_{\{\tau > n\}}$$

where

$$E^{\mathcal{F}_n}(X_{\tau \vee m})1_{\{\tau > n\}} = E^{\mathcal{F}_n}E^{\mathcal{F}_\tau}(X_m)1_{\{n < \tau \leq m\}} + 1_{\{\tau > m\}}E^{\mathcal{F}_n}(X_\tau)$$
$$\leq E^{\mathcal{F}_n}(X_\tau),$$

hence $E^{\mathcal{F}_n}(Y_m) \leq E^{\mathcal{F}_n}(X_{\tau \vee n}) = Y_n$. $\qquad\square$

On a filtered probability space $(\Omega, \mathcal{F}, (\mathcal{F}_n)_{n \geq 0}, P)$, let $M = (M_n)_{n \geq 0}$ be a square integrable martingale and let τ be a stopping time τ. The variance of a square integrable martingale M_m conditionally on $\mathcal{F}_n$ such that $n < m$ is $E\{(M_m - M_n)^2 | \mathcal{F}_n\} = E(M_m^2 | \mathcal{F}_n) - M_n^2$. If M is a martingale, for every stopping time τ and for every integer n, we have

$$\text{Var}(M_{n \wedge \tau} | \mathcal{F}_{n-1}) = E\{M_{n \wedge \tau}^2 | \mathcal{F}_n) - M_{(n-1) \wedge \tau}^2,$$
$$E(M_{n \wedge \tau}^2 | \mathcal{F}_{n-1}) = E(M_n^2 1_{\{\tau \geq n\}} + M_\tau^2 1_{\{\tau < n\}} | \mathcal{F}_{n-1})$$
$$= E(M_n^2 | \mathcal{F}_{n-1})1_{\{\tau > n-1\}} + M_\tau^2 1_{\{\tau \leq n-1\}},$$

where $\{\tau \leq n-1\}$, $\{\tau \geq n\}$ and $M_\tau^2 1_{\{\tau < n\}}$ are $\mathcal{F}_{n-1}$-measurable, therefore $\text{Var}(M_{n \wedge \tau} | \mathcal{F}_{n-1}) = \text{Var}(M_n | \mathcal{F}_{n-1})1_{\{\tau > n-1\}}$. For an increasing sequence

$(\tau_n)_n$ of stopping times tending a.s. to infinity it follows that the variable $\mathrm{Var}(M_{n \wedge \tau_n} \mid \mathcal{F}_{n-1})$ converges a.s. to $\mathrm{Var}(M_n \mid \mathcal{F}_{n-1})$. In the same way, for $n < m$

$$\mathrm{Var}(M_{m \wedge \tau} 1_{\{\tau \le m\}} \mid \mathcal{F}_n) = \mathrm{Var}(M_\tau \mid \mathcal{F}_n) 1_{\{n < \tau \le m\}},$$

and

$$\mathrm{Var}(M_{m \wedge \tau} \mid \mathcal{F}_n) = \mathrm{Var}(M_m \mid \mathcal{F}_n) 1_{\{\tau > m\}} + \mathrm{Var}(M_\tau) 1_{\{\tau \le m\}}$$

then for an increasing sequence $(\tau_n)_n$ of stopping times tending a.s. to infinity, $\mathrm{Var}(M_{m \wedge \tau_n} \mid \mathcal{F}_n)$ converges a.s. to $\mathrm{Var}(M_m \mid \mathcal{F}_n)$.

Proposition 4.21. *Let $(X_n)_{n \ge 1}$ be a positive super-martingale of L^p, for $p > 1$, and let $X_n^* = \max_{k=0,\ldots,n} X_k$ then*

$$\|X_n^*\|_p \le \frac{p}{p-1} \sup_{k \le n} \|X_k\|_p.$$

Proof. For $x > 0$, let $\tau = \inf\{n : X_n \ge x\}$ then $X_n^* \ge x$ is equivalent to $\tau \le n$. On the set $\{\tau \le n\}$ we have $X_n^* \ge X_\tau \ge x$ and

$$x P(X_n^* \ge x) \le E(X_\tau 1_{\{X_n^* \ge x\}}).$$

Integrating by parts like in the proof of Theorem 1.5 implies

$$
\begin{aligned}
E(X_n^{*p}) &= p \int_0^\infty x^{p-1} P(X_n^* \ge x) \, dx \\
&\le p E\left(\int_0^\infty x^{p-2} X_\tau 1_{\{X_n^* \ge x\}} \, dx \right) \\
&\le \frac{p}{p-1} \sup_{k \le n} E\{ X_k (X_n^*)^{p-1} \} \\
&\le \frac{p}{p-1} E\{ X_n^{*p} \}^{\frac{p-1}{p}} \sup_{k \le n} \|X_k\|_p
\end{aligned}
$$

and the result follows, with $p^{-1} + p^{-1}(p-1) = 1$. □

Corollary 4.1. *For every real $p > 1$ there exists a constant c_p such that for every martingale $X = (X_n)_{n \ge 0}$ of $\mathcal{M}^p$ and for every stopping time N*

$$E(X_N^p) \le E(X_N^{*p}) \le \left(\frac{p}{p-1} \right)^p \sup_{n \le N} E(X_n^p).$$

Corollary 4.2. *For every real $p > 1$, there exists a constant k_p such that for every martingale $X = (X_n)_{n \ge 0}$ of $\mathcal{M}_0^p$ and for every stopping time N*

$$E(A_N^p) \le E(X_N^{*2p}) \le k_p E(A_N^p).$$

Proof. The inequality $E(X_N^{*2p}) \leq k_p \sup_{n \geq 1} E(A_n^p)$ is a consequence of Propositions 4.13 and 4.21, for the increasing process A_n.   □

Let $\tau = \inf\{n : |M_n| \geq \lambda\}$, $\lambda > 0$, τ is a stopping time with respect to $\mathbb{F}$. The events $\{M_n^* \geq \lambda\}$ and $\{n \geq \tau\}$ are equivalent and for every $\lambda > 0$

$$\lambda P(M_n^* \geq \lambda) \leq E(|M_\tau| 1_{\{M_n^* \geq \lambda\}}),$$

this inequality generalizes to $\mathcal{M}^p$

$$\lambda^p P(M_n^* \geq \lambda) \leq E(|M_\tau|^p 1_{\{M_n^* \geq \lambda\}}).$$

Proposition 4.22. *Let $(X_n)_{n \geq 1}$ be a positive super-martingale of L^p, for $p > 1$, and for stopping times A and B let $X_{AB}^* = \max_{A \leq m < n \leq B} |X_n - X_m|$ then*

$$\|X_{AB}^*\|_p \leq \frac{p}{p-1} \max_{A \leq m < n \leq B} \|X_n - X_m\|_p.$$

Proof. For $x > 0$ and $A < n \leq B$, let

$$\tau_{1n} = \min\{A \leq m < n : |X_n - X_m| \geq x\},$$
$$\tau_2 = \min\{\tau_{1n} < n \leq B : |X_n - X_{\tau_{1n}}| \geq x\},$$

and let $\tau_1 = \tau_{1\tau_2}$, then

$$xP(X_n^* \geq x) \leq E(|X_{\tau_2} - X_{\tau_1}| 1_{\{X_n^* \geq x\}}).$$

Integrating by parts and using the Cauchy-Schwarz inequality like in the proof of Proposition 4.21 yields the result.   □

Corollary 4.3. *For every real $p > 1$, there exists a constant k_p such that for every martingale $X = (X_n)_{n \geq 0}$ of $\mathcal{M}_0^p$ and for all positive stopping times $M < N$*

$$E\{A_N - A_M)^p\} \leq E(X_{MN}^{*2p}) \leq k_p E\{(A_N - A_M)^p\}.$$

The bounds are deduced from Proposition 4.22 as a consequence of Proposition 4.16.

4.5 Inequalities for p-order variations

Let $(Z_n)_{n \geq 1}$ be a sequence of random variables and let $(F_n)_{n \geq 0}$ be an increasing sequence of σ-algebras such that Z_n is F_n-measurable. Moment inequalities for the conditional mean variables $z_n = E(Z_n \mid F_{n-1})$, $n \geq 1$, are obtained by Hölder's inequality and the following lemma in the Hilbert spaces L^p, $p \geq 2$.

Lemma 4.3. *Let f be a function of L^p for a real number $p > 1$ then*

$$\|f\|_p = \sup_{\|h\|_q = 1} \int f(x)h(x)\,dx,$$

where $p^{-1} + q^{-1} = 1$.

Proof. The inequality $\sup_{\|h\|_q=1} \int f(x)h(x)\,dx \leq \|f\|_p$ is a consequence of Hölder's inequality. Let $g = \|f\|_p^{-\frac{p}{q}} |f|^{p-1}$, its L^q-norm is

$$\|g\|_q^q = \|f\|_p^{-p} \int |f|^{(p-1)q}(x)\,dx = \|f\|_p^{-p} \int |f|^p(x)\,dx = 1$$

where $(p-1)q = p$, and

$$\|f\|_p = \|f\|_p^{-\frac{p}{q}} \int |f|^p(x)\,dx = \int f(x)g(x)\,dx$$

$$\leq \sup_{\|h\|_q=1} \int f(x)h(x)\,dx.$$

$\square$

Let $(X_n)_{n\geq 1}$ and $(Y_n)_{n\geq 1}$ be sequences of F_n-measurable random variables, Lemma 4.3 applies to the step-wise random constant functions $f = \sum_{n\geq 0} X_n 1_{]n,n+1]}$ and $h = \sum_{n\geq 0} Y_n 1_{]n,n+1]}$, with the L^p-norm, $p \geq 2$

$$E\|(X_n)_n\|_{l_p} = E\Big\{\Big(\sum_{n\geq 1}|X_n|^p\Big)^{\frac{1}{p}}\Big\} = \sup_{E\|(Y_n)_n\|_{l_q}=1} E\Big\{\sum_{n\geq 1} X_n Y_n\Big\}.$$

Lépingle (1978) proved the inequality

$$E\Big[\Big\{\sum_{n\geq 1} E^2\Big(Z_n \mid F_{n-1}\Big)\Big\}^{\frac{1}{2}}\Big] \leq 2E\Big\{\Big(\sum_{n\geq 1} Z_n^2\Big)^{\frac{1}{2}}\Big\},$$

applying it to a sequence of variables $(X_n)_{n\geq 1}$ such that $X_0 = 0$, X_n is $\mathcal{F}_n$ measurable and belongs to L^p, we obtain

$$E\Big(\Big[\sum_{n\geq 1}\{E(X_n \mid F_{n-1}) - X_{n-1}\}^2\Big]^{\frac{1}{2}}\Big) \leq 2E\Big[\Big\{\sum_{n\geq 1}(X_n - X_{n-1})^2\Big\}^{\frac{1}{2}}\Big]$$

$$\leq 2\sqrt{2}E\Big[\Big\{\sum_{n\geq 1} X_n(X_n - X_{n-1})\Big\}^{\frac{1}{2}}\Big]$$

and

$$E\left(\left[\sum_{n\geq 1}\{X_n - E(X_n \mid F_{n-1})\}^2\right]^{\frac{1}{2}}\right) \leq \sqrt{2}E\left(\left[\sum_{n\geq 1}\{X_n^2 + E^2(X_n \mid F_{n-1})\}\right]^{\frac{1}{2}}\right)$$

$$\leq 2\sqrt{2}E\left\{\left(\sum_{n\geq 1}X_n^2\right)^{\frac{1}{2}}\right\}.$$

If $(X_n)_{n\geq 1}$ is a martingale, then $E(X_nX_{n-1}) = E(X_{n-1}^2)$ and by concavity of the function square root, the second inequality is replaced by

$$E\left(\left[\sum_{n\geq 1}\{X_n - E(X_n \mid F_{n-1})\}^2\right]^{\frac{1}{2}}\right) \leq \left[E\sum_{n\geq 1}\{X_n(X_n - X_{n-1})\}\right]^{\frac{1}{2}}.$$

Lépingle's inequality generalizes to the expectation of L^p-norms, $p > 1$.

Proposition 4.23. *Let $(Z_n)_{n\geq 1}$ be a sequence of random variables such that $Z_0 = 0$, Z_n is $\mathcal{F}_n$ measurable and belongs to L^p, for a real number $p \geq 1$, then*

$$E\left[\left\{\sum_{n\geq 1}|E(Z_n \mid F_{n-1})|^p\right\}^{\frac{1}{p}}\right] \leq E\left\{\left(\sum_{n\geq 1}|Z_n|^p\right)^{\frac{1}{p}}\right\}.$$

Proof. By Lemma 4.3, the variables $z_n = E(Z_n \mid F_{n-1})$ satisfy

$$E\left\{\left(\sum_{n\geq 1}|z_n|^p\right)^{\frac{1}{p}}\right\} = \sup_{H_n} E\left\{\sum_{n\geq 1}z_nH_n\right\}$$

where $(H_n)_n$ is a sequence of random variables with norm

$$E\|(H_n)_n\|_{l_q} = E\left\{\left(\sum_{n\geq 1}|H_n|^q\right)^{\frac{1}{q}}\right\} = 1.$$

Let $h_n = E(H_n \mid F_{n-1})$ and let $Y_n = \sum_{k=1}^n |h_k|^q$ and $X_n = \sum_{k=1}^n |Z_k|^p$, with $Y_0 = X_0 = 0$, we have

$$E\left\{\left(\sum_{n\geq 1}|z_n|^p\right)^{\frac{1}{p}}\right\} = \sup_{E\|(H_n)_n\|_{l_q}=1} E\left(\sum_{n\geq 1}Z_nh_n\right),$$

by definition of the conditional expectation. By Hölder's inequality

$$E\left(\sum_{n\geq 1}Z_nh_n\right) \leq E\left\{\left(\sum_{n\geq 1}|Z_n|^p\right)^{\frac{1}{p}}\right\}E\left\{\left(\sum_{n\geq 1}|h_n|^q\right)^{\frac{1}{q}}\right\},$$

where the convexity of the power functions implies

$$E\left\{\left(\sum_{n\geq 1}|h_n|^q\right)^{\frac{1}{q}}\right\} \leq E\left[\left\{\sum_{n\geq 1}|E(H_n \mid F_{n-1})|^q\right\}^{\frac{1}{q}}\right]$$

$$\leq E\left[\left\{\sum_{n\geq 1}E(|H_n|^q \mid F_{n-1})\right\}^{\frac{1}{q}}\right]$$

$$\leq \left[E\left\{\sum_{n\geq 1}|H_n|^q\right\}\right]^{\frac{1}{q}} = 1.$$

$\square$

Proposition 4.23 applies to a sequence of variables $(X_n)_{n\geq1}$ such that X_n is $\mathcal{F}_n$ measurable and belongs to L^p, denoting $Z_n = X_n - X_{n-1}$ we deduce $E(Z_n \mid \mathcal{F}_{n-1}) = E(X_n \mid \mathcal{F}_{n-1}) - X_{n-1}$ and

$$E\Big[\Big\{\sum_{n\geq1}|E(X_n \mid \mathcal{F}_{n-1}) - X_{n-1}|^p\Big\}^{\frac{1}{p}}\Big] \leq E\Big\{\Big(\sum_{n\geq1}|X_n - X_{n-1}|^p\Big)^{\frac{1}{p}}\Big\}. \quad (4.9)$$

Corollary 4.4. *Let $(X_n)_{n\geq1}$ be a sequence of random variables such that X_n is $\mathcal{F}_n$ measurable, belongs to L^p and $E(|X_n|^p)$ is constant, then*

$$E\Big[\Big\{\sum_{n\geq1}|E(X_n \mid \mathcal{F}_{n-1}) - X_{n-1}|^p\Big\}^{\frac{1}{p}}\Big] \leq 2\Big\{E\Big(\sum_{n\geq1}|X_n|^p\Big)\Big\}^{\frac{1}{p}}.$$

Proof. By concavity, the upper bound (4.9) satisfies

$$E\Big\{\Big(\sum_{n\geq1}|X_n - X_{n-1}|^p\Big)^{\frac{1}{p}}\Big\} \leq E\Big[\Big\{\sum_{n\geq1}(|X_n| + |X_{n-1}|)^p\Big\}^{\frac{1}{p}}\Big]$$

$$\leq \Big\{E\Big(\sum_{n\geq1}\sum_{k=1}^{p}\binom{p}{k}|X_n|^p|X_{n-1}|^{p-k}\Big)\Big\}^{\frac{1}{p}},$$

under the assumptions and by Hölder's inequality,

$$E(|X_n|^p|X_{n-1}|^{p-k}) \leq \{E|X_n|^p\}^{\frac{k}{p}}\{E|X_{n-1}|^p\}^{\frac{p-k}{p}} = E(|X_n|^p),$$

with $\sum_{k=1}^{p}\binom{p}{k} = 2^p$. $\qquad\square$

By concavity, a sequence of variables $(X_n)_{n\geq1}$ such that X_n is $\mathcal{F}_n$ measurable and belongs to L^p satisfies

$$E\Big(\Big[\sum_{n\geq1}|X_n - E(X_n \mid \mathcal{F}_{n-1})|^p\Big]^{\frac{1}{p}}\Big)$$

$$\leq 2^{\frac{p-1}{p}}E\Big(\Big[\sum_{n\geq1}\{|X_n|^p + |E(X_n \mid \mathcal{F}_{n-1})|^p\}\Big]^{\frac{1}{p}}\Big)$$

and for a martingale sequence $(X_n)_{n\geq1}$

$$E\Big\{\Big(\sum_{n\geq1}|X_n - X_{n-1}|^p\Big)^{\frac{1}{p}}\Big\} \leq 2^{\frac{p-1}{p}}E\Big[\Big\{\sum_{n\geq1}(|X_n|^p + |X_{n-1}|^p)\Big\}^{\frac{1}{p}}\Big].$$

For a martingale $(X_n)_{n\geq1}$ of $\mathcal{M}^p$, the increasing process

$$V_{n,p} = \sum_{k=1}^{n}|X_k - X_{k-1}|^p$$

is a p-order variation process of $(X_n)_n$ and the predictable process

$$\widetilde{V}_{n,p} = \sum_{k=1}^{n} E(|X_k - X_{k-1}|^p \mid F_{k-1})$$

is such that the $(V_{n,p} - \widetilde{V}_{n,p})_{n \geq 0}$ is a martingale of $\mathcal{M}$.

Proposition 4.24. *On* $(\Omega, \mathcal{F}, \mathcal{F}_n, P)$, *let* $(X_n)_{n \geq 0}$ *be a martingale of* $\mathcal{M}^p$, *for a real number* $p \geq 1$, *then for every integer* n

$$E|X_n| \leq E(V_{n,p}^{\frac{1}{p}}),$$

$$\|X_n\|_p \leq \{E(V_{n,p})\}^{\frac{1}{p}} \leq \frac{p}{p-1} \sup_{k \geq 1} \|X_k\|_p.$$

Proof. By convexity we have

$$n^{-1} \sum_{k=1}^{n} |X_k - X_{k-1}|^p \geq \left| n^{-1} \sum_{k=1}^{n} (X_k - X_{k-1}) \right|^p = n^{-p} |X_n|^p$$

therefore $V_{n,p} \geq |X_n|^p$ and the inequalities for $|X_n|$ and $V_{n,p}^{\frac{1}{p}}$ are deduced. To prove the last inequality, let $\lambda > 0$ and $\tau = \min\{n : X_n > \lambda\}$ then

$$\lambda P(V_{n,p}^{\frac{1}{p}} > \lambda) \leq E(X_\tau 1\{V_{n,p}^{\frac{1}{p}} > \lambda\}),$$

it follows that

$$\begin{aligned}
E(V_{n,p}) &= \int_0^\infty \lambda^p \, dP(V_{n,p}^{\frac{1}{p}} \leq \lambda) \\
&= p \int_0^\infty \lambda^{p-1} P(V_{n,p}^{\frac{1}{p}} > \lambda) \, d\lambda \\
&\leq p \int_0^\infty \lambda^{p-2} E\left(X_\tau 1\{V_{n,p}^{\frac{1}{p}} > \lambda\} \right) d\lambda \\
&= \frac{p}{p-1} E\left\{ X_\tau (V_{n,p})^{\frac{p-1}{p}} \right\}.
\end{aligned}$$

Let q such that $p^{-1} + q^{-1} = 1$ so $q = \frac{p}{p-1}$, by Hölder's inequality we obtain

$$E\left(X_\tau V_{n,p}^{\frac{1}{q}} \right) \leq \|X_\tau\|_p \{E(V_{n,p})\}^{\frac{1}{q}}$$

and $E(V_{n,p}) \leq q^p E(|X_\tau|^p)$. $\qquad\square$

Proposition 4.24 generalizes to positive convex functions h

$$E\{h(n^{-1} X_n)\} \leq n^{-1} \sum_{k=1}^{n} E\{h(X_k - X_{k-1})\} \leq \sup_{k=1,\dots,n} E\{h(X_k - X)\}.$$

Proposition 4.25. *On* $(\Omega, \mathcal{F}, \mathcal{F}_n, P)$, *let* $(X_n)_{n \geq 0}$ *be a martingale of* $\mathcal{M}^p$, *for a real number* p *in* $]0, 1[$, *then for every integer* n

$$E|X_n| \geq E(V_{n,p}^{\frac{1}{p}}),$$
$$\|X_n\|_p \geq \{E(V_{n,p})\}^{\frac{1}{p}}.$$

Proof. By concavity of the function x^p we have

$$n^{-1} \sum_{k=1}^{n} |X_k - X_{k-1}|^p \leq \left| n^{-1} \sum_{k=1}^{n} (X_k - X_{k-1}) \right|^p = n^{-p} |X_n|^p$$

therefore $V_{n,p} \leq |X_n|^p$, the inequalities for $|X_n|$ and $V_{n,p}^{\frac{1}{p}}$ are deduced. $\square$

For every positive concave functions h it follows

$$E\{h(n^{-1} X_n)\} \geq n^{-1} \sum_{k=1}^{n} E\{h(X_k - X_{k-1})\} \leq \sup_{k=1,\dots,n} E\{h(X_- X_{k-1})\}.$$

The predictable variation process of the submartingale $(|X_n|^p)_{n \geq 1}$ is defined by the Doob decomposition as

$$\widetilde{A}_{n,p} = \sum_{k=1}^{n} \{E(|X_k|^p \mid \mathcal{F}_{k-1}) - |X_{k-1}|^p\},$$

the difference process $M_{n,p} = |X_n|^p - \widetilde{A}_{n,p}$ is a martigale with respect to $\mathcal{F}$, by definition

$$E(|X_n|^p \mid \mathcal{F}_{n-1}) - |X_{n-1}|^p = \widetilde{A}_{n,p} - \widetilde{A}_{n-1,p}$$

and $M_{n,p}$ satisfies the inequalities for the martigales.

4.6   Weak convergence of discrete martingales

On a probability space $(\Omega, \mathcal{F}, P)$, a sequence of processes $(X_n)_n$ in a metric space $(\mathbb{X}, \mathcal{X}, d)$ converges weakly if the sequence of its distributions P_{X_n} is relatively compact and if $\int h \, dP_{X_n}$ converges for every function h of $C_b(\mathbb{X})$.

The weak convergence of a sequence of real random variables $(X_n)_{n \geq 1}$ is equivalent to the convergence of its distribution function F_n to a distribution function F at each point of continuity of F, it is also equivalent to the convergence of its Laplace or Fourier transform.

The limit distribution of $(X_n)_n$ is the probability distribution of a variable X, it is tight: for every $\varepsilon > 0$ there exists a compact set K_ε such that $P(X \in K_\varepsilon) > 1 - \varepsilon$. A sequence variables $(X_n)_{n \geq 1}$ on a measurable metric space $(\mathcal{S}, \mathcal{B})$ is tight if for every $\varepsilon > 0$ there exists a compact set K_ε of $\mathcal{S}$ such that for every n

$$P(X_n \in K_\varepsilon) \geq 1 - \varepsilon.$$

By the Prohorov's theorem, on a separable and complete metric space $(\mathbb{X}, \mathcal{X}, d)$, the tightness of $(X_n)_n$ is equivalent to the relative compacity of $(P_{X_n})_n$.

Theorem 4.7. *On a probability space* $(\Omega, \mathcal{F}, (\mathcal{F}_n)_{n\geq 0}, P)$, *let* $(X_n)_{n\geq 1}$ *be a centered martingale with an predictable process of quadratic variations* A_n *such that* A_n *converges in probability to a constant* σ^2 *and for every* $\varepsilon > 0$

$$\lim_{n\to\infty} \sum_{k=0}^{n} E\{(X_k - X_{k-1})^2 1_{|X_k - X_{k-1}| > \varepsilon} \mid \mathcal{F}_{k-1}\} = 0,$$

then X_n *converges weakly to a Gaussian variable* $\mathcal{N}(0, \sigma^2)$.

The second condition is similar to the Lindeberg condition in the central limit theorem for independent variables $X_k - X_{k-1}$ with distinct variances, it implies the convergence to zero

$$\lim_{n\to\infty} \sum_{k=0}^{n} E\{|X_k - X_{k-1}| 1_{|X_k - X_{k-1}| > \varepsilon} \mid \mathcal{F}_{k-1}\} = 0$$

and therefore

$$\sup_{k=1,\dots,n} P(|X_k - X_{k-1}| > \varepsilon \mid \mathcal{F}_{k-1}) = 0$$

which is a tightness property.

The convergence in probability of A_n to a constant σ^2 and the second condition imply

$$\lim_{n\to\infty} \sum_{k=0}^{n} E\{(X_k - X_{k-1})^2 1_{|X_k - X_{k-1}| \leq \varepsilon} \mid \mathcal{F}_{k-1}\} = \sigma^2$$

and they provide the convergence of the variance of X_n to the expectation of A_n, as n tends to infinity.

Chapter 5

Inequalities for Time-Continuous Martingales

5.1 Introduction

On a filtered probability space $(\Omega, \mathcal{F}, (\mathcal{F}_t)_{t \geq 0}, P)$, let $(X_t)_{t \geq 0}$ be a measurable square integrable process with independent increments, such that for all $s < t < u$ in $\mathbb{R}_+$

$$E\{(X_u - X_t)(X_t - X_s)\} = E(X_u - X_t)E(X_t - X_s).$$

Gaussian processes, Poisson and renewal processes are process with independent increments. Results proved in the previous chapter for sums of independent random variables apply to martingales with independent increments.

Gaussian martingales have continuous sample paths and they are characterized by their covariances. For a k dimensional process $(X_t)_{t \geq 0}$, there is equivalence between

(1) $X = (X_t)_{t \geq 0}$ a martingale with continuous sample paths such that $< X_i, X_j > (t) = v_i(t)\delta_{ij}$, with increasing functions v_i with value zero at $t = 0$,

(2) X is a Gaussian process with independent components and with independent increments, such that $E(X_i(t)) = 0$ for every $j = 1, \ldots, k$ and $\mathrm{Var}\{X_i(t)\} = v_i(t)$ for every $t \geq 0$.

Let $(M_t)_{t \geq 0}$ be a positive supermartingale and let $\tau = \inf\{t \in \mathbb{R}_+ : M_t > a\}$ then $M_{\tau \wedge t}$ is a supermartingale, it is bounded by a for every t, if τ is infinite.

Proposition 5.1. *Let $(M_t)_{t \geq 0}$ be a positive supermartingale, for every positive stopping time τ, the process $(M_{\tau \vee t})_{t \geq 0}$ is a supermartingale a.s. bounded by M_t.*

133

Proof. Let $Y_t = M_{\tau \vee t}$, the sets $\{t < \tau\}$ and $\{\tau \leq t\}$ belong to $\mathcal{F}_t$ for every $t > 0$, and

$$Y_t = E^{\mathcal{F}_t}(M_t 1_{\{\tau \leq t\}}) + E^{\mathcal{F}_t}(M_\tau 1_{\{\tau > t\}})$$
$$= M_t 1_{\{\tau \leq t\}} + E^{\mathcal{F}_t}(M_\tau) 1_{\{\tau > t\}} \leq M_t.$$

For all $0 < s < t$, we have

$$E^{\mathcal{F}_s}(Y_t) = E^{\mathcal{F}_s}(M_{\tau \vee t} 1_{\{\tau \leq s\}}) + E^{\mathcal{F}_s}(M_{\tau \vee t} 1_{\{\tau > s\}}),$$

where $E^{\mathcal{F}_s}(M_{\tau \vee t} 1_{\{\tau \leq s\}}) = E^{\mathcal{F}_s}(M_t) 1_{\{\tau \leq s\}} \leq M_s 1_{\{\tau \leq s\}}$ and

$$E^{\mathcal{F}_s}(M_{\tau \vee t}) 1_{\{\tau > s\}} = E^{\mathcal{F}_s} E^{\mathcal{F}_\tau}(M_t) 1_{\{s < \tau \leq t\}} + E^{\mathcal{F}_s}(M_\tau) 1_{\{\tau > t\}}$$
$$\leq E^{\mathcal{F}_s}(M_\tau) 1_{\{\tau > s\}},$$

therefore $E^{\mathcal{F}_s}(Y_t) \leq M_{\tau \vee s} = Y_s$. □

5.2   Inequalities for martingales indexed by $\mathbb{R}_+$

Let $(M_t)_{t \geq 0}$ be in $\mathcal{M}^2$, it is written as the sum of a continuous process M^c and a jump process $M_t^d = \sum_{0 < s \leq t} \Delta M_s$, with jumps $\Delta M_s = M_s - M_{s-}$. The increasing predictable process of its quadratic variations is the predictable projection $V_t = \widetilde{[M]}_t$ of

$$[M]_t = <M^c>_t + \sum_{0 < s \leq t} (\Delta M_s)^2$$

such that $M_t^{c2} - <M^c>_t$ and $\sum_{0 < s \leq t}(\Delta M_s)^2 - <M^d>_t$ belong to $\mathcal{M}$, so $V_t = <M^c>_t + <M^d>_t$. The martingale properties entail

$$E(M_t^{d2} - M_s^{d2}) = E\left\{ \sum_{s < u \leq t} (\Delta M_u)^2 \right\}$$

and the following inequalities are deduced by convexity for every $p \geq 2$

$$\|M_t\|_p \leq 2\|M_t^d\|_p + 2\|M_t^c\|_p,$$

$$E(M_t^{d\,p}) \geq E[\{E(M_t^{d\,2} \mid \mathcal{F}_{t-})\}^{\frac{p}{2}}] = E\{<M^d>_t^{\frac{p}{2}}\}$$
$$\geq \{E(<M^d>_t)\}^{\frac{p}{2}},$$

$$E(M_t^{c\,p}) = E[\{(M_t^{c\,2} \mid \mathcal{F}_{t-})\}^{\frac{p}{2}}] = E\{<M^c>_t^{\frac{p}{2}}\}$$
$$\geq \{E(<M^c>_t)\}^{\frac{p}{2}}).$$

Proposition 4.12 is generalized to martingales indexed by $\mathbb{R}_+$ with the following lemma.

Lemma 5.1. *For real numbers $p \geq 2$ and $q \geq 2$, let f be a real function in $L^p(\mathbb{R}) \cap L^q(\mathbb{R})$. For every $x > 0$, there exist $C_{1,x}$ and $C_{2,x}$ depending on p and q such that $C_{1,x}\|f 1_{[0,x]}\|_p \leq \|f 1_{[0,x]}\|_q \leq C_{2,x}\|f 1_{[0,x]}\|_p$.*

Proof. Let $t > 0$, $f(t) = \lim_{n \to \infty} \sum_{i \le n} a_i r_i(t)$ with indicators functions r_i on disjoint intervals, the result is a consequence of the Hölder inequality with $\| \sum_{i \le n} r_i 1_{[0,x]} \|_p \le x^{\frac{1}{p}}$. $\qquad\square$

Inequalities for time-continuous martingales are deduced from those proved for the discrete martingales by similar arguments. For $(M_t)_{t \ge 0}$ in $\mathcal{M}^p$, let $\pi_n = (t_i)_{i \le k_n}$ be a partition of $[0,t]$ in subintervals $I_{n,i}$ with length $t_{n,i} - t_{n,i-1} = h_n(t)$ and let $m_n(t) = [h_n^{-1}]$ tend to infinity, M_t is approximated by a sum of k_n conditionally independent variables $X_{n,i} = M_{t_{n,i}} - M_{t_{n,i-1}}$

$$M_t = \lim_{n \to \infty} \sum_{i=1}^{m_n(t)} X_{n,i}, \tag{5.1}$$

$$V_t = \lim_{n \to \infty} \sum_{i=1}^{m_n(t)} E(X_{n,i}^2 | \mathcal{F}_{t_{n,i-1}}). \tag{5.2}$$

Proposition 5.2. *On a filtered probability space* $(\Omega, \mathcal{F}, (\mathcal{F}_t)_{t \ge 0}, P)$, *let* $(M_t)_{t \ge 0}$ *be a continuous process in* $\mathcal{M}^p$, $p \ge 2$, *having a process of quadratic variations* $(< M >_t)_{t \ge 0}$. *There exist functions* $C_{1,x} > 0$ *and* $C_{2,x}$ *such that for every positive random stopping time* T *of* M

$$E(C_{1,T} < M >_T^{\frac{p}{2}}) \le E(|M_T|^p) \le E(C_{2,T} < M >_T^{\frac{p}{2}}).$$

Proof. As the difference $M_t^2 - < M, M >_t$ belongs to $\mathcal{M}$ on $(\Omega, \mathcal{F}, (\mathcal{F}_t)_{t \ge 0}, P)$, the equality $E(|M_T|^2) = E < M >_T$ is true for $p = 2$. For $p > 2$, Equations (5.1) and (5.2) apply to the inequalities (4.4). From Lemma 5.1, the constant in the upper bound of inequality (4.4) does not depend on the number of variables in the sums but only on the exponent p and on t, therefore the inequality is also true as n tends to infinity and by replacing the index t by a positive stopping time of the martingale M using Doob's theorem. $\qquad\square$

Proposition 4.12 extends to martingales of $\mathcal{M}^p$, on $\mathbb{R}_+$ with constants depending only on p and not on the martingale or its index. The next result improves the constant of the lower bound in the Burkholder's inequality for martingales on $\mathbb{R}_+$.

Proposition 5.3. *For every* $p \ge 1$, *there exists a constant* $C_p > 1$ *such that for every martingale* $(M_t)_{n \ge 1}$ *of* $\mathcal{M}^{2p}$ *and for every positive stopping time* T

$$E(V_T^p) \le E(M_T^{2p}) \le C_p E(V_T^p).$$

Proof. The convexity of the function x^p, for every $p \geq 1$, implies

$$E(M_t^{2p} \mid \mathcal{F}_{t-}) \geq \{E(M_t^2 \mid \mathcal{F}_{t-})\}^p = V_t^p$$

the expectation of both sides of this inequality yields the lower bound. The upper bound is deduced from Burkholder's inequality (1.19) applied to the discrete martingale sequence

$$X_n = M_{t_n \wedge T} = \sum_{k=1}^{m_n} (M_{t_{m_k} \wedge T} - M_{t_{m_k-1} \wedge T})$$

based on a partition of $[0, T]$. By the limit as n tends to infinity, it generalizes to every stopping time T. $\qquad\square$

Proposition 4.13 extends to the variations of martingales.

Proposition 5.4. *For every real $p \geq 1$, for every martingale $(X_t)_{t \geq 0}$ of $\mathcal{M}^{2p}$, and for all $t > s > 0$*

$$E[\{E(V_t \mid \mathcal{F}_s) - V_s\}^p] \leq E\{(X_t - X_s)^{2p}\}$$
$$\leq \left(\frac{2p}{p-1}\right)^p E[\{E(V_t \mid \mathcal{F}_s) - V_s\}^p].$$

Proof. The martingale property of $X^2 - V$ implies that for every $t > s > 0$, $E(X_t \mid \mathcal{F}_s) = X_s$ and $E(X_t^2 \mid \mathcal{F}_s) = X_s^2 + E(V_t \mid \mathcal{F}_s) - V_s$ hence

$$E\{(X_t - X_s)^2 \mid \mathcal{F}_s\} = E(X_t^2 + X_s^2 - 2X_s X_t \mid \mathcal{F}_s)$$
$$= E(V_t - V_s \mid \mathcal{F}_s),$$

therefore $E\{(X_t - X_s)^2\} = E(V_t - V_s)$. By convexity of the power function x^p, for every real $p \geq 1$ we obtain

$$E\{(X_t - X_s)^{2p} \mid \mathcal{F}_s\} \geq [E\{(X_t - X_s)^2 \mid \mathcal{F}_s\}]^p$$
$$= \{E(V_t - V_s \mid \mathcal{F}_s)\}^p.$$

For the upper bound, for every $\lambda > 0$ let $B_{st}(\lambda) = \{(X_t - X_s)^2 > \lambda\}$, then

$$\lambda P((X_t - X_s)^2 > \lambda) \leq E\{(X_t - X_s)^2 1_{B_{st}(\lambda)}\}$$

where the conditional expectation $E\{(X_t - X_s)^2 1_{B_{st}(\lambda)} \mid \mathcal{F}_s\}$ is approximated by (5.2) on a partition $\pi_n = (t_i)_{i \leq k_n}$ of $[0, t]$ in $m_n(t)$ subintervals

$I_{n,i} =]t_{n,i-1}, t_{n,i}]$, as n tend to infinity, we have

$$E\{(X_t - X_s)^2 1_{B_{st}(\lambda)} \mid \mathcal{F}_s\}$$

$$= \lim_{n \to \infty} E\left[\left\{\sum_{i=m_n(s)+1}^{m_n(t)} (X_{n,i+1} - X_{n,i})\right\}^2 1_{B_{st}(\lambda)} \mid \mathcal{F}_s\right]$$

$$= \lim_{n \to \infty} E\left\{\sum_{i=m_n(s)+1}^{m_n(t)} (X_{n,i+1} - X_{n,i})^2 1_{B_{st}(\lambda)} \mid \mathcal{F}_s\right\}$$

$$+ 2E\left\{\sum_{k \neq j = m_n(s)+1}^{m_n(t)} (X_{n,i+1} - X_{n,i})(X_{n,j+1} - X_{n,j})^2 1_{B_{st}(\lambda)} \mid \mathcal{F}_s\right\}$$

$$\leq 2 \lim_{n \to \infty} E\left\{\sum_{i=m_n(s)+1}^{m_n(t)} (X_{n,i+1} - X_{n,i})^2 1_{B_{st}(\lambda)} \mid \mathcal{F}_s\right\}$$

$$= 2E\{(V_t - V_s) 1_{B_{st}(\lambda)}\},$$

by the Cauchy-Schwarz inequality. Arguing like in the proof of Doob's inequality, this entails

$$\|(X_t - X_s)^2\|_p \leq K_p \|V_t - V_s\|_p = K_p \|E(A_t \mid \mathcal{F}_s) - V_s\|_p$$

with the constant $K_p = 2p(p-1)^{-1}$ depending only on p. $\qquad\square$

Proposition 5.5. *For every p in $]0,1[$, there exists a constant $c_p < 1$ such that for ever martingale $(M_t)_{t \geq 0}$ of $\mathcal{M}^{2p}$ and for every positive stopping time T*

$$c_p E(V_T^p) \leq E(M_T^{2p}) \leq E(V_T^p).$$

The upper bound is a consequence of the concavity of the function x^p and the lower bound is due to Burkholder's inequality.

Proposition 5.6. *For every p in $]0,1[$, there exists a constant K_p such that for every martingale $(X_t)_{t \geq 0}$ of $\mathcal{M}^{2p}$, and for all $t > s > 0$*

$$K_p E[\{E(A_t \mid \mathcal{F}_s) - A_s\}^p] \leq E\{(X_t - X_s)^{2p}\}$$

$$\leq E[\{E(A_t \mid \mathcal{F}_s) - A_s\}^p].$$

Proof. By concavity of the power function x^p, for p in $]0,1[$, and by the martingale property we have $E\{(X_t - X_s)^2 \mid \mathcal{F}_s\} = \{E(V_t - V_s \mid \mathcal{F}_s)\}$, for $t > s$, then

$$E\{(X_t - X_s)^{2p} \mid \mathcal{F}_s) \leq [E\{(X_t - X_s)^2 \mid \mathcal{F}_s\}]^p$$

$$= \{E(V_t - V_s \mid \mathcal{F}_s)\}^p$$

$$= \{E(A_t \mid \mathcal{F}_s) - A_s\}^p.$$

The lower bound is established using the proof of Proposition 4.16 and the approximation (5.2) of $X_t - X_s$ and $V_t - V_s$ on a partition $\pi_n = (t_i)_{i \leq k_n}$ of $[0, t]$ in $m_n(t)$ subintervals $I_{n,i} =]t_{n,i-1}, t_{n,i}]$, as n tend to infinity. Let

$$E\{(V_t - V_s)^p\} = \lim_{n \to \infty} E\{(V_{m_n(t)} - V_{m_n(s)})^p\}$$

$$E\{(V_{m_n(t)} - V_{m_n(s)})^p\} = E\Big[\Big\{ \sum_{i=m_n(s)+1}^{m_n(t)} (X_{t_{n,i}} - X_{t_{n,i-1}})^2 \Big\}^p\Big]$$

$$\leq 2^{p-1} E[\{X_{m_n(t)} - X_{m_n(s)}\}^{2p}]$$

taking the limits as n tend to infinity, it follows that

$$E\{(V_t - V_s)^p\} = \lim_{n \to \infty} E\{(V_{m_n(t)} - V_{m_n(s)})^p\} \leq 2^{p-1} E\{(X_t - X_s)^{2p}\}.$$

$$\square$$

The next result is a generalization of Proposition 4.17 using a partition of $[0, T]$.

Proposition 5.7. *Let $(M_t)_{t \geq 0}$ belong to $\mathcal{M}^p$, for every real $p > 1$ and for every stopping time T of $(M_t)_{t \geq 0}$*

$$\big\| M_T^2 - [M]_T \big\|_{L^{\frac{p}{2}}} < \big\| [M]_T \big\|_{L^{\frac{p}{2}}}.$$

The inequalities established for martingales extend to local martingales under the same conditions by the limit as n tends to infinity on their stopping times.

On $(\Omega, \mathcal{F}, (\mathcal{F}_t)_{t \geq 0}, P)$, let $(M_t)_{t \geq 0}$ be in $\mathcal{M}^k$, for every integer $k > 1$. For $0 < s < t$, the variations of the moment generating function of the martingale are

$$E\{e^{\lambda M_t} - e^{\lambda M_s} | \mathcal{F}_s\} = \varphi_{M_s}(\lambda) E\{\exp^{\lambda(M_t - M_s)} - 1 | \mathcal{F}_s\}$$

$$= \varphi_{M_s}(\lambda) \sum_{k=2}^{\infty} \lambda^k E\{(M_t - M_s)^k | \mathcal{F}_s\},$$

from Proposition 5.3, they are bounded by the sum

$$\varphi_{M_s}(\lambda) \sum_{k=2}^{\infty} \lambda^k C_k E\{([M]_t - [M]_s)^{\frac{k}{2}} | \mathcal{F}_s\}.$$

A continuous version of Chernoff's theorem for martingales is deduced from the result for discrete martingales. Proposition 5.3 is not sufficient to establish an inequality of the same kind as Bennett's inequality and it is proved using the same argument as Lenglart's inequality.

Theorem 5.1. *On a probability space* $(\Omega, \mathcal{F}, (\mathcal{F}_t)_{t\geq 0}, P)$, *let* $(M_t)_{t\geq 0}$ *be in* $\mathcal{M}^k$, *for every integer* $k > 1$. *For all* $a > 0$ *and* $\eta > 0$

$$P\left(\sup_{t\in[0,T]} t^{-\frac{1}{2}} M_t > a\right) \leq \exp\left\{-\phi\left(\frac{a}{2\sqrt{\eta T}}\right)\right\} + P([M]_T > \eta T).$$

Proof. The constant in the upper bound of Burkholder's inequality is such that $C_p^{\frac{1}{p}} = O(q^{\frac{1}{2p}} p^{\frac{1}{p}})$, this is a decreasing function of p and C_p is strictly bounded by 2^p. From Proposition 5.3 it follows that

$$\{E(M_t^k)\}^{\frac{1}{k}} < 2\{E([M]_t^{\frac{k}{2}})\}^{\frac{1}{k}}, \tag{5.3}$$

for every $k \geq 2$ and for every t in $[0, T]$. For all $t > 0$ and $\lambda > 0$, the moment generating function of M_t is

$$\varphi_{M_t}(\lambda) = E\left\{\exp^{\lambda M_t}\right\} = 1 + \sum_{k=2}^{\infty} \frac{\lambda^k}{k!} E(M_t^k)$$

$$\leq 1 + \sum_{k=2}^{\infty} \frac{\lambda^k}{k!} 2^k E([M]_t^{\frac{k}{2}}),$$

the bound is increasing with t since $[M]$ is an increasing process. By the same arguments, the restriction of the process $X_t = t^{-\frac{1}{2}} M_t$ to the set $\{[X]_T \leq \eta\} = \{[M]_T \leq \eta T\}$ has the generating function

$$\varphi_{X_t}(\lambda) \leq e^{\lambda 2\eta^{\frac{1}{2}} T^{\frac{1}{2}}} - \lambda 2\eta^{\frac{1}{2}} T^{\frac{1}{2}}$$

$$\leq \exp(e^{\lambda 2\eta^{\frac{1}{2}} T^{\frac{1}{2}}} - 1 - \lambda 2\eta^{\frac{1}{2}} T^{\frac{1}{2}})$$

where the function

$$I_T(\lambda) = e^{\lambda 2\eta^{\frac{1}{2}} T^{\frac{1}{2}}} - 1 - \lambda 2\eta^{\frac{1}{2}} T^{\frac{1}{2}} - a\lambda$$

has the minimum value $\exp\{-\phi((2\sqrt{\eta T})^{-1}a)\}$ under the inequality (5.3) and the bound follows. $\qquad\square$

Corollary 5.1. *Under the conditions of Theorem 5.1, for all strictly positive* a *and* η *we have*

$$P(\sup_{t\in[0,T]} t^{-\frac{1}{2}} M_t > a) \leq \exp\left\{-\frac{a}{2\sqrt{\eta T}} \log\left(1 + \frac{a}{6\sqrt{\eta T}}\right)\right\} + P([M]_T > \eta T)$$

$$\leq \exp\left\{-\frac{a}{4\sqrt{\eta T}} \log\left(1 + \frac{a}{2\sqrt{\eta T}}\right)\right\} + P([M]_T > \eta T).$$

The proof is deduced from Theorem 5.1 and the inequalities

$$\exp\{-\phi(x)\} \leq \exp\left\{-x \log\left(1 + \frac{x}{3}\right)\right\} \leq \exp\left\{-\frac{x}{2} \log(1 + x)\right\}.$$

Proposition 5.8. *Let $M = (M_t)_{t \geq 0}$ be in $\mathcal{M}^k$, for every integer k and let $T > 0$ be a stopping time $T > 0$ such that $\max_{k \geq 2}(ET^{-1} \int_0^T d[M]^{\frac{k}{2}})^{\frac{1}{k}}$ is bounded by a constant C. For every $x > 0$*

$$P\Big(\sup_{t \in [0,T]} |M(t)| > x \Big) \leq \exp\Big\{ -\phi\Big(\frac{x}{CT} \Big) \Big\}.$$

Proof. The moment generating function of M_t has the bound

$$\varphi_{M_t}(\lambda) = Ee^{\lambda M_t} \leq 1 + \sum_{k \geq 2} \frac{\lambda^k}{k!} \Big\{ E\Big(\int_0^T d[M]^{\frac{k}{2}} \Big) \Big\}^k$$

$$\leq \exp\{ e^{\lambda CT} - 1 - \lambda CT \}$$

using the same expansion as in the proof of Theorem 5.1, and the proof ends using the same argument as in this proof. □

The inequalities of Corollary 5.1 extend to Proposition 5.8. Proposition 5.8 applies to the stochastic integral Y of a predictable process $A = (A_t)_{t \geq 0}$ with respect to a martingale $M = (M_t)_{t \geq 0}$, under the required integrability conditions for Y. The Bienaymé-Chebychev inequality for $Y_t = \int_0^t A_s \, dM_s$ in $\mathcal{M}^2$ is written for the supremum of the martingale over a random interval $[0, T]$ determined by an integrable stopping time of the process Y

$$P\Big(\sup_{0 \leq t \leq T} t^{-\frac{1}{2}} Y_t > \lambda \Big) \leq E \frac{1}{\lambda^2 T} \int_0^T A_u^2 \, d[M]_u \qquad (5.4)$$

$$\leq \frac{1}{\lambda^2} E\Big(\sup_{0 \leq t \leq T} A_t^2 \frac{[M]_T}{T} \Big).$$

If $[M]_t$ has a derivative, it is written as $[M]_t = \int_0^t B_s \, ds$ and the bound in inequality (5.4) can be precised by the Mean Value Theorem. For every integer $p \geq 1$, there exists θ in $]0, 1[$ such that

$$P\Big(\sup_{0 \leq t \leq T} t^{-\frac{1}{2}} Y_t > \lambda \Big) \leq \frac{1}{\lambda^{2p}} E\Big\{ \Big(\frac{1}{T} \int_0^T A_u^2 \, d[M]_u \Big)^p \Big\}$$

$$\leq \frac{1}{\lambda^{2p}} E\{ (A_{\theta T}^2 B_{\theta T})^p \}.$$

For stopping times $0 < T_1 < T_2$ of the process Y and for all $\lambda_1 > 0$ and $\lambda_2 > 0$, there exist θ_1 and θ_2 in $]0, 1[$ such that

$$P\Big(\sup_{0 \leq t \leq T_1} t^{-\frac{1}{2}} \int_0^t A_u \, dM_u > \lambda_1, \ \sup_{T_1 \leq t \leq T_2} (t - T_1)^{-\frac{1}{2}} \int_{T_1}^t A_u \, dM_u > \lambda_2 \Big)$$

$$\leq \frac{1}{(\lambda_1 \lambda_2)^2} E\Big\{ \frac{1}{T_1} \int_0^{T_1} A_u^2 \, d[M]_u \Big\} E\Big\{ \frac{1}{T_2 - T_1} \int_{T_1}^{T_2} A_u^2 \, d[M]_u \Big\}$$

$$= \frac{1}{(\lambda_1 \lambda_2)^2} E\{ A_{\theta_1 T_1}^2 B_{\theta_1 T_1} \} E\{ A_{\theta_2 (T_2 - T_1)}^2 B_{\theta_2 (T_2 - T_1)} \}.$$

Proposition 5.9. *Let $M = (M_t)_{t \geq 0}$ and $Y_t = \int_0^t A_s \, dM_s$ be in $\mathcal{M}^k$, for every integer k, where $A = (A_t)_{t \geq 0}$ is a predictable process. Let $T > 0$ be a stopping time $T > 0$ such that $\sup_{t \in [0,T]} |A|$ is a.s. bounded by a constant B and there exists a constant C such that $\max_{k \geq 2}(ET^{-1}[M]_T^{\frac{k}{2}})^{\frac{1}{k}} < C$. For every $x > 0$*

$$P\left(\sup_{t \in [0,T]} |Y(t)| > x \right) \leq \exp\left\{ -\phi\left(\frac{x}{BCT} \right) \right\}.$$

Proof. Under the condition $|A(t)| \leq B$ for every t in $[0,T]$ and $E|\int_0^t A \, dM|^k \leq B^k E \int_0^t d|M|^k$ is finite for every $t > 0$ and for every integer k, the moment generating function of Y_t is written

$$\varphi_{Y_t}(\lambda) = E e^{\lambda \int_0^t A_s \, dM_s} \leq 1 + \sum_{k \geq 2} \frac{\lambda^k B^k}{k!} \int_0^T d[M]_T^{\frac{k}{2}}$$

$$\leq \exp\left\{ e^{\lambda BCT} - 1 - \lambda BCT \right\}$$

using the same expansion as in the proof of Theorem 5.1. The proof ends by minimizing $\log L_{Y_t}(\lambda) - \lambda a$ for $\lambda > 0$ as above. $\qquad \square$

5.3 Inequalities for the maximum

On a filtered probability space $(\Omega, \mathcal{F}, P, \mathbb{F})$, let $M = (M_t)_{t \geq 0}$ be a martingale of $\mathcal{M}^2$. The inequalities for martingales apply to the maximum process $M_t^* = \sup_{0 \leq s \leq t} |M_s|$ like in the Burkholder-Davis-Gundy inequalities and the inequalities of the previous section. Let

$$\tau = \inf\{t : M_t \geq \lambda\}, \quad \lambda > 0,$$

τ is a stopping time with respect to $\mathbb{F}$ and the events $\{M_t^* \geq \lambda\}$ and $\{t \geq \tau\}$ are equivalent.

Proposition 5.10. *Let $(M_t)_{t \geq 0}$ be a positive supermartingale and for $a > 0$ let $\tau_a = \inf\{t \geq 0 : M_t^* \geq a\}$, then*

$$P(M_t^* \geq a) \leq a^{-1} E(M_{\tau_a} 1_{\{M_t^* \geq a\}}),$$

$$E(\tau_a) = a^{-1} \int_0^\infty t \, dP(M_t^* > a).$$

Proof. The equalities $\{\tau_a > t\} = \{M_t^* < a\}$ and $\{\tau_a \leq t\} = \{M_t^* \geq a\}$ is $\mathcal{F}_t$ measurable hence τ_a is a stopping time such that $M_{\tau_a} \geq a$ and on

$[0, \tau_a[$, $M_t^* < a$, it follows that

$$aP(M_t^* \geq a) = aP(\tau_a \leq t) \leq E(M_{\tau_a} 1_{\{\tau_a \leq t\}})$$
$$\leq E\{E^{\mathcal{F}_{\tau_a}}(M_t) 1_{\{\tau_a \leq t\}}\}.$$

The expectation of τ_a is

$$E(\tau_a) = \int_0^\infty t\, dP(\tau_a \leq t)$$
$$= a^{-1} \int_0^\infty t\, dP(M_t^* \geq a).$$

$\square$

Theorem 5.2. *Let* $M = (M_t)_{t \geq 0}$ *be a martingale of* $\mathcal{M}^p$, $p > 1$, *for every* $t > 0$

$$\|M_t\|_p \leq \|M_t^*\|_p \leq \frac{p}{p-1} \sup_{s \leq t} \|M_s\|_p \leq \frac{pC_p}{p-1} \left\{ E(V_t^{\frac{p}{2}}) \right\}^{\frac{1}{p}}.$$

Proof. The proof is similar to the proof of Theorem 4.21 using Proposition 5.10 and Proposition 5.3. For $x > 0$, let $\tau = \inf\{t : X_t \geq x\}$, then $X_t^* \geq x$ is equivalent to $\tau \leq t$, then Proposition 5.10 implies

$$E(M_t^{*p}) = \int_0^\infty \lambda^p\, dP(M_t^* \leq \lambda)$$
$$= p \int_0^\infty \lambda^{p-1} P(M_t^* \geq \lambda)\, d\lambda,$$
$$\leq p \sup_{s \leq t} \int_0^\infty \lambda^{p-2} E(M_s 1_{\{M_t^* > \lambda\}})\, d\lambda,$$

by integration it follows

$$E(M_t^{*p}) \leq \frac{p}{p-1} \sup_{s \leq t} E(M_t^{*p-1} M_s)$$

and by Hölder's inequality with $q = (p-1)^{-1} p$

$$E\{M_t^{*p-1} M_s\} \leq \left\{ E(M_t^{*p}) \right\}^{\frac{p-1}{p}} \{E(M_s^p)\}^{\frac{1}{p}}$$

therefore

$$\|M_t^*\|_p \leq \frac{p}{p-1} \sup_{s \leq t} \|M_s\|_p.$$

Proposition 5.3 implies $\sup_{s \leq t} E(M_s^p) \leq C_p E(V_t^{\frac{p}{2}})$ for the increasing process V and the result follows. $\square$

Theorem 5.3. *Let $M = (M_t)_{t \geq 0}$ be a martingale of $\mathcal{M}^p$, $p > 1$, for all s and t in a real interval I*

$$E[\{E(V_t \mid \mathcal{F}_s) - V_s\}^{\frac{p}{2}}] \leq E\left\{\left(\sup_{s,t \in I} |M_t - M_s|\right)^p\right\}$$

$$\leq \left(\frac{p}{p-1}\right)^p \sup_{s,t \in I} E[\{E(V_t \mid \mathcal{F}_s) - V_s\}^p].$$

Proof. The lower bound is deduced from Proposition 5.4. For the upper bound, let τ_1 and τ_2 be stopping times defined as the smallest random times such that $\sup_{s,t \in I} |M_t - M_s| = |M_{\tau_1} - M_{\tau_2}|$. For every $a > 0$

$$P\left(\sup_{s,t \in I} |M_t - M_s| > a\right) \leq a^{-1} E\left(|M_{\tau_1} - M_{\tau_2}| 1_{\{\sup_{s,t \in I} |M_t - M_s| > a\}}\right)$$

and the result follows by the same arguments as in the proof of Doob's Theorem 1.4. $\square$

The stochastic integral of a predictable process A with respect to a martingale M

$$Y_t = \int_0^t A_s \, dM_s$$

has the process of quadratic variations

$$[Y]_t = \int_0^t A_s^2 \, d[M]_s = \int_0^t A_s^2 \, d < M^c >_s + \sum_{0 < s \leq t} A_s^2 \Delta M_s^2$$

where the integral of A^2 with respect to the increasing process of the quadratic variations of M is the Stieltjes integral. If the increasing process $\int_0^t A_s^2 \, d < M >_s$ is locally finite, the process $Y^p - < Y >^{\frac{p}{2}}$ is locally integrable and

$$\|E Y_t^*\|_p \leq \frac{p C_p}{p-1} \left\{E(< Y >_t^{\frac{p}{2}})\right\}^{\frac{1}{p}}$$

$$\leq \frac{p C_p}{p-1} \|A\|_\infty \left\{E(V_t^{\frac{p}{2}})\right\}^{\frac{1}{p}}.$$

5.4 Inequalities for p-order variations

Let $(M_t)_{t \geq 0}$ be in $\mathcal{M}^p$, with a real number $p > 1$, there exists a process of p order variations

$$[M]_{t,p} = < M^c >_{t,p} + \sum_{0 < s \leq t} (\Delta M_s)^p$$

such that $E(M_t^p) = E([M]_{t,p})$, the process $< M^c >_{t,p}$ is predictable and $X_{t,p}^c = M_t^{cp} - < M^c >_{t,p}$ belongs to $\mathcal{M}$. The martingale property implies

$$E(M_t^{dp}) = \sum_{0 < s \leq t} E(\Delta M_s)^p$$

and there exists a predictable process $< M^d >_{t,p}$ such that

$$X_{t,p}^d = \sum_{0 < s \leq t} E(\Delta M_s)^p - < M^d >_{t,p}$$

is a martingale. The process $[M]_{t,p}$ has the predictable compensator $V_{t,p} = < M^c >_{t,p} + < M^d >_{t,p}$ and $M_t^p - V_{t,p}$ is a martingale $\mathcal{M}$. The inequalities of the discrete martingales extend to these processes

$$E(M_t^{cp} - M_s^{cp} \mid \mathcal{F}_s) = E(< M^c >_{t,p} \mid \mathcal{F}_s) - < M^c >_{s,p},$$

$$E\Big\{ \sum_{s < u \leq t} (\Delta M_u)^p \mid \mathcal{F}_s \Big\} = E(< M^d >_{t,p} \mid \mathcal{F}_s) - < M^d >_{s,p}.$$

For $p \geq 2$, by convexity $(M_t^p)_{t \geq 0}$ is a submartingale and

$$E(M_t^p) \geq \{E(M_t^2 \mid \mathcal{F}_{t-})\}^{\frac{p}{2}} = \{E(< M >_t)\}^{\frac{p}{2}}.$$

Proposition 5.11. *Let $(Z_t)_{t \geq 0}$ be a measurable and centered process of L^p, $p > 1$, then for all positive $s < t$*

$$E\Big[\Big\{ \int_0^t |E(Z_t \mid F_s)|^p \, ds \Big\}^{\frac{1}{p}}\Big] \leq E\Big\{ \Big(\int_0^t |Z_t|^p \, ds \Big)^{\frac{1}{p}} \Big\}.$$

Proof. Following the same arguments as in the proof of Proposition 4.23, and by Lemma 4.3 the process $z_{st} = E(Z_t \mid F_s)$ is such that

$$E\Big\{ \Big(\int_0^t |z_{st}|^p \, ds \Big)^{\frac{1}{p}} \Big\} = \sup_{\|H_{st}\|_q} E\Big\{ \int_0^t z_{st} H_{st} \, ds \Big\}$$

where $(H_{st})_{0 \leq s \leq t}$ is a process with unit norm on $[0, t]$

$$E\|H_{st}\|_q = E\Big\{ \Big(\int_0^t |H_{st}|^q \, ds \Big)^{\frac{1}{q}} \Big\} = 1.$$

Let $h_{st} = E(H_{st} \mid F_s)$ and let

$$Y_{st} = \int_0^t |h_{st}|^q \, ds, \qquad X_{st} = \int_0^t |Z_{st}|^p \, ds$$

with $Y_0 = X_0 = 0$. By definition of the conditional expectation, we obtain

$$E\Big\{ \Big(\int_0^t |z_{st}|^p \, ds \Big)^{\frac{1}{p}} \Big\} = \sup_{\|(H_{st})_{st}\|_q = 1} E\Big(\int_0^t Z_{st} h_{st} \, ds \Big),$$

and by Hölder's inequality

$$E\Big(\int_0^t Z_{st}h_{st}\,ds\Big) \le E\Big\{\Big(\int_0^t |Z_{st}|^p\,ds\Big)^{\frac{1}{p}}\Big\}E\Big\{\Big(\int_0^t |h_{st}|^q\,ds\Big)^{\frac{1}{q}}\Big\}$$

$$\le E\Big\{\Big(\int_0^t |Z_{st}|^p\,ds\Big)^{\frac{1}{p}}\Big\},$$

where the convexity of the power functions implies

$$E\Big\{\Big(\int_0^t |h_{st}|^q\,ds\Big)^{\frac{1}{q}}\Big\} = E\Big[\Big\{\int_0^t [E(H_{st}\mid F_s)|^q\,ds\Big\}^{\frac{1}{q}}\Big]$$

$$\le E\Big[\Big\{\int_0^t E(|H_{st}|^q\mid F_s)\Big\}^{\frac{1}{q}}\Big]$$

$$\le \Big[E\Big\{\int_0^t |H_{st}|^q)\,ds\Big\}\Big]^{\frac{1}{q}} = 1.$$

$\square$

Corollary 5.2. *Let $(X_t)_{t\ge 0}$ be a measurable and centered process of L^p, $p > 1$, then for all positive $s < t$*

$$E\Big[\Big\{\int_0^t |E(X_t\mid F_s) - X_s|^p\,ds\Big\}^{\frac{1}{p}}\Big] \le E\Big\{\Big(\int_0^t |X_t - X_s|^p\,ds\Big)^{\frac{1}{p}}\Big\}.$$

Proposition 4.23 applies directly to a measurable process $(X_t)_{t\ge 0}$ of L^p, with jumps $(\Delta X_t)_{t\ge 0}$ and, using a partition of $[0,t]$ with a path tending to zero, (4.9) becomes

$$E\Big[\Big\{\sum_{s\le t}|E(X_s\mid F_{s-}) - X_{s-}|^p\Big\}^{\frac{1}{p}}\Big] \le E\Big\{\Big(\sum_{s\le t}|\Delta X_s|^p\Big)^{\frac{1}{p}}\Big\}.$$

Proposition 4.24 generalizes using the same arguments.

Proposition 5.12. *Let $M = (M_t)_{t\ge 0}$ be a martingale of $\mathcal{M}^p$, for $p > 1$, then for every $t \ge 0$*

$$E|M_t| \le E(V_{t,p}^{\frac{1}{p}}),$$

$$\|M_t\|_p \le \{E(V_{t,p})\}^{\frac{1}{p}} \le q\sup_{t\ge 0}\|M_t\|_p.$$

Proposition 5.13. *Let $M = (M_t)_{t\ge 0}$ in $\mathcal{M}^p$, $p > 1$, then for all $\alpha \le p$ and $t \ge 0$*

$$\|M_t^*\|_p \le \frac{p}{p-1}\{E(V_{t,p})\}^{\frac{1}{p}}$$

and there exists a constant $C_{\alpha,p}$ such that

$$E(M_t^{*\alpha}) \le C_{\alpha,p}\{E(V_{t,p})\}^{\frac{\alpha}{p}}.$$

Proof. Theorem 5.2 implies

$$\|M_t^*\|_p \leq \frac{p}{p-1}\|M_t\|_p$$

where $E(M_t^p) = E(V_{t,p})$. Then for every $\alpha \leq p$, there exists a constant $C_{\alpha,p}$ such that

$$
\begin{aligned}
E(|M_t^*|^\alpha) &= E\{(|M_t^*|^p)^{\frac{\alpha}{p}}\} \\
&\leq \{E|M_t^*|^p\}^{\frac{\alpha}{p}} \\
&\leq C_{\alpha,p}\{E(V_{t,p})\}^{\frac{\alpha}{p}}.
\end{aligned}
$$

$\square$

5.5   Weak convergence of martingales and point processes

Let C_I be the space continuous real functions on a subinterval I of $\mathbb{R}$ with the uniform norm $\|x\| = \sup_{t \in I}|x(t)|$ for x in C_I, and let the distance $d(x,y) = \|x - y\|$, $x, y \in C_I$, then (C_I, d) is a complete separable metric space. On a probability space $(\Omega, \mathcal{F}, P)$, a sequence of processes $(X_n)_n$ with continuous sample paths is tight if

(1) for every t in I, the sequence $(X_n(t))_n$ is tight,
(2) for all $\eta > 0$ and $\varepsilon > 0$, there exists $\delta > 0$ such that

$$\limsup_{n \to \infty} P\{\sup_{|t-s|<\delta}\|X_n(t) - X_n(s)\| \geq \varepsilon\} < \eta.$$

According to Prohorov's theorem, the weak convergence of the process X_n is equivalent to its tightness and to the weak convergence of its finite dimensional distributions $(X_n(u_1), \ldots, X_n(u_j))$, for every integer j and for all $u_1, \ldots, u_j$ in I. The convergence of all moments μ_{jn} of the process X_n implies the weak convergence of its finite dimensional distributions. Billingsley (1968) proved that a sufficient condition of tightness for a process X_n of $C[0,1]$ is the existence of an exponents $\alpha > 0$ and $\beta > 1$, and an increasing continuous function h such that such that

$$E|X_n(t_1) - X_n(t_2)|^\alpha \leq |h(t_1) - h(t_2)|^\beta. \tag{5.5}$$

Let D_I be the space of right-continuous real functions with left limits on I, D_I with the uniform distance is not complete and it endowed with the Skorohod distance which allows a small shift of the functions at their jumps. Let x and y in $D_{[0,1]}$, the Skorohod distance is defined as the smallest $\varepsilon > 0$ such that there exists a strictly increasing continuous map g from $[0,1]$ onto $[0,1]$ such that $g(0) = 0$, $g(1) = 1$ and

$$\sup_{t \in [0,1]}|g(t) - t| \leq \varepsilon, \quad \sup_{t \in [0,1]}|x(t) - y(g(t))| \leq \varepsilon. \tag{5.6}$$

If a sequence $(x_n)_n$ in $D_{[0,1]}$ converges to x in $C_{[0,1]}$ for the Skorohod distance, then x_n converges uniformly to x.

Jacod and Mémin (1979b) proved the next equivalence for the weak convergence of a point process, it may be compared to Helly's theorem for the weak convergence of distribution functions.

Proposition 5.14. *A necessary and sufficient condition for the weak convergence of a point process is the weak convergence of their finite dimensional distributions in a dense set of* $\mathbb{R}_+$.

It follows that the weak convergence of a point process is equivalent to the convergence of its Fourier transform in a dense set of $\mathbb{R}_+$.

A Poisson process N with intensity $\lambda > 0$ has independent increments, the expectation and the variance of $N(t_1) - N(t_2)$ are $\lambda(t_1 - t_2)$, for all $t_1 > t_2$. The characteristic function of $N(t)$ is

$$\phi(x) = \sum_{k \geq 0} e^{ixk} P(N(t) = k) = \exp\{(e^{ix} - 1)\lambda t\}.$$

On increasing intervals $[0, t_n]$ such that $t_n = nt$, the sequence of Poisson processes defined as

$$N_n(t) = n^{-1} N(t_n)$$

has the characteristic function $\phi_n(x) = \exp\{(e^{in^{-1}x} - 1)\lambda t_n\}$. Expanding it as n tends to infinity yields

$$\phi_n(x) = \exp\left\{ ix\lambda t - \frac{x^2 \lambda t}{2n} + o(n^{-1}) \right\}$$

and it converges uniformly to $e^{ix\lambda t}$, as n tends to infinity. The process

$$\mathcal{U}_n(t) = \frac{N_n(t) - \lambda t}{(\lambda t)^{\frac{1}{2}}}$$

is a martingale and it has a convergent characteristic function, as n tends to infinity. By the independence of the increments of Poisson processes, the characteristic function of the variable $(\mathcal{U}_n(t_1) - \mathcal{U}_n(t_2), \mathcal{U}_n(t_2) - \mathcal{U}_n(t_3), \dots, \mathcal{U}_n(t_k) - \mathcal{U}_n(t_{k+1}))$, for all $t_1 > t_2 > \cdots > t_{k+1}$, is the product of the marginal characteristic functions so it converges, as n tends to infinity. Due to the martingale property, for every M in $\mathcal{M}^2(\mathcal{F}, P)$, the covariance of $M(s)$ and $M(t)$ is

$$E\{M(s)M(t)\} = E\{M^2(s \wedge t)\}$$

and the variance of $\mathcal{U}_n(t) - \mathcal{U}_n(s)$ is $\frac{2}{\lambda\sqrt{t}}(\sqrt{t} - \sqrt{s})$.

The modulus of continuity of a process X of $D(\mathbb{R}_+)$ is

$$w_D^N(X,\delta) = \inf_{\pi_N} \max_{i \leq k} \sup_{s,t \in [t_{i-1},t_i[} |X(t) - X(s)|,$$

where π_N is a partition $0 = t_0 < t_1 < \ldots < t_k = N$ of $[0, N]$ such that $t_i - t_{i-1} = \delta$, for an integer N. A sequence of processes $(X_n)_{n \geq 0}$ is tight in $D(\mathbb{R}_+)$ if for every integer N and every $t \leq N$, the sequence $(X_n(t))_n$ is tight and $w_D^N(X_n, \delta)$ converges to zero in probability as n tends to infinity and δ tends to zero. The second condition is fulfilled if for every integer N and for all $\eta, \varepsilon > 0$, there exists $\delta > 0$ such that for every n large enough

$$P\left\{ \sup_{|t_2-t_1| \leq \delta} \sup_{t_1 \leq t \leq t_2 \leq N} \min\{|X_n(t) - X_n(t_1)|, |X_n(t_2) - X_n(t)|\} \geq \varepsilon \right\} < \eta.$$

Proposition 5.15. *Let $(M_n)_{n \geq 0}$ belong to $\mathcal{M}^2(\mathcal{F}, P)$, the weak convergence of the finite dimensional distribution of M_n to the finite dimensional distribution of a Gaussian process and the convergence in probability of the predictable process V_n of the quadratic variations of M_n to a continuous function imply the weak convergence of M_n to a centered Gaussian process with independent increments and with variance function the limit of V_n.*

Proof. To prove the weak convergence of $(M_n)_{n \geq 0}$ to a process with continuous sample paths, it is sufficient to prove its C-tightness. For all $\delta > 0$ and $\varepsilon > 0$, we have

$$P\left\{ \sup_{|t-s| < \delta} |M_n(t) - M_n(s)| > \varepsilon \right\} \leq \frac{1}{\varepsilon^2} E\left[\sup_{|t-s| < \delta} \{M_n(t) - M_n(s)\}^2 \right]$$

and by Theorem 5.3

$$E\left[\sup_{|t-s| < \delta} \{M_n(t) - M_n(s)\}^2 \right] \leq \frac{p}{p-1} \sup_{|t-s| < \delta} E\{V_n(t) - V_n(s)\}$$

where M_n^2 has the predictable compensator V_n. If $(V_n)_{n \geq 0}$ is a sequence of continuous processes, there exists $\eta > 0$ such that for every n we have $\sup_{|t-s| < \delta} E(|V_n(t) - V_n(s)|) \leq \eta\varepsilon^2 p^{-1}(p-1)$ hence

$$P\{ \sup_{|t-s| < \delta} |M_n(t) - M_n(s)| > \varepsilon \} \leq \eta.$$

If the processes V_n are not continuous, their convergence in probability to a continuous function implies $\sup_{|t-s| < \delta} E(|V_n(t) - V_n(s)|)$ converges to zero as n tends to infinity, this proves the tightness of $(M_n)_{n \geq 0}$ $\qquad\square$

For every $\varepsilon > 0$, let M^ε be the sum of the jumps of M larger than ε so M is the sum

$$M = M^\varepsilon + M_\varepsilon$$

where the second term is local martingale with jumps smaller than ε. Rebolledo's weak convergence theorem for a sequence $(M_n)_{n \geq 1}$ in $\mathcal{M}^2_{0,loc}(\mathcal{F}, P)$ to a Gaussian martingale relies on the convergence in probability to $(< M_n, M_n >)_{n \geq 1}$ to a continuous function v and on a condition for $(M_n^\varepsilon)_{n \geq 1}$ that implies the C-tightness of the sequences $(M_n)_{n \geq 1}$ and $(M_{n,\varepsilon})_{n \geq 1}$, i.e. $\sup_{t \leq T} |M_n(t)- < M_{n,\varepsilon}, M_{n,\varepsilon} > (t)|$ converges to zero in probability for every T. The characterization of Gaussian processes determines the limiting distribution of $(M_n)_{n \geq 1}$ as the distribution of a centered Gaussian process with variance function v. It extends a local martingales $M_n = (M_{1n}, \ldots, M_{kn})$ of $\mathcal{M}^2_{0,loc}(\mathcal{F}, P)$.

Theorem 5.4. *Let $(M_n(t))_{t \geq 0}$ be a sequence of $\mathcal{M}^2_{0,loc}(\mathcal{F}, P)$ with values in $\mathbb{R}^k_+$ such that for every $\varepsilon > 0$ and for every $t > 0$, $< M^\varepsilon_{in}, M^\varepsilon_{in} > (t)$ converges to zero in probability as n tends to infinity, for $i = 1, \ldots, k$, the covariance function of M_n converges in probability to an increasing variance function $v = (v_1, \ldots, v_k)$ and*

$$\lim_{n \to \infty} < M_{in}, M_{jn} >= \delta_{ij} v_i(t),$$

then M_n converges weakly to a centered Gaussian process with independent components, independent increments, and with covariance function $v(s \wedge t)$.

The next proposition provides a condition on M_n^ε which ensures the tightness of $(M_n)_{n \geq 0}$ in $D(\mathbb{R}_+)$ and a simple proof of Rebolledo's convergence Theorem 5.4.

Proposition 5.16. *Let $(M_n)_{n \geq 0}$ belong to $\mathcal{M}^2_{0,loc}(\mathcal{F}, P)$ such that $< M_n >$ is discontinuous. If for every integer N, $\sup_{|t-s| < \delta, s, t \leq N} |M_n^\varepsilon(t) - M_n^\varepsilon(s)|$ converges to zero in probability as n tends to infinity, ε and δ converge to zero, then M_n is C-tight on $\mathbb{R}_+$.*

Proof. For every integer N and for all $\alpha > 0$, $\delta > 0$ and $\varepsilon > 0$, we have

$$P\Big(\sup_{|t-s| < \delta s, t \leq N} |M_n(t) - M_n(s)| > \alpha \Big)$$

$$\leq P\Big(\sup_{|t-s| < \delta s, t \leq N} |M_{n,\varepsilon}(t) - M_{n,\varepsilon}(s)| > \alpha \Big)$$

$$+ P\Big(\sup_{|t-s| < \delta s, t \leq N} |M_n^\varepsilon(t) - M_n^\varepsilon(s)| > \alpha \Big)$$

and by assumption, the last term converges to zero as n tends to infinity. Applying Theorem 5.3 implies

$$P\left(\sup_{|t-s|<\delta s, t\leq N} |M_{n,\varepsilon}(t) - M_{n,\varepsilon}(s)| > \alpha\right)$$

$$\leq \frac{1}{\alpha^2} E\left[\sup_{|t-s|<\delta, s, t\leq N} \{M_{n,\varepsilon}(t) - M_{n,\varepsilon}(s)\}^2\right]$$

$$\leq \frac{p^2}{\alpha^2(p-1)^2} \sup_{|t-s|<\delta, s, t\leq N} E\Big|[M_{n,\varepsilon}](t) - [M_{n,\varepsilon}](s)\Big|$$

$$\leq \frac{p^2}{\alpha^2(p-1)^2} \sup_{|t-s|<\delta, s, t\leq N} E\Big\{|< M_n^c >_t - < M_n^c >_s| + \sum_{u\in(s,t)} \Delta^2 M_{n,\varepsilon,u}^d\Big\}$$

with $[M_{n,\varepsilon}](t) = < M_n^c >_t + \sum_{s\leq t} \Delta^2 M_{n,\varepsilon,s}^d$. By continuity, for every $\eta > 0$ there exists $\delta > 0$ such that for every n

$$\sup_{|t-s|<\delta, s, t\leq N} E\Big|< M_{n,\varepsilon}^c > (t) - < M_{n,\varepsilon}^c > (s)\Big| \leq \eta.$$

For every $\eta > 0$ there exist $\varepsilon > 0$ and $\delta = \delta(\varepsilon, \eta) > 0$ such that for n large enough

$$\sup_{|t-s|<\delta, s, t\leq N} E\Big\{\sum_{u\in(s,t)} \Delta^2 M_{n,\varepsilon,u}^d\Big\} \leq \eta$$

this proves the $D(\mathbb{R}_+)$-tightness of $(M_n)_{n\geq 0}$. $\qquad\square$

Let N be a Poisson process with intensity λ on intervals $[0, T_n]$ and let $(a_n)_{n\geq 1}$ such that a_n tends to infinity and $a_n^{-1}T_n$ converges to a limit T, as n tends to infinity. The sequence of Poisson processes

$$N_n(t) = N(a_n t), \ t \in [0, T]$$

has the predictable compensator $A_n(t) = a_n \lambda t$ and the martingale $M_n = a_n^{-\frac{1}{2}}(N_n - A_n)$ has the variance function

$$E\{M_n^2(t)\} = a_n^{-1} E\{M_n^2(t)\} = \lambda t$$

on $[0, T]$. The sum of the jumps of M_n is $a_n^{-\frac{1}{2}} N_n$ and

$$M_n^\varepsilon(t) = \sum_{s\leq t} \Delta N_n(s) 1_{\{\Delta N_n(s)>\varepsilon\}}$$

where $\Delta N_n(s) > \varepsilon$ is equivalent to $a_n^{-\frac{1}{2}}\lambda s > \varepsilon$ and $a_n < \lambda^2 s^2 \varepsilon^{-2}$, hence M_n^ε converges a.s. to zero as n tends to infinity, for every t in $[0, T]$, then Theorem 5.4 applies to M_n.

On probability space $(\Omega, \mathcal{F}, P, \mathbb{F})_{n \geq 0}$ let $(X_n)_n$ be a semi-martingale sequence $X_n(t) = X_n(0) + A_n(t) + M_n(t)$ defined by sequences of a right-continuous process $(A_n)_n$ with integrable variations and a martingale $(M_n)_n$ of L^2. The tightness of $(X_n)_{n \geq 0}$ follows from the tightness of $(M_n)_{n \geq 0}$ and $(A_n)_{n \geq 0}$. For a sequence of point processes $(N_n)_{n \geq 0}$ with increasing predictable compensators A_n such that $M_n = N_n - A_n$ belongs to $\mathcal{M}^2_{0,loc}(\mathcal{F}_n, P)$. The tightness of $(N_n)_{n \geq 0}$ follows from the tightness of $(M_n)_{n \geq 0}$ and from the weak convergence of $(A_n)_{n \geq 0}$. If A_n is continuous then $< M_n, M_n >= A_n$ and Proposition 5.15 applies. If A_n is not continuous then Proposition 5.16 applies with

$$[M_n, M_n](t) = A_n^c(t) + \sum_{s \leq t} \{\Delta M_n(s)\}^2.$$

More generally, if the processes A_n, and respectively $[M_n, M_n]$, converge weakly to processes A, and respectively $[M, M]$, then the process $(N_n)_{n \geq 0}$ converges weakly to a point process with predictable compensator A. Jacod and Mémin (1979a) proved the convergence of semi-martingales under conditions for the convergence of the exponential of A_n.

Rebolledo (1979) proved that a semi-martingale sequence $(X_n)_{n \geq 0}$ is tight in $D(\mathbb{R}_+)$ if for every integer N

(1) for every $t \leq N$, the sequence $(X_n(t))_n$ is tight,
(2) for all $\eta > 0$ and $\varepsilon > 0$, there exists $\delta > 0$ such that for every sequence of $\mathcal{F}_n$-stopping times $T_n \leq N$

$$\limsup_{n \to \infty} P\{ \sup_{T_n \leq s \leq N \wedge (T_n + \delta)} |X_n(s) - X_n(T_n)| > \varepsilon \} < \eta.$$

Let $X_n = M_n + A_n$ be a sequence of semi-martingales such that X_n has a process of quadratic variations V_n and $A_n = A_n^+ - A_n^-$ is the difference of increasing processes of L^2.

Proposition 5.17. *Under the condition (1) and if for every integer N and for all $\eta_1, \eta_2, \eta_3 > 0$, there exists $\delta > 0$ such that for every sequence of $\mathcal{F}_n$-stopping times $T_n \leq N$*

$$\limsup_{n \to \infty} E\{V_n(T_n + \delta) - V_n(T_n)\} \leq \eta_1,$$

$$\limsup_{n \to \infty} E\{A_n^+(T_n + \delta) - A_n^+(T_n)\} \leq \eta_2,$$

$$\limsup_{n \to \infty} E\{A_n^-(T_n + \delta) - A_n^-(T_n)\} \leq \eta_3$$

then $(X_n)_{n \geq 0}$ is tight in $D(\mathbb{R}_+)$.

Proof. For all $\eta > 0$ and $\varepsilon > 0$ and for n large, there exist $\eta_1, \eta_2, \eta_3 > 0$ such that $\eta = 4\varepsilon^{-2}\{\eta_1 + 2(\eta_2 + \eta_3)\}$ and by the Cauchy-Schwarz inequality and Theorem 5.2

$$P\Big\{\sup_{T_n \leq s \leq N \wedge (T_n + \delta)} |X_n(s) - X_n(T_n)| > \varepsilon \Big\}$$

$$\leq \varepsilon^{-2} E\Big\{\sup_{T_n \leq s \leq N \wedge (T_n + \delta)} |X_n(s) - X_n(T_n)|^2 \Big\}$$

$$\leq 4\varepsilon^{-2}[E\{V_n(T_n + \delta) - V_n(T_n)\} + E\{A_n^+(T_n + \delta) - A_n^+(T_n)\}$$

$$+ 2E\{A_n^-(T_n + \delta) - A_n^-(T_n)\}] \leq \eta$$

and Rebolledo's condition (2) is fulfilled. □

5.6   Poisson and renewal processes

On a probability space $(\Omega, \mathcal{F}, P)$, the Poisson process $(N_t, t \geq 0)$ is a right-continuous process with left-hand limits defined as $N(t) = \sum_{i \geq 1} 1\{T_i \leq t\}$, $t > 0$, for a sequence of random variables $0 = T_0 < T_1 < \cdots < T_i < \cdots$. Its natural filtration $(\mathcal{F}_t)_{t \geq 0}$ is defined by the σ-algebras $\mathcal{F}_t = \sigma(T_i \leq t, i \geq 1)$. A homogeneous Poisson process has a constant intensity $\lambda > 0$, its probability distribution satisfies

$$P(N(t+h) - N(t) = 1) = h\lambda + o(h) = p(h),$$
$$P(N(t+h) - N(t) > 1) = o(h),$$

for every $t > 0$, when h tends to zero. These properties imply that for all $k \geq 1$ and $t > 0$

$$P_k(t) := P(N_t = k) = e^{-\lambda t}(\lambda t)^k (k!)^{-1}$$

and $P(N_t = 0) = e^{-\lambda t}$. The probabilities P_k are solutions of the differential equation

$$P_k'(t) = -\lambda P_k(t) + \lambda P_{k-1}(t)$$

with $P_k(0) = 0$, $k \leq 1$.

The Laplace transform of a Poisson process N_t with intensity λ is

$$L_{N_t}(x) = E(e^{-xN_t}) = \sum_{k \geq 0} P(N_t = k)e^{-xk}$$

$$= \exp\{-\lambda t(1 - e^{-x})\},$$

it is solution of the differential equation

$$\frac{L_{N_t}''(x)}{L_{N_t}'(x)} + \lambda t e^{-x}\frac{L_{N_t}'(x)}{L_{N_t}(x)} = \lambda t e^{-x}$$

with initial values $L_{N_t}(0) = 1$ and $L'_{N_t}(0) = -\lambda t$.

Let $T_n = \inf\{t : N_t \geq n\}$, for an integer $n > 0$, T_n is a stopping time of N and its value at T_n is n therefore

$$L_n(x) = E \exp\{-\lambda T_n(1 - e^{-x})\}$$

and by the change of variable $x = \log \frac{\lambda}{\lambda - y}$ for $y < \lambda$, the Laplace transform of T_n at y is

$$L_{T_n}(y) = E(e^{-T_n y}) = L_{N_{T_n}}\left(\log \frac{\lambda}{\lambda - y}\right)$$
$$= \left(\frac{\lambda - y}{\lambda}\right)^{N_{T_n}} = \left(\frac{\lambda - y}{\lambda}\right)^n.$$

The variable T_1 is therefore an exponential variable with parameter λ and the distribution of the variable T_n is the convolution of n independent exponential distributions $\mathcal{E}_\lambda$, with parameter λ so T_n is the sum of n independent exponentially distributed variables $X_i = T_i - T_{i-1}$, $i = 1, \ldots, n$, with $X_0 = 0$. The exponential variables are characterized by their lack of memory, for strictly positive s and t

$$P(T > t + s \mid T > s) = P(T > t).$$

The Poisson process is also represented in terms of the ordered statistics of the sample of a variable having the uniform distribution. For a Poisson process with parameter 1, the variable $\xi_{n:i} = T_i T_{n+1}^{-1}$ is the i-th order statistics of a vector $(\xi_i)_{i=1,\ldots,n}$ of independent and uniform variables on $[0, 1]$ and $(T_1 T_{n+1}^{-1}, \ldots T_n T_{n+1}^{-1})$ is independent of T_{n+1} (Breiman, 1968). The same result holds for a Poisson process with parameter λ.

For n independent Poisson processes $N_1, \ldots, N_n$ with parameters $\lambda_1, \ldots, \lambda_n$ respectively, and for every $t = (t_1, \ldots, t_n)$ in $\mathbb{R}_+^n$, the variable $S_{n,t} = N_{1t_1} + \cdots + N_{nt_n}$ has the distribution of a Poisson variable with parameters $\mu_{n,t} = \lambda_1 t_1 + \cdots + \lambda_n t_n$.

A renewal process has independent increments and identically distributed inter-arrival times, it has the properties of a Poisson process with an increasing functional intensity $\Lambda(t) = EN(t) > 0$. Its distribution function is defined by

$$P(N_t = k) = e^{-\Lambda(t)} \Lambda^k(t)(k!)^{-1}, t > 0$$

for every integer $k \geq 1$, and $P(N_t = 0) = e^{-\Lambda_t}$. It is a process with independent increments and the joint density at $(x_1, \ldots, x_n)$ of the independent waiting times $X_k = T_k - T_{k-1}$, $k = 1, \ldots, n$, is the product of the exponential densities $\prod_{k=1,\ldots,n} e^{-\{\Lambda_{x_k} - \Lambda_{x_{k-1}}\}} = e^{-\Lambda_{x_n}}$, with $x_0 = 0$. The transformed times $\Lambda(T_k)$ of the process are the time variables of a Poisson process with intensity one which satisfy Breiman's properties. If the function Λ has a derivative λ

$$P(N(t+h) - N(t) = 0) = 1 - e^{-\Lambda(t+h) - \Lambda(t)} = h\lambda(t) + o(h),$$

for every $t > 0$, when h tends to zero. This entails

$$P_k'(t) = -\lambda(t)P_k(t) + \lambda(t)P_{k-1}(t), \ k \leq 1.$$

Its generating function is

$$G_{N_t}(u) = e^{-\Lambda_t} \sum_{k>0} \frac{u^k \Lambda_t^k}{k!} = \exp\{\Lambda_t(u-1)\}$$

and its Laplace transform

$$L_{N_t}(u) = E e^{-uN_t} = \exp\{\Lambda_t(e^{-u} - 1)\},$$

is an decreasing function of u and its maximum for $u > 0$ is reached as u tends to zero.

Theorem 5.5. *Let Λ be an increasing function on $\mathbb{R}_+$ and let $(T_k)_{k\geq 0}$ be the jump times of a renewal process such that $P(T_k - T_{k-1} \geq x) = e^{-\Lambda(x)}$, then the variables $S_k = \Lambda(T_k)$ are the jump times of a Poisson process with intensity 1 if and only if Λ is stationary.*

Proof. Let $T_k = \Lambda^{-1}(S_k)$ such that $P(T_k - T_{k-1} \geq x) = e^{-\Lambda(x)}$, then

$$P(\Lambda^{-1}(S_k) - \Lambda^{-1}(S_{k-1}) \geq x) = e^{-\Lambda(x)},$$

if Λ is stationary it follows that for $y = \Lambda(x)$

$$P(\Lambda^{-1}(S_k - S_{k-1}) \geq x) = P(S_k - S_{k-1} \geq y) = e^{-y}$$

and $(S_k)_{k\geq 0}$ is a Poisson process with intensity 1.

Reversely if $(S_k)_{k\geq 0}$ is a Poisson process with intensity 1, its distribution is determined by the probabilities $P(S_k - S_{k-1} \geq y) = e^{-y}$ then

$$P(T_k - T_{k-1} \geq x) = P(\Lambda^{-1}(S_k) - \Lambda^{-1}(S_{k-1}) \geq x) = e^{-\Lambda(x)}$$
$$= P(\Lambda^{-1}(S_k - S_{k-1}) \geq x)$$

and Λ is stationary. $\qquad\square$

Theorem 5.6. *Let Λ be an increasing left-continuous process with right limits and let $(T_k)_{k\geq 0}$ be the jump times of a renewal process such that $P(T_k - T_{k-1} \geq x \mid \Lambda) = e^{-\Lambda(x)}$. Then the variables $S_k = \Lambda(T_k)$ are the jump times of a Poisson process with intensity 1 if and only if Λ is a stationary process.*

Proof. The proof is similar to the proof of Theorem 5.5. If Λ is a stationary process, for all integers k and n and for every $t > 0$ we have

$$\Lambda(kt) = \Lambda(t) + \Lambda((k-1)t) = k\Lambda(t),$$

$$n\Lambda\left(\frac{kt}{n}\right) = nk\Lambda\left(\frac{t}{n}\right) = k\Lambda(t),$$

by limit it follows that for all $s < t$ in $\mathbb{R}_+$

$$\Lambda(t-s) = (t-s)\Lambda(1),$$

$\Lambda(0) = 0$ a.s. and $\lim_{x\to\infty} \Lambda(x)$ is a.s. infinite. Let $\theta = \Lambda(1)$, the inverse process is $\Lambda^{-1}(t) = \theta^{-1}t$. Conditional on Λ, the variables $X_k = T_k - T_{k-1}$, $k \geq 1$, have the exponential distribution with parameter θ. $\qquad\square$

The Laplace transform of the variables $X_k = T_k - T_{k-1}$ is deduced from Theorem 5.6

$$\begin{aligned}
L_{X_k}(u_k) &= E[e^{-u_k\{\Lambda^{-1}(S_k)-\Lambda^{-1}(S_{k-1})\}}] \\
&= E[e^{-u_k(S_k-S_{k-1})\theta^{-1}}] = E[\{L_{\theta^{-1}}(u_k)\}^{S_k-S_{k-1}}] \\
&= \int_{\mathbb{R}_+} \{L_{\theta^{-1}}(u_k)\}^s e^{-s}\, ds \\
&= \int_{\mathbb{R}_+} e^{-s\{1-\log L_{\theta^{-1}}(u_k)\}}\, ds \\
&= \frac{1}{1-\log L_{\theta^{-1}}(u_k)}.
\end{aligned}$$

The Laplace transform of the variable T_0 is also L_{X_k} and the vector $(T_k)_{k=1,\ldots,n}$ has the Laplace transform

$$\begin{aligned}
L_T(u) &= E(e^{-u_0 T_0}) \prod_{k=1}^{n} E(e^{-u_k(T_k-T_{k-1})}) \\
&= \prod_{k=0}^{n} \frac{1}{1-\log L_{\theta^{-1}}(u_k)}.
\end{aligned}$$

Theorem 5.7. *Let $(N_t)_{t\geq 0}$ be a Poisson process with intensity λ and with jump $\Delta N_t = N_t - N_t^-$ at t, then for all $u > 0$ and $s < t$*

$$E(e^{-uN_t} - e^{-uN_s} \mid \mathcal{F}_s) = (e^{-u} - 1)\lambda \int_s^t E(e^{-uN_x^-} \mid \mathcal{F}_s)\, dx.$$

Proof. Let $\Delta N_t = N_t - N_t^-$ denote the jump of N at t, for $u > 0$ and $s < t$ we have

$$e^{-uN_t} - e^{-uN_s} = \sum_{s < x \le t} (e^{-uN_x} - e^{-uN_x^-})$$

$$= \sum_{s < x \le t} e^{-uN_x^-}(e^{-u} - 1)\Delta N_x$$

$$= (e^{-u} - 1)\int_s^t e^{-uN_x^-} \, dN_x,$$

and for $s \le x \le t$, $E(N_x \mid \mathcal{F}_s) = N_s + \lambda(x - s)$. $\qquad\square$

Proposition 5.18. *The process defined on $[0,1]$ as*

$$\mathcal{L}_T(t) = \frac{N_{Tt} - \lambda Tt}{\sqrt{\lambda Tt}}, \tag{5.7}$$

converges weakly to a centered Gaussian process with variance 1 and covariance $(st)^{-\frac{1}{2}}(s \wedge t)$, as T tends to infinity.

Proof. The characteristic function of $\mathcal{L}_1(t)$ is

$$\widehat{f}_t(u) = e^{-iu(\lambda t)^{\frac{1}{2}}} E e^{iuN_t(\lambda t)^{-\frac{1}{2}}}$$

$$= e^{-iu(\lambda t)^{\frac{1}{2}}} e^{-\lambda t} \sum_{n \ge 0} \frac{\left\{ e^{iu(\lambda t)^{-\frac{1}{2}}} \lambda t \right\}^n}{n!}$$

$$= e^{-iu(\lambda t)^{\frac{1}{2}}} e^{\lambda t \left\{ e^{iu(\lambda t)^{-\frac{1}{2}}} - 1 \right\}}.$$

Replacing t by Tt and expanding $\lambda Tt \left\{ e^{iu(\lambda Tt)^{-\frac{1}{2}}} - 1 \right\}$ as T tends to infinity yields

$$\lambda tT \left\{ e^{iu(\lambda Tt)^{-\frac{1}{2}}} - 1 \right\} = iu(\lambda Tt)^{\frac{1}{2}} - \frac{u^2}{2} + o(1)$$

and the characteristic function of $\mathcal{L}_T(t)$ converges to the characteristic function of the normal variable, for every t. By the independence of the increments of the process $\mathcal{L}_T$, its finite dimensional distributions converge to those of a Gaussian process. The tightness of the process $\mathcal{L}_T$ follows from Theorem 5.4. $\qquad\square$

On a probability space $(\Omega, \mathcal{F}, (\mathcal{F}_t)_{t \ge 0}, P)$, let N be a point process with independent increments and a continuous intensity function Λ, the difference $M_t = N_t - \Lambda_t$ belongs to $\mathcal{M}_{0,loc}^2$ and the function of quadratic variations of M_t is Λ_t.

Theorem 5.8. *For the point process N with independent increment and intensity function $\Lambda(t)$ and for every $x > 0$*

$$t^{-1} \log P(t^{-1} N_t > a) = \inf_{u>0} \{t^{-1} \log \varphi_{N_t}(u) - au\}$$
$$= a\{1 - \log a + \log(t^{-1}\Lambda_t)\} - t^{-1}\Lambda_t,$$

its limit as t tends to infinity is finite if $\lim_{t \to \infty} t^{-1}\Lambda_t$ is finite.

Proof. The minimum of $h_a(u) = t^{-1} \log \varphi_{N_t}(u) - au = t^{-1}\Lambda_t(e^u - 1) - au$ is reached at $u_a = \log(t\Lambda_t^{-1}a)$ where $h_a'(u_a) = t^{-1}\Lambda_t e^{u_a} - a = 0$ and

$$h_a(u_a) = a - t^{-1}\Lambda_t + a\log\{(at)^{-1}\Lambda_t\}.$$

$\square$

Proposition 5.19. *If there exist functions h and μ such that $h(T)$ tends to infinity and $h^{-1}(Tt)\Lambda(Tt)$ converges to $\mu(t)$ for every t, as T tends to infinity, then the process*

$$\mathcal{L}_T(t) = \frac{N_{Tt} - \Lambda_{Tt}}{\sqrt{\Lambda_{Tt}}}, t \in [0,1], \tag{5.8}$$

converges weakly to a centered Gaussian process $\mathcal{L}_0$ with variance 1 and covariance $\{\mu(s)\mu(t)\}^{-\frac{1}{2}}\mu(s \wedge t)$, as T tends to infinity.

Proof. The proof is the same as for Proposition 5.18 where a second order expansion of $\Lambda_{Tt}(e^{iu\Lambda_{Tt}^{-\frac{1}{2}}} - 1)$ as T and Λ_{Tt} tend to infinity yields

$$\Lambda_{Tt}\left\{e^{iu\Lambda_{Tt}^{-\frac{1}{2}}} - 1\right\} = iu\Lambda_{Tt}^{\frac{1}{2}} - \frac{u^2}{2} + o(1)$$

and the characteristic function of $\mathcal{L}_T(t)$ converges to the characteristic function of the normal variable, for every t. $\square$

The change of variable by the inverse of $\Lambda_T(t) = T^{-1}\Lambda_{Tt}$ provides an asymptotically free process $W_T = \mathcal{L}_T \circ \Lambda_T^{-1}$.

By Proposition 5.2, for every random stopping time T of N and for every $\alpha \geq 2$, there exist constants $0 < c_\alpha < C_\alpha$ such that the moments of M_t satisfy

$$c_\alpha E(\Lambda_T^{\frac{\alpha}{2}}) \leq E(|M_T|^\alpha) \leq C_\alpha E(\Lambda_T^{\frac{\alpha}{2}}).$$

The exponential submartingale related to the Poisson process is defined with its moment generating function φ_{N_t} as

$$Y_t(u) = \exp\{uN_t - \log \varphi_{N_t}(u)\} = \exp\{uN_t + \Lambda_t(1 - e^u)\}.$$

Let $(N_t, t \geq 0)$ be a Poisson process with a functional cumulative intensity $\Lambda(t) > 0$ and let T be a stopping time of N. For every $x > 0$ and for every bounded interval I

$$\sup_{T \in I} P(\sup_{0 \leq t \leq T} |N_t - \Lambda(t)| > x) \leq 2x^{-2} E\{\sup_{T \in I} \Lambda(T)\}.$$

By the martingale property, $E\{(M_t - M_s)^2 | \mathcal{F}_s\} = \Lambda_t - \Lambda_s$. For stopping times $0 < T_1 < T_2$ of the process N and for all $x_1 > 0$ and $x_2 > 0$, the independence of the increments of the Poisson process implies that for all $x_1 > 0$ and $x_2 > 0$

$$P(\sup_{0 \leq s \leq T_1} |N_s - \Lambda_s| > x_1, \sup_{T_1 \leq t \leq T_2} |N_t - \Lambda_t| > x_2)$$

$$\leq \frac{4}{x_1^2 x_2^2} E\{\Lambda_{T_2} - \Lambda_{T_1}\} E\{\Lambda_{T_1}\}.$$

For the process $\mathcal{L}_T$ defined by (5.7)

$$P(\sup_{t \in [0,1]} |\mathcal{L}_T(t)| > x) \leq 2x^{-2}.$$

Let $(N_{ij})_{j=1,\ldots,J_{in}; i=1,\ldots,I_n}$ be an array of point processes on $\mathbb{R}$ such that $I_n = O(n^{1-\alpha})$ and $J_{in} = O(n^\alpha)$ for all $i = 1, \ldots, I_n$, with a total number $n = \sum_{i=1,\ldots,I_n} J_{in}$. The predictable compensator of the processes N_{ij} are supposed to have the same form $\widetilde{N}_{ij}(t) = \int_0^t h(Y_{ij}(s)) \, d\Lambda(s)$, where $(Y_{ij})_{j=1,\ldots,J_{in}, i=1,\ldots,I_n}$ is an array of predictable processes and h is a left-continuous positive function with right-hand limits. Moreover $E\{N_{ij}(t)N_{ik}(t)\} = E < N_{ij}(t), N_{ik}(t) >$ develops as

$$\frac{1}{2}(< N_{ij} + N_{ik} >_t - < N_{ij} >_t - < N_{ik} >_t)$$

$$= \frac{1}{2} \int_0^t \{h_i(Y_{ij}(s) + Y_{ik}(s)) - h_i(Y_{ij}(s)) - h_i(Y_{ik}(s))\} \, d\Lambda(s),$$

denoted $\int_0^t k_i(Y_{ij}, Y_{ik}) \, d\Lambda$, and $E\{N_{ij}(t)N_{i'k}(t)\} = \int_0^t k_{ii'}(Y_{ij}, Y_{i'k}) \, d\Lambda$. The processes

$$X_{ij}(t) = \int_0^t \{h(Y_{ij})\}^{-1} 1_{\{h(Y_{ij}) > 0\}} \, dN_{ij},$$

$$\widetilde{X}_{ij}(t) = \int_0^t \{h(Y_{ij})\}^{-1} 1_{\{h(Y_{ij}) > 0\}} \, d\Lambda,$$

define an array of centered local martingales of L^2, $M_{ij} = X_{ij} - \widetilde{X}_{ij}$ with variances $EM_{ij}^2(t) = E \int_0^t \{h_i(Y_{ij})\}^{-1} 1_{\{h_i(Y_{ij}) > 0\}} \, d\Lambda$ and covariances

$$EM_{ij}(t)M_{ik}(s) = E \int_0^{s \wedge t} \frac{k_i(Y_{ij}, Y_{ik})}{h_i^2(Y_{ij}) h_i^2(Y_{ik})} 1_{\{h_i(Y_{ij}) h_i(Y_{ik}) > 0\}} \, d\Lambda,$$

and

$$EM_{ij}(t)M_{i'k}(s) = E \int_0^{s \wedge t} \frac{k_{ii'}(Y_{ij}, Y_{i'k})}{h_i^2(Y_{ij})h_{i'}^2(Y_{i'k})} 1_{\{h_i(Y_{ij})h_{i'}(Y_{i'k})>0\}} \, d\Lambda.$$

The martingale $S_n = (S_{in})_{i=1,\dots,I_n}$ with components $S_{in} = \sum_{j=1,\dots,J_{in}} M_{ij}$ has a variance matrix Σ_n with components $\Sigma_{ii'} = \sum_{j,j'=1}^{J_{in}} E(M_{ij}M_{i'j'})$, for i and i' in $\{1,\dots,I_n\}$. For every $t = (t_1,\dots,t_{I_n})$

$$P\left(\sup_{t \in [0,T]} S_n(t) > x \right) \leq \frac{E\|\Sigma_n(T)\|_2}{\|x\|_2^2}.$$

For a Poisson process N with a continuous cumulative intensity function Λ on $\mathbb{R}_+$, the change of time $t \mapsto \Lambda^{-1}(t)$ maps the process N_t into a Poisson process $N \circ \Lambda^{-1}(t)$ with parameter 1. Let

$$\Lambda_T(t) = T^{-1}\Lambda(Tt), \ N_T(t) = T^{-1}N(Tt), \ t \in [0,1],$$

the change of time $y = \Lambda_T^{-1}(t)$ i.e. $\Lambda_T(y) = t$ implies that $N_T \circ \Lambda^{-1}(t)$ satisfies the ergodic property

$$\sup_{t \in [0,1]} |E\{N_T \circ \Lambda^{-1}(Tt)\} - t| = 0,$$

for every continuous and bounded function φ on $\mathbb{R}_+$,

$$ET^{-1} \int_0^T \varphi(s) \, dN_s = \int_0^1 \varphi(Tx) \, d\Lambda_T(x)$$

and

$$ET^{-1} \int_0^{\Lambda_T^{-1}(1)} \varphi(s) \, dN_s = \int_0^1 \varphi \circ \Lambda^{-1}(Tx) \, dx.$$

Let $(Z_k)_{k \geq 1}$ be a sequence of independent and identically distributed real random variables and let $N = (T_k)_{k \geq 1}$ be a Poisson process with intensity λ, independent of $(Z_k)_{k \geq 1}$. The renewal process

$$Y_t = \sum_{k \geq 1} Z_k 1_{\{T_k \leq t\}} \tag{5.9}$$

is also written as

$$Y_t = \sum_{k=1}^{N_t} Z_k.$$

By the independence of the variables Z_k and the process N with independent increments, the process Y has also independent increments. Its expectation and its variance are $E(Y_t) = \lambda t \, E(Z_1)$ and $\text{Var}(Y_t) = \lambda t \, \text{Var}(Z_1)$.

Let $U_n = \sum_{k=1}^n Z_k$ and let φ_Z be the Fourier transform of the variables Z_k, then Y_t has the Fourier transform

$$\varphi_{Y_t}(u) = E\{\exp(iuY_t)\} = e^{-\lambda t} \sum_{k \geq 1} E\{\exp(iuU_k)\} P(N_t = k)$$

$$= e^{-\lambda t} \sum_{k \geq 1} E\{\exp(iuU_k)\} \frac{(\lambda t)^k}{k!}$$

$$= e^{-\lambda t} \sum_{k \geq 1} \frac{\{\lambda t \varphi_Z(u)\}^k}{k!}$$

$$= \exp[-\lambda t \{1 - \varphi_Z(u)\}] = L_{N_t} \circ \varphi_Z(u),$$

it is the Laplace transform of the Poisson process N_t at $\varphi_Z(u)$.

Let $\tau_a = \inf\{t : Y_t \geq a\}$, for $a > 0$, τ_a is a stopping time of the process Y and, with real variables U_i, the value of Y at τ_a is a therefore

$$\varphi_a(u) = E \exp[-\lambda \tau_a \{1 - \varphi_Z(u)\}].$$

By the change of variable $y = \lambda\{1 - \varphi_Z(u)\}$ i.e. $\varphi_Z(u) = 1 - \lambda^{-1} y$, the Laplace transform of τ_a at y is

$$L_{\tau_a}(y) = E(e^{-y\tau_a}) = \varphi_{Y_{\tau_a}} \circ \varphi_Z^{-1}\left(\frac{\lambda - y}{\lambda}\right)$$

$$= E \exp\left\{-N_{\tau_a}\left(\frac{\lambda}{\lambda - y}\right)\right\}$$

$$= E \exp\{-\lambda \tau_a (1 - e^{-\frac{\lambda}{\lambda - y}})\},$$

the variable τ_a is also a stopping time of the Poisson process N but N_{τ_a} differs from a.

Let a and b be strictly positive, the independence of the increments of the process Y implies that the times T_a and $T_{a+b} - T_a$ are independent and the density function of T_{a+b} is the convolution of the densities of T_a and T_b.

Proposition 5.20. *The process defined on $[0, 1]$ as*

$$\mathcal{Y}_T(t) = \frac{Y_{Tt} - \lambda T t E(Z)}{\sqrt{\lambda T t \operatorname{Var}(Z)}}, \tag{5.10}$$

converges weakly to a centered Gaussian process with independent increments, as T tends to infinity.

Proof. Let μ_Z be the expectation of the variables Z_k and let σ_Z^2 be their variance, the expectation of Y_t is

$$\mu = \sum_{n \geq 1} n \mu_Z P(N_t = n) = \lambda t \mu_Z$$

and the variance of Y_t is $\sigma^2 = \lambda t \sigma_Z^2$. The characteristic function of $\mathcal{Y}_1(t)$ is

$$\widehat{f}_t(u) = e^{-iu(\lambda t)^{\frac{1}{2}}\mu_Z \sigma_Z^{-1}} E\{e^{iuY_t(\lambda t)^{-\frac{1}{2}}\sigma_Z^{-1}}\}$$

where

$$E\{e^{iuY_t(\lambda t)^{-\frac{1}{2}}\sigma_Z^{-1}}\} = e^{-\lambda t} \sum_{n\geq 0} E\{e^{iuU_n(\lambda t)^{-\frac{1}{2}}\sigma_Z^{-1}}\} \frac{(\lambda t)^n}{n!}$$

$$= e^{-\lambda t} \sum_{n\geq 0} \frac{\{\lambda t \varphi_Z(u(\lambda t)^{-\frac{1}{2}}\sigma_Z^{-1})\}^n}{n!}$$

$$= \exp\left[\lambda t \left\{\varphi_Z\left(\frac{u}{\sqrt{\lambda t}\sigma_Z}\right) - 1\right\}\right].$$

Replacing t by Tt and expanding the characteristic function of Z and the exponential as T tends to infinity yields

$$\lambda T t \left\{\varphi_Z\left(\frac{u}{\sqrt{\lambda t T}\sigma_Z}\right) - 1\right\} = i\sqrt{\lambda t T}\frac{uE(Z)}{\sigma_Z} - \frac{u^2 E(Z^2)}{2\sigma_Z^2} + o(T^{-2})$$

therefore

$$\widehat{f}_{Tt}(u) = \exp\left\{-\frac{u^2 E(Z^2)}{2\sigma_Z^2}\right\}\{1 + +o(T^{-2})\}$$

and the characteristic function of $\mathcal{Y}_T(t)$ converges, as T tends to infinity, to the characteristic function of a centered Gaussian variable with a constant variance, for every t. By the independence of the increments of the process T, its finite dimensional distributions converge to those of a Gaussian process with independent increments. The tightness of the process $\mathcal{Y}_T$ follows from Theorem 5.4. $\qquad\square$

For a renewal process (5.9) and for $a > 0$, let

$$X_a(t) = \sum_{k=1}^{N_t} 1_{\{Z_k > a\}}, \quad t > 0. \tag{5.11}$$

Let F be the distribution function of the variables Z_k and let $\bar{F} = 1 - F$, the expectation of $X_a(t)$ is $E\{X_a(t)\} = \lambda t \bar{F}(a)$ and its variance is $\mathrm{Var}\{X_a(t)\} = \lambda t\, F(a)\bar{F}(a)$.

Proposition 5.21. *Let $(a_t)_{t\geq 0}$ be a positive real function such that $tN_t\{1 - F(a_t)\}$ converges in probability to $\mu > 0$ as t tends to infinity, then the process $(X_{a_t}(t)_{t\geq 0}$ is asymptotically equivalent to a Poisson process with intensity μ as t tends to infinity.*

Proof. Conditionally on N_t, the variable $X_a(t)$ has a random Binomial distribution with parameters $N(t)$ and $\bar{F}(a_t)$

$$P_k(t) = P(X_a(t) = k) = \binom{N(t)}{k} \bar{F}^k(a_t) F^{N(t)-k}(a_t).$$

Under the condition of convergence for $\bar{F}(a_t)$ and by the convergence of a Binomial distribution $B(n, p_n)$ such that $\lim_{n \to \infty} np_n = \mu$, for every positive sequence t_n tending to infinity, the probability $P_k(t_n)$ is asymptotically equivalent to the probability of a Poisson process with intensity μ. This convergence extends to $P_k(t)$, as t tends to infinity. □

The moment generating function of the variables $1_{\{Z_k > a\}}$ is $\phi_Z(u) = e^u + (1 - e^u)F(a)$ therefore $X_a(t)$ has the moment generating function

$$\phi_{a,t}(u) = e^{\lambda\{\phi_a(u)-1\}}$$
$$= e^{-\lambda t(1-e^u)\bar{F}(a)}.$$

Proposition 5.22. *For every $x > 0$*

$$P(X_a(t) > x) \le x^{-x}\{\lambda t \bar{F}(a)\}^x \exp\{x - \lambda t \bar{F}(a)\}.$$

Proof. For every $x > 0$, Chernoff's Theorem (1.28) entails

$$P(X_a(t) > x) \le \inf_{u>0} \exp\{-ux - \lambda t(1 - e^u)\bar{F}(a)\}$$

where the exponential is minimum as $x = e^u \lambda t \bar{F}(a)$ hence

$$P(X_a(t) > x) \le \exp\left\{-x \log \frac{x}{\lambda t \bar{F}(a)} + x - \lambda t \bar{F}(a)\right\}.$$

□

5.7   Brownian motion

Let $B = (B_t)_{t \ge 0}$ denote the Brownian motion on a filtered probability space $(\Omega, \mathcal{F}, (\mathcal{F}_t)_{t \ge 0}, P)$ with its natural filtration. Writing $B_1 - B_t = B_1 - B_s - (B_t - B_s)$, for $s < t$ in $[0, 1]$, the expectation of $B_1 - B_t$ conditionally on B_s is $B_1 - B_s$, and we deduce the equivalence in distributions

$$B_t - B_s \sim \frac{t-s}{1-s}(B_1 - B_s).$$

The independence of the variables $B_t - B_s$ and B_s, for $s < t$, implies the independence of the variables B_s and

$$B_t - B_s - \int_s^t \frac{B_1 - B_u}{1-u}\,du,$$

it follows that the process $B_t - \int_0^t \frac{B_1 - B_u}{1 - u} \, du$ is a martingale with respect to the filtration generated by B.

Proposition 5.23. *For every stopping time τ for the filtration $(\mathcal{F}_t)_{t \geq 0}$ the process $(B_{t+\tau} - B_t)_{t \geq 0}$ is a Brownian motion independent of $\mathcal{F}_s$ for every $s \leq \tau$.*

Proof. For every $x > 0$, the variable $B_{t+(\tau \wedge x)} - B_t$ is $\mathcal{F}_{t+x}$-measurable and it has the same centered Gaussian distribution as the Brownian motion $B_{\tau \wedge x}$, with covariance $E(B_{\tau \wedge x} B_{\tau \wedge y}) = \tau \wedge x \wedge y$. The variable $B_{t+\tau} - B_t$ has the same Gaussian distribution as B_τ which depends only on τ. $\square$

Proposition 5.3 applies to the Brownian motion, for every $p > 1$, there exists a constant $C_p > 1$ such that for every positive stopping time T

$$E(T^p) \leq E(B_T^{2p}) \leq C_p E(T^p).$$

Let $p > 1$, Theorem 5.2 implies that there exists a constant C such that for every positive stopping time T of B and for every $t \leq T$

$$E(|B_t^*|^p) \leq C E(T^{\frac{p}{2}}). \tag{5.12}$$

For products of the increments of the process on disjoint intervals, the inequalities are factorized and for every increasing sequence of stopping times $(T_j)_{j \leq k}$, for integers $0 < m_k < n_k$ and $p_k \geq 2$, $k > 1$

$$\prod_{j=1}^{k} E(|B_{T_j} - B_{T_{j-1}}|^{p_j}) \leq C E \left\{ \prod_{j=1}^{k} (T_j - T_{j-1})^{\frac{p}{2}} \right\} \tag{5.13}$$

and the same inequality for the maximum value of the Brownian motion on the intervals $]T_{j-1}, T_j]$.

From Proposition 5.5, for every real number α in $]0, 2[$ and for all $t > 0$ and $x > 0$, the Brownian motion has the bound

$$E(|B_t^2 - t|^{\frac{\alpha}{2}}) \leq t^{\frac{\alpha}{2}},$$

$$P(|B_t^2 - t| > x) \leq 2 e^{-\frac{(t+x)^2}{t}}$$

and for $x = 0$ this probability is strictly positive.

Theorem 5.9. *For all $a > 0$ and $x > 0$*

$$P\left(\sup_{t \in [0,x]} B_t \geq ax \right) \leq 4 C_2 e^{-\frac{a^2 x}{2}}.$$

Proof. For every $\theta > 0$, the process $Y_\theta(t) = \exp\{\theta B_t - \frac{1}{2}\theta^2 t\}$ is a martingale with respect to the filtration generated by the Brownian motion. As $B_x \geq ax$ implies $Y_\theta^*(x) \geq \exp\{\theta ax - \frac{1}{2}\theta^2 x\}$, we have

$$P(B_x^* > ax) \leq P\left(Y_\theta^*(x) > \exp\left\{\theta ax - \frac{1}{2}\theta^2 x\right\}\right)$$
$$\leq \exp\{-2\theta ax + \theta^2 x\} E\{Y_\theta^{*2}(x)\}.$$

Proposition 5.3 and Theorem 5.2 imply

$$E\{Y_\theta^{*2}(x)\} \leq 4 \sup_{t \leq x} E\{Y_\theta^2(t)\} \leq 4C_2 E\{V_\theta(x)\}$$

where

$$E\{V_\theta(x)\} = E\{Y_\theta^2(x)\} = e^{-\theta^2 x} E(e^{2\theta B_x}) = e^{\theta^2 x}$$

therefore

$$P(B_x^* > ax) \leq 4C_2 \exp\{-2\theta ax + 2\theta^2 x\}.$$

Moreover $-\theta a + \theta^2$ is minimum at $\theta = \frac{a}{2}$ where the bound is $4C_2 e^{-\frac{a^2 x}{2}}$. $\square$

Let β be a right-continuous function locally in $L^2(\mathbb{R})$ and with left limits, the process

$$Y_t = \int_0^t \beta_s \, dB_s$$

is a transformed Brownian motion, the increasing function of its quadratic variations is $V_{Y_t} = \int_0^t \beta^2(s) \, ds$ and its Laplace transform is

$$L_{Y_t}(\lambda) = \exp\left\{\frac{1}{2}\lambda^2 V_Y(t)\right\}.$$

Proposition 5.3 and Theorem 5.2, imply there exists a constant C such that

$$\|\sup_{t \in [0,T]} Y_t^p\|_{L^p} \leq C \|V_{Y_T}\|_{L^{\frac{p}{2}}}$$

for every real $p \geq 2$ and for every stopping time T. The results of the previous sections apply to the process Y, Proposition 5.3 provides the following bounds.

Proposition 5.24. *Let $p \geq 2$, there exist a constant C such that for every positive stopping time T*

$$E\left\{\left(\int_0^T \beta^2 \, ds\right)^{\frac{p}{2}}\right\} \leq E\left(\left|\int_0^T \beta \, dB\right|^p\right) \leq E\left\{C\left(\int_0^T \beta_t^2 \, dt\right)^{\frac{p}{2}}\right\}.$$

For $\lambda > 0$, let

$$Z_\lambda(t) = \exp\left\{\lambda \int_0^t \beta(s)\, dB_s - \frac{\lambda^2}{2} \int_0^t \beta^2(s)\, ds\right\}.$$

Proposition 5.25. *For every $\lambda > 0$, the process Z_λ is a martingale with respect to $(\mathcal{F}_t)_{t\geq 0}$.*

Proof. For all $s < t$, the process Z_λ is such that

$$\begin{aligned}
E\{Z_\lambda^{-1}(s)Z_\lambda(t) \mid B_s\} &= e^{-\frac{1}{2}\lambda^2(t-s)} E[\exp\{\lambda(B_t - B_s) \mid B_s\}]\\
&= e^{-\frac{1}{2}\lambda^2(t-s)} E[\exp\{\lambda B_{t-s}\}]\\
&\geq e^{-\frac{1}{2}\lambda^2(t-s)}
\end{aligned}$$

by convexity and the properties of the Brownian motion, therefore

$$E\{Z_\lambda(t) \mid B_s\} \geq Z_\lambda(s)e^{-\frac{1}{2}\lambda^2(t-s)}$$

where the exponential is lower than one. As its expectation is $E\{Z_\lambda(t)\} = 1$, we have $E\{Z_\lambda(t)\} = E\{Z_\lambda(s)\}$ for all s and t, it follows that for $s < t$

$$E\{Z_\lambda(t) \mid B_s\} = Z_\lambda(s).$$

$\square$

Corollary 5.3. *For all $a > 0$ and $x > 0$*

$$P(\sup_{t\in[0,x]} Y_t \geq ax) \leq 4C_2 e^{-\frac{a^2 x^2}{2V_{Y_x}}}.$$

Proof. Propositions 5.9 and 5.24 entail that for all $a > 0$ and $x > 0$

$$\begin{aligned}
P(Y_x^* \geq ax) &= P\left(Z_\theta^*(x) > \exp\left\{\theta ax - \frac{\theta^2}{2}V_{Y_x}\right\}\right)\\
&\leq \exp\{-2\theta ax + \theta^2 V_{Y_x}\} E\{Z_\theta^{*2}(x)\}
\end{aligned}$$

where

$$\begin{aligned}
E\{Z_\theta^{*2}(x)\} &\leq 4E\{Z_\theta^2(x)\} \leq 4C_2 E\{V_\theta(x)\},\\
E\{V_\theta(x)\} &= E\{Z_\theta^2(x)\} = e^{-\theta^2 V_{Y_x}} E(e^{2\theta Y_x}) = e^{\theta^2 V_{Y_x}}
\end{aligned}$$

therefore

$$P(Y_x^* \geq ax) \leq 4C_2 \exp\{-2\theta ax + 2\theta^2 V_{Y_x}\}).$$

The minimum of this bound is reached at $\theta = \frac{xa}{2V_{Y_x}}$ where its value is $4C_2 e^{-\frac{a^2 x^2}{2V_{Y_x}}}$.

$\square$

Proposition 5.26. *Let $a > 0$, the variable $T_{Y,a} = \inf\{s : Y(s) = a\}$ is a stopping time for the process Y and*

$$Ee^{\lambda T_{Y,a}} \geq \exp\left\{\frac{a\sqrt{2}}{\sqrt{\lambda}\|\beta\|_\infty}\right\}.$$

Proof. The martingale property of the process Y_λ and a convexity argument provide a lower bound for the moment generating function of $T_{Y,a}$

$$
\begin{aligned}
e^{a\lambda} &= E \exp\left\{\frac{1}{2}\lambda^2 V_Y(T_{Y,a})\right\} \\
&\leq E\left[\frac{1}{T_{Y,a}}\int_0^{T_{Y,a}} \exp\left\{\frac{T_{Y,a}}{2}\lambda^2\beta^2(s)\right\} ds\right] \\
&\leq E\left[\exp\left\{\frac{T_{Y,a}}{2}\lambda^2\|\beta\|_\infty^2\right\}\right].
\end{aligned}
$$

$\square$

Let $T > 0$ be a random variable and let $x > 0$, several inequalities can be written for the tail probability of $\sup_{0\leq t\leq T} t^{-\frac{1}{2}}Y_t$. The Bienaymé-Chebychev inequality implies

$$P\left(\sup_{t\in[0,T]} t^{-\frac{1}{2}}Y_t > x\right) \leq x^{-2}\left\{E\sup_{t\in[0,T]} |\beta(t)|^2\right\}.$$

From Lenglart's theorem, for every $\eta > 0$ we have

$$P\left(\sup_{t\in[0,T]} Y_t > x\right) \leq \frac{\eta}{x^2} + P\left(\int_0^T \beta^2(s)\,ds > \eta\right)$$

and the previous inequality is a special case with $\eta = E\sup_{0\leq t\leq T}|\beta(t)|^2$. The Gaussian distribution of the process Y allows us to write Chernoff's theorem in the form of

$$P(t^{-\frac{1}{2}}Y_t > x) = E\inf_{\lambda>0} \exp\left\{\frac{\lambda^2}{2}\int_0^t \beta^2(s)\,ds - \lambda x\right\}$$

$$= E\exp\left\{-\frac{x^2 t}{2\int_0^t \beta^2(s)\,ds}\right\}$$

$$P\left(\sup_{0\leq t\leq T} Y_t > x\right) \leq E\exp\left\{-\frac{x^2}{2\sup_{0\leq t\leq T}|\beta(t)|^2}\right\}.$$

For the Brownian motion, it is simply written as

$$P\left(\sup_{t\in[0,T]} t^{-\frac{1}{2}}B_t > x\right) = \exp\left(-\frac{x^2}{2}\right),$$

for every stopping time $T > 0$, where the upper bounds tend to zero as x tends to infinity. With a varying boundary $f(t)$, function on $\mathbb{R}$, Chernoff's theorem implies

$$P\{|B_t| > f(t)\} = 2\exp\Big\{-\frac{f^2(t)}{2t}\Big\},$$

it tends to 1 as t tends to infinity if $t^{-\frac{1}{2}}f(t)$ tends to zero as t tends to infinity, and it tends to zero as t tends to infinity if $|t^{-\frac{1}{2}}f(t)|$ tends to infinity with t.

Proposition 5.27. *Let* $Y(t) = \int_0^t \beta(s)\, dB_s$ *and* $T = \arg\max_{t\in[0,1]} Y_t$

$$P(T \leq t) = \frac{2}{\pi}\arcsin\Big(\frac{V_t}{V_1}\Big)^{\frac{1}{2}},$$

where $V_t = \int_0^t \beta^2(s)\, ds$.

Proof. The variable T is a stopping time of the process Y with independent increments, and its distribution is a generalization of (1.24) for the Brownian motion. The variance of $Y_{t-u} - Y_t$ is V_u and the variance of $Y_{t+v} - Y_t$ is $V_{t+v} - V_t$. The normalized processes $X_0 = V_u^{-\frac{1}{2}}(B_{t-u} - B_t)$ and $X_1 = V_v^{-\frac{1}{2}}(B_{t+v} - B_t)$ are independent and identically distributed, with a standard distribution which does not depend on t, u and v, therefore

$$P(T \leq t) = P(\sup_{u\in[0,t]} Y_{t-u} - Y_t \geq \sup_{v\in[0,1-t]} Y_{t+v} - Y_t)$$

$$= P\{V_t^{\frac{1}{2}}X_1 \geq (V_1 - V_t)^{\frac{1}{2}}X_2\}$$

and

$$P(T \leq t) = P\Big\{\frac{X_2}{(X_1^2 + X_2^2)^{\frac{1}{2}}} \leq \Big(\frac{V_t}{V_1}\Big)^{\frac{1}{2}}\Big\} = \frac{2}{\pi}\arcsin\Big(\frac{V_t}{V_1}\Big)^{\frac{1}{2}}.$$

$\square$

Proposition 5.28. *Let* $Y_t = \int_0^t (\int_0^s \beta_y^2\, dy)^{-1}\beta_s\, dB_s$, *where* β *is a positive function of* $C_b(I)$, *with a bounded interval* I *of* $\mathbb{R}_+$. *For very interval* $[S, T]$ *of* $\mathbb{R}_+$

$$\sup_{S\leq s\leq t\leq T} P(|Y_t - Y_s| > x) = 2\exp\Big\{-\frac{x^2}{2(T-S)}\Big\}.$$

Proof. The martingale $(Y_t)_{t\in[0,T]}$ has an increasing process of quadratic variations V such that $V_t - V_s = E(Y_t^2 - Y_s^2|\mathcal{F}_s) = t - s$ for every positive $s \leq t$, the process Y has therefore the distribution of a Brownian motion.

The variance $\mathrm{Var}(Y_t - Y_s) = t - s$ is bounded by $T - S$ on the interval $[S, T]$. Let $x > 0$, for every $t \geq s \geq 0$

$$P(Y_t - Y_s \geq x) \leq \exp\left\{-\frac{x^2}{2(T - S)}\right\}$$

and the bound is reached at (S, T).   □

The Brownian bridge $W_t = B_t - tB_1$ has the covariance function

$$E(W_s W_t) = s \wedge t - st$$

and the integral $I_t = \int_0^t A_s \, dW_s$ of a predictable process A with respect to the Brownian bridge has the variance function

$$E(I_t^2) = E\left\{\int_0^t A_s^2 \, ds - \left(\int_0^t A_s \, ds\right)^2\right\}.$$

The variance function

$$E(I_{t_1} I_{t_2}) = E\left\{\int_0^{t_1 \wedge t_2} A_s^2 \, ds - \left(\int_0^{t_1} A_s \, ds\right)\left(\int_0^{t_2} A_s \, ds\right)\right\}.$$

By a change of time, for every $\lambda > 0$, $B_{\lambda t} = \lambda B_t$ and $W_{\lambda t} = \lambda W_t$, and the covariance of the Brownian bridge is stationary.

Let $(X_i)_{i \geq 1}$ be a sequence of independent random variables defined on a probability space $(\Omega, \mathcal{F}, P)$ and with values in a separable and complete metric space $(\mathcal{X}, \mathcal{B})$ provided with the topology of the space $D(\mathcal{X})$. When the variables have the same distribution function F, the empirical process

$$\nu_n(t) = n^{-\frac{1}{2}}\{S_n(t) - \mu_n(t)\},$$

t in $\mathbb{R}$, defined by

$$S_n(t) = \sum_{i=1}^n 1_{\{X_i \leq t\}}, \quad \mu_n(t) = ES_n(t) = nF(t),$$

converges weakly to the process $W_F = W \circ F$ with covariance function $C(s, t) = F(s \wedge t) - F(s)F(t)$, where W is the standard Brownian bridge on $[0, 1]$. For every real number $x > 0$,

$$P(\sup_{0 \leq t \leq T} \nu_n(t) > x) \leq x^{-2} F(T).$$

Let $V_n(t) = \mathrm{Var} S_n(t) = nV(t)$, with $V(t) = F(t)\{1 - F(t)\}$. The normalized process $W_n(t) = \{V_n(t)\}^{-\frac{1}{2}}\{S_n(t) - \mu_n(t)\}$, t in $\mathbb{R}$, converges weakly to $V^{-\frac{1}{2}} W_F$ with covariance function $\rho(s, t) = \{V(s)V(t)\}^{-\frac{1}{2}} C(s, t)$. The

Laplace transform of the standard Brownian bridge W and the transformed Brownian bridge W_F at t are

$$L_{W_t} = \exp\left\{-\frac{u^2 t(1-t)}{2}\right\}, \quad L_{W_F(t)} = \exp\left\{-\frac{u^2 V(t)}{2}\right\}$$

and by Chernoff's theorem

$$\lim_{n\to\infty} P(|\nu_n(t)| > x) = 2\exp\left\{-\frac{x^2}{2V(t)}\right\}.$$

The integral $Y_n(t) = \int_0^t A_s \, d\nu_n(s)$ of a predictable process A with respect to the empirical process ν_n converges weakly to a process Y with covariance

$$E\{Y(t_1)Y(t_2)\} = E\left\{\int_0^{t_1 \wedge t_2} A_s^2 \, dF(s) - \left(\int_0^{t_1} A_s \, dF(s)\right)\left(\int_0^{t_2} A_s \, dF(s)\right)\right\}$$

and variance function $V_A(t)$, by Chernoff's theorem

$$\lim_{n\to\infty} P(|Y_n(t)| > x) = 2\exp\left\{-\frac{x^2}{2V_A(t)}\right\}.$$

A change of time by the inverse of the empirical process $\widehat{F}_n(t) = n^{-1}S_n(t)$ implies the weak convergence of the process $\nu_n \circ \widehat{F}_n^{-1}$ to the Brownian bridge and the same inequalities applied to this process are free from the distribution of the variables X_i.

Proposition 5.29. *For all $\alpha > 0$ and $\beta > \frac{1}{2}\alpha$, the Brownian bridge satisfies the inequality*

$$P(Z_t < \alpha(1-t) + \beta t, 0 \le t \le 1) \le \frac{1}{2(\alpha+\beta)}.$$

Proof. The Gaussian variable $Y_t = \{\alpha(1-t) + \beta t\}^{-1} Z_t$ has the variance

$$v_t = \frac{t(1-t)}{\{\alpha(1-t) + \beta t\}^2},$$

its first derivatives are

$$v_t' = \frac{\alpha - (\alpha+\beta)t}{\{\alpha(1-t) + \beta t\}^3},$$

$$v_t'' = \frac{2\alpha(\alpha - 2\beta)}{\{\alpha(1-t) + \beta t\}^4},$$

such that $v_t'' < 0$ hence the variance is concave and maximum at $(\alpha+\beta)^{-1}\alpha$ where it is

$$v_M = \frac{1}{2(\alpha+\beta)},$$

it follows that the variance of $\sup_{t\in[0,1]} Y_t$ has the upper bound v_M and by the inequality (1.13) we obtain

$$P(Z_t > \alpha(1-t) + \beta t, 0 \leq t \leq 1) = P\left(\sup_{0 \leq t \leq 1} Y_t > 1\right)$$

$$\leq \frac{1}{2(\alpha+\beta)}.$$

$\square$

Proposition 5.30. *For all $\alpha > 0$ and $\beta > \frac{1}{2}\alpha$, the Brownian bridge satisfies the inequality*

$$P\left(\sup_{a \leq t \leq 1} Z_t > \lambda t\right) \leq \frac{1-a}{a\lambda^2}.$$

Proof. The variable $t^{-1}Z_t$ is Gaussian, centered with a decreasing variance $t^{-1}(1-t)$ on $[a,1]$, its maximum is $a^{-1}(1-a)$ then the variance of $\sup_{a \leq t \leq 1} t^{-1}Z_t$ is bounded by $a^{-1}(1-a)$. The inequality is deduced from the inequality (1.13). $\square$

5.8   Diffusion processes

On a probability space $(\Omega, \mathcal{F}, P)$, let $X = (X_t)_{t \leq 0}$ be a continuous diffusion process with sample paths in $\mathbb{R}_+$, defined by a random initial value X_0, $\mathcal{F}_0$-measurable and belonging to $L^2(P)$, and by the differential equation

$$dX_t = \alpha(X_t)\, dt + \beta(X_t)\, dB_t \tag{5.14}$$

where B is the Brownian motion, α and β are real functions on $\mathbb{R}_+$ such that the drift function α belongs to $L^1(\mathbb{R}_+)$ and the variance function β belongs to $L^2(\mathbb{R}_+)$. Equivalently

$$X_t - X_0 = \int_0^t \alpha(X_s)\, ds + \int_0^t \beta(X_s)\, dB_s.$$

The Ornstein-Uhlenbeck process is defined by the differential equation

$$dX_t = \alpha X_t\, dt + \beta\, dB_t$$

initial value X_0, so

$$X_t = -X_0 e^{\alpha t} + \beta \int_0^t e^{\alpha(t-s)}\, dB_s.$$

The process defined with real valued drift and variance

$$X_t - X_0 = \int_0^t \alpha_s\, ds + \int_0^t \beta_s\, dB_s$$

has a Gaussian distribution with expectation $E(X_t - X_0) = \int_0^t \alpha_s \, ds$ and variance $\mathrm{Var}(X_t - X_0) = \int_0^t \beta_s^2 \, ds$.

The distribution of the Brownian motion is invariant by translation $B_t \sim B_{T+t} - B_T$ for every $T > 0$, it follows that periodic functions α and β having the same period determine a periodic diffusion process by (5.14). On a circle or a ball, complex periodic exponential functions α and β determine a periodic diffusion and a periodic diffusion inside a ball is defined by variation of the radius with functions $\alpha(r\theta, \varphi)$ and $\beta(r\theta, \varphi)$, $r \leq R$, with angles θ and φ.

More generally, the sample paths of a Gaussian process solution of a stochastic differential equation belong to functional sets determined by the expression of the drift of the equation. With a linear drift

$$dX(t) = \alpha(t)X(t) \, dt + \beta(t) \, dB(t), \tag{5.15}$$

the stochastic integral

$$X_t = X_0 e^{A(t)} + \int_0^t e^{A(t)-A(s)} \beta(s) \, dB(s), \tag{5.16}$$

defined with

$$A(t) = \int_0^t \alpha(s) \, ds,$$

is the solution of the diffusion Equation (5.15). Let $\mathcal{F}_t$ be the σ-algebra generated by $\{B_s, 0 \leq s \leq t\}$, the Brownian motion B is a $(\mathcal{F}_t)_{t \geq 0}$-martingale with independent increments and with expectation zero and the variance $EB_t^2 = t$. The initial value X_0 is supposed to be independent of $B(t) - B(s)$ for every $t > s > 0$, then the diffusion process has the expectation

$$E(X_t) = e^{A(t)} E(X_0)$$

and the centered process

$$M_t = X_t - X_0 e^{A(t)} = \int_0^t e^{A(t)-A(s)} \beta(s) \, dB(s)$$

is a transformed Brownian motion on $(\Omega, \mathcal{F}, (\mathcal{F}_t)_{t \geq 0}, P)$. It satisfies the properties of Section 5.7, in particular

$$E(M_t^2) = \int_0^t e^{2\{A(t)-A(s)\}} \beta^2(s) \, ds,$$

for every $0 < s < t < u$, $E\{(M_u - M_t)(M_t - M_s)\} = 0$ and

$$E(M_t^2 | \mathcal{F}_s) = M_s^2 + \int_s^t e^{2\{A(t)-A(u)\}} \beta^2(u) \, du.$$

The predictable process of its quadratic variations is

$$< M >_t = \int_0^t e^{2\{A(t)-A(s)\}}\beta^2(s)\,ds.$$

The process $X = M + X_0 e^A$ has the variance

$$v_t := \text{Var}(X_t) = \text{Var}(X_0)\,e^{2A(t)} + \int_0^t e^{2\{A(t)-A(s)\}}\beta^2(s)\,ds$$

and the quadratic variations

$$E\{(X_t - X_s)^2|\mathcal{F}_s\} = EX_0^2\,(e^{A(t)} - e^{A(s)})^2 + \int_s^t e^{2\{A(t)-A(s)\}}\beta^2(s)\,ds,$$

the predictable process of its quadratic variations is

$$< X >_t = X_0^2\,e^{2A(t)} + < M >_t.$$

For each t such that $\text{Var}(X_t)$ is strictly positive, the variable X_t has a Gaussian distribution with expectation $e^{A(t)}E(X_0)$ and variance v_t, its Laplace transform is $L_{X_t}(\lambda) = \exp\{\frac{v_t\lambda^2}{2} - e^{A(t)}E(X_0)\}$.

The Brownian process M defines the martingale $(M_t^2 - < M >_t)_{t\geq 0}$ such that

$$E\{(M_t^2 - < M >_t)^2\} = 2 < M >_t^2.$$

Propositions 5.3 and 5.5 imply that for every real $p \geq 1$ and for every $t > 0$

$$2^p < M >_t^{2p} \leq E\{(M_t^2 - < M >_t)^{2p}\} \leq 2^p C_p < M >_t^{2p},\ p \geq 1,$$
$$c_p 2^p < M >_t^{2p} \leq E\{(M_t^2 - < M >_t)^{2p}\} \leq 2^p < M >_t^{2p},\ p \in\ ]0,1[.$$

If α and β are strictly positive and bounded, the process $< M >$ exponentially increasing. The Bienaymé-Chebychev inequality applies to M and X, for every positive stopping time T independent of X_0

$$P\left(\sup_{t\in[0,T]} t^{-\frac{1}{2}}|M_t| > \lambda\right) \leq \lambda^{-2}\sup_{T>0} E\left[T^{-1}\int_0^T e^{2\{A(t)-A(s)\}}\beta^2(s)\,ds\right]$$

$$\leq \lambda^{-2}E\left[\sup_{t\in[0,T]}\beta^2(t)\,T^{-1}\int_0^T e^{2\{A(t)-A(s)\}}\,ds\right],$$

$$P\left(\sup_{t\in[0,T]} t^{-\frac{1}{2}}|X_t - EX_t| > 2\lambda\right) \leq P\left(\sup_{t\in[0,T]} t^{-\frac{1}{2}}|M_t| > \lambda\right)$$

$$+ P\left(\sup_{t\in[0,T]} t^{-\frac{1}{2}}e^{A(t)}|X_0 - EX_0| > \lambda\right),$$

where the last term has the bound

$$\lambda^{-2}\mathrm{Var}(X_0)E\Big\{\sup_{t\in[0,T]} t^{-1}e^{A(t)}\Big\}.$$

Other inequalities are deduced from the results of Sections 5.2 and 5.3, in particular Theorem 5.1 applies and

$$\|M_T^*\|_2 \le 2E(<M>_T)^{\frac12} = 2E\Big\{\int_0^t e^{2\{A(t)-A(s)\}}\beta^2(s)\,ds\Big\}^{\frac12}.$$

Exponential inequalities are consequences of the properties of the Brownian motion. For every strictly positive x and for every stopping time $T>0$

$$P\Big(\sup_{t\le T} M_T > x\Big) \le 4C_2\exp\Big(-\frac{x^2}{2<M>_t}\Big).$$

Consider a diffusion process with a drift function α depending on the time index and on the sample path of the process, determined by the differential equation

$$dX(t) = \alpha(t,X_t)\,dt + \beta(t)\,dB(t) \tag{5.17}$$

and the initial value X_0. Its expectation is $A(t) = \int_0^t E\alpha(s,X_s)\,ds$ and its variance is

$$V_t = \int_0^t \beta^2(s)\,ds + \mathrm{Var}\int_0^t \alpha(s,X_s)\,ds.$$

Under the condition that $E\Big|\int_0^T \alpha(s,X_s)\,ds\Big|$ is finite, the stochastic integral

$$X_t = X_0 e^{A(t,X_t)} + e^{A(t,X_t)}\int_0^t e^{-A(s,X_s)}\beta(s)\,dB(s), \tag{5.18}$$

$$A(t,X_t) = \int_0^t X_s^{-1}1_{\{X_s\ne0\}}\alpha(s,X_s)\,ds$$

is an implicit solution of the diffusion Equation (5.17), with the convention $\frac00 = 0$, equivalently

$$dX(t) = X_t\,dA(t,X_t) + \beta(t)\,dB(t)$$

and $A(0,X_0) = 0$. The expectation of X is $E(X_t) = E\{X_0 e^{A(t,X_t)}\}$ and its variance is

$$\mathrm{Var}(X_t) = \mathrm{Var}\{X_0 e^{A(t,X)}\}$$
$$+ \int_0^t E\Big[e^{2\{A(t,X_t)-A(s,X_s)\}}\Big]\beta^2(s)\,ds$$

where $\mathrm{Var}\{X_0 e^{A(t,X)}\} = E[X_0^2 \mathrm{Var}\{e^{A(t,X)}|\mathcal{F}_0\}] + \mathrm{Var}[X_0 E\{e^{A(t,X)}|\mathcal{F}_0\}]$. The centered process

$$M_t = X_t - X_0 e^{A(t,X_t)} = \int_0^t e^{e^{A(t,X_t)-A(s,X_s)}} \beta(s)\, dB(s)$$

is a transformed Brownian motion and the predictable process of its quadratic variations is

$$< M >_t = \int_0^t e^{2\{A(t,X_t)-A(s,X_s)\}} \beta^2(s)\, ds.$$

For every positive stopping time T independent of X_0 we have

$$P\Big(\sup_{t\in[0,T]} t^{-\frac{1}{2}}|M_t| > \lambda \Big) \le \lambda^{-2} E\Big[T^{-1} \int_0^T e^{2\{A(t,X_t)-A(s,X_s)\}} \beta^2(s)\, ds \Big]$$

$$\le \frac{\sup_{t\in[0,T]} \beta^2(t)}{\lambda^2} E\Big[T^{-1} \int_0^T e^{2\{A(t,X_t)-A(s,X_s)\}}\, ds \Big]$$

if the function $|\beta|$ is bounded. The inequalities for the martingales are satisfied and by Chernoff's theorem we have

$$\lim_{n\to\infty} P(|X_t| > x) = 2\exp\Big[-\frac{\{X_t - E\int_0^t \alpha(s,X_s)\, ds\}^2}{2\mathrm{Var}(X_t)} \Big].$$

5.9   Martingales in the plane

Walsh (1974) and Cairoli and Walsh (1975) presented inequalities for martingales and stochastic integrals with respect to (local) martingales in the plane. A partial order is defined as the order with respect to both coordinates and the notion of martingale is defined for this order, it implies that the martingale property holds with respect to both marginal filtrations. For every $z = (s,t)$ in $\mathbb{R}_+^2$, let $(\mathcal{F}_z)_{z>0}$ be a family of increasing σ-algebra and let $(\mathcal{F}_z^k)_{z>0}$ be the marginal defined σ-algebras defined by $\mathcal{F}_z^1 = \mathcal{F}_{s,\infty}$ and $\mathcal{F}_z^2 = \mathcal{F}_{\infty,t}$. A martingale M with respect to the filtration on $(\Omega, \mathcal{F}, P)$ is an integrable real process, adapted to $(\mathcal{F}_z)_{z>0}$ such that $E(M_z|\mathcal{F}_{z'}) = M_z$ for every $z' \le z$ in the total order of $\mathbb{R}_+^2$. The increment of the martingale on the rectangle $R_{z',z} = [z', z]$ is

$$\Delta M_{[z',z]} = M_z + M_{z'} - M_{(s',t)} - M_{(s,t')} \qquad (5.19)$$

and a weak martingale is defined by the property $E(M_{[z',z]}|\mathcal{F}_{z'}) = 0$. The process M is a strong martingale if

$$E(M_{[z',z]}|\mathcal{F}_{z'}^1 \vee \mathcal{F}_{z'}^2) = 0.$$

Let $R_z = [0, z]$ and $R_{z'} = [0, z']$ be rectangles from 0 to z and from 0 to z' respectively, then $R_z \cap R_{z'} = [0, z \wedge z']$ with the minimum in both coordinates and

$$E(M_{[z \wedge z', z \vee z']} | \mathcal{F}_{z \wedge z'}) = M_{R_{z \wedge z'}}.$$

If a weak martingale satisfies the condition $M(0) = M(0, t) = M(s, 0) = 0$ for every $z = (s, t)$, where the values of M are considered as its increments on empty rectangles, then

$$E(M_z | \mathcal{F}_{(0,t)}) = E(M_z | \mathcal{F}_{(s,0)}) = 0.$$

Under the condition of conditionally independent marginal σ-algebras, a local square integrable strong martingale M has a predictable compensator $[M]$ such that $M^2 - [M]$ is a martingale. Lenglart's theorem applies to a strong martingale in the plane like on $\mathbb{R}_+$.

On $(\Omega, \mathcal{F}, P)$, a counting process N in $\mathbb{R}_+^d$ is defined by the cardinal $N(A)$ of N in the subsets A of $\mathbb{R}_+^d$. The notions of martingales (respectively weak and strong martingales) are generalized to $\mathbb{R}^n$ with similar properties. Point processes with independent increments are defined as d-dimensional stochastic discrete measures. The cumulative intensity function Λ of a Poisson process N is a positive measure in $\mathbb{R}_+^d$ such that $N - \Lambda$ is a weak martingale with respect to the filtration generated by N on the rectangles of $\mathbb{R}_+^d$ and the function $\Lambda(A) = \int_A \lambda(s) \, ds$ is a positive measure in $\mathbb{R}_+^d$. The process $N - \Lambda$ has independent rectangular increments distributed like Poisson variables having as parameters the measure of the sets, with respect to Λ, then the property of conditional independence of the marginal σ-algebras is satisfied and the process $N - \Lambda$ is a martingale. Moreover, $\Lambda(A)^{-\frac{1}{2}} \{N(A) - \Lambda(A)\}_{A \in \mathbb{R}^d}$ converges weakly to a Gaussian process as the volume of A tends to infinity, when $\sup_{A \in \mathbb{R}^d} \|A\|^{-1} \Lambda(A)$ is bounded. From the martingale property, the variance of $N(A)$ is $\Lambda(A)$ and the covariance of $N(A)$ and $N(B)$ is

$$E\{N(A)N(B)\} - \Lambda(A)\Lambda(B) = E(N - \Lambda)^2(A \cap B) = \Lambda(A \cap B),$$

the normalized process has therefore a covariance equal to the correlation of N on the subsets of $\mathbb{R}_+^d$

$$R(A, B) = \frac{\Lambda(A \cap B)}{\{\Lambda(A)\Lambda(B)\}^{\frac{1}{2}}}.$$

The independence of the increments is a property of the Poisson process, and every point process with independent increments has a similar behaviour with a stochastic or deterministic compensator instead of the function Λ. More generally, every point process N has a predictable compensator $\widetilde{N}$ such that $M = N - \widetilde{N}$ is a weak martingale, it is not a martingale if the increments of the process N are dependent.

Because of the independence of the increments of N, as $\|A\|$ tends to infinity, the variable $N(A)$ is the sum of the number of points of N in an increasing number of nonintersecting subsets A_i of A with bounded volume, and the variables $N(A_i)$ are independent. It follows that $\|A\|^{-1}N(A)$ converges a.s. to $\Lambda(A)$ and $\|A\|^{-1}\widetilde{N}(A)$ converges a.s. to the same limit. The normalized variable

$$X(A) = \|A\|^{-\frac{1}{2}}\{N(A) - \widetilde{N}(A)\}$$

has the expectation zero and it converges weakly to a Gaussian variable as $\|A\|$ tends to infinity. Its covariance is

$$E\{X(A)X(B)\} = \frac{E\{N(A) - \widetilde{N}(A)\}\{N(B) - \widetilde{N}(B)\}}{\{\|A\|\,\|B\|\}^{\frac{1}{2}}}.$$

Under the condition of independent increments, the covariances and the variances of the counting process N and of the centered process X are $E\{N(A)N(B)\} = EN(A)\,EN(B) + \mathrm{Var}N(A \cap B)$, therefore

$$E\{X(A)X(B)\} = \frac{1}{\{\|A\|\,\|B\|\}^{\frac{1}{2}}}\,\mathrm{Var}\{N(A \cap B) - \widetilde{N}(A \cap B)\}.$$

The inequalities of the Poisson process on $\mathbb{R}_+$ extend to the spatial Poisson process, for every bounded set B

$$P\Big(\sup_{A \subset B} |N(A) - \Lambda(A)| > x\Big) \leq 2x^{-2}E\Lambda(B)$$

and the maximum inequality on sets with empty intersection factorizes

$$P\Big(\sup_{A_1 \subset B} |N(A_1) - \Lambda(A_1)| > x_1, \sup_{A_2 \subset B : A_1 \cap A_2 =} |N(A_2) - \Lambda(A_2)| > x_2\Big)$$

$$\leq \frac{4}{x_1^2 x_2^2} E\{\Lambda_{A_2}\}E\{\Lambda_{A_1}\}.$$

For every bounded set A, Theorem 5.8 implies

$$\|A\|^{-1}\log P(\|A\|^{-1}N(A) > a) = a[1 - \log a + \log\{\|A\|^{-1}\Lambda(A)\}]$$

$$- \|A\|^{-1}\Lambda(A)\}$$

and its limit as $\|A\|^{-1}$ tends to infinity is finite if $\lim_{\|A\| \to \infty} \|A\|^{-1}\Lambda(A)$ is finite.

Chapter 6

Stochastic Calculus

6.1 Stochastic integration

The stochastic integral of a predictable process U with respect to an increasing process V on $[0, t]$ is the limit, as n tends to infinity, of a series $\sum_{i=1}^{n} U(t_i)\{V(t_{i+1}) - V(t_i)\}$ approximating the integral on a regular partition $\pi_n = (t_i)_{i=1,\ldots,n}$ of $[0, t]$. It has been extended to the integral of a predictable process U with respect to a martingale as a limit in L^2 of the integral of U^2 with respect to the increasing process $[M]$. Much of the stochastic relies on the same kind of approximations and on Taylor expansions for Itô's formula of change of variable and for exponential martingales (Meyer, 1976). The first sections of this chapter present the main results without proofs, and applications.

The integration by parts formula for two right-continuous increasing processes U and V with left limits on a real interval $[a, b]$ is

$$U_{(b)}V(b) - U_-(a)V_-(a) = \int_{[a,b]} U_- \, dV + \int_{[a,b]} V \, dU$$

$$= 2\int_{[a,b]} U_- \, dV + \sum_{s \in [a,b]} \Delta(UV),$$

it is obtained from a decomposition of the integral on n sub-intervals $]t_i, t_{i+1}]$ of π_n, as n tends to infinity.

Let F be a function of $C^1(\mathbb{R})$ and let X be a continuous increasing process, the differentiation of $F(X_t)$ is written like for F on $\mathbb{R}$, $dF(X_t) = F'(X_t) \, dX_t$ and

$$F(X_t) = F(X_0) + \int_0^t F'(X_s) \, dX_s.$$

Proposition 6.1. *Let F be a function on $\mathbb{R}$ having a right-continuous derivative f with left limits, and let X be a right-continuous increasing process with left limits, then for every $t \geq 0$*

$$F(X_t) = F(X_0) + \int_0^t f(X_{s-}) \, dX_s$$

$$+ \sum_{0 \leq s \leq t} \{\Delta F(X_s) - \Delta X_s f(X_{s-})\}$$

$$= F(X_0) + \int_0^t f(X_{s-}) \, dX_s^c + \sum_{0 \leq s \leq t} \Delta F(X_s).$$

For a point process $N_t = \sum_{i \geq 1} 1_{\{T_i \leq t\}}$ on $\mathbb{R}_+$, with an increasing predictable compensator A such that $X = N - A$ belongs to $\mathcal{M}_{0,loc}^2$, the stochastic integrals of a predictable process Y with respect to X is $I_t = \int_0^t Y_s \, dN_s - \int_0^t Y_s \, dA_s$ where $\int_0^t Y_s \, dN_s = \sum_{i \geq 1} 1_{\{T_i \leq t\}} Y_{T_i}$. Proposition 6.1 applies to differentiable functions of X and I, as differences of two increasing processes therefore

$$X_t^2 = -2 \int_0^t X_{s-} \, dA_s^c + \sum_{0 \leq s \leq t} \Delta(X_s^2),$$

$$I_t^2 = -2 \int_0^t Y_s I_{s-} \, dA_s^c + \sum_{0 \leq s \leq t} \Delta(I_s^2).$$

The proof of Proposition 6.1 relies on a discretization of $F(X_t) - F(X_0)$ on a partition π_n of $[0, t]$ and a first order expansion of $F(X_{t_{i+1}}) - F(X_{t_i})$, as n tends to infinity. It extends to functions of several increasing processes and to functions of martingales.

Proposition 6.2. *Let F be a real function on $\mathbb{R}^n$ with right-continuous partial derivatives $D_i F$, $i = 1, \dots, n$, having left limits, and let X be a right-continuous increasing process with left limits and locally bounded variations in $\mathbb{R}^n$, then*

$$F(X_t) = F(X_0) + \sum_{i \leq n} \int_0^t D_i F(X_{s-}) \, dX_s^i$$

$$+ \sum_{0 \leq s \leq t} \{\Delta F(X_s) - \sum_{i \leq n} D_i F(X_{s-}) \Delta X_s^i\}.$$

Theorem 6.1. *Let F be a function of $C^2(\mathbb{R})$ and let X in $\mathcal{M}^2_{loc}$, then*

$$F(X_t) = F(X_0) + \int_0^t F'(X_{s-})\,dX_s + \frac{1}{2}\int_0^t F''(X_{s-})\,d<X^c>_s$$

$$+ \sum_{0\leq s\leq t}\{\Delta F(X_s) - \Delta X_s F'(X_{s-})\}$$

$$= F(X_0) + \int_0^t F'(X_{s-})\,dX_s + \frac{1}{2}\int_0^t F''(X_{s-})\,d[X]_s$$

$$+ \sum_{0\leq s\leq t}\{\Delta F(X_s) - \Delta X_s F'(X_{s-}) - \frac{1}{2}(\Delta X_s)^2 F''(X_{s-})\}.$$

For a right-continuous local martingale M with left limits in $\mathcal{M}^2_{0,loc}$, Theorem 6.1 implies

$$M_t^2 = 2\int_0^t M_{s-}\,dM_s^c + <M^c>_t + \sum_{0\leq s\leq t}\Delta(M_t^2),$$

if M is continuous in $\mathcal{M}^2_{0,loc}$, it reduces to $M_t^2 - <M>_t = 2\int_0^t M_s\,dM_s$ which belongs to $\mathcal{M}_{0,loc}$.

The Brownian motion B_t has the same distribution as $t^{\frac{1}{2}}X$ where X is a normal variable. Itô's formula for B_t^2, with a function $F(x) = x^2$ reduces to $d(B_t^2) = 2\int_0^t B_s\,dB_s + dt$, equivalently

$$\int_0^t B_s\,dB_s = \frac{1}{2}(B_t^2 - t).$$

Applying Theorem 6.1 to the stochastic integral $U.B$ of a measurable process U with respect to the Brownian motion, it is written as

$$F(U.B_t) = F(0) + \int_0^t F'(U.B_s)U_s\,dB_s + \frac{1}{2}\int_0^t F''(U.B_s)U_s^2\,ds.$$

A point process $(N_t)_{t\geq 0}$ with an increasing predictable compensator $(A)_{t\geq 0}$ defines $X = N - A$ with increasing process of predictable variation

$$<X>_t = \int_0^t (1 - \Delta A_s)\,dA_s$$

and $[X]_t = <X^c>_t + \sum_{s\leq t}(\Delta X_s)^2$. Theorem 6.1 is written as

$$F(X_t) = F(0) + \int_0^t F'(X_{s-})\,dX_s^c + \frac{1}{2}\int_0^t F''(X_{s-})\,d<X^c>_s$$

$$+ \sum_{0\leq s\leq t}\Delta F(X_s),$$

in particular

$$X_t^2 = 2 \int_0^t X_{s-} \, dX_s^c + < X^c >_t + \sum_{0 \leq s \leq t} \Delta(X_s^2).$$

If A is a continuous process, we have $< X^c >_t = A$ and $\Delta X_t = \Delta N_t$, $\sum_{s \leq t} \Delta(X_s^2) = N_t - 2 \int_0^t A_s \, dN_s$ hence

$$X_t^2 = A_t - 2 \int_0^t X_{s-} \, dA_s + N_t - 2 \int_0^t A_s \, dN_s$$
$$= A_t(1 + A_t^2) + N_t - 2N_t A_t.$$

Let U be a predictable process and let I_t be its stochastic integral with respect to N, $I_t = \int_0^t U_s \, dN_s$, with predictable compensator $\widetilde{I}_t = \int_0^t U_s \, dA_s$. The increasing process of predictable variations of $Y = I - \widetilde{I}$ is

$$< Y >_t = \int_0^t U_s^2 (1 - \Delta A_s) \, dA_s$$

then $< Y^c >_t = \int_0^t U_s^2 \, dA_s^c$ and $[Y]_t = < Y^c >_t + \int_0^t U_s^2 \, dN_s$. By Proposition 6.1, it follows that

$$Y_t^2 = \int_0^t U_s^2 \, dA_s^c - 2 \int_0^t Y_{s-} U_s \, dA_s^c + \sum_{0 \leq s \leq t} \Delta(Y_s^2)$$

and, if A is continuous, $\Delta(Y_s^2) = U_s^2 \Delta N_s$ hence

$$Y_t^2 = \int_0^t U_s^2 \, d(N_s + A_s) - 2 \int_0^t Y_{s-} U_s \, dA_s.$$

Theorem 6.2. *Let $F(t, x)$ be a function continuously differentiable with respect to t on $\mathbb{R}^+$ and twice continuously differentiable with respect to x and let X be a martingale, then*

$$F(t, X_t) = F(X_0) + \int_0^t F_s'(s, X_s) \, ds + \int_0^t F_x'(s, X_{s-}) \, dX_s$$
$$+ \frac{1}{2} \int_0^t F_x''(s, X_{s-}) \, d[X]_s + \sum_{0 \leq s \leq t} \left\{ F(s, X_s) - F(s, X_{s-}) \right.$$
$$\left. - \Delta X_s F_x'(s, X_{s-}) - \frac{1}{2} (\Delta X_s)^2 F_x''(s, X_{s-}) \right\}.$$

Theorem 6.2 generalizes to a function $F(V_t, X_t)$ where V_t is a right-continuous increasing process with left limits and locally bounded variations

$$F(V_t, X_t) = F(V_0, X_0) + \int_0^t F'_v(V_s, X_s)\, dV_s + \int_0^t F'_x(V_s, X_{s-})\, dX_s$$

$$+ \frac{1}{2} \int_0^t F''_x(V_s, X_{s-})\, d[X]_s + \sum_{0 \le s \le t} \Big\{ F(V_s, X_s) - F(V_{s-}, X_s)$$

$$- \Delta V_s\, F'_v(V_s, X_s) \Big\} + \sum_{0 \le s \le t} \Big\{ F(V_s, X_s) - F(V_s, X_{s-})$$

$$- \Delta X_s F'_x(V_s, X_{s-}) - \frac{1}{2}(\Delta X_s)^2 F''_x(V_{s-}, X_{s-}) \Big\}.$$

It generalizes also to a function F on $\mathbb{R}^n$ and a multidimensional process $X = (X^1, \ldots, X^n)$, with sums of the partial derivatives. Let $< X^i >_t$ be the predictable process of the quadratic variation of X^i then

$$< X^i, X^j >_t = \frac{1}{2}\{ < X^i + X^j >_t - < X^i >_t - < X^j >_t \},$$

$$[X^i, X^j]_t = \frac{1}{2}\{ [X^i + X^j]_t - [X^i]_t - [X^j]_t \}$$

$$= [X^i, X^j]_t^c + \sum_{s \le t} \Delta X_s^i \Delta X_s^j.$$

Itô's formula applies in the same form to semi-martingales $X_t = X_0 + A_t + M_t$ with X_0 bounded, A increasing continuous and locally bounded process such that $A_0 = 0$, and M in $\mathcal{M}^2$

$$F(X_t) = F(X_0) + \int_0^t F'(X_s)\,(dM_s + dA_s) + \frac{1}{2}\int_0^t F''(X_s)\, d < M >_s$$

$$+ \sum_{0 \le s \le t} \{ \Delta F(X_s) - \Delta X_s F'(X_{s-}) \}$$

$$= F(X_0) + \int_0^t F'(X_{s-})\, dX_s + \frac{1}{2}\int_0^t F''(X_{s-})\, d[X]_s$$

$$+ \sum_{0 \le s \le t} \{ \Delta F(X_s) - \Delta X_s F'(X_{s-}) - \frac{1}{2}(\Delta X_s)^2 F''(X_{s-}) \}.$$

Let M and N be semi-martingales without common jump times, then $\Delta(M_s N_s) = M_{s-}\,\Delta N_s + N_{s-}\,\Delta M_s$ and

$$M_t N_t = \int_0^t M_{s-}\, dN_s^c + \int_0^t N_{s-}\, dM_s^c + < M, N >_t^c.$$

A real diffusion process $X_t = X_0 + \int_0^t \alpha(s, X_s)\, ds + \int_0^t \beta(s, X_s)\, dB_s$ defined with a continuous increasing process $A_t = \int_0^t \alpha(s, X_s)\, ds$ and a function β such that $E \int_0^t \beta^2(s, X_s)\, ds$ is finite, for every real t, is a continuous semi-martingale such that

$$X_t^2 = X_0^2 + 2 \int_0^t X_s\, dX_s + \int_0^t \beta^2(s, X_s)\, ds.$$

Its expectation is deduced

$$E(X_t^2) = E(X_0^2) + 2E \int_0^t X_s\, dX_s + \int_0^t E\{\beta^2(s, X_s)\}\, ds.$$

6.2 Exponential solutions of differential equations

For a right-continuous distribution function F with left-limits, the survival function

$$\bar{F}(t) = \int_{[t,\infty[} dF$$

determines a hazard function

$$\Lambda(t) = \int_{]0,t]} \frac{dF}{1 - F_-},$$

reciprocally, the function $\Lambda = \Lambda^c + \Lambda^d$ determines uniquely the function $\bar{F}$ by the inversion formula

$$\bar{F}(t) = e^{-\Lambda^c(t)} \prod_{s \leq t} \{1 - \Delta\Lambda(s)\}.$$

The discrete survival function $\bar{F}^d(t) = e^{\Lambda^c(t)} \bar{F}(t)$ has the product representation

$$\bar{F}^d(t) = \prod_{s \leq t} \{1 - F^d(s^-)\}^{-1} \{1 - F^d(s)\}$$

and $\Delta F(t) = \Delta\Lambda(t)\{1 - F^d(t^-)\}$.

For a Brownian motion $(B_t)_{t \geq 0}$ on $(\Omega, \mathcal{F}, P, (F_t)_{t \geq 0})$ and a predictable process $(\beta_t)_{t \geq 0}$, the Gaussian process $Y_t = \int_0^t \beta_s\, dB_s$ defines the exponential martingale

$$Z_t = \exp\left\{ Y_t - \frac{1}{2} \int_0^t \beta_s^2\, ds \right\}$$

by Theorem 6.1 it is the unique solution of the equation

$$Z_t = 1 + \int_0^t Z_s \beta_s \, dB_s.$$

In the same way, some differential equations have an exponential solution generalized to functions and processes with jumps. Let A be a right-continuous increasing process with left-limits and let a be a function of $L^1(A)$. The stochastic implicit equation

$$Z_t = Z_0 + \int_0^t a_s Z_s^- \, dA_s \qquad (6.1)$$

has the unique solution

$$Z_t = Z_0 e^{\int_0^t a_s \, dA_s^c} \prod_{s \leq t} \{1 + a_s \Delta A(s)\}$$

$$= Z_0 e^{\int_0^t a_s \, dA_s} \prod_{s \leq t} \{1 + a_s \Delta A(s)\} e^{-a_s \Delta A_s}$$

with a continuous part $Z_t^c = Z_0^{-1} e^{-\int_0^t a_s \, dA_s^c}$ and a discrete part

$$Z_t^d = \prod_{s \leq t} \{1 + a_s \Delta A(s)\}$$

such that

$$\Delta Z^d(s) = Z^d(s^-) a_s \Delta A(s).$$

The proof is straightforward from the inversion formula of the process $\int_0^t a_s \, dA_s$. This implicit equation generalizes to a function a on a measurable space $(\mathbb{F}, \mathcal{Z})$ where the process Z is defined by the integral with respect to a right-continuous increasing process A with left-limits

$$a(Z_t) = a(Z_0) + \int_0^t a(Z_s^-) \, dA_s,$$

it has the unique solution

$$a(Z_t) = a(Z_0) e^{A_t} \prod_{s \leq t} \{1 + \Delta A(s)\} e^{-\Delta A_s}$$

with a continuous part $Z_t^c = a(Z_0) e^{A_t^c}$ and a discrete part with jumps $\Delta a(Z_t) = 1 + \Delta A(t)$.

More generally, the exponential process (1.27) of a semi-martingale X was deduced from Theorem 6.2, it is the unique solution of the equation $Z_t = 1 + \int_0^t Z_- \, dX$.

6.3 Exponential martingales, submartingales

On a filtered probability space $(\Omega, \mathcal{F}, P, \mathbb{F})$, let $M = (M_n)_{n \geq 0}$ be a local martingale with respect to $\mathbb{F} = (F_n)_{n \leq 0}$, with $M_0 = 0$. For every stopping time τ, $\{\tau \leq n\}$ is an element of F_n, the variable $M_{n \wedge \tau}$ is F_n measurable and $(M_{n \wedge \tau})_{n \geq 0}$ is a $\mathbb{F}$-martingale

$$
\begin{aligned}
E(M_{n \wedge \tau} \mid F_{n-1}) &= E(M_n 1_{\{n \leq \tau\}} + M_\tau 1_{\{n > \tau\}} \mid F_{n-1}) \\
&= E(M_n \mid F_{n-1}) 1_{\{\tau > n-1\}} + M_\tau 1_{\{\tau \leq n-1\}} \\
&= M_{(n-1) \wedge \tau}.
\end{aligned}
$$

Proposition 6.3. *Let $(M_n)_{n \geq 1}$ be a sequence of $\mathcal{M}_{0loc}^2$ and let $(A_n)_{n \geq 1}$ be the predictable process of its quadratic variations, then $Z_n = \exp(M_n^2 - A_n)$ is a local submartingale.*

Proof. The process $X_n = \sum_{k=1}^n (M_n - M_{n-1})^2 - (A_n - A_{n-1})$ is a centered local martingale such that, by convexity of the exponential and for every stopping time τ

$$
\begin{aligned}
E\left(\frac{Z_{n \wedge \tau}}{Z_{(n-1) \wedge \tau}} \mid \mathcal{F}_{(n-1) \wedge \tau} \right) &= \exp\left\{ M_{n \wedge \tau}^2 - M_{(n-1) \wedge \tau}^2 - (A_{n \wedge \tau} - A_{(n-1) \wedge \tau}) \right\} \\
&\geq \exp\{ E(X_{n \wedge \tau} - X_{(n-1) \wedge \tau} \mid \mathcal{F}_{(n-1) \wedge \tau}) \} = 1
\end{aligned}
$$

therefore $E(Z_{n \wedge \tau} \mid \mathcal{F}_{(n-1) \wedge \tau}) = Z_{(n-1) \wedge \tau}$ and $(Z_n)_{n \geq 1}$ is a submartingale with respect to $(\mathcal{F}_n)_{n \geq 1}$. $\square$

For every $n \geq 1$ and for $\lambda \neq 0$, the logarithm of the moment generating function $\varphi_t(\lambda) = E\{e^{\lambda M_n}\}$ of M_n is

$$
\phi_n(\lambda) = \log E \prod_{k=1}^n E\{e^{\lambda(M_k - M_{k-1})} \mid F_{k-1}\},
$$

the functions ϕ_n are concave. For every λ such that $\phi_n(\lambda)$ is finite, we consider the process

$$
Z_n(\lambda) = \exp\{\lambda M_n - \phi_n(\lambda)\}.
$$

Proposition 6.4. *If M is a local martingale with independent increments, for every $\lambda \neq 0$ such that $\phi_n(\lambda)$ is finite, the sequence $(Z_n(\lambda))_{n \geq 0}$ is a $\mathbb{F}$-local martingale with expectation $EZ_n(\lambda) = 1$ and a strictly positive variance $\exp\{\phi_n(2\lambda) - 2\phi_n(\lambda)\} - 1$.*

Proof. For a process X with independent increments and $X_0 = 0$, the equality $X_n = \sum_{k=1}^{n}(X_k - X_{k-1})$ entails

$$E(e^{\lambda X_n}) = E\left\{\prod_{k=1}^{n} e^{\lambda(X_k - X_{k-1})}\right\} = \prod_{k=1}^{n} E\left\{e^{\lambda(X_k - X_{k-1})}\right\}$$

therefore

$$\phi_n(\lambda) = \log \prod_{k=1}^{n} E\left\{e^{\lambda(X_k - X_{k-1})}\right\}$$

$$= \sum_{k=1}^{n} \log E\left\{e^{\lambda(X_k - X_{k-1})}\right\},$$

$$e^{\phi_n(\lambda)} = \prod_{k=1}^{n} E\left\{e^{\lambda(X_k - X_{k-1})}\right\} = E(e^{\lambda X_n})$$

hence $E\{Z_n(\lambda)\} = 1$. For every integrable stopping time τ

$$E\{Z_{(n+1)\wedge\tau}(\lambda) \mid F_n\} = Z_{n\wedge\tau}(\lambda)E\left[\frac{\exp\{\lambda(X_{(n+1)\wedge\tau} - X_{n\wedge\tau})\}}{e^{\phi_{n+1}(\lambda) - \phi_n(\lambda)}} \mid F_n\right]$$

$$= \frac{E[\exp\{\lambda(X_{n+1} - X_n)\}]}{e^{\phi_{n+1}(\lambda) - \phi_n(\lambda)}} 1_{\{n+1 \leq \tau\}}$$

$$+ 1_{\{n \geq \tau\}},$$

and by convexity

$$E[\exp\{\lambda(X_{n+1} - X_n)\} \mid F_n] \geq \exp[E\{\lambda(X_{n+1} - X_n)\}] = 1$$

therefore $E\{Z_{(n+1)\wedge\tau}(\lambda) \mid F_n\} \geq Z_{n\wedge\tau}(\lambda)$ and $(Z_n(\lambda))_{n\geq 0}$ is a local submartingale. Since $E\{Z_n(\lambda)\} = 1$, $(Z_n(\lambda))_{n\geq 0}$ is a local martingale. Moreover

$$EZ_n^2(\lambda) = E\exp\{2\lambda X_n - 2\phi_n(\lambda)\}$$

$$= E\{Z_n(2\lambda)\}\exp\{\phi_n(2\lambda) - 2\phi_n(\lambda)\}$$

$$= \exp\{\phi_n(2\lambda) - 2\phi_n(\lambda)\},$$

by the concavity of the logarithm, the variance $\exp\{\phi_n(2\lambda) - 2\phi_n(\lambda)\} - 1$ of Z_n is strictly positive.

For every stopping time τ

$$E\{Z_{(n+1)\wedge\tau}(\lambda) \mid F_n\} = Z_{n\wedge\tau}(\lambda)E\left[\frac{\exp\{\lambda(M_{(n+1)\wedge\tau} - M_{n\wedge\tau})\}}{e^{\phi_{(n+1)\wedge\tau}(\lambda) - \phi_{n\wedge\tau}(\lambda)}} \mid F_n\right]$$

where

$$E\left[\frac{e^{\lambda(M_{(n+1)\wedge\tau} - M_{n\wedge\tau})}}{e^{\phi_{(n+1)\wedge\tau}(\lambda) - \phi_{n\wedge\tau}(\lambda)}} \mid F_n\right] = E\left[\frac{\exp\{\lambda(M_{n+1} - M_n)\}}{e^{\phi_{n+1}(\lambda) - \phi_n(\lambda)}}\right]1_{\{n+1 \leq \tau\}}$$

$$+ 1_{\{n \geq \tau\}} = 1$$

therefore $E\{Z_{(n+1)\wedge\tau}(\lambda) \mid F_n\} = Z_{n\wedge\tau}(\lambda)$ and its expectation is 1

$$EZ_n^2(\lambda) = E\{Z_n(2\lambda)\}\exp\{\phi_n(2\lambda) - 2\phi_n(\lambda)\} = \exp\{\phi_n(2\lambda) - 2\phi_n(\lambda)\},$$

by concavity of the functions ϕ_n, $EZ_n^2(\lambda) > 1$. □

Proposition 6.4 leads to a generalization of Chernoff's theorem (Theorem 1.7) to the local supremum M_n^* of a martingale $(M_n)_{n\geq 1}$.

Theorem 6.3. *There exists a constant C such that for every $(M_n)_{n\geq 1}$ of $\mathcal{M}_0^{2loc}$, for every $a > 0$ and for every integer $n \geq 1$ we have*

$$P(M_n^* \geq a) \leq 4C_2 \inf_{\lambda>0} \exp\{-2\lambda a + \phi_n(2\lambda)\}.$$

For every $a > 0$ and for every $n \geq 1$

$$P(M_n^* \geq a) \leq \inf_{\lambda>0} \exp\{-\lambda a + \phi_n(\lambda)\}.$$

Proof. For all $a > 0$, $\lambda > 0$ and $n > 0$

$$P(M_n^* \geq a) = P(Z_n^*(\lambda) > \exp\{\lambda a - \phi_n(\lambda)\})$$
$$\leq \exp\{-\lambda a + \phi_n(\lambda)\}E\{Z_n^*(\lambda)\}$$

and the first inequality follows from the expectation $E\{Z_n^*(\lambda)\} = 1$. The second inequality is a consequence of Kolmogorov's inequality for $Z_n^*(\lambda)$

$$P(M_n^* \geq a) = P(Z_n^*(\lambda) > \exp\{\lambda a - \phi_n(\lambda)\})$$
$$\leq \exp\{-2\lambda a + 2\phi_n(\lambda)\}E\{Z_n^{*2}(\lambda)\}$$

where Propositions 4.13 and 6.4 imply

$$E\{Z_n^{*2}(\lambda)\} \leq 4E\{Z_\lambda^2(n)\}$$
$$= 4C_2 \exp\{\phi_n(2\lambda) - 2\phi_n(\lambda)\}, \tag{6.2}$$

it follows that for every $\lambda > 0$

$$P(M_n^* \geq an) \leq 4C_2 \exp\{-2\lambda a + \phi_n(2\lambda)\}.$$

□

Similar properties are obtained for martingales on $\mathbb{R}_+$.

Proposition 6.5. *For every $(M_t)_{t\geq 0}$ of $\mathcal{M}_{0loc}^2$, $Z_t = \exp(M_t^2 - <M>_t)$ is a local submartingale.*

Proof. The result is a consequence of the convexity of the exponential function if M is a continuous local martingale. If M is discrete, M_t^d is the limit of $\sum_{k=1}^n (X_{t_k} - X_{t_k-1})$ as $t_n - t_{n-1} = \delta_n$ tends to zero, where $(t_k)_{k=0,\ldots,n}$ is a partition of $[0, t]$, then the result is a consequence of Proposition 6.3. □

Let $\mathbb{F} = (\mathcal{F}_t)_{t \geq 0}$ be a filtration and let $M = (M_t)_{t \geq 0}$ be in $\mathcal{M}^2_{0,loc}(\mathbb{F})$. On a partition $\pi_n = (t_k)_{k=0,\dots,n}$ of $[0, t]$ such that $t_n - t_{n-1} = \delta_n$, $t_0 = 0$ and $t_n = t$, the martingale M is approximated by the sequence of variables $X_k = M_{t_k}$, $k = 1, \dots, n$, and its increments on π_n define the function $\phi_n(\lambda) = \log E \prod_{k=1}^{n} E\{e^{\lambda(X_k - X_{k-1})} \mid \mathcal{F}_{t_{k-1}}\}$ such that

$$E\{e^{\lambda(X_k - X_{k-1})} \mid \mathcal{F}_{t_{k-1}}\} \geq \exp\{\lambda E(X_k - X_{k-1} \mid \mathcal{F}_{t_{k-1}})\} = 1 \qquad (6.3)$$

and $\phi_n(\lambda) \geq 0$.

For a sequence $(t_k)_{k \geq 1}$, for $t > 0$ such that $t_n = t$ and for every $\lambda > 0$, the sequence of variables

$$\zeta_n(\lambda) = \prod_{k \leq n} E\{e^{\lambda(X_k - X_{k-1})} \mid \mathcal{F}_{t_{k-1}}\}$$

is a submartingale. If $\sup_{n \geq 1} E|\zeta_n(\lambda)|$ is finite, the submartingale $(\zeta_n(\lambda))_{n \geq 1}$ converges, as n tends to infinity, to a finite limit $\zeta_\infty(\lambda)$ such that $\zeta_n(\lambda) = E\{\zeta_\infty(\lambda) \mid \mathcal{F}_n\}$. The transform $\varphi_t(\lambda)$ of M_t is defined as $\zeta_\infty(\lambda)$, for a partition of $[0, t]$. Let $\phi_t(\lambda) = \log \varphi_t(\lambda)$.

Proposition 6.6. *Let $(M_t)_{t \geq 0}$ belong to $\mathcal{M}^2_{0loc}$, then for every $\lambda > 0$ such that $\sup_{\pi_n, n \geq 1} E|\zeta_n|$ is finite, the process*

$$Z_t(\lambda) = \exp\{\lambda M_t - \phi_t(\lambda)\}$$

is a local martingale of L^1.

Proof. For every stopping time τ and for all (t_k, t_{k+1}) of a partition π_n of $[0, t]$

$$\frac{E\{e^{\lambda(M_{t_{k+1} \wedge \tau} - M_{t_k \wedge \tau})} \mid F_k\}}{e^{\phi_{t_{k+1} \wedge \tau}(\lambda) - \phi_{t_k \wedge \tau}(\lambda)}}$$
$$= \frac{E[\exp\{\lambda(M_{t_{k+1} \wedge \tau} - M_{t_k \wedge \tau})\} \mid F_k]}{e^{\phi_{t_{k+1}}(\lambda) - \phi_{t_k}(\lambda)}} 1_{\{t_{k+1} \leq \tau\}} + 1_{\{t_k \geq \tau\}}$$

and it is larger than 1. From the limit as n tends to infinity, $(Z_{t_k \wedge \tau}(\lambda))_k$ is a local submartingale. Applying the same argument as in the proof of Proposition 6.4, as δ_n tends to zero implies that $E(Z_t) = 1$ so Z is a local martingale. $\qquad \square$

Theorem 6.4. *There exists a constant C such that for every $(M_t)_{t \geq 0}$ of $\mathcal{M}^{2loc}_0$, for every $a > 0$ and for every $t \geq 0$ we have*

$$P(M_t^* \geq a) \leq 4C_2 \inf_{\lambda > 0} \exp\{-2\lambda a + \phi_t(2\lambda)\}.$$

For every $a > 0$ and for every $t \geq 0$

$$P(M_t^* \geq a) \leq \inf_{\lambda > 0} \exp\{-\lambda a + \phi_t(\lambda)\}.$$

Proof. For all $a > 0$, $\lambda > 0$ and $t > 0$

$$P(M_t^* \geq a) = P(Z_t^*(\lambda) > \exp\{\lambda a - \phi_t(\lambda)\})$$
$$\leq \exp\{-\lambda a + \phi_t(\lambda)\}E\{Z_t^*(\lambda)\},$$

the first inequality follows from the expectation of the exponential martingale $E\{Z_t^*(\lambda)\} = E\{Z_\tau(\lambda)\} = 1$ for $\tau = \inf\{s; Z_s(\lambda) = Z_t^*(\lambda)\}$. The second inequality is a consequence of Kolmogorov's inequality for $Z_t^*(\lambda)$

$$P(M_t^* \geq a) = P(Z_t^*(\lambda) > \exp\{\lambda a - \phi_t(\lambda)\})$$
$$\leq \exp\{-2\lambda a + 2\phi_t(\lambda)\}E\{Z_t^{*2}(\lambda)\}$$

where Propositions 5.3 and 6.6 imply

$$E\{Z_n^{*2}(\lambda)\} \leq 4E\{Z_\lambda^2(n)\}$$
$$= 4C_2 \exp\{\phi_n(2\lambda) - 2\phi_n(\lambda)\}, \qquad (6.4)$$

it follows that for every $\lambda > 0$

$$P(M_n^* \geq an) \geq 4C_2 \exp\{-2\lambda a + \phi_n(2\lambda)\}.$$

$\square$

Let N be a point process with independent increments, with a real intensity $\Lambda_t > 0$, for every $t > 0$, and with the probabilities $P(N_t = k) = e^{-\Lambda_t}\Lambda_t^k(k!)^{-1}$, the local martingale of its compensated jumps $M_t = N_t - \Lambda_t$ has the moment generating function

$$\varphi_t(\lambda) = Ee^{\lambda M_t} = \exp\{\Lambda_t(e^\lambda - 1 - \lambda)\}$$

for $\lambda > 0$. Let $\phi_t(\lambda) = \log \varphi_t(\lambda)$, it follows from Proposition 6.6 that the process

$$Z_t(\lambda) = \exp\{\lambda M_t - \Lambda_t(e^\lambda - 1 - \lambda)\} = \exp\{\lambda N_t - \Lambda_t(e^\lambda - 1)\} \qquad (6.5)$$

is a local martingale. Theorem 6.4 implies that for every $a > 0$ and for every $t > 0$

$$P(M_t^* \geq a) \leq \inf_{\lambda > 0} \exp\{-\lambda a + \Lambda_t(e^\lambda - 1 - \lambda)\}$$

and the function $h_{at}(\lambda) = \Lambda_t(e^\lambda - 1 - \lambda) - \lambda a$ is minimum as $\Lambda_t(e^\lambda - 1) = a$ where its value is

$$\inf_{\lambda > 0} h_{at}(\lambda) = a - \sup_{\lambda > 0} \lambda(a + \Lambda_t)$$
$$= a - (a + \Lambda_t) \log \frac{\Lambda_t}{a + \Lambda_t}.$$

Equation (6.5) is also written as $Z_t(\lambda) = \exp\{\lambda N_t - \Lambda_t(e^\lambda - 1)\}$ and it is the exponential martingale of N_t.

Freedman (1973, 1975) determined several exponential sub and super-martingales on a probability space $(\Omega, \mathcal{F}, (\mathcal{F}_n)_{n\geq 0}, P)$ for a sequence of $\mathcal{F}_n$-adapted variables X_n such that $|X_n| \leq 1$, $n \geq 1$. They are obtained from a larger bound of the exponential. Let $Z_n = E(X_n \mid \mathcal{F}_{n-1})$ and, for every $t > 0$, let

$$M_n = \exp\{tX_n - Z_n(e^{t-1})\} \tag{6.6}$$

then $(M_n)_{n\geq 1}$ is a $(\mathcal{F}_n)_{n\geq 1}$ supermartingale. This property extends under the same conditions to a process on $\mathbb{R}_+$, $(X_t)_{t\geq 0}$ with predictable compensator $(A_t)_{t\geq 0}$, by discretization of the interval $[0,t]$ on a partition. Let $(\mathcal{F}_t)_{t\geq 0}$ be a filtration and let $N_n(t) = \sum_{i\leq n} 1_{\{T_i \leq t\}}$ be a point process with $T_0 = 0$ such that $N_n(t)$ is $\mathcal{F}_t$-measurable, its predictable compensator with respect to the filtration $(\mathcal{F}_t)_{t\geq 0}$ is

$$A_n(t) = \sum_{i\leq n} 1_{\{T_{i-1} < t \leq T_i\}} \int_0^t \{1 - \Delta\Lambda(s + T_i)\}\, d\Lambda(s + T_i).$$

For all s and $t > 0$, let

$$M_n(t) = \exp\{sN_n(t) - A_n(t)(e^{ns} - 1)\},$$

then by (6.6), $(M_n(t), \mathcal{F}_t)$ is a supermartingale for every n.

Let $V_n = E(X_n^2 \mid \mathcal{F}_{n-1})$, $n \geq 1$, be the predictable quadratic variations of a sequence of variables $(X_n)_{n\geq 1}$ such that $|X_n| \leq 1$ and $EX_n = 0$ for every n and, for every $t > 0$, the functions $e(t) = e^t - 1 - t$ and $f(t) = e^{-t} - 1 + t$, define the processes

$$Y_n = \exp\{tX_n - e(t)V_n\}, \tag{6.7}$$
$$Z_n = \exp\{tX_n - f(t)V_n\},$$

Freedman (1975) proved that $(Y_n, \mathcal{F}_n)_{n\geq 1}$ is a supermartingale and $(Z_n, \mathcal{F}_n)_{n\geq 1}$ is a submartingale.

A point process $N_n(t) = \sum_{i\leq n} 1_{\{T_i \leq t\}}$ with predictable compensator $A_n(t)$ defines the centered martingale $M_n = (N_n(t) - A_n(t))_{t\geq 0}$ with predictable quadratic variations $V_n(t) = A_n^c(t) + \widetilde{[A]}_n(t)$, for every integer n. The processes

$$Y_n(t) = \exp\{sM_n(t) - e(ns)V_n(t)\},$$
$$Z_n(t) = \exp\{sM_n(t) - f(ns)V_n(t)\},$$

are such that $(Y_n(t), \mathcal{F}_t)_{t \geq 0}$ is a supermartingale and $(Z_n(t), \mathcal{F}_t)_{t \geq 0}$ is a submartingale.

By Proposition 6.6, the process Z_t defined by (6.5) is a local martingale if and only if the point process N_t is a Poisson process. This is an example where the process Y_t studied by Freedman is a local martingale.

Let $S_n = \sum_{i=1}^{n} X_i$ for the sum of n independent and identically distributed real variables X_i, and let $Y_t = \sum_{i=1}^{N_t} X_i = S_{N_t}$ be a point process defined by an independent point process $N_t = \sum_{i \geq 1} 1_{\{T_i \leq t\}}$ with independent increments and with a deterministic intensity Λ_t. The process Y_t has the expectation $E(X)\Lambda_t$ and its moment generating function is determined by the moment generating function φ_X of the variables X_i

$$\varphi_{Y_t}(u) = E\{\exp(uY_t)\} = e^{-\Lambda_t} \sum_{k \geq 1} E\{\exp(uS_k)\}P(N_t = k)$$

$$= e^{-\Lambda_t} \sum_{k \geq 1} \frac{\{\Lambda_t \varphi_X(u)\}^k}{k!}$$

$$= \exp[-\Lambda_t\{1 - \varphi_X(u)\}].$$

The process $U_t = Y_t - E(X)\Lambda_t$ is a local martingale with independent increments with respect to the filtration $\mathcal{F}_t$ generated by N, such that the variables X_i are $\mathcal{F}_0$-measurable. Its moment generating function is

$$\varphi_t(u) = e^{-uE(X)\Lambda_t} E\{e^{uY_t}\} = \exp[\Lambda_t\{\varphi_X(u) - 1 - uE(X)\}]$$

and, by Proposition 6.6, for every $u > 0$ the process

$$Z_t(u) = \exp\{u(Y - EX\Lambda_t) - \log \varphi_t(u)\}$$

is a local martingale. Theorem 6.4 implies that for $a > 0$ and $t > 0$

$$P(U_t^* \geq a) \leq \inf_{u > 0} \exp[-ua + \Lambda_t\{\varphi_X(u) - 1 - uE(X)\}].$$

This example extends the cases already considered. Let N be a point process with a predictable compensator $\tilde{N}_t$, if N does not have independent increments then the process

$$Z_t(s) = \exp\{s(N_t - \tilde{N}_t) - \log \varphi_t(s)\}$$

defined by the generating function $\varphi_t(s) = Ee^{s(N_t - \tilde{N}_t)_t}$ is a local submartingale for every $s > 0$, by the same argument as in (6.3).

For a Brownian motion B, the variable $\tau_a = \sup\{t > 0; B_t \leq a\}$ is a stopping time and the events $\{\tau_a \leq t\}$ and $\{B_t^* > a\}$ are identical. For all $a > 0$ and $s > 0$, the process

$$Y(t \wedge \tau_a) = \exp\{sB(t \wedge \tau_a) - e(as)(t \wedge \tau_a)\}$$

is a $(\mathcal{F}_t)_{t \geq 0}$ supermartingale and

$$Z(t \wedge \tau_a) = \exp\{sB(t \wedge \tau_a) - f(as)(t \wedge \tau_a)\}$$

is a $(\mathcal{F}_t)_{t \geq 0}$ submartingale. These properties apply to the stochastic integrals of a predictable process H with respect to the Brownian motion $X_t = \int_0^t H_s \, dB_s$ or the martingale related to a point processes $X_t = \int_0^t H_s \, dM_s$. For every $a > 0$, the variable $\tau_a = \sup\{t > 0; X_t \leq a\}$ is a stopping time for the stochastic integrals. The processes Y and Z are now defined for the Brownian motion as

$$Y(t \wedge \tau_a) = \exp\left\{sX(t \wedge \tau_a) - e(as) \int_0^{t \wedge \tau_a} H_s^2 \, ds\right\},$$

$$Z(t \wedge \tau_a) = \exp\left\{sX(t \wedge \tau_a) - f(as) \int_0^{t \wedge \tau_a} H_s^2 \, ds\right\},$$

they have the same form for the martingale related to the point process, with a quadratic variations process $V(t) = \int_0^{t \wedge \tau_a} H_s^2 \, d\widetilde{[M]}(t)$.

The k-th moment of a variable Y is the value at zero of the k-th derivatives of the moment generating function $\varphi_Y(t) = Ee^{tY} = \sum_{k \geq 0} \frac{t^k}{k!} E(Y^k)$, its k-th cumulent κ_k is the coefficient of t^k in the expansion in series of the function $\varphi_Y = \log L_Y$. The cumulents depend on the moments of Y, $\kappa_1 = EY$, $\kappa_2 = \operatorname{Var} Y$, $\kappa_3 = E(Y - EY)^3$, $\kappa_4 = E(Y - EY)^4 - 3(\operatorname{Var} Y)^2$.

For a Poisson process N_t with function Λ_t, the k-th moment of $M_t = N_t - \Lambda_t$ is given by the k-th derivative with respect to s of the function $Ee^{s(N_t - \Lambda_t)} = \exp\{\Lambda_t(e^s - 1 - s)\}$, we get

$$E\{(N_t - \Lambda_t)^2\} = E\{(N_t - \Lambda_t)^3\} = \Lambda_t,$$

$E\{(N_t - \Lambda_t)^4\} = \Lambda_t + 2\Lambda_t^2$ and so on. Expansions of the functions exponential and h_s is another method to calculate the moments of M_t.

6.4 Gaussian processes

The sample paths of the Brownian motion are continuous but not differentiable, the next proposition concerns their regularity.

Proposition 6.7. *For the Brownian motion X on $[0, 1]$, there exists δ in $]0, \frac{1}{2}[$ such that $h^{\delta - \frac{1}{2}} \sup_{|t-s| \leq h} |X_t - X_s|$ converges a.s. to zero as h tends to zero.*

Proof. The L^2 norm of $X_t - X_s$ is $E|X_t - X_s|^2 = |t - s|$ and $h^{2\delta - 1} \sup_{|t-s| \leq h} E|X_t - X_s|^2 = h^{2\delta}$ converges to zero. Let $(t_j)_{j \leq J_h}$ in $[0, 1]$ such that $|t_i - t_{i+1}| \leq h$ and $t_0 = 0 < t_1 < \cdots < t_i < t_{i+1} < \cdots < t_{J_h} = 1$. It follows that for every $\varepsilon > 0$

$$\sum_{t_i \leq J_h} P(h^{\delta - \frac{1}{2}} |X_{t_i} - X_{t_{i+1}}| > \varepsilon) \leq \frac{h J_h}{h^{2\delta - 1} \varepsilon^2}$$

where J_h has the order h^{-1} hence

$$\frac{h J_h}{h^{2\delta - 1} \varepsilon^2} = \frac{O(h^{1-2\delta})}{\varepsilon^2}$$

and it converges to zero as h tends to zero, it follows that $h^{\delta - \frac{1}{2}} \sup_{|t-s| \leq h} |X_t - X_s|$ converges a.s. to zero. $\square$

Applying Proposition 5.3 with $p > 1$ to the martingales B_t, such that $E(B_t^4) = 3t^2$, and $B_t^2 - t$ yields that for all positive $s < t$

$$E\{(B_t - B_s)^{2p}\} \geq (t - s)^p,$$
$$E\{(B_t^2 - t)^{2p}\} \geq 2^p t^{2p}$$

and for every convex function g, $E\{g((B_t^2 - t)^2)\} \geq g(2t^2)$.

Let $a > 0$ and let P_a be the probability distribution of the Brownian motion B starting from a. The density $f_{t,a}$ of B under P_a has the limit zero as t tends to infinity and it tends to infinity as $t \to 0^+$. It is solution of the differential equation

$$\left\{ 2\frac{\partial}{\partial t} - \frac{\partial^2}{\partial a^2} \right\} f_{t,a} = 0$$

with initial value $f_{a,a} = 0$. Let

$$\tau_a = \min\{t : B_t > a\}$$

the properties of the Brownian motion at t apply at τ_a and at every stopping time, in particular $EB(\tau_a) = a$ and by the Laplace transform

$$e^{-\lambda a} = E_0 e^{-\frac{\lambda^2 \tau_a}{2}}, \quad a > 0.$$

This property extends to the waiting time for the Brownian motion to reach a positive and continuous function h on $\mathbb{R}_+$. The moment generating function of B at the stopping time $\tau = \min\{t : B_t \geq h_t\}$ is

$$\varphi_\tau(\lambda) = E_0(e^{\lambda h_\tau}) = E_0 \exp\left\{-\frac{\lambda^2 \tau}{2}\right\}$$

for every $\lambda > 0$, and the Laplace transform of τ is

$$E_0 e^{-s\tau} = E_0 e^{\sqrt{2s}h(\tau)}.$$

The exponential of a centered martingale is generally a submartingale, the martingale property of the process $Z_t = \exp\{B_t - \frac{t}{2}\}$ extends to other processes.

Proposition 6.8. *On a filtered probability space* $(\Omega, \mathcal{A}, (\mathcal{F}_t)_{t\geq 0}, P)$, *let* $(X_t)_{t\geq 0}$ *be a martingale with independent increments such* $EX_0 = 0$. *For all* u *and* t *such that the moment generating function* $\varphi_{X_t}(u)$ *of* X_t *is finite, the process* $Y_t(u) = \exp\{uX_t - \log \varphi_{X_t}(u)\}$ *is a* $(\mathcal{F}_t)_{t\geq 0}$-*submartingale if* $\varphi_{X_s}\varphi_{X_{t-s}} \geq \varphi_{X_t}$ *for every* $0 < s < t$ *and it is a supermartingale if* $\varphi_{X_s}\varphi_{X_{t-s}} \leq \varphi_{X_t}$ *for every* $0 < s < t$.

Proof. For all $u > 0$ and $t > s > 0$, $\varphi_{X_t}(u) - \varphi_{X_s}(u) = E(e^{uX_t} - e^{uX_s})$ and

$$E\{(Y_t - Y_s)(u)|\mathcal{F}_s\} = Y_s(u)E\left\{\frac{Y_t}{Y_s}(u) - 1|\mathcal{F}_s\right\}$$

$$= Y_s(u)\left[\frac{\varphi_{X_s}}{\varphi_{X_t}}(u)E\exp\{u(X_t - X_s)|\mathcal{F}_s\} - 1\right]$$

$$= Y_s(u)\left\{\frac{\varphi_{X_s}\varphi_{X_{t-s}}}{\varphi_{X_t}}(u) - 1\right\},$$

therefore $E\{Y_t(u)|\mathcal{F}_s\} = Y_s(u)\varphi_{X_s}(u)\varphi_{X_{t-s}}(u)\varphi_{X_t}^{-1}(u)$ and the result follows. $\square$

Corollary 6.1. *Let* $(X_t)_{t\geq 0}$ *be a martingale with independent increments and* $EX_0 = 0$ *and let* φ_{X_t} *be the moment generating function of* X_t. *If for every* u *the process* $(Y_t(u))_{t\geq 0} = (\exp\{uX_t - \log \varphi_{X_t}(u)\})_{t\geq 0}$ *is a martingale with respect to the filtration generated by the process* $(X_t)_{t\geq 0}$, *then* $(X_t)_{t\geq 0}$ *is a stationary process.*

By Proposition 6.8, the process $(Y_t(u))_t$ is such that $\log \varphi_{X_t}(u)$ is stationary in t, for every u, and by an expansion of the exponential with respect to u, all moments of X_t are stationary.

Example 6.1. Let $(X_n)_n$ be a random walk where the variables X_n have the moment generating function φ, let $S_n = \sum_{i=1}^{n} X_i$ and let $\mathcal{F}_n$ be the σ-algebra generated by $(X_i)_{1 \leq i \leq n}$. For every t such that $\varphi(t)$ is finite, the process $Y_n(t) = \exp\{tS_n - n \log \varphi(t)\}$, is such that

$$\frac{E\{Y_{n+1}(t) \mid \mathcal{F}_n\}}{Y_n(t)} = E\left\{ \frac{Y_{n+1}(t)}{Y_n(t)} \mid \mathcal{F}_n \right\} = \frac{E \exp(tX_{n+1})}{\varphi(t)} = 1$$

therefore $(Y_n)_{n \geq 0}$ is a martingale.

On a filtered probability space $(\Omega, \mathcal{A}, \mathcal{F}_{(t)t>0}, P)$, let B be a Brownian motion and let h be a function of $L^2(\mathbb{R}_+)$. The translated Brownian motion $B'_t = B_t - \frac{1}{2} \int_0^t h_s \, ds$ has the expectation $-\frac{1}{2} \int_0^t h_s \, ds$ and the variance t, and the process

$$Z_t = \exp\left(\int_0^t h_s \, dB_s - \frac{1}{2} \int_0^t h_s^2 \, ds \right) = \exp\left(\int_0^t h_s \, dB'_s \right) \qquad (6.8)$$

is a martingale with respect to the filtration $\mathcal{F}_{(t)t>0}$, its expectation is 1 and $E(Z_t^2) = \exp\{\int_0^t h_s^2 \, ds\}$. It determines a probability P' having the Radon-Nikodym derivative Z_t with respect to P, on $\mathcal{F}_t$

$$Z_t = E\left(\frac{dP'}{dP} \mid \mathcal{F}_t \right).$$

By the change of probability and the martingale property, we have $E(Z_\infty) = 1$, $Z_t = E(Z_\infty \mid \mathcal{F}_t)$ for $t > 0$ and $E(Z_{t+s} \mid \mathcal{F}_t) = Z_t$ for s and $t > 0$

$$E_{P'}(Z_{t+s} \mid \mathcal{F}_t) = E_P\{Z_t(Z_{t+s} \mid \mathcal{F}_t)\} = Z_t^2.$$

Under the probability P', Z_t has the expectation $E_{P'} Z_t = E_p(Z_t^2)$ and the variance $\text{Var}_{P'} Z_t = E_p(Z_t^3) - E_p^2(Z_t^2)$, the Brownian motion B_t has the expectation $E_{P'} B_t = E_p(Z_t B_t)$ and the variance $\text{Var}_{P'} Z_t = E_p(Z_t B_t^2) - E_p^2(Z_t B_t)$.

Let $(X_t, \mathcal{F}_t)_{t \geq 0}$ be continuous in $\mathcal{M}_{0,loc}^2$ and let $(< X >_t)t \geq 0$ be the predictable process of its quadratic variations. Let h be a function of $L^2(d < X >_t)$, for a generalization of the martingale Z_t defined by (6.8), the question is to find conditions under which the process

$$Z_t(h) = \exp\left(\int_0^t h_s \, dX_s - \frac{1}{2} \int_0^t h_s^2 \, d < X >_s \right)$$

is a martingale. Now Z_t is solution of the differential equation of a diffusion

$$dZ_t(h) = h_t Z_t \, dX_t - \frac{1}{2} h_t^2 Z_t \, d < X >_t \qquad (6.9)$$

where $-\frac{1}{2}h_t^2 Z_t \, d < X >_t$ may be stochastic. Let

$$\zeta_{s,t}(h) = \int_s^t h_u \, dX_u$$

it is centered and its variance function is

$$v_{s,t}(h) = \int_s^t h_u^2 \, dE < X >_u .$$

For all $0 < s < t$, we have

$$E\left(\frac{Z_t(h)}{Z_s(h)} \mid \mathcal{F}_t\right) = E \exp\left\{\zeta_{s,t}(h) - \frac{1}{2}v_{s,t}(h)\right\}$$

$$= \exp\left\{\log \varphi_{s,t}(h) - \frac{1}{2}v_{s,t}(h)\right\}$$

where $\varphi_{s,t}(h) = E \exp\left\{\zeta_{s,t}(h)\right\}$.

Let $(B_{<X>_t})_{t \geq 0}$ be the transformed Brownian motion with variance $< X >_t$.

Proposition 6.9. *The process* $Z_t = \exp\left(B_{<X>_t} - \frac{1}{2} < X >_t\right)$ *is a submartingale and it is a martingale if and only if* $< X >_t$ *is deterministic.*

Proof. For every stopping time τ for X_t, the process $M_\tau = B_{<X>_\tau}$ is a stopped Brownian motion, where M_t is a martingale conditionally on $< X >_t$, with variance $EM_t^2 = E(< X >_t)$ and its moment generating function is

$$\varphi_t(\lambda) = E \exp(\lambda B_{<X>_t}) = E \exp\left(\frac{\lambda^2 < X >_t}{2}\right).$$

Let $\widetilde{Z}_t(\lambda) = \exp\left\{\lambda B_{<X>_t} - \frac{1}{2}\log \varphi_t(\lambda)\right\}$, $\lambda > 0$, the process Z_t is related to $\widetilde{Z}_t$ by the relation

$$Z_t = \exp\left(M_t - \frac{1}{2} < X >_t\right)$$

$$= \widetilde{Z}_t(1) \exp\left[-\frac{1}{2}\{< X >_t - \log \varphi_t(1)\}\right]$$

$$= \widetilde{Z}_t(1) \exp\left(-\frac{< X >_t}{2}\right) E\left\{\exp\left(\frac{< X >_t}{2}\right)\right\}.$$

By convexity, its expectation is larger than 1 and it is 1 if and only if $< X >_t$ is deterministic. The same argument proves that $E(Z_s^{-1} Z_t \mid \mathcal{F}_s) \leq 1$, for $s < t$, which implies Z_t is a submartingale. $\qquad\square$

Proposition 6.9 applies to the process $Z(h)$ for every function h such that $v_{0,t}(h)$ is locally bounded.

Proposition 6.10. *The process $Z_t(h)$ is a martingale if and only if X is the Brownian motion or the integral of a real function with respect to the Brownian motion.*

The moment generating function of a non Gaussian process Z depends on the moments of X higher than two and Z is not a martingale.

The definition of a diffusion $Y_t = \int_0^t a_s \, ds + \int_0^t b_s \, dB_s$ absolutely continuous with respect to $X_t = \int_0^t \alpha_s \, ds + \int_0^t \beta_s \, dB_s$ requires a to be absolutely continuous with respect to α and b is absolutely continuous with respect to β, with the same ratio $dY_t = \varphi_s \, dX_t$. According to the Gaussian distributions of X_t and Y_t, the logarithm of the ratio of the distributions of Y_t and X_t is

$$l_t = -\frac{1}{2}\left\{ \frac{(Y_t - \int_0^t a_s \varphi_s \, ds)^2}{\int_0^t \beta_s^2 \varphi_s^2 \, ds} - \frac{(X_t - \int_0^t \alpha_s \, ds)^2}{\int_0^t \beta_s^2 \, ds} + \log \frac{\int_0^t \beta_s^2 \varphi_s^2 \, ds}{\int_0^t \beta_s^2 \, ds} \right\}.$$

For a sequence of probabilities $(P_{\varphi_n})_{n\geq 1}$ contiguous to P, such that φ_n converges uniformly to 1 on $[0, t]$, or $(P_{\varphi_T})_{T>0}$ such that ϕ_T converges uniformly to 1 on an interval $[0, T]$, as T tends to infinity, the log-likelihood ratio l_t has an expansion depending on the difference of the processes $Y_T(t) - X(t)$ under P_{φ_T} and P. Let $\varphi_T(t) = 1 + T^{-\frac{1}{2}}\theta_T(t)$ where the sequence $(\theta_T)_{T>0}$ converges uniformly on $\mathbb{R}_+$ to a limit θ such that $\eta_T = T^{\frac{1}{2}}(\theta_T - \theta)$ converges uniformly to a function η as T tends to infinity, then $2l_T(t)$ has a second-order expansion converging weakly to a quadratic limit from the convergence properties of the normalized process X_T.

Consider a Gaussian diffusion

$$dX_t = \alpha(X_t)\, dt + C_t \, dB_t,$$

with initial value X_0 and with random coefficients defined by a $L^2(P)$ process $A_t = \int_0^t \alpha(X_s)\, ds$ and a $L^2(P \otimes \mu)$ predictable process C_t, the process X has the expectation $E(X_0) + E\{\int_0^t \alpha(X_s)\, ds\}$ and the variance

$$v_t = \mathrm{Var}\left\{ X_0 + \int_0^t \alpha(X_s)\, ds \right\} + \int_0^t E(C_s^2)\, ds.$$

The condition $\int_0^t E(C_s^2)\, ds$ finite is equivalent to $\mathrm{Var}\{X_t - X_0 - \int_0^t \alpha(X_t)\, dt\}$ finite. The process

$$Z_t(\lambda) = \exp\left[\lambda\left\{ X_t - X_0 - \int_0^t \alpha(X_s)\, ds \right\} - \frac{\lambda^2 \int_0^t E(C_s^2)\, ds}{2} \right]$$

is a martingale with respect to filtration $F^X = (F_t^X)_{t>0}$ generated by X and $EZ_t(\lambda) = 1$ for all $t > 0$ and $\lambda > 0$.

The convergence of $t^{-1} \int_0^t E(C_s^2) \, ds$ to a finite limit $v > 0$ implies the convergence to zero of $\exp\{-\frac{1}{2}\lambda^2 \int_0^t E(C_s^2) \, ds\}$, as t tends to infinity. The condition of a finite integral $\int_0^\infty E(C_s^2) \, ds$ implies

$$EZ_\infty(\lambda) \geq \exp\left\{-\frac{\lambda^2 \int_0^\infty E(C_s^2) \, ds}{2}\right\} > 0$$

and $Z_t = E(Z_\infty \mid \mathcal{F}_t)$.

Let h be a function such that $E \int_0^t h_s^2 C_s^2 \, ds$ is finite, by the integration of h with respect to $C \, dB$, the process

$$Z_t(\lambda, h) = \exp\left[\lambda\left\{\int_0^t h_s \, dX_s - \int_0^t h_s \alpha(X_s) \, ds\right\} - \frac{\lambda^2 \int_0^t h_s^2 E(C_s^2) \, ds}{2}\right]$$

is a martingale with expectation 1 for all $t > 0$ and $\lambda > 0$.

6.5 Processes with independent increments

Let $(X_t)_{t \geq 0}$ be a process with independent and stationary increments on a filtered probability space $(\Omega, \mathcal{F}, P, \mathbb{F})$, for all $s < t$ and for every stopping time τ

$$E(X_{t \wedge \tau} \mid F_s) = X_{s \wedge \tau}.$$

For $\lambda > 0$, the function $\varphi_t(\lambda) = E(e^{\lambda X_t})$ factorizes as

$$\varphi_t(\lambda) = E\{e^{\lambda X_s}\} E\{e^{\lambda(X_t - X_s)}\} = \varphi_s(\lambda)\varphi_{t-s}(\lambda)$$

for all $s < t$, and $\phi_t(\lambda) = \log E(e^{\lambda X_t})$ is the sum

$$\phi_t(\lambda) = \phi_s(\lambda) + \phi_{t-s}(\lambda).$$

Proposition 6.11. *Let $(X_t)_{t \geq 0}$ be a right-continuous process with independent and stationary increments and such that $X_0 = 0$, then the function $\phi_t = \log E(e^{X_t})$ is such that $\phi_t = \phi_s + \phi_{t-s}$ for all $t > s > 0$.*

Proof. The function $\varphi_t = E(e^{X_t})$ factorizes as

$$\varphi_t = E\{e^{X_t}\} E\{e^{X_t - X_s}\} = \varphi_s \varphi_{t-s}$$

for all $s < t$, and $\phi_t = \log E(e^{X_t})$ is the sum $\phi_t = \phi_s + \phi_{t-s}$. $\qquad \square$

For a process X_t with independent increments, the factorization of the function φ_t implies that the process

$$Z_t(\lambda) = \exp\{\lambda X_t - \phi_t(\lambda)\}$$

is a martingale. For all $s < t$, it satisfies

$$E\Big\{Z_t(\lambda) - Z_s(\lambda) \mid \mathcal{F}_s\Big\} = Z_s(\lambda)\Big[E\Big\{\frac{Z_t}{Z_s}(\lambda) \mid \mathcal{F}_s\Big\} - 1\Big],$$

$$E\Big\{\frac{Z_t}{Z_s}(\lambda) \mid \mathcal{F}_s\Big\} = E\{e^{\lambda(X_t - X_s)}\}\frac{\varphi_s}{\varphi_t}(\lambda) = 1.$$

Let $A = [0, a]$ be a positive interval, and let

$$\tau_a = \min\{t : X_t \notin A\}$$

τ_a is a stopping time for X_t such that

$$\{\tau_a > t\} = \{X_s \in A, \forall s \leq t\}.$$

Proposition 6.12. *The function $\psi_{a,t} = \log P_a(\tau_a > t)$ is additive, for all $a > 0$ and $t > s > 0$*

$$\psi_{a,t} = \psi_{a,s} + \psi_{a,t-s},$$

$P_a(\tau_a > 0) = 1$ *and τ_a is independent of X_{τ_a}.*

Proof. For $0 < s < t$, the properties if X_t imply

$$
\begin{aligned}
P_a(\tau_a > t + s) &= P_a(X_u \in A, \forall u \leq t + s)\\
&= P_a(X_u \in A, \forall u \leq t, X_v \in A, \forall t < v \leq t + s),\\
&= P_a(X_u \in A, \forall u \leq t)P_a(X_v \in A, \forall t < v \leq t + s)\\
&= P_a(\tau_a > t)P_a(\tau_a > s)
\end{aligned}
$$

and by the same proof as for φ_t, there exists a stationary function u_t, satisfying $u_t = u_s + u_{t-s}$ for $t > s > 0$ for all $t > s \geq 0$, and such that the distribution function $P_a(\tau_a > t)$ is an exponential distribution e^{ut} with $u \geq 0$ and $e^{u_0} = 1$, under P_a. Then $P_a(\tau_a > 0) = e^{u_0} = 1$.

Moreover, $\tau_a > t$ implies $\tau_a = t + \sigma_{a,t}$ and for every borelian set E

$$
\begin{aligned}
P_a(\tau_a > t, X_{\tau_a} \in E) &= P_a(X_s \leq a, \forall s \leq t, X_{t+\sigma_{a,t}} \in E)\\
&= P_a(\tau_a > t)P_{X_t}(X_{\sigma_{a,t}} < b - X_t)\\
&= P_a(\tau_a > t)P_a(X_{X_{\tau_a}} \in E).
\end{aligned}
$$

$\square$

The function $P_a(\tau_a > t)$ satisfies

$$P_a(\tau_a > t) = \exp\left\{-\int_0^t \frac{dP_a(\tau_a \leq s)}{P_a(\tau_a > s)}\right\} = \exp(\psi_{a,t}),$$

where $\psi_{a,t} < 0$ for all $a > 0$ and $t > 0$, if $\lim_{t\to\infty} \psi_{a,t} = -\infty$, it follows that $\lim_{t\to\infty} t P_a(\tau_a > t) = 0$ and $E_a \tau_a = \int_0^\infty P_a(\tau_a > t)\, dt$.

Let $A = [0, a]$ and $B = [0, b]$ be positive intervals such that $0 < a < b$ is finite, the last exit time from A for X_t belonging to B is

$$\gamma = \sup\{t > 0 : X_t \in A, t < \tau_b\}.$$

Let $\tau_{1,a} = \tau_a$ and for $n > 1$ let

$$\tau_{n,a} = \inf\{t : \tau_{n-1,a} < t \leq \gamma, X_t \notin A\},$$

γ belongs to the increasing sequence of stopping times $(\tau_{n,a})_{n\geq 1}$ or it is its limit. From Proposition 6.12, $P_a(\gamma > 0) = 1$ and the stopping time

$$\gamma = \max_{n\geq 1}\{\tau_{n,a} \leq \tau_b\},$$

is such that

$$\{\gamma > t\} = \cup_{n\geq 1}\{\tau_{n,a} : t < \tau_{n,a} \leq \tau_b\}.$$

For every $t > \tau_a$, the tail probability of τ_b starting from a is

$$\begin{aligned}
P_a(\tau_b > t + s) &= P_a(X_u \in B \setminus A, \forall u \leq t + s) \\
&= P_a(X_u \in B \setminus A, \forall u \leq t, X_v \in B \setminus A, \forall t < v \leq t + s) \\
&= P_a(\tau_b > t) P_a(\tau_b > s),
\end{aligned}$$

by the same arguments as for Proposition 6.12, $\psi_{a,b,t} = \log P_a(\tau_b > t)$ is additive

$$\psi_{a,b,t} = \psi_{a,b,s} + \psi_{a,b,t-s}$$

for $t > s > 0$, $P_a(\tau_b > 0) = 1$ and τ_b is independent of X_{τ_b}.

Proposition 6.13. *For all $a > 0$, $s > 0$ and $t > 0$*

$$P_a(\gamma > t + s) = P_a(\gamma > t) P_a(\gamma > s)$$

$P_a(\gamma > 0) = 1$ *and γ is independent of X_γ.*

Proof. Let $a > 0$, $P_a(\tau_a > 0) = 1$ and $P_a(\gamma > 0) \geq P_a(\tau_a > 0) = 1$. Since γ is the maximum of the stopping times $\tau_{n,a} \leq \tau_b$, either γ is one of them and Proposition 6.12 applies or it is their limit, then the distribution function of γ under P_a satisfies

$$P_a(\gamma > t) = \lim_{n \to \infty} P_a(\tau_{n,a} > t)$$

where $P_a(\tau_{n,a} > t) = \exp\{\psi_{n,a,t}\}$. The function $\psi_{n,a,t} = \log P_a(\tau_{n,a} > t)$ is additive and it converges uniformly to an additive function $\phi_{a,t}$, the properties of the stopping times $\tau_{n,a}$ extend to γ. $\qquad\square$

From the property $P_a(\gamma > t) = \lim_{n\to\infty} P_a(\tau_{n,a} > t)$, the function

$$P_a(\gamma > t) = \exp\left\{ -\int_0^t \frac{dP_a(\gamma \leq s)}{P_a(\gamma > s)} \right\}$$

satisfies the property $\lim_{t\to\infty} t P_a(\gamma > t) = 0$ if the stopping times $\tau_{n,a}$ satisfy it, then γ has the expectation $E_a\gamma = \int_0^\infty P_a(\gamma > t)\, dt$.

For a diffusion $dX_t = \alpha_t\, dt + \beta_t\, dB_t$ such that $\sigma_t^2 = E \int_0^t \beta_s^2\, ds$, the process $Y_t = X_t - \int_0^t \alpha_s\, ds$ is a transformed Brownian motion $W_t = B \circ \sigma_t^2$, it has independent increments. If the function σ_t^2 is stationary, the same propositions apply to the stopping times of the process W_t, by the properties of its exponential martingale.

6.6 Semi-martingales

On a probability space $(\Omega, \mathcal{F}, (\mathcal{F}_t)_{t\geq 0}, P)$, a semi-martingale is defined as a sum

$$X = X_0 + A + M$$

of an initial value X_0, a $(\mathcal{F}_t)_t$-adapted process A with locally bounded variations, such that $E \int_0^\infty |dA_s|$ is finite for every $t > 0$, and a local martingale M of $\mathcal{M}_{0,loc}^2(P)$. For all $s < t$, it follows that $E(X_t - X_s \mid \mathcal{F}_s) = E(A_t - A_s \mid \mathcal{F}_s)$ and, if A is predictable and belongs to $L_{0,loc}^2(P)$

$$E\{(X_t - X_s)^2 \mid \mathcal{F}_s\} = E\{(A_t - A_s)^2 + (M_t - M_s)^2 \mid \mathcal{F}_s\}$$

i.e. $< X >_t\, =\, < A >_t + < M >_t$ where

$$< A >_t\, =\, < A >_s + E\{(A_t - A_s)^2 \mid \mathcal{F}_s\},$$
$$< M >_t\, =\, < M >_s + E\{(M_t - M_s)^2 \mid \mathcal{F}_s\}$$

and the process $X_t^2 - < X >_t$ is a local martingale.

The stochastic integral of a predictable process Y with respect to a semi-martingale of L^2 is the sum of the stochastic integrals of Y with respect to A and to M. For a point process N with a continuous increasing predictable process A, $[N]_t - <N>_t$ is identical to $N_t - A_t$. Let $Y.N_t = \int_0^t Y_s \, dN_s$ be the integral of a predictable process Y of $L^{2,loc}(P)$ with respect to a point process N with predictable compensator

$$\widetilde{N}_t = \sum_{n \geq 1} 1_{\{T_{n-1} \leq t < T_n\}} \int_{T_{n-1}}^{t \wedge T_n} \frac{dF_n(s)}{1 - F_n(s^-)}$$

where $F_n(t) = P(T_n \leq t \mid \mathcal{F}_{T_{n-1}})$, the process $<Y.N>_t = \int_0^t Y_s \, \widetilde{N}_s$ is such that $(A_t - \widetilde{A}_t)^2 - \widetilde{A}_t$ is a $(\mathcal{F}_t)_t$-local martingale.

Let $A_t = \sum_{n=1}^{N_t} H_{T_n}$ be a stationary point process with independent increments, its transform is expressed according to the conditional transforms of its jumps

$$Ee^{\lambda A_t} = E\left\{ \prod_{n \leq N_t} E(e^{\lambda H_{T_n}} \mid \mathcal{F}_{T_{n-1}}) \right\}$$

$$= E \exp\left\{ \sum_{0 < s \leq t} \log E(e^{\lambda \Delta A_s} \mid \mathcal{F}_{s-}) \right\} = \exp\left\{ \sum_{s \leq t} \log \varphi_s(\lambda) \right\}$$

where $\varphi_s(\lambda) = E(e^{\lambda \Delta A_s} \mid \mathcal{F}_{s-}) = E(e^{\lambda \Delta A_{T_n}} \mid \mathcal{F}_{T_{n-1}})$ at $s = T_n$. Its predictable compensator has independent increments and a transform

$$Ee^{\lambda \widetilde{A}_t} = E \exp\left(\sum_{n \geq 1} 1_{\{T_{n-1} \leq t < T_n\}} \int_{T_{n-1}}^{t \wedge T_n} H \, d\widetilde{N} \right).$$

The local martingale $A_t - \widetilde{A}_t$ has independent increments and an exponential transform $\varphi_t = Ee^{\lambda(A_t - \widetilde{A}_t)}$ such that

$$\varphi_t = E \exp\left\{ \sum_{n \leq N_t} \log \varphi_{T_{n-1}}(\lambda) - \lambda \widetilde{A}_t \right\}$$

$$= \exp\left\{ \sum_{s \leq t} \log \varphi_s(\lambda) + \log Ee^{-\lambda \widetilde{A}_t} \right\}.$$

If X_t is a continuous martingale, the exponential martingale

$$Z_t(\lambda) = \exp\{\lambda X_t - \phi_t(\lambda)\}$$

is solution of the diffusion equation

$$dZ_t(\lambda) = \lambda Z_t \, dX_t - d\phi_t(\lambda) Z_t.$$

The process Z_t is also the product of the independent processes Z_s and $\exp\{\lambda(X_t - X_s) - \phi_t(\lambda) + \phi_s(\lambda)\}$, for every $s < t$. By stationarity, Z_t has the same distribution as the product of independent processes Z_s and $\widetilde{Z}_{t-s}$ which has the same distribution as Z_{t-s}. Let $t > 0$, for every sequence $(t_0, \ldots, t_n)$ of $\mathbb{R}_+$ such that $t_0 = 0$ and $t_n = t$, there exist n independent and identically distributed variables $\widetilde{Z}_{t,k}$ having respectively the same distribution as $Z_{t_k - t_{k-1}}$, $k = 1, \ldots, n$, and such that the variables Z_t and $Z_{t_1} \prod_{1 < k \leq n} \widetilde{Z}_{t,k}$ have the same distribution.

A martingale with independent and stationary increments, sum of a continuous process X^c and a discrete process X^d has an exponential martingale Z_t solution of the equations

$$Z_t = Z_t^c \prod_{s \leq t} \Delta Z_s^d,$$

$$Z_t^c = \lambda \int_0^t Z_s \, dX_s^c - \int_0^t \frac{\phi_s^{c\prime}}{\phi_s^c}(\lambda) Z_s \, ds,$$

$$\Delta Z_t^d(\lambda) = e^{\lambda \Delta X_t^d - \Delta \phi_t^d}$$

where the function $\phi_t(\lambda) = \log E(e^{\lambda X_t})$ is the sum $\phi_t = \phi_t^c + \sum_{s \leq t} \Delta \phi_s^d$. The function ϕ_t^d is the discrete part of the expectation of Z^d. For a process

$$X_t = X_0 + X_t^c + N_t,$$

with a point process N_t, ϕ_t is continuous if the predictable compensator of N_t is continuous.

6.7   Level crossing probabilities

The previous sections provide bounds for the tail probabilities of sums of variables and martingales. The exact tail probabilities of distributions of empirical laws have been computed by many authors who established tables used in goodness-of-fit testing for samples of variables, such as tables for the Kolmogorov-Smirnov statistics. They give exact or asymptotic expansions for the tail probabilities of empirical distributions, calculated from the order statistics of the variable sample. Other results concern the limiting probabilities of the number of level crossings. The first result in this field concerns sums of independent and identically distributed variables with expectation zero. Let

$$N_n = \min\{k : S_k \geq S_m, 1 \leq k, m \leq n\}$$

be the first index where the maximum of the partial sums $S_k = X_1 + \cdots + X_k$ is achieved, for $k = 1, \ldots, n$. Considering the binomial distribution with values -1 and 1 with probabilities $\frac{1}{2}$. Lévy (1939) established that for every α in $[0, 1]$

$$\lim_{n \to \infty} P(n^{-\frac{1}{2}} N_n < \alpha) = \frac{2}{\pi} \arcsin(\alpha^{\frac{1}{2}}).$$

Erdös and Kac (1947) extended this result to other distributions. These results are easily proved from the weak convergence of the process $n^{-\frac{1}{2}} N_n$ to the standard Brownian motion and from the arcsine law (1.24) for the time when the Brownian motion reaches its maximum.

Consider the sum S_n of n exponential variables with parameter 1, $\lim_{n \to \infty} n^{-1} S_n = 1$ a.s. and the variables S_n are the increasing times of a Poisson process. For a threshold $t > 0$, the number of sums such that $S_n < t$ is N_{t^-}, where $N_t = \sum_{k \geq 1} 1_{\{S_k \leq t\}}$, therefore $N_t = n$ when t belongs to the random interval $[S_n, S_{n+1}[$. The process N_t has a homogeneous Poisson distribution with intensity t and expectation $E N_t = t$, and the process $t^{-1} N_t$ converges a.s. to one. Applying Proposition 1.2 to the Poisson process implies that for every x in $]0, 1[$

$$\lim_{t \to \infty} P\{t^{-1}(N_t - t) > x - 1\} = \lim_{t \to \infty} P\{t^{-1} N_t > x\} \leq \frac{1}{x}$$

and the limit is given by Chernoff's theorem.

These results are generalized to Markov processes. Lamperti (1958) studied the limit of the probability $G_n(t) = P(n^{-1} N_n \leq t)$ for the number of transitions in a recurrent state s of a Markov process having occurrence probabilities p_n. Under the necessary and sufficient conditions that the limits $\alpha = \lim_{n \to \infty} E n^{-1} N_n$ and $\delta = \lim_{x \to 1}(1 - x) F'(x) \{1 - F(x)\}^{-1}$ are finite, where F is the generating function of the occurrence probabilities, the probability function $G_n(t)$ converges to a distribution function $G(t)$ with a density generalizing the arcsine law and depending on the parameters α and δ. Other limits of interest are those of the current time $\gamma(t) = t - S_{N_t}$ and the residual time $\delta(t) = S_{N_t+1} - t$ of sums of independent and identically distributed variables X_i with distribution function F. As $t \to \infty$, the distribution function of the current time $\gamma(t)$ converges to the distribution function

$$F_{\gamma_\infty}(y) = \lim_{t \to \infty} P(\gamma(t) \leq y) = \frac{1}{EX} \int_y^\infty \{1 - F(t^-)\} \, dt$$

and the distribution function of the residual time δ_t converges to the distribution function F_{δ_∞}

$$\bar{F}_{\delta_\infty}(y) = \lim_{t \to \infty} P(\delta(t) \leq y) = \frac{1}{EX} \int_0^y \{1 - F(t^-)\} \, dt$$

therefore

$$\lim_{t \to \infty} P(S_{N_t} \leq t + y) = F_{\gamma_\infty}(y),$$
$$\lim_{t \to \infty} P(S_{N_t} \geq -y) = F_{\delta_\infty}(y).$$

Karlin and McGregor (1965) provided expressions for the limits $\lim_{t \to \infty} P(t^{-1}\gamma(t) \leq x)$ and $\lim_{t \to \infty} P(t^{-1}\delta(t) \leq x)$, under the condition that $1 - F(t) \sim t^{-\alpha}L^{-1}(t)$ as t tends to infinity, where L is a slowly varying function and $0 < \alpha < 1$.

By the weak convergence of the normalized process $(t^{-\frac{1}{2}}(N_t - t))_{t \geq 0}$ to a standard Brownian motion $(B_t)_{t \geq 0}$, the same limit applies to B and to $t^{-\frac{1}{2}}(N_t - t)$. For every x in $[0, 1]$

$$\lim_{t \to \infty} P\{t^{-\frac{1}{2}}B_t > x\} = \lim_{t \to \infty} P\{t^{-\frac{1}{2}}B_t < -x\}$$
$$= \lim_{t \to \infty} P\{t^{-1}N_t < 1 - x\} = \exp\left\{-\frac{x^2}{2}\right\}$$

from the probability of large deviations for the Brownian motion. Replacing x by a function h defined from $\mathbb{R}$ to $\mathbb{R}_+$ implies

$$P\{B_t > h(t)\} = \exp\left\{-\frac{h^2(t)}{2t}\right\}.$$

Assuming $\lim_{t \to \infty} t^{-\frac{1}{2}}h(t) = 0$, it follows that $\lim_{t \to \infty} P\{B_t > h(t)\} = 1$ and the Brownian motion crosses every function $h(t) = o(\sqrt{t})$ with probability converging to one, as t tends to zero.

Almost sure limits for the Brownian motion are usually expressed by the law of iterated logarithm (Kiefer, 1961)

$$\limsup_{t \to \infty} \frac{B_t}{\sqrt{2t \log \log t}} = 1, \qquad \liminf_{t \to \infty} \frac{B_t}{\sqrt{2t \log \log t}} = -1, \quad \text{a.s.}$$

This inequality proved for the discrete process extends to the Brownian motion indexed by $\mathbb{R}$ by limit as n tends to infinity. It is also true for other functions than the iterated logarithm. Section A.2 presents some examples from Robbins and Siegmund (1970). The next boundary is more general.

Proposition 6.14. *For every real function h such that $\int_0^\infty h_t^{-1}\, dt$ is finite, the Brownian motion satisfies*

$$\limsup_{t \to \infty} \frac{|B_t|}{\sqrt{th(t)}} \leq 1, \quad \text{a.s.}$$

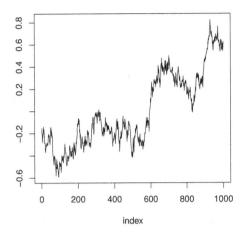

Fig. 6.1 A sample path of the Brownian motion.

Proof. For every real function h and for every $t > 0$ the inequality

$$P(B_t > \sqrt{th(t)}) \leq \frac{1}{h(t)}$$

implies that for every sequence $(t_n)_{n \leq N}$ tending to infinity as N tends to infinity and such that $t_{n+1} - t_n = \delta_N$ tends to zero, we have

$$\lim_{N \to \infty} \delta_N \sum_{n \leq N} P(B_t > \sqrt{t_n h(t_n)}) < \lim_{N \to \infty} \delta_N \sum_{n \leq N} \frac{1}{h(t_n)}$$

and it is finite. The result follows by the Borel-Cantelli lemma. $\square$

For example, a function such that $h(t) \leq \exp\{t^{-\alpha}\}$, $\alpha > 1$, fulfills the summability condition of Proposition 6.14. Consider the function $h_t = \alpha(t+s)^{\frac{1}{2}}$ on $\mathbb{R}_+$, with $\alpha > 0$ and $s > 0$, $t^{-\frac{1}{2}} h_t$ converges to α as t tends to infinity and it tends to infinity as t converges to zero. Let $\tau = \inf\{t : B_t \geq h_t\}$ and let $F_t = P(\tau \leq t)$, the expectation of the exponential martingale of B_t at τ is $\int_0^\infty e^{\lambda h_t - \frac{\lambda^2 t}{2}} \, dF_t = 1$, for every $\lambda > 0$, equivalently

$$e^{-\frac{\lambda^2 s}{2}} = \int_0^\infty e^{\lambda \alpha(t+s)^{\frac{1}{2}} - \frac{\lambda^2(t+s)}{2}} \, dF_t.$$

Integrating with respect to λ, we get

$$\int_0^\infty e^{-\frac{\lambda^2 s}{2}}\, d\lambda = \int_0^\infty \int_0^\infty e^{\lambda \alpha (t+s)^{\frac{1}{2}} - \frac{\lambda^2 (t+s)}{2}}\, d\lambda\, dF_t,$$

$$s^{-\frac{1}{2}} = \frac{2}{\sqrt{2\pi}} e^{\frac{\alpha^2}{2}} \int_0^\infty e^{-\frac{(y-\alpha)^2}{2}}\, dy \int_0^\infty \frac{dF_t}{(t+s)^{\frac{1}{2}}},$$

where $y = \lambda(t+s)^{\frac{1}{2}}$, this implies an inequality for the Cauchy transform of F, for every $\alpha > 0$

$$\int_0^\infty \frac{dF_t}{(t+s)^{\frac{1}{2}}} \geq s^{-\frac{1}{2}} e^{-\frac{\alpha^2}{2}}.$$

The properties of envelop functions of B_t extend to transformed Brownian motions.

Proposition 6.15. *Let* $Y_t = A_t^{-1} \int_0^t A_s\, dB_s$, *where* A *is a.s. decreasing and positive process. For very real function* h *on* $\mathbb{R}$ *such that* $\int_0^x h_t^{-1}\, dt$ *is finite as* x *tends to infinity, the process* Y *satisfies*

$$\limsup_{t\to\infty} \frac{|Y_t|}{\sqrt{th(t)}} \leq 1, \quad a.s.$$

Indeed the variance of Y_t is $E \int_0^t (A_t^{-1} A_s)^2\, ds \geq t$ and the proof follows the same arguments as in the proof for the Brownian motion, using Chernoff's theorem

$$\int_0^x P\{Y_t > \sqrt{th_t}\}\, dt < \int_0^x \frac{1}{h_t}\, dt$$

which tends to infinity with x. With an a.s. increasing positive process A and a function h such that $\int_0^\infty h_t^{-1}\, dt$ is finite

$$\limsup_{t\to\infty} \frac{|Y_t|}{\sqrt{th(t)}} \leq 1, \quad a.s.$$

Freedman (1975) extended the results to a martingale $S_n = \sum_{i=1}^n X_i$, with the quadratic variations $V_n = \sum_{i=1}^n \text{Var}(X_n|\mathcal{F}_{n-1})$ and such that $|X_n| \leq 1$

$$\limsup_{n\to\infty} (2V_n \log\log V_n)^{-\frac{1}{2}} |S_n| \leq 1 \text{ a.s. on } \{V_n = \infty\}.$$

Under similar conditions, for a supermartingale S_n such that $|X_n| \leq 1$, we have $\limsup_{n \to \infty} (2V_n \log \log V_n)^{-\frac{1}{2}} |S_n| \leq 1$, a.s.

Proposition 6.14 also applies to the martingale S_n with increasing threshold functions φ. By the a.s. asymptotic equivalence of V_n and EV_n, for every function φ such that $\sum_{n \geq 1} \exp\{-\frac{1}{2}\varphi_n^2\}$ or $\int_1^\infty \exp\{-\frac{1}{2}\varphi_t^2\} \, dt$ are finite

$$\limsup_{n \to \infty} V_n^{-\frac{1}{2}} \varphi_n^{-1} S_n \leq 1, \quad \text{a.s.}$$

$$\liminf_{n \to \infty} V_n^{-\frac{1}{2}} \varphi_n^{-1} S_n \geq -1, \quad \text{a.s.}$$

Replacing S_n by the sum of the independent transformed increments of the martingale $Y_n = \sum_{i=1}^n \varphi(X_i)$, with an $L^2(\mathbb{R})$ function φ such that the empirical expectation $\bar{\mu}_n = n^{-1} \sum_{i=1}^n E\varphi(X_i)$ converges a.s. to a finite limit, the law of iterated logarithm for the process $(Y_n)_{n \geq 0}$ is written in terms of the quadratic variations $T_n = \sum_{i=1}^n \text{Var}\{\varphi(X_i)|\mathcal{F}_{i-1}\}$

$$\limsup_{n \to \infty} (2T_n \log \log \varphi_n)^{-\frac{1}{2}} (Y_n - \bar{\mu}_n) \leq 1, \quad \text{a.s.}$$

$$\liminf_{n \to \infty} (2T_n \log \log \varphi_n)^{-\frac{1}{2}} (Y_n - \bar{\mu}_n) \geq -1, \quad \text{a.s.}$$

It is generalized to varying threshold functions like for S_n.

6.8 Local times

Let B be a Brownian motion process on $(\Omega, \mathcal{F}, (\mathcal{F}_t)_{t \geq 0}, P)$ and let τ_a be a stopping time defined for $a > 0$ as

$$\tau_a = \inf\{s > 0 : B_s > a\}.$$

The event $\{\tau_a > t\}$ is equivalent to $\{B_t^* < a\}$ where $B_t^* = \sup_{0 \leq s \leq t} B_s$ hence

$$P(\tau_a > t) = P(B_t^* < a)$$

and $P(\tau_a < \infty) = P(B_\infty^* > a) = 1$. The mapping $a \mapsto \tau_a$ and, for every $t > 0$, $1_{B_s < a} \, ds$ are increasing with respect to a. The probability $P(\tau_a < t)$ is strictly positive if and only if $P(D_{t,a}) < 1$ where $D_{t,a} = \cap_{s \leq t} \{B_s < a\}$. The cumulated probability

$$\int_0^t P(B_s < a) \, ds = E \int_0^t 1_{\{B_s < a\}} \, ds$$

is the expectation of the sojourn time of the process under a during the time interval $[0, t]$, it is the integral of $P(B_s < a) = \Phi(s^{-\frac{1}{2}}a)$, where Φ is the normal distribution. Integrating by parts, it follows that for all $a > 0$ and $t > 0$

$$\int_0^t P(B_s < a)\, ds = t\Phi(t^{-\frac{1}{2}}a) + \frac{a}{2}\int_0^t \frac{e^{-\frac{a^2}{2s}}}{\sqrt{2\pi s}}\, ds$$

$$= t\Phi(t^{-\frac{1}{2}}a) + \frac{a^2}{\sqrt{2\pi}}\int_{\frac{a}{\sqrt{t}}}^\infty u^{-2}e^{-\frac{u^2}{2}}\, du,$$

$$\int_0^t P(B_s < -a)\, ds = t\Phi(-t^{-\frac{1}{2}}a) - \frac{a}{2}\int_0^t \frac{e^{-\frac{a^2}{2s}}}{\sqrt{\pi s}}\, ds,$$

then for every $\mu > 0$

$$\int_0^t P(|B_s - \mu| < a)\, ds = \int_0^t P(B_s < \mu + a)\, ds - \int_0^t P(B_s < \mu - a)\, ds$$

$$= t\{\Phi(t^{-\frac{1}{2}}(\mu + a))) - \Phi(t^{-\frac{1}{2}}(\mu - a))\}$$

$$+ \frac{2a^2}{\sqrt{2\pi}}\int_{\frac{a-\mu}{\sqrt{t}}}^\infty u^{-2}e^{-\frac{u^2}{2}}\, du. \tag{6.11}$$

As a tends to zero, the first term has a first order expansion

$$t\{\Phi(t^{-\frac{1}{2}}(\mu + a)) - \Phi(t^{-\frac{1}{2}}(\mu - a))\} = 2a\sqrt{t}(2\pi)^{-\frac{1}{2}}e^{-\frac{\mu^2}{2t}} + o(a^2)$$

and the second term is a $O(a^2)$.

Let $X_t = \int_0^t \beta_s\, dB_s$ with a function β locally in L^2 and with variance the increasing function $v_t = \int_0^t \beta_s^2\, ds$, $X_t = B \circ v_t$ is a transformed Brownian motion. For $a > 0$ the stopping time $\tau_a = \min\{s; X_s > a\}$ satisfies

$$P(\tau_a > t) = P(B_{v_t}^* < a)$$

and $P(\tau_a < \infty) = 1$ if v_t tends to infinity with t. Let P_μ be the probability for the processes starting from $\mu > 0$, under P_μ the mean sojourn time of the process under a up to time $[0, t]$ is

$$\int_0^t \{\Phi((\mu + a)v_s^{-\frac{1}{2}}) - \Phi(\mu - a)v_s^{-\frac{1}{2}})\}\, ds.$$

Integrating by parts it is written as

$$t\{\Phi((\mu + a)v_t^{-\frac{1}{2}}) - \Phi(\mu - a)v_t^{-\frac{1}{2}})\}$$

$$- \int_0^t s\{\Phi((\mu + a)v_t^{-\frac{1}{2}}) - \Phi(\mu - a)v_t^{-\frac{1}{2}})\}'\, ds$$

$$= t\{\Phi((\mu + a)v_t^{-\frac{1}{2}}) - \Phi((\mu - a)v_t^{-\frac{1}{2}})\} + \frac{a}{2\sqrt{2\pi}}\int_0^t se^{-\frac{(\mu+a)^2}{2v_s}}v_s^{-\frac{3}{2}}v_s'\, ds$$

$$+ \frac{a}{2\sqrt{2\pi}}\int_0^t se^{-\frac{(\mu-a)^2}{2v_s}}v_s^{-\frac{3}{2}}v_s'\, ds.$$

By a first order expansion as a tends to zero, the first term is asymptotically equivalent to $2at(2\pi v_t)^{-\frac{1}{2}}e^{-\frac{\mu^2}{2v_t}} + o(a^2)$ and the second term is a $O(a)$.

For a diffusion process $dX_t = \alpha_t\,dt + \beta_s\,dB_s$, with a positive function α, the hitting time T_a of a by X_t is smaller than τ_a defined for $\int_0^t \beta_s\,dB_s$, therefore $P(T_a < \infty) = 1$. With a negative function α, T_a is larger than τ_a and $P(T_a < \infty) \leq 1$. Denoting $A_t = \int_0^t a_s\,ds$, the mean sojourn time of the process under a during the time interval $[0, t]$ is

$$\int_0^t \Phi(v_s^{-\frac{1}{2}}(a - A_s))\,ds = t\Phi(v_t^{-\frac{1}{2}}(a - A_t)) + \frac{a - A_t}{2}\int_0^t \frac{e^{-\frac{(a - A_s)^2}{2v_s}}}{\sqrt{2v_s\pi}}\,ds.$$

Let $(N_t)_{t\geq 0}$ be a Poisson process with parameter $\lambda > 0$, for every $k > 0$, the random times of the process satisfy

$$\int_0^\infty P(N_s = k)\,ds = E\int_{T_k}^{T_{k+1}} ds = \lambda^{-1}$$

and the sojourn time of the process in k during the time interval $[0, t]$ is the variable $Y_t = \int_0^t 1_{\{N_s = k\}}\,ds = (T_{k+1} \wedge t) - T_k$, it has the expectation $\lambda^{-1} - E\{(T_{k+1} - t)1_{\{t < T_{k+1}\}}\} \leq \lambda^{-1}$.

For a Brownian motion, the occupation time of the interval $[x - \varepsilon, x + \varepsilon]$ for B_s during $[0, t]$ is

$$m_\varepsilon^x(t) = \int_0^t 1_{\{|B_s - x| \leq \varepsilon\}}\,ds, \tag{6.12}$$

as ε tends to zero, $m_\varepsilon^x(t)$ converges a.s. to the number of hitting times of zero by the process $B - x$, during the time interval $[0, t]$, this is the local time of the Brownian motion at x

$$L_t^x = \lim_{\varepsilon \to 0} (2\varepsilon)^{-1}m_\varepsilon^x(t).$$

The expectation of L_t^x is

$$\mu_t^x = \lim_{\varepsilon \to 0} (2\varepsilon)^{-1}\int_0^t P(|B_s - x| \leq \varepsilon)\,ds = \frac{\sqrt{t}}{\sqrt{2\pi}}e^{-\frac{x^2}{2t}}. \tag{6.13}$$

The density of occurrence formula implies that for every real measurable function f on $\mathbb{R}$

$$\int_0^t f(B_s)\,ds = \int_\mathbb{R} f(x)L_t^x\,dx.$$

The expectation of $m_\varepsilon(t)$, at $x = 0$, is

$$Em_\varepsilon(t) = \int_0^t \{P(B_s \leq \varepsilon) - P(B_s \leq \varepsilon)\} \, ds$$

$$= t\{\Phi(t^{-\frac{1}{2}}\varepsilon) - \Phi(t^{-\frac{1}{2}}\varepsilon)\} + 2\frac{\varepsilon^2}{\sqrt{2\pi}} \int_{\frac{\varepsilon}{\sqrt{t}}}^\infty u^{-2} e^{-\frac{u^2}{2}} \, du,$$

it is a $O(\varepsilon)$ and the first term is asymptotically equivalent to $2\varepsilon t^{\frac{1}{2}}(2\pi)^{-\frac{1}{2}}$, hence

$$\mu_t = t^{\frac{1}{2}}(2\pi)^{-\frac{1}{2}}. \tag{6.14}$$

Lemma 6.1. *The variance of $m_\varepsilon(t)$, $t > 0$, is*

$$v_\varepsilon(t) = \int_0^t E\{2m_{2\varepsilon}(t - v) - m_\varepsilon(t)\} \, dm_\varepsilon(v)$$

and the covariance of $m_\varepsilon(s)$ and $m_\varepsilon(t)$, $t > s > 0$, is

$$c_\varepsilon(s, t) = 2E\left\{ m_{2\varepsilon} * m_\varepsilon(t) - \int_0^s m_{2\varepsilon}(t - v) \, dm_\varepsilon(v) + m_{2\varepsilon} * m_\varepsilon(s) \right\}.$$

For every integer $k \geq 1$ and for $0 < s < t$, there exists constants C_1 and C_2 such that

$$0 \leq (\mu_t - \mu_s)^k \leq C_1(t - s)^k,$$

$$0 \leq \{v_\varepsilon^{\frac{1}{2}}(t) - v_\varepsilon^{\frac{1}{2}}(s)\}^k \leq C_2 \varepsilon^k (t - s)^k.$$

Proof. For every integer N and for $0 < s < t \leq N$, by (6.14) and the inequality

$$t^{\frac{1}{2}} - s^{\frac{1}{2}} = s^{\frac{1}{2}}\left\{ \left(1 + \frac{t - s}{s}\right)^{(\frac{1}{2})} - 1 \right\} \leq s^{\frac{1}{2}}\frac{t - s}{2s}$$

we deduce $\mu_t - \mu_s = O(t - s)$. The variance of $m_\varepsilon(t)$ is $v_\varepsilon(t) = E\{m_\varepsilon^2(t)\} - m_\varepsilon^2(t)$ with

$$E\{m_\varepsilon^2(t)\} = 2 \int_0^t \int_v^t P(|B_u| \leq \varepsilon, |B_v| \leq \varepsilon) \, du \, dv$$

$$= 2 \int_0^t P(|B_v| \leq \varepsilon) \int_v^t P(|B_{u-v}| \leq 2\varepsilon) \, du \, dv$$

$$= 2 \int_0^t P(|B_v| \leq \varepsilon) \int_0^{t-v} P(|B_x| \leq 2\varepsilon) \, dx \, dv$$

$$= 2E \int_0^t m_{2\varepsilon}(t - v) \, dm_\varepsilon(v),$$

it is finite for every t and strictly positive. For $s < t$ we have

$$v_\varepsilon(t) - v_\varepsilon(s) = 2E \int_0^s \{m_{2\varepsilon}(t-v) - m_{2\varepsilon}(s-v)\}\, dm_\varepsilon(v)$$

$$+ 2E \int_s^t m_{2\varepsilon}(t-u)\, dm_\varepsilon(u) = O(\varepsilon^2(t-s)),$$

and for every integer $k \geq 1$

$$(v_t^{\frac{1}{2}} - v_s^{\frac{1}{2}})^k \leq v_s^{\frac{k}{2}}\left\{\left(1 + \frac{v_t - v_s}{v_s}\right)^{\frac{1}{2}} - 1\right\}^k$$

$$\leq \frac{(v_t - v_s)^k}{2 v_s^{\frac{k}{2}}},$$

it is therefore a $O(\varepsilon^{2k}(t-s)^k)$. The covariance of $m_\varepsilon(t)$ and $m_\varepsilon(s)$ is $c_\varepsilon(s,t) = v_\varepsilon(t) - 2E\{m_\varepsilon(t)m_\varepsilon(s)\} + v_\varepsilon(s)$ where

$$E\{m_\varepsilon(t)m_\varepsilon(s)\} = \int_0^s \int_v^t P(|B_v| \leq \varepsilon)P(|B_{u-v}| \leq 2\varepsilon)\, du\, dv$$

$$= \int_0^s P(|B_v| \leq \varepsilon) \int_0^{t-v} P(|B_x| \leq 2\varepsilon)\, dx\, dv$$

$$= E \int_0^s m_{2\varepsilon}(t-v)\, dm_\varepsilon(v).$$

$\square$

Proposition 6.16. *As ε tends to zero, the process*

$$\mathcal{T}_\varepsilon(t) = \frac{m_\varepsilon(t) - 2\varepsilon\mu_t}{\sqrt{v_t}}$$

converges weakly in $C(\mathbb{R}_+)$ to a centered Gaussian process.

Proof. The expectation of $\mathcal{T}_\varepsilon(t)$ converges to zero as ε tends to zero and by Lemma 6.1, $(2\varepsilon)^{-2}v_\varepsilon(t)$ converges to

$$\sigma^2(t) = \int_0^t 2\mu_{t-v}\, d\mu_v - \mu_t^2.$$

Let $\varepsilon_n = n^{-1}$ converge to zero as n tends to infinity, let $(s_k)_{k \leq n}$ be a regular partition of $[0,t]$ and let $m_{k,n} = \int_{s_{k-1}}^{s_k} 1_{\{|B_s| \leq n^{-1}\}}\, ds$, then the local time μ_t is the limit of $\sum_{k=1}^n m_{k,n}$ and by the central limit theorem $(n \sum_{k=1}^n m_{k,n} - \mu_t)\{\text{Var}(n \sum_{k=1}^n m_{k,n})\}^{-\frac{1}{2}}$ converges weakly to a centered Gaussian variable, as n tends to infinity. Equivalently, $v_\varepsilon^{-\frac{1}{2}}(t)\{m_\varepsilon(t) - 2\varepsilon\mu_t\}$ converges weakly to a centered Gaussian variable as ε tends to zero.

Using the same arguments as for $\mathcal{T}_\varepsilon(t)$ and with the convergence of $\varepsilon^{-2}c_\varepsilon(s,t)$ for the covariance of $m_\varepsilon(t)$ and $m_\varepsilon(t)$, it follows that for every integer m and for positive $(t_1,\ldots,t_m)$, the vector $(\mathcal{T}_\varepsilon(t_1),\ldots,\mathcal{T}_\varepsilon(t_m))$ converges weakly to the finite dimensional distributions at $(t_1,\ldots,t_m)$ of a centered Gaussian process, as ε tends to zero.

It remains to prove the tightness of the process $\mathcal{T}_\varepsilon$ in the space of processes with continuous sample paths on $[-M,M]$, for every $M > 0$. For $s < t$ in $[-M,M]$, we have $\{v_\varepsilon^{\frac{1}{2}}(t) - v_\varepsilon^{\frac{1}{2}}(t)\}^2 = O((t-s)^2)$ by Lemma 6.1.

Moreover $A_\varepsilon(s,t) = E[\{m_\varepsilon(t) - m_\varepsilon(s) - 2\varepsilon\mu_t + 2\varepsilon\mu_s\}^2] \leq 2E[\{m_\varepsilon(t) - m_\varepsilon(s)\}^2] + 4\varepsilon^2(\mu_t - \mu_s)^2$ where $E\{m_\varepsilon(t) - m_\varepsilon(s)\}^2$ and $(\mu_t - \mu_s)^2$ are $O(\varepsilon^2(t-s)^2)$ by Lemma 6.1 then $A_\varepsilon(s,t) = O(\varepsilon^2(t-s)^2)$ and

$$E(\mathcal{T}_\varepsilon(t) - \mathcal{T}_\varepsilon(s))^2 \leq 2\frac{A_\varepsilon(s,t)}{v_\varepsilon(t)} + 2\frac{\{m_\varepsilon(s) - 2\varepsilon\mu_s\}^2\{v_\varepsilon^{\frac{1}{2}}(t) - v_\varepsilon^{\frac{1}{2}}(t)\}^2}{v_\varepsilon(s)v_\varepsilon(t)},$$

it is a $O((t-s)^2)$. The tightness condition (5.5) is therefore fulfilled by the process $\mathcal{T}_\varepsilon$. $\qquad\square$

Proposition 6.16 extends to m_ε^x, for every x, the process

$$\mathcal{T}_\varepsilon^x(t) = \frac{m_\varepsilon^x(t) - 2\varepsilon\mu_t^x}{\sqrt{v_t^x}}$$

converges weakly in $C(\mathbb{R}_+)$ to a centered Gaussian process, as ε tends to zero.

Let C_t be a predictable process and let $X_t = \int_0^t C_s\,dB_s$, under the condition that the variance $\phi_t = \int_0^t EC_s^2\,ds$ of X_t is locally finite, the local time of X at zero is

$$L_X(t) = \lim_{\varepsilon\to 0}(2\varepsilon)^{-1}\int_0^t \mathbb{1}_{[-\varepsilon,\varepsilon]}(X_s)\,ds,$$

its expectation is $\mu_{X,t}$. The process X_t is a transformed Brownian motion with independent increments written as $X_t = B_t \circ \phi_t^{\frac{1}{2}}$ and, for every real x

$$\mu_{X,t}^x = \int_0^t P_0(|X_s - x| = 0)\,ds$$

is determined by the probability $h_x(s) = P_0(X_s = x)$ of sojourn at zero.

Proposition 6.17. *Let* $v_{X,\varepsilon}^x(t) = \int_0^t\{2m_{X,2\varepsilon}^x(t-v) - m_{X,\varepsilon}^x(t)\}\,dm_{X,\varepsilon}^x(v)$, *for* $t > 0$. *For every* x, *the process*

$$\mathcal{Y}_\varepsilon^x(t) = \frac{m_{X,\varepsilon}^x(t) - 2\varepsilon\mu_{X,t}^x}{\sqrt{v_{X,t}^x}}$$

converges weakly in $C(\mathbb{R}_+)$ *to a centered Gaussian process as* ε *tends to zero.*

Proof. The expectation of the process $\mathcal{Y}_\varepsilon^x$ converges to zero as ε tends to zero and its variance function $v_{X,\varepsilon}^x$ is increasing, locally bounded and strictly positive. The weak convergence of Y_ε^x relies on the same arguments as in Proposition 6.16 for the Brownian motion. $\square$

Let C_1 and C_2 be independent predictable strong Markov processes such that, for $k = 1, 2$, $X_k(t) = \int_0^t C_{ks}\, dB_s$ belong to $\mathcal{M}_{0,loc}^2$ and have the variance $\phi_{kt} = \int_0^t E(C_{ks}^2)\, ds$. For every t, let

$$m_{k\varepsilon}^x(t) = \int_0^t 1_{\{|X_k(s)-x|\leq\varepsilon\}}\, ds,$$

$$\mu_{kt}^x = \lim_{\varepsilon\to 0}(2\varepsilon)^{-1}\int_0^t P(|X_k(s) - x| \leq \varepsilon)\, ds = \int_0^t P(|X_k(s) - x| = 0)\, ds,$$

they satisfy the inequalities

$$|\phi_{1t}^{\frac{1}{2}} - \phi_{2t}^{\frac{1}{2}}| \leq \frac{|\phi_{1t} - \phi_{2t}|}{2\phi_{2t}^{\frac{1}{2}}},$$

$$E(\mu_{1t}^x - \mu_{2t}^x)^2 \leq \int_0^t P(\{|X_1(s) - x| \leq \varepsilon\}\Delta\{|X_2(s) - x| \leq \varepsilon\})\, ds$$

$$E|\mu_{1t}^x - \mu_{2t}^x| \leq \int_0^t \max\{P(|X_1(s) - x| \leq \varepsilon), P(|X_2(s) - x| \leq \varepsilon)\}\, ds.$$

Proposition 6.17 applies to the processes $Y_{k\varepsilon}$.

Proposition 6.18. *Let C_1 and C_2 be independent predictable processes such that the variance of $X_1 - X_2$ is locally bounded, then for every real x, $|\mu_{1t}^x - \mu_{2t}^x|$ is finite for every t and the process $Y_{1\varepsilon}^x - Y_{2\varepsilon}^x$ converges weakly to a centered Gaussian process as ε tends to zero.*

Proof. Under the conditions, the expectation and the variance $\int_0^t E(C_{1s} - C_{2s})^2\, ds$ of $X_1(t) - X_2(t)$ are finite for every t, the expectation is a $O(\varepsilon t)$ and the variance is a $O(\varepsilon^2 t)$. The weak convergence of the process $Y_{1\varepsilon}^x - Y_{2\varepsilon}^x$ is obtained like in Proposition 6.16. $\square$

Let M be a martingale of $\mathcal{M}^2$ with a predictable process A of quadratic variations, the occupation time of the interval $[x - \varepsilon, x + \varepsilon]$ by M during $[0, t]$ is defined like in 6.12, with respect to the continuous part of A

$$m_\varepsilon^x(t) = \int_0^t 1_{\{|M_s-x|\leq\varepsilon\}}\, dA_s^c.$$

As A^c is predictable, the expectation of $m_\varepsilon^x(t)$ is

$$\mu_\varepsilon^x(t) = E\int_0^t P(|M_s - x| \leq \varepsilon \mid \mathcal{F}_{s-})\, dA_s^c$$

and it is a $O(\varepsilon)$. Right and left local times of M at x are

$$L^x(t) = \lim_{\varepsilon \to 0} \frac{1}{2\varepsilon} \int_0^t 1_{\{M_s \in [x, x+\varepsilon]\}} \, dA_s^c,$$

$$L^{x-}(t) = \lim_{\varepsilon \to 0} \frac{1}{2\varepsilon} \int_0^t 1_{\{M_s \in [x-\varepsilon, x+\varepsilon]\}} \, dA_s^c,$$

and

$$\frac{1}{2}\{L^x(t) + L^{x-}(t)\} = \lim_{\varepsilon \to 0} \frac{1}{2\varepsilon} \int_0^t 1_{\{|M_s - x| \le \varepsilon\}} \, dA_s^c = \lim_{\varepsilon \to 0} \frac{1}{2\varepsilon} m_\varepsilon^x(t).$$

For every real measurable function f on $\mathbb{R}$, the density of occurrence formula implies

$$\int_0^t f(M_s) \, dA_s^c = \int_{\mathbb{R}} f(x) \bar{L}_t^x \, dx.$$

For all $\varepsilon > 0$ and x, the process $(m_\varepsilon^x(t))_{t \ge 0}$ is increasing and there exists a predictable process $\tilde{m}_\varepsilon^x$ such that $m_\varepsilon^x - \tilde{m}_\varepsilon^x$ is a centered martingale.

Proposition 6.19. *Let A^c be a process with independent increments of $L^2(\mathbb{R}_+)$, then for every x, the process $\mathcal{M}_\varepsilon^x(t) = \varepsilon^{-1}\{m_\varepsilon^x(t) - \tilde{m}_\varepsilon^x(t)\}$ converges weakly in $C(\mathbb{R}_+)$ to a centered Gaussian process as ε tends to zero.*

Proof. Arguing as in the proof of Proposition 6.16, and by the martingale property

$$E\{m_\varepsilon^{x2}(t)\} = 2E \int_0^t 1_{\{|M_v - x| \le \varepsilon\}} \int_v^t 1_{\{|M_u - M_v| \le 2\varepsilon\}} \, dA_u^c \, dA_v^c,$$

the variance of $m_\varepsilon^x(t)$ is therefore $0(\varepsilon^2)$ such that $\varepsilon^{-2}\mathrm{Var}\{m_\varepsilon^x(t)\}$ converges to a continuous function v. Choosing $\varepsilon = n^{-1}$, as n tends to infinity, by Theorem 5.4 the process $\mathcal{M}_\varepsilon^x$ converges weakly to a centered Gaussian process with independent increments, and with variance function v. □

Barlow and Yor (1982) obtained bounds for the difference of local times of semi-martingales under conditions for the norms of their difference. Let $X_t = X_0 + M_t + A_t$ where M is a martingale of $\mathcal{M}^2$ and A is a processes with bounded variations. The right and left local times of X at x are defined with respect to the continuous part of $< X >$.

Chapter 7

Functional Inequalities

7.1 Introduction

Analytic inequalities for operators in functional spaces are often deduced from their derivatives, such as bounds for the entropy or the information of classes (Section 4.3). The derivative of an operator T on a vector space $(\mathcal{F}, \| \cdot \|)$ at f is

$$T'(f)(h) = \lim_{t \to \infty} \frac{T(f + th) - T(f)}{t},$$

where h belongs to $\mathcal{F}$ and t to $\mathbb{R}$ where the limit for the topology of the norm is the convergence to zero of $\|t^{-1}\{T(f + th) - T(f)\} - T'(f)(h)\|$.

Let $(X_1, \ldots, X_n)$ be a sequence of random and identically distributed variables defined on a probability space $(\Omega, \mathcal{A}, P)$ and with values in a separable Banach space $(\mathcal{X}, \mathcal{B})$ provided with the Borel σ-algebra $\mathcal{B}$. Let $\mathcal{P}_\mathcal{X}$ denote the space of probability distributions on $\mathcal{X}$ and let P in $\mathcal{P}_\mathcal{X}$ be the probability distribution of the variables X_i, $\widehat{P}_n(t) = \sum_{i=1}^{n} 1_{\{X_i \leq t\}}$ is their empirical distribution, $\nu_{n,P} = \sqrt{n}(\widehat{P}_n - P)$ is the normalized empirical process and G_P is the Brownian bridge associated to P.

Let $N(\mathcal{X}, \varepsilon)$ be the minimum number of sets covering the space $\mathcal{X}$ with ball of radius less than ε for the metric of $\mathcal{X}$, the ε-entropy of $\mathcal{X}$ is

$$H(\mathcal{X}, \varepsilon) = \log N(\mathcal{X}, \varepsilon).$$

Under the condition that $\int_0^1 H(\mathcal{X}, \varepsilon) \, d\varepsilon$ is finite and under regularity conditions for the variables, the normalized sum $n^{-\frac{1}{2}} S_n$ converges weakly to a centered Gaussian variable (Dudley, 1974; Giné, 1974). This result was applied to the sums of continuous processes, with the uniform topology.

Processes are also defined on functional families of $\mathcal{L}_P^2 = \{f : \mathcal{X} \mapsto \mathbb{R}; \int_\mathcal{X} f^2 \, dP < \infty\}$. Pollard's conditions and notations for the functional empirical process are adopted. A pseudo-distance

$$\rho_P(f, g) = \sigma_P(f - g)$$

is used on $\mathcal{F}$, with

$$\sigma_P^2(f) = P(f^2) - P^2(f).$$

The space $\ell^\infty(\mathcal{F}) = \{h : \mathcal{F} \mapsto \mathbb{R}; h \in C_b\}$ is a measurable space provided with the uniform norm on $\mathcal{F}$ and the Borel σ-algebra for this norm, the sample paths of the Brownian bridge belong to $\ell^\infty(\mathcal{F})$. The space $\mathcal{U}_b(\mathcal{F}, P)$ of the uniformly continuous and bounded real functions on $(\mathcal{F}, \rho_P)$ is a separable Banach space in $\ell^\infty(\mathcal{F})$ (Dudley, 1984). We suppose that $\mathcal{F}$ has a measurable envelope function $F = \sup\{f \in \mathcal{F}\}$ in L^2. Let

$$N_F(\varepsilon, \mathcal{F}) = \sup_{Q \in \mathcal{P}_\mathcal{X}} \max \ \{m : \exists \text{ distinct } f_1, \ldots, f_m \in \mathcal{F} \cap C_b(\mathcal{X});$$

$$Q(|f_i - f_j|^2) > \varepsilon^2 Q(F^2), \forall i \neq j\}$$

be the Pollard entropy function of $(\mathcal{F}, F)$, a set of functions $(f_j)_j$ bracketing all functions of a class $\mathcal{F}$ for a constant ε is called an ε-net of $\mathcal{F}$. The entropy dimension of $\mathcal{F}$ is

$$d_F^{(2)}(\mathcal{F}) = \inf \left\{ \delta > 0; \limsup_{\varepsilon \to 0} \varepsilon^\delta N_F(\varepsilon, \mathcal{F}) < \infty \right\}$$

and its exponent entropy is

$$e_F^{(2)}(\mathcal{F}) = \inf \left\{ \delta > 0; \limsup_{\varepsilon \to 0} \varepsilon^\delta \log N_F(\varepsilon, \mathcal{F}) < \infty \right\}.$$

The supremum of the Brownian bridge W on a class $(\mathcal{F}, F)$ such that $e_F^{(2)}(\mathcal{F}) < 2$ and $\sup_{f \in \mathcal{F}} \sigma^2(f) < \sigma^2$ satisfies

$$P(\|W\|_\mathcal{F} > t) \leq C_\sigma \exp\left\{ -\frac{t^2}{2\sigma^2} \right\}.$$

Similar inequalities are proved in the next section for $\sup_{f \in \mathcal{F}} |S_n(f)|$ under entropy and integrability conditions, generalizing Bennett's inequality. For $s \geq 2$, the entropy function of order s for $(\mathcal{F}, F)$ is defined as

$$N_F(\varepsilon, \mathcal{F}, s) = \sup_{Q \in \mathcal{P}_\mathcal{X}} \max \ \{m : \exists \text{ distinct } f_1, \ldots, f_m \in \mathcal{F} \cap C_b(\mathcal{X});$$

$$Q(|f_i - f_j|^s) > \varepsilon^s Q(F^s), \forall i \neq j\}.$$

7.2 Exponential inequalities for functional empirical processes

A subset $\mathcal{C}$ of $\mathcal{X}$ is a Vapnik-Cervonenkis' class if $\|\widehat{P}_n - P\|_{\mathcal{C}}$ converges a.s. to zero as n tends to infinity. It is a P-Donsker class if there exists a sequence of uniformly continuous Brownian bridges $G_P^{(n)}$ such that for every $\varepsilon > 0$ there exists an integer n_0 such that $P(\|\nu_n - G_P^{(n)}\|_{\mathcal{C}} > \varepsilon) \leq \varepsilon$, for every $n \geq n_0$. Let $\mathcal{F}$ be a class of functions on a metric space $(\mathbb{X}, \mathcal{X}, \|\cdot\|)$, measurable with respect to the σ-algebra generated by the closed balls of bounded functions for the uniform topology, centered at uniformly continuous functions in $L^2(\mathbb{X})$. Pollard (1981) proved that sufficient conditions implying that $\mathcal{F}$ is a P-Donsker class are the existence of an envelope function F of $\mathcal{F}$ belonging to $L^2(\mathbb{X}, P)$ and

$$\int_0^1 (\log N_F(\varepsilon, \mathcal{F}))^{\frac{1}{2}} \, d\varepsilon < \infty.$$

From Pollard (1982), this integral is finite if the envelope F belongs to $L^2(P)$ and $N_F(\varepsilon, \mathcal{F}) \leq A\varepsilon^{-2k}$ on $[0, 1]$, with constants A and k. Assouad (1981), Dudley (1984), Massart (1983), van der Vaart and Wellner (1996) present examples of Vapnik-Cervonenkis and Donsker classes.

The moments of $\sup_{f \in \mathcal{F}} |S_n(f)|$ satisfy inequalities similar to those established for the sum of variables, with $V_n(f) = \sum_{i=1}^n f^2(X_i)$. The bounds depends on the same constants $c_{\alpha,n}$ and $C_{\alpha,n}$ as in Proposition 4.4.

Proposition 7.1. *Let $(X_i)_{i=1,\ldots,n}$ be a vector of independent random and identically distributed variables on a probability space $(\Omega, \mathcal{A}, P)$, with values in a metric space $(\mathbb{X}, \mathcal{X}, \|\cdot\|)$. Let $\alpha \geq 2$ and let $\mathcal{F}$ be a class of measurable $L^\alpha(P)$ functions on $(\mathbb{X}, \mathcal{X}, \|\cdot\|)$ such that $P(f) = 0$ for every f in $\mathcal{F}$ and $E_P\|f(X_i)\|_{2,\mathcal{F}} = \sigma_P^2(F)$, then*

$$E\{\sup_{f \in \mathcal{F}} V_n^{\frac{\alpha}{2}}(f)\} \leq E(\sup_{f \in \mathcal{F}} |S_n(f)|^\alpha) \leq C_{\alpha,n} E\{\sup_{f \in \mathcal{F}} V_n^{\frac{\alpha}{2}}(f)\}.$$

Proof. Writing $E \sup_{f \in \mathcal{F}} |S_n(f)| \leq \{E \sup_{f \in \mathcal{F}} |S_n(f)|^2\}^{\frac{1}{2}}$ and using the inequality $|S_n(f)|^2 \leq C_{2,n} V_n(f)$ for every f of $\mathcal{F}$ and we obtain

$$E \sup_{f \in \mathcal{F}} |S_n(f)| \leq \{C_{2,n} E \sup_{f \in \mathcal{F}} V_n(f)\}^{\frac{1}{2}} = \{C_{2,n} E V_n(F)\}^{\frac{1}{2}}.$$

For $\alpha \geq 2$, the inequalities are proved by the same arguments, as in the proof of Proposition 4.4. $\qquad\square$

The constant $C_{\mathcal{F}}$ that appears in the next functional inequalities is an integral function of the dimension of the class $\mathcal{F}$. Here the main constant of dimension is

$$C_{\mathcal{F}} = \left\{ \int_0^1 N_F(\varepsilon, \mathcal{F}) \, d\varepsilon \right\}^{\frac{1}{2}}. \tag{7.1}$$

Proposition 7.2. *Let $(X_i)_{i=1,\ldots,n}$ be a sequence of independent and identically distributed random variables on $(\Omega, \mathcal{A}, P)$, with values in a metric space $(\mathbb{X}, \mathcal{X}, \|\cdot\|)$. Let $\mathcal{F}$ be a class of measurable functions on $\mathbb{X}$ such that $\|f(X_i)\|_{\mathcal{F}}$ belongs to $L^2(P)$ and $C_{\mathcal{F}}$ is finite. For every $t > 0$*

$$E \exp\left\{ t \sup_{f \in \mathcal{F}} S_n(f) \right\} \leq C_{\mathcal{F}} L_{F(X)}^{\frac{n}{2}}(2t).$$

Proof. From the equivalence of the norms L^p of random variables, $\|f(X_i)\|_{\mathcal{F}}$ belongs to L^p for every $p \geq 2$. Let $t > 0$, the independence of the variables $(X_i)_{i=1,\ldots,n}$ implies

$$E \exp\{ t \sup_{f \in \mathcal{F}} S_n(f) \} \leq [E \exp\{ t \sup_{f \in \mathcal{F}} f(X_i) \}]^n.$$

Let ε be in $]0, \delta[$, with $\delta < 1$. Let $\mathcal{F}_\varepsilon$ be an ε-net of $\mathcal{F}$ adapted to P and let π_ε be the projection from $\mathcal{F}$ to $\mathcal{F}_\varepsilon$, then for every f in $\mathcal{F}$ and for $p \geq 2$

$$\|(f - \pi_\varepsilon(f))(X)\|_{L^2} \leq \varepsilon \|F(X)\|_{L^2},$$

$$E \exp\left\{ t \sup_{f \in \mathcal{F}} f(X) \right\} \leq \left[E \exp\left\{ 2t \sup_{f_\varepsilon \in \mathcal{F}_\varepsilon} f_\varepsilon(X) \right\} \right]^{\frac{1}{2}}$$

$$\left[E \exp\left\{ 2t \sup_{f \in \mathcal{F}} (f - \pi_\varepsilon(f))(X) \right\} \right]^{\frac{1}{2}}, \tag{7.2}$$

by the Cauchy-Schwarz inequality. Moreover

$$E \left[\exp\left\{ 2t \sup_{f_\varepsilon \in \mathcal{F}_\varepsilon} f_\varepsilon(X) \right\} \right] \leq E \exp\{ 2t F(X) \} \int_0^\delta N_F(\varepsilon, \mathcal{F}) \, d\varepsilon$$

and the exponential function in $E \exp\{ 2t \sup_{f \in \mathcal{F}} (f - \pi_\varepsilon)(X) \}$ is expanded as a sum bounded, using Proposition 7.1, as $1 + \frac{1}{2}\varepsilon^2 \|F(X)\|_{L^2}^2 + o(\varepsilon^2)$ when δ tends to zero.

Since the bounds are valid for every $\delta < 1$, the integral over $[0, 1]$ gives an upper bound for the product of (7.2). $\square$

Functional Bienaymé-Chebychev inequalities for the sum $S_n(f)$ are consequences of Proposition 7.1, they are proved like Proposition 7.2, by projections on each function of $\mathcal{F}$ in an ε-net of the class, a bound for the supremum is obtained by summing a uniform bound of the projections over all functions of the ε-net

$$P\Big(\sup_{f\in\mathcal{F}}|n^{-\frac{1}{2}}S_n(f)|\geq t\Big)\leq 2C_{\mathcal{F}}\frac{\sigma_P^2(F)}{t^2},$$

$$P\Big(\sup_{f\in\mathcal{F}}|S_n(f)S_m(f)|\geq (n\wedge m)t\Big)\leq 2C_{\mathcal{F}}\frac{\sigma_P^2(F)}{t},$$

and for every $p>1$

$$P\Big(\sup_{f\in\mathcal{F}}|n^{-\frac{1}{2}}S_n(f)|\geq t\Big)\leq 2C_{\mathcal{F}}\frac{E\{V_n^p(F)\}}{t^{2p}}. \tag{7.3}$$

Similar results hold uniformly in n, for every stopping time N of S_n

$$P\Big(\sup_{1\leq n\leq N}\sup_{f\in\mathcal{F}}|n^{-\frac{1}{2}}S_n(f)|\geq t\Big)\leq 2C_{\mathcal{F}}E(N)\frac{\sigma_P^2(F)}{t^2},$$

$$P\Big(\sup_{1\leq n\leq N}\sup_{f\in\mathcal{F}}|n^{-\frac{1}{2}}S_n(f)|\geq t\Big)\leq \frac{2C_{\mathcal{F}}}{t^{2p}}E\Big\{\sum_{n=1}^{N}n^{-p}V_n^p(F)\Big\}.$$

Let $\|F_2-F_1\|_p=[\frac{1}{n}\sum_{i=1}^n E\{(F_2-F_1)(X_i)\}^p]^{\frac{1}{p}}$.

Proposition 7.3. *Let X be a random variable on a probability space $(\Omega,\mathcal{A},P)$, with values in a metric space $(\mathbb{X},\mathcal{X},\|\cdot\|)$. Let $\mathcal{F}$ be a class of measurable functions on $(\mathbb{X},\mathcal{X},\|\cdot\|)$ such that there exist functions F_1 and F_2 in $\mathcal{F}$, belonging to $L^p(\mathbb{X})$ and satisfying $F_1\leq f\leq F_2$ for every f of $\mathcal{F}$. For every $p\geq 2$*

$$P\Big(\sup_{f\in\mathcal{F}}|f(X)-\frac{1}{2}\{F_1(X)+F_2(X)\}|\geq t\Big)\leq 2C_{\mathcal{F}}\frac{\|F_2-F_1\|_p^p}{(2t)^p},$$

$$P\Big(\sup_{f\in\mathcal{F}}|f(X)-Ef(X)|\geq t\Big)\leq 2C_{\mathcal{F}}\frac{\|F_2-F_1\|_p^p}{t^p}.$$

Proof. The variable $Y(f)=|f(X)-\frac{1}{2}\{F_1(X)+F_2(X)\}|$ is uniformly bounded on $\mathcal{F}$ by $\frac{1}{2}(F_2-F_1)(X)$ and $\sup_{f\in\mathcal{F}}|f(X)-Ef(X)|$ is bounded by $(F_2-F_1)(X)$ which belongs to $L^p(\mathbb{X})$, the functional Bienaymé-Chebychev inequalities yield the result. $\qquad\square$

Under the conditions of Proposition 7.3, the empirical process of a non centered sample $(X_i)_{i=1,\ldots,n}$ of measurable random variables with values in $\mathbb{X}$ satisfies similar inequalities, due to (7.3)

$$P\Big(\sup_{f\in\mathcal{F}}|\nu_n(f)-\frac{1}{2}\{F_1(X)+F_2(X)\}|>t\Big)\leq 2C_{\mathcal{F}}\frac{\|F_2-F_1\|_2^2}{4t^2},$$

$$P\Big(\sup_{f\in\mathcal{F}}|\nu_n(f)|>t\Big)\leq 2C_{\mathcal{F}}\frac{\|F_2-F_1\|_2^2}{t^2}.$$

The first inequality is due to the bound of the empirical process by the empirical mean of the envelope of the variable $Y(f)$

$$E\Big\{\sup_{f\in\mathcal{F}}|\nu_n(f)|\Big\}^2 \leq 2C_{\mathcal{F}}\frac{1}{2n}\sum_{i=1}^{n}E\{(F_2-F_1)(X_i)\}^2.$$

The second inequality is a consequence of the uniform bounds on $\mathcal{F}$ $X(F_1) \leq X(f) \leq X(F_2)$, which implies

$$P(X(F_2) > x) \leq P(X(f) > x) \leq P(X(F_1) > x)$$

and $E|\nu_n(f)|^2 \leq \|F_2 - F_1\|_2^2$, from Proposition 7.1.

Bennett's inequality for independent random variables is extended to the sum $S_n(f)$, with a uniform norm over a class of functions $\mathcal{F}$.

Theorem 7.1. *Let $(X_i)_{i=1,...,n}$ be a sequence of independent and identically distributed random variables on a probability space $(\Omega, \mathcal{A}, P)$, with values in a metric space $(\mathbb{X}, \mathcal{X}, \|\cdot\|)$. Let $\mathcal{F}$ be a class of measurable functions on $(\mathbb{X}, \mathcal{X}, \|\cdot\|)$ with envelope F in L^p for every $p \geq 1$, and such that $P(f) = 0$ in $\mathcal{F}$, $C_{\mathcal{F}}$ is finite and there exists a constant M for which $f(X_i)| \leq \sigma_P(F)M$, a.s. in $\mathcal{F}$. For every $t > 0$*

$$P\Big(\sup_{f\in\mathcal{F}}|S_n(f)| \geq t\Big) \leq 2C_{\mathcal{F}}\exp\Big\{-n\phi\Big(\frac{t}{n\sigma_P(F)M}\Big)\Big\}$$

where $\phi(x) = (1+x)\log(1+x) - x$.

Proof. From Chernoff's theorem

$$P\Big(\sup_{f\in\mathcal{F}}S_n(f) \geq t\Big) = E\exp\Big\{\lambda \sup_{f\in\mathcal{F}}S_n(f) - \lambda t\Big\}.$$

For every $t > 0$ and ε in $]0, \delta[$, with $\delta < 1$, let $\mathcal{F}_\varepsilon$ be an ε-net of $\mathcal{F}$ adapted to P. A bound for the moment generating function of $F(X)$ is obtained from an expansion of the exponential function, $\varphi_{F(X)} \leq \exp\{\exp(b\lambda) - 1 - b\lambda\}$, with the bound $b = \sigma_P(F)M$ for the uniform L^p-moments of the variables $f(X_i)$. By Proposition 7.2

$$P\Big(\sup_{f\in\mathcal{F}}S_n(f) \geq t\Big) \leq C_{\mathcal{F}}\inf_{\lambda>0}\exp\{\psi_t(2\lambda)\},$$

where the function $\psi_t(\lambda) = n(\exp(b\lambda) - 1 - b\lambda) - \lambda t$ is

$$\inf_{\lambda>0}\psi_t(\lambda) = -n\phi\Big(\frac{t}{nb}\Big)$$

therefore

$$\inf_{\lambda>0}\exp\Big\{\frac{1}{2}\psi_t(2\lambda)\Big\} = \exp\Big\{-\frac{n}{2}\phi\Big(\frac{t}{nb}\Big)\Big\}.$$

$\square$

Corollary 7.1. *Under the conditions of Theorem 7.1, for a sequence of independent, identically distributed and non centered random variables* $(X_i)_{i=1,\ldots,n}$, *and for a class $\mathcal{F}$ of functions, for every $x > 0$*

$$P\Big(\sup_{f \in \mathcal{F}} |S_n(f)| \geq x + n \sup_{f \in \mathcal{F}} |P(f)|\Big) \leq 2C_{\mathcal{F}} \exp\Big\{-n\phi\Big(\frac{x}{n\sigma_P(F)M}\Big)\Big\}.$$

This result is a consequence of Theorem 7.1 and of the inequality $|\sup_{\mathcal{F}} |S_n(f)| - n \sup_{\mathcal{F}} |P(f)|| \leq \sup_{\mathcal{F}} |S_n(f) - nP(f)|$, hence

$$P\Big(\sup_{\mathcal{F}} |S_n(f)| \geq x + n \sup_{\mathcal{F}} |P(f)|\Big) \leq P\Big(\sup_{\mathcal{F}} |S_n(f) - nP(f)| \geq x\Big).$$

Theorem 7.1 is also written without boundedness condition for $\sup_{f \in \mathcal{F}} |f(X_i)|$ and with a condition for the upper bound of $f^2(X_i)$. For every $x > 0$

$$P\Big(\sup_{f \in \mathcal{F}} |n^{-\frac{1}{2}} S_n(f)| \geq x\Big) \leq 2C_{\mathcal{F}} \exp\Big\{-n\phi\Big(\frac{x}{\sqrt{n\eta}}\Big)\Big\} + P(F^2(X) > \eta).$$

With non identically distributed random variables, the boundedness condition implies that the moment generating functions $\varphi_{|f(X_i)|}$ satisfy the condition of convergence of $n^{-1} \sum_{i=1}^{n} \log \varphi_{|f(X_i)|}$ to a limit, $\log \varphi_{|f(X)|}$. The bound in the expression of the moment generating function is $b_n = M \max_{i=1,\ldots,n} \sigma_{P,i}$ with $\sigma_{P,i}^2 = E_P \|X_i\|_{\mathcal{F}}$, and the upper bound of the limit $\varphi_{|f(X)|}(t)$ is similar to the bound for the Laplace transform of X with i.i.d. variables, replacing σ_P by $\sigma_{P,n}^* = \max_{i=1,\ldots,n} \sigma_{P,i}$. The constant depending on the entropy function is unchanged by considering a maximum over the probability distributions $(P_{X_i})_{i=1,\ldots,n}$.

Let $(X_i)_{i=1,\ldots,n}$ be a sequence of independent variables on $(\Omega, \mathcal{A}, P)$ and let $\mathcal{F}$ be a class of measurable functions on $(\mathbb{X}, \mathcal{X}, \| \cdot \|)$ such that the variables $f(X_i)$ have the respective expectations $E_P f(X_i) = P_i(f)$ and variances $E_P f^2(X_i) - E_P f(X_i) = \sigma_{P,i}^2(f)$, and such that $\|f(X_i)\|_{\mathcal{F}}$ belongs to L^p for every integer p. For every $x > 0$

$$P\Big(\sup_{f \in \mathcal{F}} |\nu_n(f)| \geq x\Big) \leq 2C_{\mathcal{F}} \exp\Big\{-n\phi\Big(\frac{x}{\sqrt{n\eta}}\Big)\Big\}$$
$$+ P(\max_{i=1,\ldots,n} |F(X_i)| > \eta).$$

The next version of Corollary 7.1 applies to functional classes having lower and upper envelopes.

Corollary 7.2. *Let $(X_i)_{i=1,\ldots,n}$ be independent variables with values in a metric space $(\mathbb{X}, \mathcal{X}, \| \cdot \|)$ and let $\mathcal{F}$ be a class of measurable functions on $\mathbb{X}$ such that $C_{\mathcal{F}}$ is finite and for every i there exist upper and lower* envelops F_{1i} and F_{2i} in $\mathcal{F}$ for X_i, with finite moment generating functions, $F_{1i}(X_i) \leq f(X_i) \leq F_{2i}(X_i)$ for every f of $\mathcal{F}$. Then, for every $t > 0$

$$P\Big(\sup_{f \in \mathcal{F}} |\nu_n(f)| \geq t\Big) \leq 2C_{\mathcal{F}} \exp\Big\{-n\phi\Big(\frac{t\sqrt{n}}{\{\sum_{i=1}^n E(F_{2i} - F_{1i})(X_i)\}^{\frac{1}{2}}}\Big)\Big\}.$$

Proof. Under the conditions of Corollary 7.2, $E\{f(X_i) - Ef(X_i)\}^2$ is bounded by

$$EF_{2i}^2(X_i) - \{EF_{1i}(X_i)\}^2 \leq E\{F_{2i}(X_i) - EF_{1i}(X_i)\}^2$$
$$\leq E\{F_{2i}(X_i) - F_{1i}(X_i)\}^2$$

and the moment generating function of $\sup_{f \in \mathcal{F}} \nu_n(f)$ is bounded using functional Burkholder-Davis-Gundy inequality of Proposition 7.1 for non identically distributed variables

$$\varphi_f(t) = E \exp\{tn^{-\frac{1}{2}}(S_n - ES_n)(f)\}$$
$$\leq \exp[t\{n^{-1}E(F_{2i} - F_{1i})(X_i)\}^{\frac{1}{2}}]$$
$$- t\{n^{-1} \sum_{i=1}^n E(F_{2i} - F_{1i})(X_i)\}^{\frac{1}{2}}.$$

The proof ends similarly to that of proving Bennett's inequality. $\square$

If the moment generating function of the variables $F_{ki}(X_i)$ is not finite, the inequality of Corollary 7.2 can be replaced by the following one, for every $\eta > 0$

$$P\Big(\sup_{f \in \mathcal{F}} |\nu_n(f)| \geq t\Big) \leq 2\,C_{\mathcal{F}} \exp\Big\{-n\phi\Big(\frac{t}{\sqrt{n\eta}}\Big)\Big\}$$
$$+ P\Big(n^{-1} \sum_{i=1}^n (F_{2i} - F_{1i})(X_i) > \eta\Big).$$

Not only the expectation of $n^{-1} \sum_{i=1}^n (F_{2i} - F_{1i})(X_i)$ enters in the bound but also its distribution, however a value of η larger than the expectation of $n^{-1} \sum_{i=1}^n (F_{2i} - F_{1i})(X_i)$ reduces the exponential term. The second term is bounded by $\eta^{-2} n^{-1} \sum_{i=1}^n E\{(F_{2i} - F_{1i})^2(X_i)\}$.

An exponential inequality for $\sup_{f \in \mathcal{F}} |\nu_n(f)|$ is proved like the inequality (A.3) with the bound $E\{F_{2i}(X_i) - EF_{1i}(X_i)\}^2$ for the variance of the variables $f(X_i)$.

Theorem 7.2. *Let* $(X_i)_{i=1,\dots,n}$ *be independent variables satisfying the conditions of Corollary 7.2. For every* $\lambda > 0$

$$P\Big(\sup_{f \in \mathcal{F}} |\nu_n(f)| \geq t\Big) \leq C_{\mathcal{F}} \exp\Big\{-\frac{t^2}{2n^{-1} \sum_{i=1}^n E(F_{2i} - F_{1i})(X_i)}\Big\}.$$

A process $(X_t)_{t \in \mathbb{R}}$ with values in a metric space $(\mathcal{X}, d)$ and with a variance function σ_t^2 which satisfies the equality in Hoefding's exponential bound

$$P(|X_t - X_t)| > x) \leq \exp\left\{-\frac{t^2}{2|\sigma_t^2 - \sigma_s^2|}\right\}$$

for every $x > 0$, has a tail behaviour similar to that of a Gaussian process, or a Brownian motion if $\sigma_t^2 = t$. The Gaussian martingales also reach the bound. With a strict inequality, it is called *sub-Gaussian* (van der Vaart and Wellner, 1996). Thus, the empirical process and other martingales are sub-Gaussian. Let $(X_t)_{t \in \mathbb{R}}$ be a separable sub-Gaussian process, in their Corollary 2.2.8 van der Vaart and Wellner (1996) proved that there exists a constant K such that for every $\delta > 0$

$$E \sup_{\|t-s\| \leq \delta} |X_t - X_s| \leq K \int_0^\delta \sqrt{\log D(x, \|\cdot\|)} \, dx,$$

where $D(\varepsilon, d)$ is the packing number, defined as the maximum number of ε-separated points of the metric space $(\mathcal{X}, d)$, and it is equivalent to the covering number $N(\varepsilon, d)$ by the inequality $N(\varepsilon, d) \leq D(\varepsilon, d) \leq N(\frac{1}{2}\varepsilon, d)$. The constant $C_{\mathcal{F}}$ is larger than the constant $\int_0^\delta \sqrt{1 + \log N(x, \|\cdot\|)} \, dx$ of their inequalities which cannot be compared with the above results.

The results apply to U-statistics considered in Section 4.2

$$U_n = \binom{n}{k}^{-\frac{1}{2}} \sum_{i_1 \neq \ldots \neq i_k = 1}^n [f(X_{i_1}, \ldots, X_{i_k}) - E\{f(X_{i_1}, \ldots, X_{i_k})\}],$$

with $\mathcal{X} = \mathbb{R}^k$, k-dimensional variables X_i, $i = 1, \ldots, n$, and f belonging to the class $\mathcal{F}$.

7.3 Inequalities for functional martingales

On a filtered probability space $(\Omega, \mathcal{G}, (\mathcal{G}_n)_{n \geq 0}, P)$, let $X = (X_n)_n$ be in $\mathcal{M}_{0,loc}^2$ and $(U_n)_{n \geq 1}$ be a predictable process. A functional inequality is established for the discrete martingales

$$Y_n(f) = \sum_{k=1}^n f(U_k)(X_k - X_{k-1}) \tag{7.4}$$

defined for functions f of a class $\mathcal{F}$ of measurable $L^2(P)$ functions on $(\mathbb{R}, \mathcal{B})$. The proofs are deduced from Propositions 4.12 and 4.13 for the process Y

with quadratic variations

$$V_n(f) = \sum_{k=1}^{n} f^2(U_k)(V_k - V_{k-1})$$

and from Propositions 5.2 and 5.3 for continuous martingales.

Proposition 7.4. *For every $\alpha \geq 2$, let $\mathcal{F}$ be a class of measurable L^α functions on $(\mathbb{R}, \mathcal{B})$. There exists a constant $C_\alpha > 0$ such that for every real centered local martingale $(X_n)_{n \geq 0}$ of L^α, on a filtered probability space $(\Omega, \mathcal{G}, (\mathcal{G}_n)_{n \geq 0}, P)$, with a process of quadratic variations $(V_n(X))_{n \geq 0}$, and for every predictable process $(U_n)_{n \geq 0}$ of L^α*

$$E\left\{\sup_{f \in \mathcal{F}} V_n(f)\right\}^{\frac{\alpha}{2}} \leq E\left|\sup_{f \in \mathcal{F}} Y_n(f)\right|^\alpha \leq C_\alpha E\left\{\sup_{f \in \mathcal{F}} V_n(f)\right\}^{\frac{\alpha}{2}}.$$

Proposition 7.4 applies to stopping times. For every $p \geq 2$ and for every stopping time N, there exists a constant C_α depending only on α such that

$$E\left[\left\{\sup_{f \in \mathcal{F}} V_N(f)\right\}^{\frac{\alpha}{2}}\right] \leq E\left\{\left|\sup_{f \in \mathcal{F}} Y_N(f)\right|^\alpha\right\} \leq C_p E\left[\left\{\sup_{f \in \mathcal{F}} V_N(f)\right\}^{\frac{\alpha}{2}}\right].$$

If $\mathcal{F}$ has an envelope F, by monotonicity of $V_N(f)$ in $\mathcal{F}$ this inequality and a projection of every function of $\mathcal{F}$ on ε-nets implies

$$E|\sup_{f \in \mathcal{F}} Y_N(f)|^\alpha \leq C_\alpha C_\mathcal{F} E|V_N(F)|^{\frac{\alpha}{2}}.$$

For $\lambda > 0$, $\alpha \leq 2$ and for every stopping time N

$$P\left(\sup_{f \in \mathcal{F}} Y_N(f) > \lambda\right) \leq C_\alpha C_\mathcal{F} \lambda^{-\alpha} E\{V_N^{\frac{\alpha}{2}}(F)\},$$

$$P\left(\sup_{f \in \mathcal{F}} Y_N^*(f) > \lambda\right) \leq C_\alpha C_\mathcal{F} \lambda^{-\alpha} E\left\{\sum_{n=1}^{N} V_n^{\frac{\alpha}{2}}(F)\right\}.$$

If $0 < \alpha < 1$, the function x^α is concave and the moment inequality of Proposition 7.4 is reversed

$$E\left[\left|\sup_{\mathcal{F}} Y_n(f)\right|^\alpha \mid \mathcal{F}_{n-1}\right] \leq E\left[\left\{\sup_{\mathcal{F}} Y_n^2(f) \mid \mathcal{F}_{n-1}\right\}^{\frac{\alpha}{2}}\right]$$

$$= E\left[\left\{\sup_{\mathcal{F}} V_n(f) \mid \mathcal{F}_{n-1}\right\}^{\frac{\alpha}{2}}\right]$$

$$= E\left\{\sup_{\mathcal{F}} A_n^{\frac{\alpha}{2}}(f)\right\},$$

where $A_n(f) = E\{V_n(f) \mid \mathcal{F}_{n-1}\}$ and $V_n(f)$ are increasing processes.

Proposition 7.5. *For every α in $]0, 1[$, there exists a constant $c_\alpha < 1$ such that for every local martingale $(X_n)_{n \geq 1}$ of $\mathcal{M}_{0,loc}^\alpha$ and for every $n \geq 2$*

$$c_\alpha E\left\{\sup_{\mathcal{F}} A_n^{\frac{\alpha}{2}}(f)\right\} \leq E\left\{\left|\sup_{\mathcal{F}} Y_n(f)\right|^\alpha\right\} \leq E\left\{\sup_{\mathcal{F}} A_n^{\frac{\alpha}{2}}(f)\right\}.$$

Proposition 7.4 generalizes to the variations of the process $Y_n(f)$.

Proposition 7.6. *For every $\alpha \geq 2$, there exists a constant $C_\alpha > 0$ such that for every real centered local martingale $(X_n)_{n\geq 0}$ of L^α with a process of quadratic variations $(V_n)_{n\geq 0}$, and for every predictable process $(U_n)_{n\geq 0}$ of L^α*

$$E\left(\left[\sup_{\mathcal{F}} E\{V_n(f) \mid \mathcal{F}_m\} - V_m(f)\right]^{\frac{\alpha}{2}}\right)$$

$$\leq E\left\{\sup_{\mathcal{F}} |Y_n(f) - Y_m(f)|^\alpha\right\}$$

$$\leq C_\alpha E\left(\left[\sup_{\mathcal{F}} E\{V_n(f) \mid \mathcal{F}_m\} - V_m(f)\}\right]^{\frac{\alpha}{2}}\right).$$

Proof. The martingale property implies that for every $n > m$

$$E\left[\sup_{\mathcal{F}}\{Y_n(f) - Y_m(f)\}^2\right] = E\left[\sup_{\mathcal{F}}\left\{\sum_{k=m+1}^{n} f(U_k)(X_k - X_{k-1})\right\}^2\right]$$

$$= E\left\{\sup_{\mathcal{F}} \sum_{k=m+1}^{n} f^2(U_k)(X_k - X_{k-1})^2\right\}$$

$$= E\left[\sup_{\mathcal{F}}\{V_n(f) - V_m(f)\}\right],$$

and the convexity of the power function x^α, for every real $\alpha \geq 2$, yields

$$E\left[\sup_{\mathcal{F}}\{Y_n(f) - Y_m(f)\}^\alpha \mid \mathcal{F}_m\right] \geq E\left[\sup_{\mathcal{F}}\{Y_n(f) - Y_m(f)\}^2 \mid \mathcal{F}_m\right]^{\frac{\alpha}{2}}$$

$$= E\left[\sup_{\mathcal{F}}\{V_n(f) - V_m(f) \mid \mathcal{F}_m\}^{\frac{\alpha}{2}}\right]$$

and the lower bound of $E \sup_{f\in\mathcal{F}} |Y_n(f) - Y_m(f)|^\alpha$ follows. The upper bound is established like in the proof of Proposition 4.14. For every $\lambda > 0$ let $B_{nm}(\lambda) = \{\sup_{\mathcal{F}}\{Y_n(f) - Y_m(f)\}^2 > \lambda\}$, we have

$$\lambda P\left(\sup_{\mathcal{F}}\{Y_n(f) - Y_m(f)\}^2 > \lambda\right) \leq E\left[\sup_{\mathcal{F}}\{Y_n(f) - Y_m(f)\}^2 1_{B_{nm}(\lambda)}\right]$$

$$= E\left[\sup_{\mathcal{F}}\{V_n(f) - V_m(f)\} 1_{B_{nm}(\lambda)}\right]$$

$$+ E\left\{1_{B_{nm}(\lambda)} \sup_{\mathcal{F}} \sum_{j\neq k=m+1}^{n} f(U_j)f(U_k)(X_j - X_{j-1})(X_k - X_{k-1})\right\}$$

$$\leq 2E\left[\sup_{\mathcal{F}}\{V_n(f) - V_m(f)\} 1_{B_{nm}(\lambda)}\right]$$

by the Cauchy-Schwarz inequality. Arguing like in the proof of Doob's inequality gives the bound with the constant $C_p = 2p(p-1)^{-1}$ depending only on p. $\square$

Proposition 7.7. *For every α in $]0,1[$, there exists a constant c_α such that for every local martingale $(Y_n(f))_{n\geq 1}$ of $\mathcal{M}_{0,loc}^\alpha$ and for all $n > m \geq 2$*

$$c_\alpha E\left(\left[\sup_{\mathcal{F}} E\{V_n(f) \mid \mathcal{F}_m\} - V_m(f)\right]^{\frac{\alpha}{2}}\right)$$

$$\leq E\left\{\sup_{\mathcal{F}} |Y_n(f) - Y_m(f)|^\alpha\right\}$$

$$\leq E\left(\left[\sup_{\mathcal{F}} E\{V_n(f) \mid \mathcal{F}_m\} - V_m(f)\}\right]^{\frac{\alpha}{2}}\right).$$

Proof. By the concavity of the power function x^α, for α in $]0,1[$, and by the martingale property of $Y_n(f)$, for all $n > m$ we have

$$E\{\sup_{\mathcal{F}} |Y_n(f) - Y_m(f)|^\alpha \mid \mathcal{F}_m\} \leq \left[E\left\{\sup_{\mathcal{F}} |Y_n(f) - Y_m(f)|^2 \mid \mathcal{F}_m\right\}\right]^{\frac{\alpha}{2}}$$

$$= \{E(V_n - V_m \mid \mathcal{F}_m)\}^{\frac{\alpha}{2}}$$

and the upper bound of $E\{\sup_{\mathcal{F}} |Y_n(f) - Y_m(f)|^\alpha\}$ follows.

The increasing process $V_n(f)$ is such that the expectations of $\sup_{\mathcal{F}}\{V_n(f) - V_m(f)\}$ and $\sup_{\mathcal{F}}\{Y_n(f) - Y_m(f)\}^2$ are equal and the convexity of the power function implies

$$E\left(\sup_{\mathcal{F}}\left[\{V_n(f) - V_m(f)\} - \{Y_n(f) - Y_m(f)\}^2\right]^p\right)$$

$$\leq \left(E \sup_{\mathcal{F}}\left[\{V_n(f) - V_m(f)\} - \{Y_n(f) - Y_m(f)\}^2\right]\right)^p = 0$$

and therefore

$$0 \leq E\left[\sup_{\mathcal{F}}\{V_n(f) - V_m(f)\}^p\right] \leq 2^{p-1} E\left[\sup_{\mathcal{F}}\{Y_n(f) - Y_m(f)\}^{2p}\right]$$

lower bound follows. $\square$

Theorem 7.3. *On a filtered probability space $(\Omega, \mathcal{G}, (\mathcal{G}_n)_{n\geq 0}, P)$, let $(X_n)_{n\geq 0}$ belong to $L_{0,loc}^p$, for every integer $p \geq 2$ and let $(A_n)_{n\geq 0}$ be a predictable process of L^p. Let $\mathcal{F}$ be a class of measurable $L^p(P)$ functions on $(\mathbb{R}, \mathcal{B})$ with envelope F such that $\sup_{f\in\mathcal{F}} Y_n(f)$ belongs to L^p for every integer p, $P(f) = 0$ in $\mathcal{F}$ and $C_{\mathcal{F}}$ is finite. If there exists a constant M such that $|f(X_i)| \leq \sigma_P(F)M$ a.s. in $\mathcal{F}$, then for every $x > 0$*

$$P\left(\sup_{f\in\mathcal{F}} Y_n(f) \geq x\right) \leq C_{\mathcal{F}} \exp\left\{-n\phi\left(\frac{x}{n\sigma_P(F)M}\right)\right\}$$

where $\phi(x) = (1+x)\log(1+x) - x$. More generally

$$P\left(\sup_{f\in\mathcal{F}} Y_n(f) \geq x\right) \leq C_{\mathcal{F}} \exp\left\{-n\phi\left(\frac{x}{n\sqrt{\eta}}\right)\right\} + P(V_n(F) > \eta).$$

Since $V_n(Y)$ is increasing over $\mathcal{F}$, $\sup_{f\in\mathcal{F}} V_n(f) = V_n(F)$. This implies the second inequality of Theorem 7.3.

For a local martingale $X = (X_t)_{t\geq 0}$ indexed by $\mathbb{R}_+$ and a predictable process $(U_t)_{t\geq 0}$, a functional local martingale

$$Y_t(f) = \int_0^t f(U_s)\, dX_s, \tag{7.5}$$

is defined for measurable functions f of a class $\mathcal{F}$ in $L^2(P)$. The process $Y_t(f)$ and its increasing predictable process $V_t(f) = \int_0^t f^2(U_s)\, d < X >_s$ satisfy similar inequalities as the discrete martingale $Y_n(f)$, uniformly in $\mathcal{F}$.

Proposition 7.8. *For every $\alpha \geq 2$, let $\mathcal{F}$ be a class of measurable L^α functions on $(\mathbb{R}, \mathcal{B})$. There exists a constant $C_\alpha > 0$ such that for every $(X_t)_{t\geq 0}$ of $\mathcal{M}^\alpha_{loc,0}$*

$$E\Big\{\sup_{f\in\mathcal{F}} < X >_t (f)\Big\}^{\frac{\alpha}{2}} \leq E\Big|\sup_{f\in\mathcal{F}} X_t(f)\Big|^\alpha \leq C_\alpha E\Big\{\sup_{f\in\mathcal{F}} < X >_t (f)\Big\}^{\frac{\alpha}{2}}.$$

Proposition 7.9. *For every $\alpha \geq 2$, let $\mathcal{F}$ be a class of measurable functions of L^α. There exists a constant $C_\alpha > 0$ such that for every X be in $\mathcal{M}^\alpha_{0,loc}$, for every predictable process U of L^α and for every stopping time T*

$$E\Big\{\sup_{f\in\mathcal{F}} \int_0^T f^2(U_s)\, d < X >_s\Big\}^{\frac{\alpha}{2}} \leq E\Big|\sup_{f\in\mathcal{F}} \int_0^T f(U_s)\, dX_s\Big|^\alpha$$

$$\leq C_\alpha E\Big\{\sup_{f\in\mathcal{F}} \int_0^T f^2(U_s)\, d < X >_s\Big\}^{\frac{\alpha}{2}},$$

and for all stopping times $S < T$

$$E\Big(\Big[\sup_{\mathcal{F}} E\{V_T(f) \mid \mathcal{F}_S\} - V_S(f)\Big]^{\frac{\alpha}{2}}\Big)$$

$$\leq E\Big\{\sup_{\mathcal{F}} |Y_T(f) - Y_S(f)|^\alpha\Big\}$$

$$\leq C_\alpha E\Big(\Big[\sup_{\mathcal{F}} E\{V_T(f) \mid \mathcal{F}_S\} - V_S(f)\}\Big]^{\frac{\alpha}{2}}\Big).$$

Proposition 7.10. *For every α in $]0,1[$, there exists a constant $c_\alpha < 1$ such that for every local martingale $(X_t)_{t\geq 0}$ of $\mathcal{M}^\alpha_{0,loc}$ and for every stopping time T*

$$c_\alpha E\Big\{\sup_{\mathcal{F}} V_T^{\frac{\alpha}{2}}(f)\Big\} \leq E\Big\{\Big|\sup_{\mathcal{F}} Y_T(f)\Big|^\alpha\Big\} \leq E\Big\{\sup_{\mathcal{F}} V_T^{\frac{\alpha}{2}}(f)\Big\}.$$

For all stopping times $S < T$, we have

$$c_\alpha E\left(\left[\sup_{\mathcal{F}} E\{V_T(f) \mid \mathcal{F}_S\} - V_S(f)\right]^{\frac{\alpha}{2}}\right)$$

$$\leq E\left\{\sup_{\mathcal{F}} |Y_T(f) - Y_S(f)|^\alpha\right\}$$

$$\leq E\left(\left[\sup_{\mathcal{F}} E\{V_T(f) \mid \mathcal{F}_S\} - V_S(f)\}\right]^{\frac{\alpha}{2}}\right).$$

Like in Theorem 7.3, the local martingale Y satisfies an exponential inequality related to a bound for its predictable process and depending on the cumulated entropy function.

Theorem 7.4. *On a filtered probability space $(\Omega, \mathcal{G}, (\mathcal{G}_t)_{t\geq 0}, P)$, let X be in $\mathcal{M}^p_{0,loc}$, $p \geq 2$ and let A be a predictable process of L^p. Let $\mathcal{F}$ be a class of measurable $L^p(P)$ functions on $(\mathbb{R}, \mathcal{B})$ with envelope F such that $\|Y_t(f)\|_{\mathcal{F}}$ belongs to L^p, for every integer p, $P(f) = 0$ in $\mathcal{F}$ and $C_{\mathcal{F}}$ is finite. If there exists a constant c such that $|Y_t(f)| \leq c \|Y_t(F)\|_{L^2}$ a.s. in $\mathcal{F}$, then for every stopping time T and for every $x > 0$*

$$P\left(\sup_{f\in\mathcal{F}} |Y_T(f)| \geq x\right) \leq C_{\mathcal{F}} E \exp\left\{-\phi\left(\frac{x}{c\|Y_T(F)\|_{L^2}}\right)\right\}.$$

More generally

$$P\left(\sup_{f\in\mathcal{F}} |Y_T(f)| \geq t\right) \leq C_{\mathcal{F}} 2 \exp\left\{-\phi\left(\frac{x}{\sqrt{\eta}}\right)\right\}$$

$$+ P\left(\int_0^T F^2(U_s)\, d < M >_s> \eta\right).$$

Extending Proposition 5.28, the Brownian distribution of the process

$$Y_t = \int_0^t \left(\int_0^s \beta_y^2\, dy\right)^{-\frac{1}{2}} \beta_s\, dB_s,$$

where β is a predictable process with sample paths in $\mathcal{C}_b(\mathbb{R}_+)$, allows to write a uniform tightness property.

Proposition 7.11. *On very finite interval $[S, T]$ of $\mathbb{R}_+$ and for every $x > 0$*

$$\lim_{\varepsilon\to 0} P\left(\sup_{S\leq s\leq t\leq T, |t-s|<\varepsilon} |Y_t - Y_s| > x\right) = 0.$$

Proof. The predictable compensator of Y is

$$< Y >_t = \int_0^t \left(\int_0^s \beta_y^2\, dy\right)^{-1} \beta_s^2\, ds$$

and its continuity implies that for every $\eta > 0$ and $x > 0$, there exists $\varepsilon > 0$ such that

$$P\left(\sup_{S \leq s \leq t \leq T, |t-s| < \varepsilon} < Y >_t - < Y >_s > x\right) \leq \eta.$$

The real ε defines an integer $k_\varepsilon = [\varepsilon^{-1}(T-S)]$, $k_\varepsilon + 1$ points $x_k = S + k\varepsilon$ such that $x_1 = S$ and $x_{k_\varepsilon+1} = T$, and variables

$$Z_k = \sup_{x_k \leq s \leq t \leq x_{k+1}} (< Y >_t - < Y >_s).$$

It follows that for every finite interval $[S, T]$ of $\mathbb{R}_+$ and $x > 0$

$$P\left(\sup_{S \leq s \leq t \leq T} < Y >_t - < Y >_s > x\right) \leq \sum_{k=1,\ldots,k_\varepsilon} P(Z_k > k_\varepsilon^{-1} x)$$

$$\leq k_\varepsilon \eta = O\left(\frac{\eta}{\varepsilon}\right).$$

The proof ends by using Lenglart's inequality of Proposition 1.3. $\quad\square$

Considering the uniform metric on the space $C(\mathbb{R}_+)$, the functional variations of the process Y defined by (7.5) satisfy a property similar to the previous proposition.

Proposition 7.12. *Let* $\mathcal{F} = \{\alpha \geq 0, \alpha \in D(\mathbb{R}_+)\}$. *On every finite interval* $[0, T]$ *of* $\mathbb{R}_+$

$$\lim_{\varepsilon \to 0} \sup_{0 \leq t \leq T} P\left(\sup_{\alpha, \beta \in \mathcal{F}, \|\alpha - \beta\|_{[0,t]} \leq \varepsilon} |Y_t(\alpha) - Y_t(\beta)| > x\right) = 0.$$

Theorem 7.5. *Let* $\mathcal{F}$ *be a class of measurable functions on* $(\mathbb{R}, \mathcal{B})$ *such that there exist functions* F_1 *and* F_2 *in* $\mathcal{F}$, *belonging to* $L^p(\mathbb{X})$ *and satisfying* $F_1 \leq \beta \leq F_2$ *for every* β *of* $\mathcal{F}$. *Then for every stopping time* T *and for every* $x > 0$

$$P\left(\sup_{\beta \in \mathcal{F}} \sup_{t \in [0,T]} |Y_t(\beta)| \geq x\right) \leq 2C_{\mathcal{F}} E \int_0^T \left(\int_0^s F_1^2(y)\, dy\right)^{-1} F_2^2(s)\, ds.$$

Proof. The predictable compensator of the process $Y(\beta)$ has a uniform bound in $\mathcal{F}$

$$< Y >_t (\beta) \leq \int_0^t \left(\int_0^s F_1^2(y)\, dy\right)^{-1} F_2^2(s)\, ds.$$

Following the proof of Theorem 7.3 with the bounds of Proposition 5.2 for the moments of the process $Y(\beta)$ yields the result, with $E\sqrt{b_T}$. $\quad\square$

Replacing the Brownian motion by a local martingale M of $\mathcal{M}_{0,loc}^p$, let Y be the process defined by

$$Y_t = \left(\int_0^t \beta_y^2 \, d < M >_y\right)^{-\frac{1}{2}} \int_0^t \beta_s \, dM_s,$$

where β is a process with sample paths in $\mathcal{C}_b(\mathbb{R}_+)$.

Theorem 7.6. *Let $\mathcal{F}$ be a class of measurable functions on $(\mathbb{R}, \mathcal{B})$ such that there exist functions F_1 and F_2 in $\mathcal{F}$, belonging to $L^p(\mathbb{X})$, for every $p \geq 2$ and satisfying $F_1 \leq \beta \leq F_2$ for every β of $\mathcal{F}$. For all stopping time T and $x > 0$*

$$P\left(\sup_{\beta \in \mathcal{F}} |Y_T(\beta)| \geq x\right) \leq 2C_{\mathcal{F}} E \exp\left\{-\phi\left(\frac{x}{\sqrt{a_T}}\right)\right\}$$

where $a_T = \int_0^T F_2^2 \, d < M > \{\int_0^T F_1^2 \, d < M >\}^{-1}$.

The proof is the same as above, with the following uniform bounds for the predictable compensator of the process $Y(\beta)$

$$\sup_{\beta \in \mathcal{F}} < Y >_t (\beta) \leq \left(\int_0^t F_1^2(y) \, dy\right)^{-1} \int_0^t F_2^2(s) \, ds = a_t.$$

The bound is replaced by an exponential inequality by Hoeffding's inequality (Theorem A.4). Under the conditions of Theorem 7.6, for every $x > 0$

$$P\left(\sup_{\beta \in \mathcal{F}} Y_T(\beta) \geq x\right) \leq C_{\mathcal{F}} E \exp\left\{-\frac{x^2}{2a_T}\right\}.$$

7.4   Weak convergence of functional processes

Let Φ be a functional defined on the set $\mathcal{P}_{\mathcal{X}}$ of the probability distributions on $(C([0,1]), \mathcal{C})$, with values in $(C([0,1]), \mathcal{C})$ and satisfying a Lipschitz condition

$$\|\Phi(X_t) - \Phi(Y_t)\|_{[0,1]} \leq K \|X_t - Y_t\|_{[0,1]},$$

for processes X and Y of $C([0,1])$. For every $x > 0$, the empirical process of independent uniform variables on $[0,1]$ and the Brownian bridge have the bound

$$P(\|\Phi(\nu_{n,t}) - \Phi(W_t)\|_{[0,1]} > x) \leq \frac{K}{x} E \|\nu_{n,t} - W_t\|_{[0,1]}$$

and it converges to zero.

The continuous differentiability of a function $\Phi : \mathbb{R} \mapsto \mathbb{R}$ implies

$$\Phi(\nu_{n,t}) - \Phi(W_t) = (\nu_{n,t} - W_t)\Phi'(W_t) + o(\|\nu_{n,t} - W_t\|),$$

and the function is lipschitzian if there exists a constant K such that $\sup_{x \in \mathbb{R}} |\Phi'(x)|^\alpha \leq K$, the remainder term of the expansion is uniform over $[0, 1]$ if the derivative of the function Φ is uniformly continuous, i.e. $\limsup_{x,x' \in \mathbb{R}, |x-x'| \to 0} |\Phi'(x) - \Phi'(x')| = 0$. Conversely, a Lipschitz function with exponent $\alpha > 1$ is continuously differentiable.

The weak convergence of the empirical processes in the space $C(\mathcal{X})$ of the continuous functions on $(\mathcal{X}, \mathcal{B})$ with the uniform metric has been proven under the condition that the class of functions has a $L^2(P)$ envelope and under conditions about the dimension of $\mathcal{F}$ that ensure the tightness of the process.

Example 7.1. On a probability space $(\Omega, \mathcal{A}, P)$, let $T > 0$ and $C > 0$ be independent real random variables with distribution functions F and G respectively and let $\delta = 1_{\{T \leq C\}}$ and let $X = T \wedge C$ be the minimum variable defined on $[0, \tau)$, $\tau \leq \infty$. Let $(T_i, C_i)_{i=1,\ldots,n}$ be a vector of independent variables having the same distribution as (T, C) and let φ be a real function of $L^2(F)$. Let $\Lambda(t) = \int_0^t \{1 - F^-\}^{-1} dF$ and let the empirical processes

$$N_n(t) = \sum_{i=1}^n 1_{\{T_i \wedge C_i \leq t\}},$$

$$Y_n(t) = \sum_{i=1}^n 1_{\{T_i \wedge C_i \geq t\}}.$$

The difference $M_n(t) = n^{\frac{1}{2}} \int_0^t 1_{\{Y_n > 0\}} Y_n^{-1} N_n - \int_0^t 1_{\{Y_n > 0\}} d\Lambda$ is a local square integrable martingale in the support of I_F of F, $EM_n(t) = 0$ and

$$EM_n^2(t) = \int_{[0, t \wedge \tau]} nEY_n^{-1} d\Lambda$$

where $\int_{[0,t \wedge \tau]} nY_n^{-1} d\Lambda$ converges to $\sigma_t^2 = \int_{[0,t \wedge \tau]} \{(1 - F^-)^2(1 - G^-)\}^{-1} dF$ a.s., uniformly in every compact subinterval of I_F, the predictable compensator of N_n is denoted $\widetilde{N}_n(t) = \int_0^t 1_{\{Y_n > 0\}} Y_n d\Lambda$. Let $\mathcal{H}$ be a family of functions with envelope H such that $\sigma_t^2(h) = \sup_{h \in \mathcal{H}} \int_0^t h^2 \{(1 - F^-)^2(1 - G^-)\}^{-1} dF$ is finite. Applying Proposition 7.9, for every $\alpha \geq 2$ and $t < \tau$

$$E|\sup_{h \in \mathcal{H}} \int_0^t h \, dM_n|^\alpha \leq C_{\alpha,t} E \left\{ \int_0^t 1_{\{Y_n > 0\}} H^2 Y_n^{-1} d\Lambda \right\}^{\frac{\alpha}{2}}.$$

If the constant of entropy dimension $C_{\mathcal{H}}$ is finite and if there exist m_n such that $N_n(t) \le m_n \int_0^t Y_n \, d\Lambda$ in I_F, Theorem 7.3 entails that for every stopping time T

$$P\left(\sup_{h \in \mathcal{H}} \sup_{t \le T} \int_0^t h \, dM_n \ge x\right) \le k_0 C_{\mathcal{H}} \exp\left\{-\phi\left(\frac{x}{\sqrt{n\eta}}\right)\right\}$$

$$+ P\left(\int_0^{T \wedge \tau} H^2 1_{\{Y_n > 0\}} Y_n^{-1} \, d\Lambda > \eta\right), \quad n \ge 1.$$

For every $\varepsilon > 0$, t and η can be chosen to be sufficiently large to ensure that the upper bound is smaller than ε. This inequality implies the tightness of the sequence $(\sup_{h \in \mathcal{H}} \sup_{t \le T} \int_0^t h \, dM_n)_{n \ge 1}$, and therefore its convergence to a centered Gaussian variable with variance $\int_0^T H_t^2 d\sigma_t^2$.

Example 7.2. Let $0 < S < T$ and $C > 0$ be real random variables on a probability space $(\Omega, \mathcal{A}, P)$, with C independent of S and T, and let $\delta = 1_{\{T \le C\}}$ and $\delta' = 1_{\{S \le C\}}$. Let $X = T - S$ and $\widetilde{X} = \delta'\{X \wedge (C - S)\}$ which takes the values X if $\delta = 1$, $C - S$ if $\delta' = 1$ and $\delta = 0$, and zero if $\delta' = 0$. Let $\tau > 0$ be included in the support $I_{T \wedge C}$ of the distribution of the variable $T \wedge C$ and let Y be the indicator process

$$Y(x) = 1_{\{T \wedge C \ge S + x\}}, \quad x \le \tau.$$

We consider the empirical process related to the right-censored variables $(S, T \wedge C)$ defined from a vector of independent variables $(S_i, T_i, C_i)_{i=1,\dots,n}$ having the same distribution as (S, T, C). Counting processes are defined for (s, x) in I_τ by

$$N_n(x) = \sum_{i=1}^n N_i(x) = \sum_{i=1}^n \delta_i 1_{\{T_i \le S_i + x\}},$$

$$Y_n(x) = \sum_{i=1}^n Y_i(x) = \sum_{i=1}^n 1_{\{T_i \wedge C_i \ge S_i + x\}},$$

$$\widetilde{N}_n(x) = \sum_{i=1}^n \widetilde{N}_i(x) = \sum_{i=1}^n \int_0^x Y_i(y) \lambda_{Y|S}(y; S_i) \, dy,$$

where $\lambda_{X|S}(x; s) = \lim_{\varepsilon \downarrow 0} \frac{1}{\varepsilon} P(x \le X < x + \varepsilon | X \ge x, S = s)$ is the hazard function of X conditionally on $S = s$. Summing the weighted differences $N_i - \widetilde{N}_i$ yields

$$M_n(x) = n^{-\frac{1}{2}} \left\{\int_0^x 1_{\{Y_n > 0\}} Y_n^{-1} \, dN_n - \sum_{i=1}^n \int_0^x 1_{\{Y_n(y) > 0\}} \lambda_{Y|S}(y; S_i) \, dy\right\}$$

is a local square integrable martingale with respect to the filtration generated by $((N_i(t), Y_i(t))_{t \geq S_i})_{i=1,\ldots,n}$, then

$$EM_n^2(x) = \int_{[0,x \wedge \tau]} E \frac{\lambda_{Y|S}(y; S)}{n^{-1} Y_n(y)} \, dy := E\widetilde{N}_n(x).$$

Let $\mathcal{H}$ be a family of functions h defined in $I_{T \wedge C}$, with an envelope H such that the integral

$$E \int_0^x \frac{H^2(s+y)}{P(T \geq (s+y) \wedge C)} \lambda_{Y|S}(y; s) \, dy \qquad (7.6)$$

is finite for every (s, x) with $s + x$ in $I_{T \wedge C}$. For every h of $\mathcal{H}$

$$E\{h(S+X)1_{\{T \leq S+x\}} 1_{\{T \leq S+x|S\}}\}$$
$$= \int_{[0,x \wedge \tau]} E\{h(S+y)Y(y)|S = s\} \lambda_{Y|S}(y; s) \, dy.$$

For every function h of $\mathcal{H}$, the process $W_n(s, x) = \int_{[0,x \wedge \tau]} h(s+y) \, dM_n(y)$, (s, x) in $I_{T \wedge C}$, converges weakly to a centered Gaussian process W_h, as an empirical process in the class of functions $\mathcal{F}_h = \{\delta h(s+y)1_{\{s+y \leq s+x\}} - \int_0^x E\{h(S+u)Y(s,u)|S = s\} \lambda_{X|S}(u; s) \, du, (s, x) \in I_{T \wedge C}$, at fixed h. For every $p > 1$ and for every x in $I_{T \wedge C}$

$$E|\sup_{h \in \mathcal{H}} \int_0^x h \, dM_n|^{2p} \leq C_{\mathcal{H}} E\{\sup_{h \in \mathcal{H}} \int_0^x h^2 \, d\widetilde{N}_n\}^p$$

and this inequality extends to a uniform inequality in subintervals $[0, T]$ of $I_{T \wedge C}$

$$E|\sup_{h \in \mathcal{H}} \sup_{x \geq T} \int_0^x h \, dM_n|^{2p} \leq C_{\mathcal{H}} E\{\int_0^T h^2 \, d\widetilde{N}_n\}^p.$$

This inequality implies Chernoff and Bennett inequalities for the variable $\sup_{h \in \mathcal{H}} \sup_{x \geq T} \int_0^x h \, dM_n$, by Theorem 7.3, hence its tightness and therefore its weak convergence.

Extending the process to a class of functions $\mathcal{H}$ defined by (7.6) and having a finite entropy integral $C_{\mathcal{H}}$, the weak convergence of the process is also uniform on $\mathcal{H}$.

7.5 Differentiable functionals of empirical processes

Let Φ be a function defined on the subset $\mathcal{P}_{\mathcal{X}}$ of the probability distributions on $(\mathcal{X}, \mathcal{B})$ and with values in a normed vector space and let $(B_2, \mathcal{B}_2)$ be the Borel σ-algebra. The function Φ is supposed to be differentiable in the following sense.

Definition 7.1. Let B_1 and B_2 be normed vector spaces provided with the Borel σ-algebra, a measurable map $\Phi : E \in B_1 \to B_2$ is *differentiable* at $x \in B_1$, *tangentially to a separable subspace* C of B_1 if there exists a linear map $d\phi(x) : B_1 \to B_2$ such that for all sequences $(t_n)_n$ in $\mathbb{R}$ and $(h_n)_n$ in B_1, such that t_n tends to zero, $x + t_n h_n$ belongs to E and h_n converges to a limit h in C as n tends to infinity, then

$$\lim_{n \to \infty} \left\| \frac{\phi(x + t_n h_n) - \phi(x)}{t_n} - d\phi(x).h \right\| = 0.$$

A measurable map $\phi : E \subset B_1 \to B_2$ is *continuously differentiable* at $x \in E$ along a sequence $(x_n)_n \in E$ that converges to x and *tangentially to a separable subspace* C of B_1 if there exists a linear map $d\phi(x) : B_1 \to B_2$, continuous on E and such that for all sequences $(t_n)_n$ in $\mathbb{R}$ and $(h)_n$ in B_1, such that t_n tends to zero, $x_n + t_n h_n$ in E and h_n converges to h in C as n tends to infinity

$$\lim_{n \to \infty} \left\| \frac{\phi(x_n + t_n h_n) - \phi(x_n)}{t_n} - d\phi(x).h \right\| = 0.$$

Proposition 7.13. *Let P be a probability on $(\mathcal{X}, \mathcal{B})$ and let $\mathcal{F}$ be a functional subset of $\mathcal{L}_P^2$ with an envelope $F \in \mathcal{L}_P^2$ and such that $\int_0^1 (\log D_F^{(2)}(x, \mathcal{F}))^{\frac{1}{2}} dx < \infty$. Let Φ be a map defined from a subset $\mathcal{P}_\mathcal{X}$ of $\ell^\infty(\mathcal{F})$ to B_2, differentiable at P tangentially to $\mathcal{U}_b(\mathcal{F}, P)$, then $\sqrt{n}\{\Phi(\widehat{P}_n) - \Phi(P)\}$ converges weakly to $d\Phi(P).G_P$.*

It is a consequence of the assumption of differentiability and of Pollard's central limit theorem.

The differentiability of a functional is applied to nonparametric models. Here, it is used to prove a triangular version of the $\sqrt{n}$-consistency of $\Phi(\widehat{P}_n)$ to $\Phi(P)$, for a differentiable function Φ. It relies on the existence of a sequence of Brownian bridges G_{P_n} related to P_n and converging to the Brownian bridge G_P as P_n converges to P, uniformly on $\mathcal{F}$.

Proposition 7.14. *Let P_n and P be probabilities on $(\mathcal{X}, \mathcal{B})$ and let $\mathcal{F}$ be a subset of $\bigcap_n \mathcal{L}_{P_n}^2 \bigcap \mathcal{L}_P^2$ with an envelope F in $\bigcap_n \mathcal{L}_{P_n}^2 \bigcap \mathcal{L}_P^2$. Suppose that $\bigcap_n \mathcal{L}_{P_n}^2 \bigcap \mathcal{L}_P^2$ has a finite entropy dimension, $\lim \|P_n - P\|_\mathcal{F} = 0$ and $\lim \|P_n - P\|_{\mathcal{F}^2} = 0$. Then, for every n, there exist uniformly continuous versions of the Brownian bridges for G_P and G_{P_n}, defined on the same space and such that for every $\varepsilon > 0$, $\lim_n P\{\|G_{P_n}^{(n)} - G_P^{(n)}\|_\mathcal{F} > \varepsilon\} = 0$.*

Proof. Let $\mathcal{F}_{\sigma_n}$ be a σ_n-net of $\mathcal{F}$ adapted to P and let π be the projection $\mathcal{F} \to \mathcal{F}_{\sigma_n}$. Since $\sigma^2_{P_n}(f - \pi(f)) \leq \sigma^2_P(f - \pi(f)) + 4\|P_n - P\|_{\mathcal{F}^2}$, $\|G_P(f - \pi(f))\|_{\mathcal{F}}$ and $|G_{P_n}(f - \pi(f))\|_{\mathcal{F}}$ tend to zero with σ_n. The restrictions of G_P and G_{P_n} to $\mathcal{F}_{\sigma_n}$ are Gaussian variables with dimension $k_n \leq C\sigma_n^{-d}$, for n large enough, if d is the entropy dimension of $\mathcal{F}$. Strassen's theorem implies the existence of a probability space where a Gaussian variable (X, Y) is defined as having the same margins as the restrictions of G_{P_n} and G_P to $\mathcal{F}_{\sigma_n}$ and such that $P\{\|X - Y\|_\infty > \Pi\} < \Pi$, where Π is the Prohorov distance for the uniform norm on $\mathbb{R}^{k_n}$. The existence of uniformly continuous versions of the Brownian bridge G_{P_n} and G_P defined on the same probability space is deduced from Berkes and Philipp's lemma (1979). Finally, from Dehling (1983), Π has a bound depending on k_n and δ_n as follows

$$\Pi \leq C\delta_n^{\frac{1}{2}}k_n^{\frac{1}{6}}\left(1 + |\log\frac{k_n}{\delta_n}|^{\frac{1}{2}}\right),$$

$$\delta_n \leq k_n^2 \sup_{f,g \in \mathcal{F}_{\sigma_n}} |\text{Cov}_{P_n}(f, g) - \text{Cov}_P(f, g)|.$$

Choosing σ_n that converges to zero with a rate such that $\lim_n \delta_n^{\frac{1}{2}}k_n^{\frac{1}{6}} = 0$ and since

$$|\text{Cov}_{P_n}(f, g) - \text{Cov}_P(f, g)| \leq \|P_n - P\|_{\mathcal{F}^2} + 2\|P_n - P\|_{\mathcal{F}} \sup_n \int F\,dP_n,$$

then Π tends to 0 as n tends to infinity. $\qquad\square$

Proposition 7.15. *Let $\mathcal{F}$ be a family of finite entropy dimension, let P_n and P be probabilities on $(\mathcal{X}, \mathcal{B})$ such that $\lim_n \|P_n - P\|_{\mathcal{F}} = 0$, $\lim_n \|P_n - P\|_{\mathcal{F}^2} = 0$, $F \in \mathcal{L}_P^{2+\delta}$ and $F \in \mathcal{L}_{P_n}^{2+\delta}$ for every n, with $\delta > 0$. Then, for every n, there exists a uniformly continuous version of the Brownian bridge G_P defined on the same space as ν_{n,P_n} and such that for every $\varepsilon > 0$, $P\left\{\|\nu_{n,P_n} - G_P^{(n)}\|_{\mathcal{F}} > \varepsilon\right\}$ tends to zero.*

Let Φ be a map defined from a subset $\mathcal{P}_\mathcal{X}$ of $\ell^\infty(\mathcal{F})$ to B_2, differentiable at P along the sequence $(P_n)_n$ and tangentially to $\mathcal{U}_b(\mathcal{F}, P)$, then

$$\sqrt{n}\left(\Phi(\widehat{P}_n) - \Phi(P_n)\right) \xrightarrow{\mathcal{D}} d\Phi(P).G_P.$$

Proof. This is a consequence of Massart (1986) that ensures for every n, the existence of a uniformly continuous Brownian bridge related to P_n and such that

$$P\{\|\nu_{n,P_n} - G_{P_n}^{(n)}\|_{\mathcal{F}} \geq \alpha_n\} \leq \beta_n,$$

with α_n and β_n depending only on the entropy dimension of $\mathcal{F}$ and converging to zero.

Let $\mathcal{F}_{\sigma_n, P_n}$ be a (σ_n, P_n)-net of $\mathcal{F}$ and let $\nu_{n, P_n}(\sigma_n)$ be the restriction of ν_{n, P_n} to $\mathcal{F}_{\sigma_n, P_n}$. By Proposition 7.14 and Berkes and Philipp's lemma, there exists a distribution on $\ell^\infty(\mathcal{F}_{\sigma_n, P_n}) \times \mathcal{U}_b(\mathcal{F}, P_n) \times \mathcal{U}_b(\mathcal{F}, P)$ having $\nu_{n, P_n}(\sigma_n)$ and G_P as margins on $\ell^\infty(\mathcal{F}_{\sigma_n, P_n})$ and $\mathcal{U}_b(\mathcal{F}, P)$. Skorohod's lemma implies the existence of a uniformly continuous version of G_P defined on Ω and such that $P\left\{ \|\nu_{n, P_n}(\sigma_n) - G_P^{(n)}\|_{\mathcal{F}} > \varepsilon \right\}$ tends to zero, then the bounds for the variations of ν_{n, P_n} end the proof. $\qquad\square$

Corollary 7.3. *Let P be a probability distribution on $(\mathcal{X}, \mathcal{B})$ and let $\mathcal{F}$ be a subset of a family $\mathcal{L}_P^2$ with finite entropy dimension, such that F belongs to $\mathcal{L}_P^{2+\delta}$, for some $\delta > 0$. Let Φ be a map defined from a subset $\mathcal{P}_\mathcal{X}$ of $\ell^\infty(\mathcal{F})$ to B_2, differentiable at P tangentially to $\mathcal{U}_b(\mathcal{F}, P)$ along every probability sequence $(P_n)_n$ such that $\lim_n \|P_n - P\|_{\mathcal{F}} = 0$, $\lim_n \|P_n - P\|_{\mathcal{F}^2} = 0$, and F belongs to $\bigcap_n \mathcal{L}_{P_n}^{2+\delta}$. Then for a.e. $(X_1, \ldots, X_n)$*

$$\sqrt{n}\left(\Phi(P_n^*) - \Phi(\widehat{P}_n) \right) \xrightarrow{\mathcal{D}} d\Phi(P).G_P$$

where P_n^* is the empirical distribution of an i.i.d. sequence of variables with distribution function P_n.

As an application, let $[0, \tau]$ be a subset of $\mathbb{R}_+^2$, let H be a continuous distribution function on $\mathbb{R}_+^2$ and Λ be defined on $\mathbb{R}_+^2$ by $\Lambda(t) = \int_{]0,t]} \bar{H}^{-1} \, dH$, for $t = (t_1, t_2)$ such that $\bar{H}(t) = \int_{[t_1, \infty[\times [t_2, \infty[} dH > 0$. For t in $\mathbb{R}_+^2$, let Q_t be the quadrant larger than t and T_t be the triangle under the diagonal and lower than t

$$Q_t = \left\{ (u, v) : u \epsilon \mathbb{R}_+^2, v \epsilon \mathbb{R}_+^2, u \geq t, v \geq t \right\},$$
$$T_t = \left\{ (u, v) : u \epsilon \mathbb{R}_+^2, v \epsilon \mathbb{R}_+^2, u \leq v, u \leq t \right\},$$

then $\mathcal{Q}_\tau = \{ Q_t, t \in [0, \tau] \}$ and $\mathcal{T}_\tau = \{ T_t, t \in [0, \tau] \}$ are Vapnik-Cervonenkis' classes, i.e. $\|\widehat{P}_n - P\|_{\mathcal{Q}_\tau}$ and $\|\widehat{P}_n - P\|_{\mathcal{T}_\tau}$ converge a.s. to zero as n tends to infinity. Let $\Lambda = \Phi(P)$, where P is the product probability distribution $P_H \times P_G$ related to continuous distribution functions H and G, then

$$\Phi(P)(t) = \int_{\mathbb{R}_+^4} \frac{1}{P(Q_u)} 1_{T_t}(u, v) \, dP(u, v).$$

On the set $E_n = \{ \widehat{P}_n(Q_\tau) > 0 \}$ having a probability that tends to 1, $\widehat{\Lambda}_n = \Phi(\widehat{P}_n)$. The process $W_{n, P} = \sqrt{n}(\widehat{\Lambda}_n - \Lambda)$ is written as

$$W_{n, P}(t) = \nu_{n, P}(g_t) - \int \frac{\nu_{n, P}(Q_u)}{\widehat{P}_n(Q_u) P(Q_u)} 1_{T_t}(u, v) \, d\widehat{P}_n(u, v),$$

with $g_t(u,v) = \frac{1}{P(Q_u)} 1_{T_t}(u,v)$. Similarly, let

$$g_{n,t}(u,v) = \frac{1}{P_n(Q_u)} 1_{T_t}(u,v)$$

Proposition 7.15 and Corollary 7.3 apply with the class $\mathcal{F} = \bigcup_n \mathcal{F}_n$ defined by $\mathcal{F}_n = \{g_{n,t}, g_t, 1_{Q_t}, 1_{T_t}; t \in [0,\tau]\}$.

Proposition 7.16. *Let P be a probability on $\mathbb{R}^4_+$ and let W be the Gaussian process defined by*

$$W(t) = G_P(g_t) - \int \frac{G_P(Q_u)}{P^2(Q_u)} 1_{T_t} dP(u,v),$$

then $W_{n,P} = \sqrt{n}\{\Phi(\widehat{P}_n) - \Phi(P)\} = \sqrt{n}(\widehat{\Lambda}_n) - \Lambda)$ converges weakly to W on every interval $[0,\tau]$ such that $\Lambda(\tau)$ and the variance of $W(\tau)$ are finite.

Its proof relies on the differentiability of the functional Φ (Pons, 1986). The functional Φ is also differentiable in the sense of the Definition 7.1 under the next conditions.

Proposition 7.17. *Let P_n and P be probability distributions on $\mathbb{R}^4_+$ such that $P(Q_\tau) > 0$, $n^\alpha \sup_{u \leq \tau} |P_n(Q_u) - P(Q_u)| \leq M$, for constants $\alpha > 0$ and $M > 0$, and such that $\lim_n ||P_n - P||_{\mathcal{F}} = 0$ and $\lim_n ||P_n - P||_{\mathcal{F}^2} = 0$. Then Φ defining Λ as $\Phi(P)$ on $D([0,\tau])$ is differentiable at P along $(P_n)_n$ and tangentially to $\mathcal{U}_b(\mathcal{F}, P)$.*

The condition $n^\alpha \sup_{u \leq \tau} |P_n(Q_u) - P(Q_u)| \leq M$ entails that $\mathcal{F}$ has a finite entropy dimension, moreover the envelope F is finite under the condition $P(Q_\tau) > 0$. Let P_n^* the empirical distribution of an i.i.d. sequence of variables with the empirical distribution function $\widehat{P}_n$.

Proposition 7.18. *Under the conditions of Proposition 7.15, the process $W_{n,P_n} = \sqrt{n}\{\Phi(\widehat{P}_n) - \Phi(P_n)\}$ converges weakly to W under P_n and the process $W_n^* = \sqrt{n}\{\Phi(\widehat{P}_n^*) - \Phi(\widehat{P}_n)\}$ converges weakly to W under $\widehat{P}_n$, conditionally on the random vector $(X_i, \delta_i)_{i \leq n}$.*

7.6 Regression functions and biased length

On a probability space $(\Omega, \mathcal{F}, P)$, let (X,Y) be a random variable with values in a separable and complete metric space $(\mathcal{X}_1 \times \mathcal{X}_2, \mathcal{B})$. For every x

in $\mathcal{X}_1$, a regression function is defined by the conditional expectation of Y given $X \leq x$,

$$m(x) = E(Y|X \leq x) = \frac{E(Y1_{\{X \leq x\}})}{P(X \leq x)}.$$

Its empirical version is defined from a sequence of independent random variables distributed like (X, Y)

$$\widehat{m}_n(x) = \frac{\sum_{i=1}^{n} Y_i 1_{\{X_i \leq x\}}}{\sum_{i=1}^{n} 1_{\{X_i \leq x\}}}, \quad x \in \mathcal{X}_1, \tag{7.7}$$

where the denominator is the empirical distribution $\widehat{F}_{X,n}(x)$, with expectation $F(x) = P(X \leq x)$, and the numerator is an empirical expectation process denoted $\mu_n(x)$ with expectation $\mu(x) = E(Y1_{\{X \leq x\}})$.

Proposition 7.19. *If Y belongs to $L^4(\mathcal{X}_2)$, for every x in a subset of I_X of $\mathcal{X}_1$ such that there exist constants for which $0 < M_1 < m(x) < M_2$ finite, then $E\widehat{m}_n(x) = m(x) + O(n^{-\frac{1}{2}})$ and*

$$\mathrm{Var}\{\widehat{m}_n(x)\} = n^{-1} F_X^{-1}(x)\{E(Y^2|X \leq x) - m^2(x)\} + o(1),$$
$$n^{\frac{1}{2}}(\widehat{m}_n - m) = F_X^{-1}\{n^{\frac{1}{2}}(\mu_n - \mu) - m\nu_{X,n}\} + r_n$$

where $\sup_{I_X} r_n = o_{L^2}(1)$, as n tends to infinity.

Proof. Let $A_n(x) = n^{-1} \sum_{i=1}^{n} Y_i 1_{\{X_i \leq x\}}$ be the numerator of $\widehat{m}_n$ and let $\mu(x) = E(Y1_{\{X \leq x\}})$ be its expectation. Under the condition, there exists a strictly positive $k(x)$ such that for every x in I_X, $k(x) \leq \widehat{F}_{Xn}(x)$ if n is large enough. For the expectation of $\widehat{m}_n(x)$

$$E\widehat{m}_n(x) - m(x) = E\frac{\mu_n - \mu}{\widehat{F}_{Xn}}(x) - m(x)E\frac{\widehat{F}_{Xn} - F_X}{\widehat{F}_{Xn}}(x)$$

$$\leq m(x)E\frac{(\widehat{F}_{Xn} - F_X)^2}{F_X \widehat{F}_{Xn}}(x) + E\frac{|(\mu_n - \mu)(\widehat{F}_{Xn} - F_X)|}{F_X \widehat{F}_{Xn}}(x)$$

$$\leq m(x)\frac{\|\widehat{F}_{Xn}(x) - F_X(x)\|_2^2}{F_X(x)k_2(x)} + \|\mu_n(x) - \mu(x)\|_2 \frac{\|\widehat{F}_{Xn}(x) - F_X(x)\|_2}{F_X(x)k_2(x)}$$

with the L^2-norm, then $\|\widehat{F}_{Xn} - F_X\|_2 = O(n^{-\frac{1}{2}})$ and $\|\mu_n - \mu\|_2 = O(n^{-\frac{1}{2}})$. For its variance, we have $\mathrm{Var}\{\widehat{m}_n(x)\} = E\{\widehat{m}_n(x) - m(x)\}^2 + O(n^{-1})$ and the first term develops as

$$F_X^2(x)E\{\widehat{m}_n(x) - m(x)\}^2 = \{E\widehat{m}_n(x)\}^2\{\mathrm{Var}\widehat{F}_{Xn}(x)$$
$$-2E\widehat{m}_n(x)\mathrm{Cov}(\mu_n(x),$$
$$\widehat{F}_{Xn}(x))\} + \mathrm{Var}\mu_n(x) + o(n^{-1}) = O(n^{-1}).$$

$\square$

Let $\widehat{F}_n$ be the empirical distribution of $(X_i, Y_i)_{i=1,\ldots,n}$ with distribution function F and let ν_n be its empirical process, then

$$n^{\frac{1}{2}}(\widehat{m}_n - m) = \Phi_n.\nu_n(x) := F_X^{-1}(x) \int_{\mathcal{X}_2} y\, \nu_n(x, dy)$$

$$-F_X^{-1}(x)\widehat{F}_{Xn}^{-1}(x)\nu_{Xn}(x) \int_{\mathcal{X}_2} y\, F(x, dy),$$

as n tends to infinity, it converges weakly in $D(\mathcal{X}_1)$ to the centered Gaussian process

$$\Phi.\nu(x) = F_X^{-1}(x) \int_{\mathcal{X}_2} y\, \nu(x, dy) - F_X^{-2}(x)\nu_X(x) \int_{\mathcal{X}_2} y\, F(x, dy)$$

with variance function $F_X^{-1}(x)\{E(Y^2|X \leq x) - m^2(x)\}$. On every bounded subset C of $\mathcal{X}_1 \times \mathcal{X}_2$ where F_X is strictly positive, the function Φ_n converges uniformly to Φ and it is a Lipschitz functional, hence for every $\lambda > 0$

$$P\left(\sup_{|x| \leq M} \|\Phi_n.\nu_n(x) - \Phi.\nu(x)\| > \lambda \right) \leq \frac{K}{\lambda} E\left\{ \sup_{|x| \leq M} |\nu_n(x) - \nu(x)| \right\}$$

with a Lipschitz constant K.

Replacing the variable Y by $f(Y)$, for a function f belonging to a class of functions $\mathcal{F}$, let

$$m_f(x) = E\{f(Y)|X \leq x\}$$

be a regression function indexed by f, the empirical regression function becomes

$$\widehat{m}_{f,n}(x) = \frac{\sum_{i=1}^n f(Y_i)1_{\{X_i \leq x\}}}{\sum_{i=1}^n 1_{\{X_i \leq x\}}}, \quad x \in \mathcal{X}_1,$$

as n tends to infinity, the variance of the normalized process

$$\zeta_{f,n} = n^{-\frac{1}{2}}(\widehat{m}_{f,n} - m_f)$$

is approximated by

$$\sigma_f^2(x) = \operatorname{Var}\zeta_{f,n}(x) = F_X^{-1}(x)\{E(f^2(Y)|X \leq x) - m_f^2(x)\}.$$

For every f such that σ_f^2 is locally bounded, the process $\zeta_{f,n}$ converges weakly in $D(\mathcal{X}_1)$ to a centered Gaussian process with variance function σ_f^2, as n tends to infinity.

If $F_1 \leq f \leq F_2$, then $\sigma_f^2(x) \leq F_X^{-1}(x)\{E(F_2^2(Y)|X \leq x) - m_{F_1}^2(x)\}$ and this bound is denoted $\sigma_{F_1, F_2}^2(x)$.

Proposition 7.20. *Let $\mathcal{F}$ be a class of measurable functions on $(\mathbb{X}_2, \mathcal{X}_2)$ such that $C_{\mathcal{F}}$ is finite and there exist envelopes $F_1 \leq f \leq F_2$ belonging to $L^p(\mathbb{X})$. Under the conditions of Proposition 7.19, for every x of I_X*

$$\lim_{n\to\infty} P\left(\sup_{f\in\mathcal{F}} |n^{-\frac{1}{2}}(\widehat{m}_{f,n} - m_f)(x)| \geq t\right) \leq 2C_{\mathcal{F}}\frac{\sigma^2_{F_1,F_2}(x)}{t^2}.$$

The odd moments of order $p \geq 2$ of the process $\zeta_{f,n}$ are $o(1)$ and its even moments are $O(1)$, from its expansion of Proposition 7.19, like those of the empirical process ν_n. From Proposition 7.1, for every $p \geq 2$ and for every x of I_X

$$\lim_{n\to\infty} P\left(\sup_{f\in\mathcal{F}} |\zeta_{f,n}(x)| \geq t\right) \leq 2C_{\mathcal{F}}\frac{\sigma^p_{F_1,F_2}(x)}{t^p}.$$

Biased length variables appear in processes observed on random intervals (Cox, 1960). Let Y be a positive random variable sampled at a uniform and independent random time variable U on $[0, 1]$. The variable Y is not directly observed and only a biased length variable $X = YU$ is observed, therefore $F_Y \leq F_X$ and $EX = \frac{1}{2}EY$. The variable $U = (Y^{-1}X) \wedge 1$ has a uniform distribution on $[0, 1]$ and its expectation is $\int_0^\infty \{x \int_x^\infty y^{-1} dF_Y(y) + F_Y(x)\} dF_X(x) = \frac{1}{2}$.

Lemma 7.1. *The distribution function of X and Y are defined for every positive x by*

$$F_X(x) = E(xY^{-1} \wedge 1) = F_Y(x) + x\int_x^\infty y^{-1} dF_Y(y), \qquad (7.8)$$

$$F_Y(y) = 1 - E(Xy^{-1} \wedge 1) = F_X(y) - y^{-1}\int_0^y x\, dF_X(x). \qquad (7.9)$$

Proof. Let $x > 0$, the distribution function of $X = UY$ is defined by

$$F_X(x) = \int_0^1 P(Y \leq u^{-1}x)\, du = \int_0^1 F_Y(u^{-1}x)\, du$$
$$= F_Y(x) + E(Y^{-1}x 1_{\{x<Y\}}) = E(xY^{-1} \wedge 1).$$

The distribution of $Y = U^{-1}X$ is written

$$F_Y(y) = \int_0^1 P(X \leq uy)\, du = \int_0^1 F_X(uy)\, du = y^{-1}\int_0^y F_X(x)\, dx$$
$$= F_X(x) - y^{-1}\int_0^y x\, dF_X(x).$$

$\square$

The expected mean lifetime function is defined as

$$m(y) = E\{X1_{X \leq y}\}.$$

It is related to the distribution functions of the variables X and Y by (7.9) in Lemma 7.1. For every $y > 0$, $m(y) = y\{F_X(y) - F_Y(y)\}$.

From the observation of n independent and identically distributed random variables X_i distributed like X, we define the empirical distribution function $\widehat{F}_{X,n}(x) = n^{-1} \sum_{i=1}^n 1_{\{X_i \leq x\}}$ of F_X and the empirical version of the function $m(y)$

$$\widehat{m}_n(y) = n^{-1} \sum_{i=1}^n X_i 1_{\{X_i \leq y\}}.$$

By plugging in (7.9), $\widehat{F}_{X,n}$ and $\widehat{m}_n$ define the empirical distribution function $\widehat{F}_{Y,n}$ of the unobserved variable Y

$$\widehat{F}_{Y,n}(y) = n^{-1} \sum_{i=1}^n \left(1 - \frac{X_i}{y}\right) 1_{\{X_i \leq y\}}.$$

The variance of the empirical process related to $\widehat{F}_{Y,n}$ is

$$\sigma_Y^2(y) = \{F_Y(1 - F_Y)\}(y) + E\left(\frac{X^2}{y^2} 1_{\{X \leq y\}}\right) - \frac{m_X(y)}{y} \leq \{F_Y(1 - F_Y)\}(y).$$

Proposition 7.21. *The estimator $\widehat{F}_{Y,n}$ converges uniformly to F_Y in probability and $n^{\frac{1}{2}}(\widehat{F}_{Y,n} - F_Y)$ converges weakly to a centered Gaussian variable with variance function σ_Y^2.*

From the inequality (1.13), for all $t > 0$ and $y > 0$

$$P(|n^{-\frac{1}{2}}\{\widehat{F}_{Y,n}(y) - F_Y(y)\}| \geq t) \leq 2\frac{\sigma_Y^2(y)}{t^2} \leq \frac{2F_Y(y)\{1 - F_Y(y)\}}{t^2}.$$

There exists a constant C such that for every $t > 0$

$$P\left(\sup_{y>0} |n^{-\frac{1}{2}}\{\widehat{F}_{Y,n}(y) - F_Y(y)\}| \geq t\right) \leq \frac{C}{t^2},$$

since $\sigma_Y^2(y) \leq \{1 - F_Y(y)\}F_Y(y) \leq \frac{1}{4}$ for every y.

A continuous multiplicative mixture model is more generally defined for a real variable U having a non uniform distribution function on $[0, 1]$. Let

F_U denote its distribution function and let F_Y be the distribution function of Y. The distribution functions of X and Y are

$$F_X(x) = \int F_U(xy^{-1}) \, dF_Y(y),$$

$$F_Y(y) = \int_0^1 F_X(uy) \, dF_U(u)$$

$$= \int_0^y \{1 - F_U(y^{-1}x)\} \, dF_X(x)$$

$$= F_X(y) - E_X\{1_{\{X \leq y\}} F_U(y^{-1}X)\} \tag{7.10}$$

and the conditional density of Y given X is

$$f_{Y|X}(y; x) = \frac{f_Y(y) f_U(y^{-1}x)}{f_X(x)}.$$

The empirical distribution of Y is deduced from (7.10) in the form

$$\widehat{F}_{Y,n}(y) = n^{-1} \sum_{i=1}^n 1_{\{X_i \leq y\}} \{1 - F_U(y^{-1}X_i)\}.$$

The variance of the empirical process related to $\widehat{F}_{Y,n}$ is

$$\sigma_{F_U}^2(y) = F_Y(y)\{1 - F_Y(y)\} + E\{F_U^2(y^{-1}X) 1_{\{X \leq y\}}\} - m_{F_U}(y)$$

$$\leq F_Y(y)\{1 - F_Y(y)\},$$

where the expected mean lifetime is now $m_{F_U}(y) = E_X\{1_{\{X \leq y\}} F_U(y^{-1}X)\}$. Applying the inequality (1.13), for every $t > 0$

$$P(|n^{-\frac{1}{2}}\{\widehat{F}_{Y,n}(y) - F_Y(y)\})| \geq t) \leq \frac{2F_Y(y)\{1 - F_Y(y)\}}{t^2} \leq \frac{1}{2t^2}.$$

The process $n^{-\frac{1}{2}}(\widehat{F}_{Y,n} - F_Y)$ is bounded in probability in $\mathbb{R}$ endowed with the uniform metric, and it converges weakly to a Gaussian process with expectation zero and variance function $\sigma_{F_U}^2$.

Let $p > 1$ and let $\mathcal{F}_U$ be a class of distribution functions on $[0, 1]$ with an envelope F of $L^p([0, 1])$, from Proposition 7.19

$$P\left(\sup_{y>0} |n^{-\frac{1}{2}}\{\widehat{F}_{Y,n}(y) - F_Y(y)\}| \geq t\right) \leq C \frac{F_U^p(1)}{t^p},$$

$$P\left(\sup_{F_U \in \mathcal{F}} \sup_{y>0} |n^{-\frac{1}{2}}\{\widehat{F}_{Y,n}(y) - F_Y(y)\}| \geq t\right) \leq C \frac{F^p(1)}{t^p}, \quad t > 0.$$

Another biased length model is defined by the limiting density, as t tends to infinity, of the variations $X(t) = S_{N(t)+1} - t$ between t and the sum of random number of independent and identically distributed random variables ξ_k having the distribution function G, $S_{N(t)+1} = \sum_{k=1}^{N(t)+1} \xi_k$, with the random number $N(t) = \sum_{i=1}^{\infty} 1_{\{S_i \leq t\}}$. This limiting density only depends on the distribution of the variables ξ_i in the form

$$f_X(x) = \mu^{-1}\{1 - G(x)\}$$

where $\mu = \{f_X(0)\}^{-1}$ (Feller, 1971). This is equivalent to

$$G(x) = 1 - f_X(x)\{f_X(0)\}^{-1}.$$

Let F_Y be the limiting density of $S_{N(t)+1} - S_{N(t)}$. The distribution functions F_X and F_Y are

$$F_X(x) = \mu^{-1} \int_0^x \{1 - G(y)\} \, dy$$
$$= \mu^{-1}x\{1 - G(x) + x^{-1}E_G(\xi 1_{\{\xi \leq x\}})\},$$
$$F_Y(y) = F_X(y) - y^{-1}E(X1_{\{X \leq y\}})$$

and the expected mean lifetime distribution function for X is

$$m_X(y) = (2\mu)^{-1}[y^2\{1 - G(y)\} - E_G(\xi^2 1_{\{\xi \leq x\}})].$$

The empirical versions of the functions m, G, F_X and F_Y are all easily calculated from a sample $(\xi_i)_{i \leq n}$.

7.7 Regression functions for processes

Let $X = (X_n)_{n \geq 0}$ be an adapted process of $L^2(P, (\mathcal{F}_n)_{n \geq 0})$, with values in a separable and complete metric space $\mathcal{X}_1$, and let $Y = (Y_n)_{n \geq 0}$ be an adapted real process. We assume that the processes have independent increments $(X_n - X_{n-1})_{n \geq 0}$, with a common distribution function F_X with a density f_X and there exists a function $m > 0$ of $C_1(\mathcal{X}_1)$, such that

$$E(Y_n - Y_{n-1}|X_n - X_{n-1} \leq x) = m(x), \, n \geq 1.$$

This implies that $(Y_n - Y_{n-1})_{n \geq 0}$ is a sequence of independent variables with a common distribution function F_Y, with a density f_Y. The empirical version of the function m is

$$\widehat{m}_n(x) = \frac{\sum_{i=1}^{n}(Y_i - Y_{i-1})1_{\{X_i - X_{i-1} \leq x\}}}{\sum_{i=1}^{n} 1_{\{X_i - X_{i-1} \leq x\}}}, \, x \in \mathcal{X}_1. \tag{7.11}$$

The empirical means

$$n^{-1} \sum_{i=1}^{n} (1_{\{X_i - X_{i-1} \leq x\}}, (Y_i - Y_{i-1})1_{\{X_i - X_{i-1} \leq x\}})$$

converge a.s. uniformly to the expectation $F_X(x)(1, m(x))$ of the variables, therefore $\lim_{n \to \infty} \sup_{\mathcal{X}_1} \|\widehat{m}_n - m\| = 0$, a.s. and Propositions 7.19 and 7.20 apply to the processes X and Y.

Let $(X, Y) = (X_n, Y_n)_{n \geq 0}$ be an ergodic sequence of $L^2(P, (\mathcal{F}_n)_n)$ with values in $\mathcal{X}^2$, there exists an invariant measure π on $\mathcal{X}$ such that for every continuous and bounded function φ on $\mathcal{X}^2$

$$\frac{1}{n} \sum_{k=1}^{n} \varphi(X_k, Y_k, X_{k-1}, Y_{k-1}) \to \int_{\mathcal{X}} \varphi(z_k, z) F_{X_k, Y_k | X_{k-1}, Y_{k-1}}(dz_k, z)) \, d\pi(z)$$

and the estimator

$$\widehat{m}_n(x) = \frac{\sum_{i=1}^{n} Y_i 1_{\{X_i \leq x\}}}{\sum_{i=1}^{n} 1_{\{X_i \leq x\}}}, \quad x \in \mathcal{X}_1,$$

converges in probability to $m(x)$, uniformly in $\mathcal{X}_1$. Under the condition (4.3) and a φ-mixing assumption, it converges in distribution to a centred Gaussian process with variance function

$$\sigma_m^2(x) = F_X^{-1}(x)\{E(Y^2 | X \leq x) - m^2(x)\}.$$

Let $\mathcal{H}$ be a class of functions, the transformed variables $h(Y_i)$ define a functional empirical regression

$$\widehat{m}_n(h, x) = \frac{\sum_{i=1}^{n} h(Y_i) 1_{\{X_i \leq x\}}}{\sum_{i=1}^{n} 1_{\{X_i \leq x\}}}, \quad x \in \mathcal{X}_1,$$

under the condition (4.3), it converges to $m(f, x) = E\{h(Y) | X = x\}$ defined as a expectation with respect to the invariant measure. If the class $\mathcal{H}$ has a finite constant $C_{\mathcal{H}}$ and an envelope H, the convergence is uniform over $\mathcal{H}$ and the process $\sup_{h \in \mathcal{H}} \sup_{x \in I} |\widehat{m}_n(f, x) - m(h, x)|$ converges to zero in every real interval where m has lower and upper bounds.

7.8 Functional inequalities and applications

Let x_t be a function in a metric space $\mathbb{X}$ of $C^2(\mathbb{R}_+)$ with a first derivative x_t' in $\mathbb{X}'$ and let $f(t, x_t, x_t')$ be a bounded functional of x_t and its derivative. Conditions for a function x_t^* to minimizes the integral over an interval $[t_0, t_1]$ of a function f of $C^2(\mathbb{R}_+ \times \mathbb{X} \times \mathbb{X}')$ are conditions for the first two derivatives of f. The first condition for the function f that ensures the existence of a function x_t^* that minimizes the integral

$$I = \int_{t_0}^{t_1} f(t, x_t, x_t') \, dt$$

with respect to x is a null first derivative of I, which is equivalent to the next Euler-Lagrange condition.

Theorem 7.7. *The first necessary condition for the existence of a function x_t^* that minimizes the finite integral $I = \int_{t_0}^{t_1} f(t, x_t, x_t')\, dt$ with respect to x is*

$$\frac{\partial f}{\partial x}(t, x_{t,\theta}^*, x_{t,\theta}^{*\prime}) - \frac{d}{dt}\frac{\partial f}{\partial x'}(t, x_{t,\theta}^*, x_{t,\theta}^{*\prime}) = 0. \tag{7.12}$$

The second condition for a minimum of I at x^* is $I'' \geq 0$. It is written in the next equivalent forms

$$\frac{\partial^2}{\partial x^2} f(t, x_{t,\theta}^*, x_{t,\theta}^{*\prime}) \geq 0,$$

$$\left\{ \frac{\partial^2}{\partial x \partial x'} f(t, x_{t,\theta}^*, x_{t,\theta}^{*\prime}) \right\}^2 \leq \left\{ \frac{\partial^2}{\partial x^2} f(t, x_{t,\theta}^*, x_{t,\theta}^{*\prime}) \right\} \left\{ \frac{\partial^2}{\partial x'^2} f(t, x_{t,\theta}^*, x_{t,\theta}^{*\prime}) \right\}.$$

A parametric family of probabilities $(P_\theta)_{\theta \in \Theta}$, with densities $(f_\theta)_{\theta \in \Theta}$ belonging to $C^2(\Theta)$ on a bounded parametric set Θ, has a maximum Radon-Nikodym derivative with respect to the Lebesgue measure under the condition that the integral $I(f_\theta, f_\theta') = \int_{\mathbb{R}} f_\theta^{-1}(t) f_\theta'^2(t)\, dt$ is finite, where f_θ' is the derivative with respect to the parameter. This integral has a bound determined from Theorem 7.7 by the equality

$$\int \frac{f'^2}{f}\, dt = 2 \int \frac{f'^2 - f f''}{f^2}\, dt$$

which must be satisfied by the maximum density for which the right-side integral is strictly positive.

Consider the minimization of an integral

$$I_\lambda(f, f') = \int_0^1 \{ (f - f_0)^2 + \lambda(f'^2 - 1) \}\, dt,$$

the first condition for the existence of a minimum is

$$\int_0^1 (f - f_0)\, dt - \lambda \int f'\, dt = 0.$$

For a density f and its derivative belonging to $L^2([0,1])$, we have $\int f'\, dt = 0$ and this condition is equivalent to $\int_0^1 f\, dt = \int_0^1 f_0\, dt = 1$, so f is a density with the same support as f_0. The derivative of $I_\lambda(f, f')$ with respect to the Lagrange multiplier λ is equivalent to the constraint $\int_0^1 f'^2\, dt = 1$. The information of $f - f_0$ with respect to f_0 is minimum at $f = f_0$ where it is $-\infty$, and the criteria of differentiability do not apply. These conditions extend to higher order functionals (Pons, 2015)

$$I(x_t, x_t^{(1)}, \ldots, x_t^{(k)}) = \int_{t_0}^{t_1} f(t, x_t, x_t^{(1)}, \ldots, x_t^{(k)}) \, dt$$

and a least square approximation in $\mathcal{F}$ of f_0 under constraints on the integrals of the squares derivatives of order k is based on the minimization in a functional class $\mathcal{F}$ of an integral

$$I(\lambda(f, f', \ldots, f^{(k)}) \, dt = \int_{\mathbb{R}} \left\{ (f - f_0)^2 + \sum_{j=1}^{k} \lambda_j (f^{(j)2} - c_j) \right\} dt,$$

which is equivalent to the minimization of $\int_{\mathbb{R}} \{(f - f_0)^2 \, dt$ in $\mathcal{F}$ under the conditions $\int_{\mathbb{R}} f^{(j)2} \, dt = c_j$ or $\int_{\mathbb{R}} f^{(j)2} \, dt \leq c_j$, for $j = 1, \ldots, k$. In the latter case, the Lagrange multiplier λ_j apply to the inequality.

For a vector of variables (X, Y) with joint distribution function F and marginal F_X for X under a probability P_0, a least square approximation of a regression function $m_0(x) = E_0(Y \mid X = x)$ by $m(x)$ in a class $\mathcal{R}$ is performed by minimization of the integral

$$I_\lambda(m, m^{(k)}) = \int_{\mathcal{X}} \{y - m(x)\}^2 \, F(dx, dy).$$

The first Euler-Lagrange condition is $\int_{\mathcal{X}} \{y - m(x)\} \, F(dx, dy) = 0$ which is equivalent to $\int_{\mathcal{X}} \{m_0(x) - m(x)\} \, F_X(x) = 0$. Under the constraint of a bounded k-th order derivative of m the integral becomes

$$I_\lambda(m, m^{(k)}) = \int_{\mathcal{X}} \{y - m(x)\}^2 \, dF_X(x) + \lambda \left\{ \int_{\mathcal{X}} m^{(k)2}(x) \, dF_X(x) - c \right\}$$

and, using the derivative with respect to λ, the condition is

$$\int_{\mathcal{X}} \{m_0(x) - m(x)\} \, dF_X(x) = \pm \lambda \left\{ \int_{\mathcal{X}} m^{(k)}(x) \, dF_X(x) - c \right\} = 0.$$

Chapter 8

Markov Processes

8.1 Ergodic theorems

Let $X = (X_t)_{t \geq 0}$ be a Markov process on a probability space $(\Omega, \mathcal{B}, P)$, with values in a Banach space $(E, \mathcal{E})$. For every $t > 0$ and for A in $\mathcal{E}$

$$P(X_t \in A \mid (X_s)_{s<t}) = P(X_t \in A \mid X_{t_-}).$$

The transition probability of the process is

$$P_t(x, A) = \int_A P_t(x, dy) = P(X_t \in A \mid X_0 = x)$$

and for every $s < t$

$$P_t(x, A) = \int_A \int_E P_t(z, dy) P_s(x, dz).$$

A Markov process is Harris-recurrent if there exists a σ-finite measure π such that for every B of $\mathcal{E}$ such that $\pi(B) > 0$

$$P_x(X_t \in B \text{ infinitely often}) = 1,$$

for every x of E.

The ergodic theorem states that if X is Harris-recurrent and has an invariant measure π on a bounded state space endowed with the borelian σ-algebra $(E, \mathcal{E})$, π is unique and for every bounded real function ϕ integrable with respect to π on E

$$\lim_{t \to \infty} \left| \frac{1}{t} \int_0^t \phi(X_s) \, ds - \int_E \phi(x) \, d\pi(x) \right| = 0.$$

The expectation of $\phi(X_t)$ conditionally on $X_0 = x$ is

$$P_t(\phi)(x) := E_x \phi(X_t) = \int_E \phi(y) P_t(x, dy),$$

247

the ergodic theorem implies the weak convergence of the measures $t^{-1} \int_0^t \int_E P_s(x, dy)\, ds$ to a measure $\pi(dy)$ on $(E, \mathcal{E})$. Let P_0 be the probability of the initial value $X_0 = x$, the ergodic measure π is written as

$$\pi(dy) = \int_E \Pi(x, dy)\, dP_0(x)$$

where $\Pi(x, dy)$ is the limit in probability of $t^{-1} \int_0^t P_t(x, dy)$, as t tends to infinity. For every integrable function ψ on $(E^2, \mathcal{E}^2, \Pi)$

$$\lim_{t \to \infty} \left| \frac{1}{t} \int_0^t \psi(X_s, X_0)\, ds - \int_{E^2} \psi(y, x)\, \Pi(x, dy)\, d\pi(x) \right| = 0$$

and by the Markov property, for every $u \geq 0$

$$\lim_{t \to \infty} \left| \frac{1}{t} \int_0^t \psi(X_{s+u}, X_u)\, ds - \int_{E^3} \psi(y, z)\, \Pi(z, dy)\, d\pi(z) \right| = 0.$$

For all $s < t$ the integral

$$E_x\{\phi(X_s)\phi(X_t)\} = \int_{E \otimes 2} \phi(y)\phi(z) P_t(y, dz) P_s(x, dy)$$

is written as

$$\begin{aligned} E_x\{\phi(X_s)\phi(X_t)\} &= E_x[\phi(X_s) E_{X_s}\{\phi(X_t)\}] \\ &= E_x\{\phi(X_s) P_t\phi(X_s)\} = P_s\{\phi(x) P_t\phi(x)\}. \end{aligned}$$

The limit as t tends to infinity of the expectation distribution function of a real valued process X_t, $t^{-1} \int_0^t P(X_s \leq x)\, ds$ is $\int_E F(x) 1_{\{y \in [0, x]\}}\, d\pi(y)$ and for $s < t$

$$\lim_{t \to \infty} \frac{1}{t} \int_0^t P(X_s \leq x, X_t \leq y)\, ds = \int_{E^2} F(x, y) 1_{\{u \leq x, v \leq y\}} \Pi(u, dv)\, d\pi(u).$$

If the process X is stationary

$$P(X_s \leq x, X_t \leq y) = \int_0^x P(X_{t-s} \leq y - z)\, dP(X_s \leq z),$$

$$\int_0^x \int_0^{y-x} F(x, y)\, \Pi(u, dv)\, d\pi(u)$$

$$= \int_{E^2} \int_0^{y-z} 1_{\{u \leq z, v \leq y-z\}} F(y - z)\, dF(z)\, \Pi(u, dv)\, d\pi(u).$$

The expectation and the variance of the integral $I_t = \int_0^t \phi(X_s)\, ds$ conditionally on $X_0 = x$ are

$$E_t\phi(x) = E_x I_t = \int_0^t P_s\phi(x)\, ds,$$

$$V_t\phi(x) = \operatorname{Var}_x I_t = 2 \int_0^t \int_0^s E_x\{\phi(X_u)\phi(X_s)\}\, du\, ds - E_x^2 I_t$$

$$= 2 \int_0^t \int_0^s P_u\{\phi(x) P_s\phi(x)\}\, du\, ds - E_x^2 I_t.$$

A second-order ergodic theorem for the variance is obtained with the function $\psi(X_s, X_u) = \{\phi(X_s) - \pi(\phi)\}\{\phi(X_u) - \pi(\phi)\}$ on E^2, with the convergence of $t^{-1}\left[\int_0^t \{\phi(X_s) - \pi(\phi)\} ds\right]^2$ to

$$V(\phi) = \lim_{t \to \infty} t \text{Var}\left\{t^{-1} \int_0^t \phi(X_s) ds\right\} \tag{8.1}$$

$$= \int_{E^3} \{\phi(y) - \pi(\phi)\}\{\phi(z) - \pi(\phi)\} \Pi(dz, dy) d\pi(x).$$

Let X be a continuous Markov process and let ϕ be $C(E)$, the process $\phi(X_t)$ has a modulus of continuity ω

$$\omega_{\phi(X)}(\eta) = \sup_{|t-s| \leq \eta} |\phi(X_t) - \phi(X_s)|,$$

the process

$$Y_t = t^{-1} \int_0^t \phi(X_s) ds$$

is related to the limit as η tends to zero of the expectation of a Markov chain defined by a partition $(t_i)_{i=0,\ldots,n}$ of $[0, t]$, such that $t_i - t_{i-1} = \delta$ for every i, $t_0 = 0$ and $t_{n+1} = t$

$$Y_n = \frac{1}{n} \sum_{i=0}^{n-1} \phi(X_{t_i}),$$

for every $t > 0$, $\lim_{\eta \to 0} |Y_t - Y_n| = \eta^{-1}\omega(\eta) + o(1)$.

If ϕ is a Lispschitz function, $\eta^{-1}\omega(\eta)$ is bounded by a constant and Y_n converges to $\pi(\phi)$ by the ergodic property of the Markov chain $\{\phi(X_{t_i})\}_{i\geq 0}$. Necessary conditions for the convergence to zero of $|Y_t - Y_n|$ is the derivability of $\phi(X_t)$ and $\omega(\eta) = o(\eta)$.

Let $X = (X_n)_{n\geq 0}$ be a recurrent and irreductible Markov chain on a probability space $(\Omega, \mathcal{A}, P)$ such that there exists a density of transition f for (X_{k-1}, X_k), $k \geq 1$, with respect to the Lebesgue measure. The density of $(X_1, \ldots, X_n)$ conditionally on $X_0 = x_0$ is

$$Y_n = f_{X_1,\ldots,X_n}(x_1, \ldots, x_n; x_0) = \prod_{k=1}^n f(x_k, x_{k-1}).$$

Let F_n be the σ-algebra generated by $(X_1, \ldots, X_n)$, the sequence $(Y_n)_{n\geq 1}$ is a martingale with respect to the filtration $F = (F_n)_{n\geq 1}$

$$E(Y_{n+1} \mid F_n) = Y_n E\{f(X_{n+1}, X_n) \mid F_n\}$$

$$= Y_n \int_E f(y, X_n) d\Pi(dy, X_n) = Y_n.$$

By the ergodic theorem, there exists an invariant measure π such that

$$n^{-1} \log f_{X_1,\ldots,X_n}(X_1,\ldots,X_n;x_0) = n^{-1} \sum_{k=1}^{n} \log f(X_k, X_{k-1})$$

converges in probability to

$$\int_{E^2} \log f(x,y)\, d\Pi(dy,x)\, d\pi(x) \leq \log E f(X_k, X_{k-1}) = \log \pi(f).$$

A sum $S_n(\psi) = \sum_{k=1}^{n} \psi(X_k, X_{k-1})$ is such that $n^{-1} S_n(\psi)$ converges in probability to a limit $\pi(\psi) = \int_{E^2} \psi(y,x)\, d\Pi(dy,x)\, d\pi(x)$ defined by the invariance measure of the transition probabilities and the normalized variable $W_n(\psi) = n^{\frac{1}{2}}\{n^{-1} S_n(\psi) - \pi(\psi)\}$ has the variance

$$V_n(\psi) = \frac{1}{n} \sum_{k=1}^{n} \psi^2(X_k, X_{k-1}) + \frac{1}{n} \sum_{j\neq k=1}^{n} \psi(X_j, X_{j-1})\psi(X_k, X_{k-1}),$$

$$\frac{1}{n} E \sum_{k=1}^{n} \psi^2(X_k, X_{k-1}) = \int_{E^2} \psi^2(y,x)\, d\Pi(dy,x)\, d\pi(x),$$

$$\frac{1}{n} E \sum_{j<k=1}^{n} \psi(X_j, X_{j-1})\psi(X_k, X_{k-1}) = \frac{1}{n} \sum_{i=2}^{n} \int_{E^4} \psi(y,x)\psi(y',x')d\Pi(dy',x')$$

$$.d\Pi^{*(i-2)}(dx',y)\, d\Pi(dy,x)\, d\pi(x)$$

where Π^{*i} is the $i = k_j$-th convolution of the transition probability Π. Under the mixing condition

$$\sup_{x\in E, B\subset E}\left|\frac{1}{n} \sum_{k=1}^{n} \Pi^{*k}(B,x) - \pi(B)\right| = 0,$$

the variance $V_n(\psi)$ converges weakly to

$$v(\psi) = \int_{E^2} \psi^2(y,x)\, d\Pi(dy,x)\, d\pi(x) - \pi^2(\psi)$$

and the variable $W_n(\psi)$ converges weakly to a centered Gaussian process with variance $v(\psi)$. The uniform functional bounds of the previous chapter apply to the process W_n defined on a class of L^2, under similar conditions of dimension.

8.2 Inequalities for Markov processes

The Bienaymé-Chebychev inequality applies to Markov processes $(X_t)_{t \geq 0}$, let $Y_t = t^{-\frac{1}{2}} \int_0^t \|\phi(X_s) - \pi(\phi)\| \, ds$, for every $a > 0$ and for every real function ϕ on $(E, \mathcal{E})$ such that the variance $V(\phi)$ of Y_t is finite

$$P(Y_t > a) \leq a^{-2} V_t \phi, \tag{8.2}$$

where $V_t \phi = t^{-1} \int_E V_t \phi(x) \, d\pi(x)$ converges to the limit $V(\phi)$ by the ergodic property (8.1). This probability tends to zero as t and a tend to infinity.

The first hitting times

$$T_{y,x} = \inf\{t : X_t \geq y\}$$

of a Markov process X_t starting at $x \leq y$ satisfies an additive property in distribution

$$T_{z,x} = T_{z,y} + T_{y,x}$$

for all $x \leq y \leq z$. Then $X_{T_{y,x}} \geq y$ and

$$E_x(X_{T_{z,x}}) = E_x(X_{T_{y,x}}) + E_y(X_{T_{z,y}} - X_{T_{y,x}}),$$

the variance has the same transitivity property.

Let $(X_n)_{n \geq 0}$ be a Markov chain on $(\Omega, \mathcal{B}, P)$, with values in $(E, \mathcal{E})$ and let $(F_n)_{n \geq 0}$ be the increasing sequence of σ-algebras generated by $\sigma(X_0, \ldots, X_n)$. For a Markov chain with independent increments, the variable $\xi_n = X_n - X_{n-1}$ is independent of F_{n-1}, with values in a space $\Gamma = \{x - y, (x, y) \in E^2\}$. For every function ϕ of $L^2(\Gamma)$, the variable $n^{-1} \sum_{i=1}^n \phi(\xi_i)$ converges to a Gaussian variable with the expectation $m_\phi = \int_E \phi(x) \, d\pi(x)$ and the variance $\sigma_\phi^2 = \int_\Gamma \phi^2(x) \, d\pi(x) - m_\phi^2$.

The inequality (8.2) applies to the variables

$$Y_n(\phi) = n^{-\frac{1}{2}} \sum_{i=1}^n \{\phi(X_i) - \pi(\phi)\},$$

for every $a > 0$ and for every real function ϕ on $(E, \mathcal{E})$ such that the sequence of the variances $V_n(\phi)$ of Y_n is bounded

$$P(Y_n > e^{\frac{a}{2}} V_n^{\frac{1}{2}}(\phi)) \leq e^{-a}.$$

Generally, the Markov property of the chain $(X_n)_{n \geq 0}$ implies

$$P\{(X_n, \xi_n) \in A \times B \mid X_{n-1}, F_{n-1}\} = P\{(X_n, \xi_n) \in A \times B \mid X_{n-1}\}$$

and the inequalities for the sums of independent variables extend to the sums of dependent variables $X_n = \sum_{i=1}^{n} \xi_i$ and $X_n(\phi) = \sum_{i=1}^{n} \phi(\xi_i)$, for every function ϕ of $L^2(E)$. By the ergodic theorem, for every real function ϕ if $L^2(E, \Pi)$, the variable $n^{-1} X_n(\phi)$ converges a.s. to

$$m_\phi = \int_{E^{\otimes 2}} \phi(y - x) \Pi(x, dy) \, d\pi(x)$$

and the variance $V_{n,\phi}$ of the variable $Y_n(\phi)$ converges to a finite limit σ_ϕ^2. The Bienaymé-Chebychev inequality for the tail probabilities of the sequence of variables $Y_n(\phi)$ is written as $P(Y_n(\phi) > a) \leq a^{-2} V_{n,\phi}$. In particular, it applies to X_n with the expectation $m = \int_0^\infty y \Pi(x, dy) \, d\pi(x)$ and the variance $V = \int_0^\infty y^2 \Pi(x, dy) \, d\pi(x) - m^2$.

8.3 Convergence of diffusion processes

Let X be a second-order stationary process on $\mathbb{R}_+$ with expectation function $\mu(t)$ and covariance function $R(s, t)$ such that for all $0 \leq s < t$

$$R(s, t) = Cov\{X(s), X(t)\} = Cov\{X(t - s), X(0)\} = R(t - s).$$

The average process

$$\bar{X}_T = T^{-1} \int_0^T X(t) \, dt$$

has the expectation

$$\bar{\mu}_T = T^{-1} \int_0^T \mu(t) \, dt$$

and the variance

$$V_T = E\bar{X}_T^2 - \mu_T^2 = T^{-2} \int_0^T \int_0^T R(t - s) \, ds \, dt$$

where $E(\bar{X}_T - \bar{\mu})^2 \leq E(\bar{X}_T - \bar{\mu}_T)^2 + (\bar{\mu}_T - \bar{\mu})^2$.

Under the condition that V_T and, respectively $\bar{\mu}_T$, converge to finite limits V and, respectively $\bar{\mu}$, as T tends to infinity, the Bienaymé-Chebychev inequality implies for every $a > 0$

$$P(|\bar{X}_T - \bar{\mu}| > aT^{\frac{1}{2}}) \leq \frac{V_T + (\mu_T - \bar{\mu})^2}{a^2 T},$$

it converges to zero and the process $T^{-\frac{1}{2}}(\bar{X}_T - \bar{\mu})$ is tight.

Let $dX_t = \alpha(t)\,dt + \beta(t)\,dB_t$, $X_0 = x_0$, be a diffusion process defined by integrable functions α and β^2 on $\mathbb{R}_+$, and by the Brownian motion B_t. The expectation and the variance functions of X_t are

$$\mu_t = \int_0^t \alpha(s)\,ds, \quad \sigma_t^2 = \int_0^t \beta^2(s)\,ds$$

and the covariance function of X_s and X_t is stationary. The integrability of α and β entails $\bar{\mu}_T = \mu(t_T)$ where $t_T \leq T$ and $V_T = T^{-1}\sigma_T^2$ has the same property. The variance σ_T^2 is increasing and it is bounded under the integrability condition, therefore V_T converges to a finite limit V. If the function μ is positive, by the same argument $\bar{\mu}_T$ converges to a finite limit $\bar{\mu}$.

A diffusion process $dX_t = \alpha X_t\,dt + \beta X_t^{\frac{1}{2}}\,dB_t$, with $X_0 = x_0$ and a drift linear in X_t, has exponential expectation and variance

$$\mu_t = x_0 e^{\alpha t}, \quad \sigma_t^2 = \beta^2 \int_0^t \mu_s\,ds = \frac{\beta^2 x_0}{\alpha} e^{\alpha t},$$

$\bar{\mu}_T$ and V_T do not converge.

A hyperbolic diffusion process $dX_t = \alpha \frac{(1+X_t^2)^{\frac{1}{2}}}{X_t}\,dt + \beta\,dB_t$, with $X_0 = x_0$ and a constant variance, has the unbounded expectation

$$\mu_t = \int_0^t \frac{X_t\,dX_t}{(1+X_t^2)^{\frac{1}{2}}} = \arg\sinh t.$$

A diffusion process X_t with integrable drift and variance functions depending on X_t, $dX_t = \alpha(X_t)\,dt + \beta(X_t)\,dB_t$, $X_0 = x_0$, has the expectation and variance

$$\mu_t = \int_0^t E\alpha(X_s)\,ds, \quad V_T = \mathrm{Var}\int_0^t \alpha(X_s)\,ds + \int_0^t E\beta^2(X_s)\,ds,$$

$\bar{\mu}_T = T^{-1}\mu_T$ and $T^{-1}V_T$ are bounded and they converge to finite limits.

8.4 Branching process

The Galton-Watson branching process is the sum of an array of independent and identically distributed (i.i.d.) random variables

$$X_n = \sum_{k=1}^{X_{n-1}} Y_{n,k}, \tag{8.3}$$

where $Y_{n,k} \geq 0$, $k \geq 1$, are i.i.d. with distribution function F on $\mathbb{R}_+$, having the expectation m and the variance σ^2, they are independent of X_{n-1} and $X_0 = 1$. The conditional expectation of X_n is

$$E\{X_n \mid X_{n-1}\} = X_{n-1}m,$$

and for every $k < n$

$$E\{X_n \mid X_k\} = X_i m^{n-k},$$

The sequence of variables $(X_n)_{n \geq 1}$ is a submartingale with respect to the increasing sequence of σ-algebras $\mathcal{F}_{n-1}$ generated by the variables $X_0, \ldots, X_{n-1}$, if $m \geq 1$, and it is a super-martingale if $m \leq 1$, moreover $(m^{-n} X_n, \mathcal{F}_n)_{n \geq 1}$ is a martingale. Their conditional generating function is

$$G_n(s) = E\{s^{X_n} \mid X_{n-1}\} = G_Y^{X_{n-1}}(s)$$

where $G_Y(s) = \int_0^\infty s^y \, dF_Y(y)$ and the generating function of X_n is the n-th composition of the generating function of a variable Y with distribution function F

$$\widehat{G}_{n+1}(s) = E\{G_Y^{X_n}(s)\} = \widehat{G}_n \circ G_Y(s) = G_Y \circ \cdots \circ G_Y(s),$$

therefore $\widehat{G}_{n+k}(s) = \widehat{G}_n \circ \widehat{G}_k(s)$ for all integers n and k.

Their conditional variance is

$$\mathrm{Var}(X_n \mid X_{n-1}) = X_{n-1} \sigma^2$$

and $E(X_n^2 \mid X_{n-1}) = X_{n-1} \sigma^2 + m^2 X_{n-1}^2$. Their variance is

$$\mathrm{Var} X_n = E X_n^2 - m^{2n} = m^{n-1} \sigma^2 + m^2 \mathrm{Var} X_{n-1}$$
$$= m^{n-1}(1 + m + \cdots + m^{n-1}) \sigma^2$$

and for all $k < n$, if $m < 1$

$$m^{-k} \mathrm{Var}(X_n \mid X_{n-k}) \sim X_{n-k} \sigma^2$$

as k tends to infinity as $n - k$ tends to infinity, if $m > 1$

$$m^{-2k} \mathrm{Var}(X_n \mid X_{n-k}) \sim X_{n-k} \sigma^2$$

as k tends to infinity.

If $m = 1$, $(X_n, \mathcal{F}_n)_{n \geq 1}$ is a martingale with constant expectation 1 and conditional variance is

$$\mathrm{Var}(X_n \mid X_{n-k}) \sim k X_{n-k} \sigma^2$$

as k tends to infinity. The properties of the martingale and the inequality

$$P(k^{-\frac{1}{2}} |X_n - X_{n-k}| > \varepsilon \mid X_{n-k}) \leq \varepsilon^{-2} \sigma^2 X_{n-k}$$

implies $n^{-\frac{1}{2}} X_n$ converges in probability to a limiting variable X.

Let $\xi_{n,k} = Y_{n,k} - m$ for $k \geq 1$ and let $Z_n = \sum_{k=1}^{X_{n-1}} \xi_{n,k}$, $n \geq 1$, the variables Z_n are centered and their variance is the variance of X_n, it tends to zero with the rate $b_n = m^n$ if $m < 1$ and to infinity with the rate $b_n = m^{2n}$ if $m > 1$. By normalization, it follows that for every $m \neq 1$

$$P(b_n^{-\frac{1}{2}}|Z_n| > a) \leq a^{-2}\sigma^2$$

tends to zero as a and n tend to infinity and the sequence $b_n^{-\frac{1}{2}}Z_n, n \geq 1$, is tight. For every $\delta > 0$

$$P\left(\max_{k \leq n} b_n^{-\frac{1}{2}}|Z_k| > ak^{\frac{1+\delta}{2}}\right) \leq \sum_{k=1}^{n} P(b_n^{-\frac{1}{2}}|Z_k| > ak^{\frac{1+\delta}{2}})$$

$$\leq a^{-2}\sigma^2 \sum_{k=1}^{n} k^{-(1+\delta)}$$

this probability tends to zero as a and n tend to infinity.

Considering a normalisation by the random size of the process X_n, the variables

$$\widehat{m}_n = X_{n-1}^{-1} \sum_{k=1}^{X_{n-1}} Y_{n,k}$$

have the expectation m and the variance $E\{\mathrm{Var}(\widehat{m}_n \mid \mathcal{F}_{n-1})\} = X_{n-1}^{-1}\sigma^2$. The centered normalized variables

$$X_{n-1}^{\frac{1}{2}}(\widehat{m}_n - m) = X_{n-1}^{-\frac{1}{2}} \sum_{k=1}^{X_{n-1}} \xi_{n,k}$$

have the variance σ^2, their conditional third moment is

$$E(X_{n-1}^{-\frac{1}{2}})E(Y_{n,k} - m)^3 = E(X_{n-1}^{-\frac{1}{2}})\mu_3,$$

by convexity, it has the lower bound $m^{-\frac{n-1}{2}}\mu_3$. If $m < 1$, the third moment $E\{X_{n-1}^{\frac{1}{2}}(\widehat{m}_n - m)\}^3$ diverges as n tends to infinity and the higher moments have the same behaviour so $X_{n-1}^{\frac{1}{2}}(\widehat{m}_n - m)$ does not have a Gaussian limiting distribution. Moreover, if $m < 1$, the expectation and the centered moments of X_n tend to zero as n tends to infinity, X_n tends to zero in probability and $X_{n-1}^{\frac{1}{2}}(\widehat{m}_n - m)$ cannot satisfy a central limit theorem.

For every $m > 0$, the Laplace transform of X_n is

$$L_{X_n}(t) = Ee^{tX_n} = EL_F^{X_{n-1}}(t) = g_{X_{n-1}} \circ L_F(t)$$

where g is the generating function of the jump process X_n. Starting from $Ee^{tX_1} = L_Y(t)$

$$\log L_{X_n}(t) = \log L_Y \circ \log L_Y \cdots \circ \log L_Y(t)$$

denoted as $(\log L_Y)^{\otimes n}(t)$ for the n-th composition of $\log L_Y$.

With the centered Gaussian distribution function F with variance σ^2, the Laplace transform of X_n is

$$\log L_{X_n}(t) = \frac{t^{2r_n}\sigma^{2r_n}}{2^{r_n}},$$

where $r_n = 2^{n-1}(2^n + 1) \sim 2^{2n-1}$ as n tends to infinity. For every $a > 0$

$$\lim_{n\to\infty} \log P(X_n > a) = -\frac{a}{\sigma\sqrt{2}}.$$

If $m > 1$, Chernoff's theorem provides the lower bound $P(X_n > a) \geq 1$, for every $a > 0$, as n tends to infinity therefore

$$\lim_{n\to\infty} P(X_n > a) = 1$$

for every $a > 0$, X_n diverges in probability and $EX_{n-1}^{-\frac{1}{2}}$ converges to zero as n tends to infinity. By the same arguments, the centered moments of order larger than 3 of the variable

$$Z_n = X_{n-1}^{\frac{1}{2}}(\widehat{m}_n - m)$$

also converge to zero as n tends to infinity. Moreover, Z_n has an asymptotically normal distribution conditionally on X_{n-1}. The martingale $m^{-n}X_n$ is integrable and its variance is constant, it converges a.s. to a limiting variable X_∞ with variance $\sigma^2 m^{-1}(m-1)^{-1}$.

With an arbitrary distribution function F for the variables $Y_{n,k}$ and with $m > 1$, for every $a > 0$, by Chernoff's theorem conditionally on X_{n-1}

$$P(X_n > aX_{n-1} \mid X_{n-1}) = \exp\{-X_{n-1}I(a)\}$$

where $I(a) = \sup_{t>0}\{at - \log L_Y(t)\}$, for every n

$$m^{-(n-1)}E\log P(X_n > aX_{n-1} \mid X_{n-1}) = -I(a)$$

and by concavity

$$m^{-(n-1)}\log P(X_n > aX_{n-1}) \geq -I(a).$$

If there exists a function

$$J(a) = \lim_{n\to\infty}\inf_{t>0}\{at - m^{-(n-1)}(\log L_Y)^{\otimes n}(t)\},$$

for every $a > 0$, the probability

$$P(X_n > aX_{n-1}) = \exp\{-m^{n-1}J(a)\}$$

converges to zero as n tends to infinity. If $m > 1$ and F is symmetric at m, for every $\varepsilon > 0$, the probability

$$P(|\widehat{m}_n - m| > \varepsilon) = 2P\{X_n > (\varepsilon + m)X_{n-1}\}$$

converges to zero and $\widehat{m}_n$ converges in probability to m, as n tends to infinity.

A spatial branching process is defined by (8.3) where the variables $Y_{n,k}$ take their values in a spatial domain $\mathcal{D}$ of $\mathbb{R}^2$ or $\mathbb{R}^3$. Varadhan's large deviation principle applies and the convergences of the variables are the same.

8.5 Renewal processes

Let $(X_n)_{n\geq 1}$ be a sequence of independent positive variables having the same distribution function F and a finite expectation $\mu = EX_n$. For $n \geq 1$, the n-th jump time of the process is $T_n = \sum_{1 \leq k \leq n} X_k$ and

$$N(t) = \sum_{n\geq 1} 1_{\{T_n \leq t\}}$$

is a renewal process with parameter μ, this is a Poisson process if F is exponential.

With a distribution function F, the expectation duration time is

$$\mu = EX_n = \int_0^\infty y\, dF(y) = \int_0^\infty \bar{F}(y)\, dy$$

and for $n > 1$

$$P(T_n \leq t) = F^{*(n)}(t) = \int_0^t F^{*(n-1)}(t)\, dF(t)$$

with $F^{*(1)} = F$. For every t, $F^{*(n)}(t)$ tends to zero, $F^{*(n)}(t)$ is decreasing with respect to n, it is lower than 1 for $t < \tau_F = \sup\{x : F(x) < 1\}$. The expectation $M(t)$ of $N(t)$ is

$$M(t) = \sum_{n=1}^\infty nP(N(t) = n) = \sum_{n=1}^\infty P(N(t) \geq n) = \sum_{n=1}^\infty F^{*(n)}(t).$$

Let L be the Laplace transform operator, $Lf(t) = \int_0^\infty e^{xt} f(x)\, dx$. For all s and $t < \tau_F$

$$LF^{*(n)}(s) = LF^{*(n-1)}(s)Lf(s) = LF(s)(Lf(s))^{n-1},$$

where $sLF(s) = Lf(s)$ and $Lf(s) \leq 1$ for every $s > 0$, these properties entail

$$LM(s) = \frac{LF(s)}{(1 - Lf(s))} = \frac{LF(s)}{1 - sLF(s)}. \tag{8.4}$$

Considering the expectations conditionally on $X_1 = x$ with $t < x$ or $t \geq x$

$$M(t) = E[E\{N(t)|X_1\}] = \int_0^t (1 + M(t - s)) \, dF(s). \tag{8.5}$$

The Laplace transform of Equation (8.5)

$$M(t) = F(t) + \int_0^t M(t - s) \, dF(s)$$

is equivalent to (8.4).

For every bounded function a with value zero at 0, the renewal equation is defined as

$$A(t) = a(t) + \int_0^t A(t - s) \, dF(s), \tag{8.6}$$

it has the unique solution

$$A(t) = a(t) + \int_0^t a(t - s) \, dM(s). \tag{8.7}$$

The expression and the unicity of this solution are a consequence of the fact that Equations (8.6) and (8.7) have the same Laplace transform. From (8.4), it is written

$$LA(t) = \frac{La(t)}{1 - tLF(t)}. \tag{8.8}$$

Proposition 8.1. *For every $t > 0$, $ET_{N_t+1} = \mu(1 + EN_t)$ and*

$$EN_t = \mu^{-1} ET_{N_t}.$$

Proof. Let $A(t) = ET_{N_t+1}$, $E(T_{N_t+1}|X_1 = x) = x$ if $t < x$ and if $x \leq t$, $E(T_{N_t+1}|X_1 = x) = x + A(t - x)$ therefore

$$A(t) = \mu + \int_0^t A(t - x) \, dF(x)$$

and the result is the expression (8.7) of the solution of (8.6) with $a(t) = \mu$ and $A(t) = \mu + \int_0^t \mu \, dM(x)$. $\qquad \square$

Let $\mu(t) = \int_0^t x \, dF(x) = E(X 1_{\{X \leq t\}})$.

Proposition 8.2. *For every* $t > 0$

$$ET_{N_t+1}^2 = E(X^2)(1 + EN_t) + 2\mu \Big\{ \mu(x) + \int_0^t \mu(t - x) \, dM(x)$$

$$+ \int_0^t \int_0^{t-x} M(t - x - s) \, d\mu(s) \, dM(x) \Big\}.$$

Proof. Let $A(t) = ET_{N_t+1}$ and $B(t) = ET_{N_t+1}^2$, $E(T_{N_t+1}^2 | X_1 = x) = x^2$ if $t < x$ and, if $x \leq t$

$$E(T_{N_t+1}^2 | X_1 = x) = x^2 + 2xA(t - x) + B(t - x),$$

therefore

$$B(t) = E(X^2) + 2 \int_0^t x A(t - x) \, dF(x) + \int_0^t B(t - x) \, dF(x).$$

It is denoted $B(t) = b(t) + \int_0^t B(t - x) \, dF(x)$ where

$$b(t) = E(X^2) + 2\mu \Big\{ \mu(t) + \int_0^t M(t - x) \, d\mu(x) \Big\}.$$

According to the equivalence of Equations (8.6) and (8.7)

$$B(t) = b(t) + \int_0^t b(t - x) \, dM(x)$$

and $B(t)$ is deduced from Proposition 8.1 as

$$B(t) = E(X^2)(1 + EN_t) + 2\mu \int_0^t \{1 + M(t - x)\} \, d\mu(x)$$

$$+ \int_0^t \Big\{ \mu(t - x) + \int_0^{t-x} M(t - x - s) \, d\mu(s) \Big\} \, dM(x)$$

which yields the result. $\qquad \square$

Let σ^2 be the variance of the variables X_n, from Proposition 8.1 the variance of T_{N_t+1} is

$$V(t) = \sigma^2(1 + EN_t) + 2\mu \Big\{ \mu(x) + \int_0^t \mu(t - x) \, dM(x)$$

$$+ 2\mu \int_0^t \int_0^{t-x} M(t - x - s) \, d\mu(s) \, dM(s) \Big\} - \mu^2 M(t) \{1 + M(t)\}.$$

The current waiting time C_t and the residual waiting time R_t of the process $(T_n)_n$ at time t are defined as

$$R_t = T_{N_t+1} - t, \quad C_t = t - T_{N_t}.$$

Then, $E(R_t + C_t) = EX$ and their expectations are

$$EC_t = \int_0^t \bar{F}_{T_{N_t}}(s)\, ds,$$

$$ER_t = \int_t^\infty \bar{F}_{T_{N_t}}(s)\, ds.$$

For a Poisson process N with intensity λ and starting at $T_0 = 0$, $M(t) = \lambda t$, the variables $T_{n+1} - T_n$ are independent with an exponential distribution $\mathcal{E}_\lambda$ with expectation $E(T_{n+1} - T_n) = \lambda^{-1}$ and variance λ^{-2}, hence $ET_n = n\lambda^{-1}$ and $\mathrm{Var}T_n = n\lambda^{-2}$. Moreover

$$P(t - T_{N_t} \le x) = P(N_t - N_{t-x} = 0) = (1 - e^{-\lambda x})1_{\{x \le t\}},$$

$$P(T_{N_t+1} - t \ge x) = P(N_{t+x} - N_t = 0) = (1 - e^{-\lambda x})1_{\{t \le x\}},$$

it follows that

$$\mu(t) = EC_t = \lambda^{-1}\{1 - e^{-\lambda t}(\lambda t + 1)\},$$

$$ER_t = \lambda^{-1}e^{-\lambda t}(\lambda t + 1).$$

Proposition 8.3. *As t tends to infinity, the normalized Poisson process $Z_t = (\lambda t)^{-\frac{1}{2}}(N_t - \lambda t)$ converges weakly to a normal variable $\mathcal{N}(0,1)$ and the renewal process $Z_t = t^{-\frac{1}{2}}(N_t - (EX)^{-1}t)$ converges weakly to a Gaussian variable $\mathcal{N}(0, (EX)^{-3}\mathrm{Var}(X))$.*

Proof. Let $a_t(x) = xt^{\frac{1}{2}} + \lambda t$ where $\lambda = (EX)^{-1}$, by the equivalence between $N_t \ge n$ and $T_n \le t$

$$P(t^{-\frac{1}{2}}(N_t - \lambda t) > x) = P(N_t > a_t(x)) = P(T_{a_t(x)} < t)$$

where $ET_{a_t(x)} = E(X)\, a_t(x) = \lambda^{-1}a_t(x)$ and $\mathrm{Var}T_{a_t(x)} = \sigma^2 a_t(x)$. By the theorem of the central limit for i.i.d. variables, $n^{\frac{1}{2}}(n^{-1}T_n - \lambda^{-1})$ converges weakly to a centered Gaussian variable with variance σ^2. It follows that

$$P(t^{-\frac{1}{2}}(N_t - \lambda t) > x)$$
$$= P(a_t^{-\frac{1}{2}}(x)\{T_{a_t(x)} - \lambda^{-1}a_t(x)\} < a_t^{-\frac{1}{2}}(x)\{t - \lambda^{-1}a_t(x)\})$$
$$= P(\sigma^{-1}a_t^{-\frac{1}{2}}(x)\{T_{a_t(x)} - \lambda^{-1}a_t(x)\} < -\sigma^{-1}\lambda^{-1}a_t^{-\frac{1}{2}}(x)xt^{\frac{1}{2}}),$$

as t tends to infinity, $t^{\frac{1}{2}}a_t^{-\frac{1}{2}}$ converges to $\lambda^{-\frac{1}{2}}$ and $P(t^{-\frac{1}{2}}(N_t - \lambda t) > x)$ converges to $P(Z > x\sigma\lambda^{-\frac{3}{2}})$ where Z is a normal variable. For a Poisson process, $\sigma^2 = \lambda^{-2}$ and the limit is $P(Z > x\lambda^{-\frac{1}{2}})$. $\qquad\square$

Let F_{R_t} be the distribution function of R_t and let $\bar{F}_{R_t} = 1 - F_{R_t}$, then $\bar{F}_{R_t}(y) = P(N_{t+y} - N_t = 0)$ and

$$\bar{F}_{R_t}(y) = P(R_t > y) = P(T_{N_t+1} > y + t)$$

$$= \int_0^\infty P(T_{N_t+1} > y + t \,|\, X_1 = x)\, dF(x)$$

$$= \bar{F}(t + y) + \int_0^t P(R_{t-x} > y)\, dF(x)$$

the survival function of R_t is therefore solution of a renewal equation and

$$\bar{F}_{R_t}(y) = \bar{F}(t + y) + \int_0^t \bar{F}(t + y - x)\, dM(x).$$

Let F_{C_t} the distribution function of C_t and $\bar{F}_{C_t} = 1 - F_{C_t}$. in the same way, for $y < t$,

$$F_{C_t}(y) = P(C_t \le y) = P(T_{N_t} \ge t - y)$$

$$= \int_0^\infty P(T_{N_t} \ge t - y \,|\, X_1 = x)\, dF(x)$$

$$= \bar{F}(t) + \int_0^t P(C_{t-x} \le y)\, dF(x), \text{ therefore}$$

$$F_{C_t}(y) = 1_{\{y<t\}}\Big\{\bar{F}(t) + \int_0^{t-y} \bar{F}(t - y - x)\, dM(x)\Big\}$$

$$= 1_{\{y<t\}}\bar{F}(t) + \int_0^t 1_{\{y<t-x\}} F(t - y - x)\, dM(x).$$

The time t may be replaced by a random stopping time.

For a finite number of n durations, let $N_n(t) = \sum_{1 \le k \le n} 1_{\{T_k \le t\}}$ and let

$$M_n(t) = EN_n(t) = \sum_{k=1}^n k P(N_n(t) = k) = \sum_{k=1}^n F^{*(k)}(t),$$

$$LM_n(t) = \frac{1 - \{tLF(t)\}^n}{1 - tLF(t)}, \tag{8.9}$$

$$M_n(t) = F(t) + \int_0^t M_n(t - s)\, dF(s) - F^{*(n+1)}(t), \tag{8.10}$$

and, from the definition of M_n, $M_n(t) = F(t) + \int_0^t M_{n-1}(t - s)\, dF(s)$. For every bounded function a with value zero at 0, a renewal equation is written

$$A_n(t) = a(t) + \int_0^t A_{n-1}(t - s)\, dF(s), \tag{8.11}$$

with the unique solution

$$A_n(t) = a(t) + \int_0^t a(t-s)\,dM_n(s). \tag{8.12}$$

Proof. From (8.9), the Laplace transform (8.12) is written

$$LA_n(t) = La(t)\{1 + tLM_n(t)\} = La(t)\frac{1 - \{tLF(t)\}^{n+1}}{1 - tLF(t)}$$

$$= \frac{La(t)LM_{n+1}(t)}{LF(t)} = \frac{tLa(t)LM_{n+1}(t)}{Lf(t)},$$

let $LA_n(t)Lf(t) = tLa(t)LM_{n+1}(t)$, which implies

$$\int_0^t A_{n-1}(t-s)\,dF(s) = \int_0^t a(t-s)\,dM_n(s)$$

and equation

$$A_n(t) = a(t) + \int_0^t A_n(t-s)\,dF(s), \tag{8.13}$$

has the unique solution

$$A_n(t) = a(t) + \int_0^t a(t-s)\,dM_{n+1}(s). \tag{8.14}$$

$\square$

The n-th current time and residual current time are

$$C_n(t) = t - T_{N_n(t)}, \quad R_n(t) = T_{N_n(t)+1} - t.$$

Let $F_{R_n(t)}$ and $F_{C_n(t)}$ be the distribution functions of the residual and current times and let $\bar{F}_{R_n(t)}$ and $F_{C_n(t)}$ be their survival functions

$$\bar{F}_{R_n(t)}(y) = \bar{F}(t+y) + \int_0^t P(R_{n-1}(t-x) > y)\,dF(x), \text{ therefore}$$

$$\bar{F}_{R_n(t)}(y) = \bar{F}(t+y) + \int_0^t \bar{F}(t+y-x)\,dM_n(x),$$

$$F_{C_n(t)}(y) = \bar{F}(t)1_{\{y<t\}} + \int_0^t P(C_{n-1}(t-x) \le y)\,dF(x), \text{ therefore}$$

$$F_{C_n(t)}(y) = \bar{F}(t)1_{\{y<t\}} + \int_0^t F(t-y-x)1_{\{y<t-x\}}\,dM_n(x)$$

$$= \bar{F}(t)1_{\{y<t\}} + \int_0^{t-y} F(t-y-x)\,dM_n(x),$$

since $P(C_t > y|X_1 = x) = 0$ if $t < y < x$.

The asymptotic distribution of the current times is determined by the renewal theorem (Feller, 1966): For a monotone and integrable function a, $A(t)$ tends to $\mu^{-1} \int_0^\infty a(s)\, ds$ as t tends to infinity if μ is finite and the limit is zero otherwise.

Theorem 8.1. *As t tends to infinity, the distribution function of the residual time R_t converges to the distribution function*

$$F_{R_\infty}(y) = \mu^{-1} \int_0^y \bar{F}(t)\, dt$$

and the distribution function of the current time F_{C_t} converges to the distribution function F_{C_∞} such that

$$F_{C_\infty}(y) = \mu^{-1} \int_y^\infty \bar{F}(t)\, dt.$$

Proof. Let $A_y(t) = \bar{F}_{R_t}(y)$ and $a(t) = \bar{F}(t+y)$, $\int_0^\infty a(t)\, dt$ is written as $\int_y^\infty \bar{F}(t)\, dt = \mu - \int_0^y \bar{F}(t)\, dt$, which implies the result for $F_{R_\infty}(y)$. The distribution function of the current time is obtained by the same arguments with the notations $A_y(t) = F_{C_t}(y)$ and $a(t) = \bar{F}(t) 1_{\{y < t\}}$. □

8.6 Maximum variables

The maximum $X_n^* = \max_{i=1,\ldots,n} X_i$ of n independent and identically distributed variables and the integer stopping times

$$N_k^* = \min\{j > N_{k-1}^* : X_j = X_k^*\}, \; k \geq 1,$$

starting from $N_1^* = 1$, define the local maximum variables as the increasing sequence $X_{N_k^*}^*$ if N_k^* is finite or X_∞^* otherwise. If the support of the variables X_k is unbounded, X_n^* and N_n^* are increasing sequences tending to infinity and for every n, there exists $k \leq n$ such that $N_{k-1}^* \leq n < N_k^*$, then

$$N_{k-1}^{*-1} X_{N_{k-1}^*}^* \leq n^{-1} X_n^* \leq N_{k-1}^{*-1} X_{N_k^*}^*$$

and

$$\lim_{k \to \infty} N_{k-1}^{*-1} X_{N_{k-1}^*}^* = \lim_{n \to \infty} n^{-1} X_n^*.$$

For every k in $\{1, \ldots, n\}$

$$P(X_n^* = X_k) = \prod_{i \neq k, i \in \{1,\ldots,n\}} E P(X_i < X_k \mid X_k)$$

$$= \int_{\mathbb{R}_+} F^{n-1}(x_k)\, dF(x_k) = n^{-1}$$

and the moment generating function of X_n^* is

$$G(s) = Ee^{sX_n^*} = \sum_{k=1}^{n} Ee^{sX_k} P(X_n^* = X_k) = Ee^{sX_1}.$$

The moments of X_n^* are the same as the moments of X_1. At x such that $0 < F(x) < 1$, the distribution function of X_n^* is $P(X_n^* \leq x) = F^n(x)$ and it tends to zero as n tends to infinity. For a uniform variable U on $[0,1]$

$$P(U_n^* \leq 1 - \frac{x}{n}) = \left(1 - \frac{x}{n}\right)^n$$

and it converges to e^{-x} as n tends to infinity. By the change of variable $X = F^{-1}(U)$, the variable X with distribution function F has a tail probability such that $P(X_n^* \leq F^{-1}(1 - \frac{x}{n}))$ converges to e^{-x} as n tends to infinity and

$$\lim_{n \to \infty} P\left\{X_n^* \leq F^{-1}\left(1 - \frac{e^{-x}}{n}\right)\right\} = \exp\{e^{-x}\}.$$

This convergence does not depend on the form of the distribution function F at infinity.

For i.i.d. variables $\max_{m < i \leq n} X_i$ is independent of X_m^* and it has the same distribution as X_{n-m}^*. By the independence of the observations, for all integers $m < n$

$$P(X_n^* > t) \leq P(X_m^* > t) + P(\max_{m < i \leq n} X_i > t)$$

$$\equiv^d P(X_m^* > t) + P(X_{n-m}^* > t),$$

so the distribution function of X_m^* and X_n^* satisfy $\bar{F}_{X_n^*} \leq \bar{F}_{X_m^*} + \bar{F}_{X_{n-m}^*}$. For all integers k and $m \geq 1$

$$\bar{F}_{X_{km}^*} \leq k\bar{F}_{X_m^*}, \tag{8.15}$$

that is also the distribution of the maximum of a sample observed in k independent blocks of size m.

Proposition 8.4. *If F is continuous, for every integer $k \leq n$*

$$P(X_{N_k^*}^* > t | N_{k-1}^*) = \frac{(n - N_k^*)(n - N_k^* + 1)}{2n} \bar{F}(t).$$

Proof. With F continuous, $P(N_k^* = j) = n^{-1}$ for all j and k in $\{1, \ldots, n\}$. The probability of $\{X_{N_k^*}^* > t\}$ conditionally on $N_{k-1}^* = j$ is

$$P_j = P\left(\max_{j < i \leq N_k^*} X_i > t | N_{k-1}^* = j\right)$$

$$= \sum_{l=j+1}^{n} P(N_k^* = l)(l - j)\bar{F}(t) = n^{-1} \sum_{l=1}^{n-j} l\bar{F}(t).$$

$\square$

As in Proposition 8.1

$$EX^*_{N^*_k} = EX\, EN^*_k.$$

The maximum variables of sums $S_1 = X_1, \ldots, S_n = \sum_{k=1}^n X_k$ of independent and identically distributed random variable X with expectation zero satisfy the above results with the n-th convolution of the distribution of X and they are recursively written for $n \geq 1$ and conditionally on X_0, as

$$S^*_{n+1} = S^*_n \vee S_{n+1} = S^*_n 1_{\{X_{n+1} \leq 0\}} + S_{n+1} 1_{\{X_{n+1} > 0\}} \qquad (8.16)$$

and recursive inequalities apply to their tail probability

$$P(S^*_{n+1} > x, X_{n+1} \leq 0) = P(S^*_n > x),$$
$$P(S^*_{n+1} > x, X_{n+1} > 0) \geq P(S^*_n > x).$$

Let

$$\nu^+_1 = \min\{k > 0 : S_k > 0\}, \quad \nu^-_1 = \min\{k > 0 : S_k < 0\},$$

and for $n > 1$, let

$$\nu^+_{n+1} = \min\{k > \nu^+_n : S_k > 0\}, \quad \nu^-_{n+1} = \min\{k > \nu^-_n : S_k < 0\},$$

they are infinite if the event does not occur.

By Proposition 6.12, for all integers k and $m < n$ the function $P(\nu^+_n > k)$ is multiplicative

$$P(\nu^+_n > k) = P(\nu^+_m > k) P(\nu^+_{n-m} > k),$$

$P(\nu^+_n > 0) = 1$ and ν^+_n is independent of $S_{\nu^+_n}$. For the stopping times ν^-_n, the probabilities $P(\nu^-_n < -k)$ have the same multiplicative property, $P_k(\nu^-_n < 0) = 1$ and ν^-_n is independent of $S_{\nu^-_n}$.

The variables $S_{\nu^+_n} - S_{\nu^+_{n-1}}$, $n \geq 1$, are independent and identically distributed, so the weak law of large numbers apply to their empirical mean $n^{-1} S_{\nu^+_n}$

$$n^{-1} S_{\nu^+_n} \xrightarrow{P} ES_{\nu^+_1} = EX\, E\nu^+_1,$$

and by the theorem of the central limit, the normalized variable

$$\sqrt{\nu^+_n}\{(\nu^+_n)^{-1} S_{\nu^+_n} - EX\}$$

converges weakly to a normal variable.

If $EX > 0$, the expectation $ES_{\nu^+_1}$ and the limit of $S_{\nu^+_n}$ as n tends to infinity are strictly positive, $P(\nu^+_n < \infty) = 1$ for every n. If $EX < 0$, $ES_{\nu^+_1}$ and the limit of $S_{\nu^+_n}$ as n tends to infinity are strictly negative,

$P(\nu_n^+ < \infty) < 1$ for every n. The sequence indexed by ν_n^- has similar properties, $ES_{\nu_1^-} < 0$ if $EX > 0$ and $ES_{\nu_1^-} > 0$ if $EX < 0$.

For every $n > 1$, the last time where the sums S_k have a strictly positive minimum after ν_n^+ is

$$\nu_n = \max\{k : S_k = \min \nu_n^+ < j < \nu_n^- S_j\}, \text{ if } \nu_n^+ < \nu_n^-,$$
$$\nu_n = \max\{k : S_k = \min \nu_n^+ < j < \nu_{n+1}^- S_j\}, \text{ if } \nu_n^- < \nu_n^+,$$

and the sums at ν_n and after ν_n are $S\nu_n > 0$ and $S\nu_n + 1 \leq 0$ for every n. By the equivalence between the existence of n such that $\nu_n^- < \nu_n^+$ and $S_1 < 0$

$$P(\nu_n = i \mid S_1 < 0) = \frac{P\left(\cap_{\nu_n^- < k < \nu_n^+}\{S_i \leq S_k\} \cap \{S_{i+1} \leq 0\}\right)}{P(S_1 < 0)}.$$

Considering independent variables with respective distribution continuous functions F_k, $k \geq 1$, and the sums $S_n = \sum_{k=1}^{n} X_k$, the expectation of S_n is the expectation $\sum_{k=1}^{n} EX_k$ and $EX_n^* = n^{-1}ES_n$. For every k in $\{1, \ldots, n\}$

$$P(X_n^* = X_k) = \prod_{i \neq k, i \in \{1,\ldots,n\}} EP(X_i < X_k \mid X_k)$$

$$= \int_{\mathbb{R}_+} \prod_{i \neq k, i \in \{1,\ldots,n\}} F_i(x_k) \, dF(x_k)$$

and $P(X_n^* = X_k)$ is generally different from n^{-1}, the moment generating function of X_n^* is determined with this probability

$$G(s) = Ee^{sX_n^*} = \sum_{k=1}^{n} Ee^{sX_k}P(X_n^* = X_k).$$

The variable $\max_{m < i \leq n} X_i$ is independent of X_m^* and for all integers $m < n$

$$P(X_n^* > t) = P(X_m^* > t) + P(\max_{m < k \leq n} X_k > t)$$

$$= \left\{1 - \prod_{k \leq m} F_k(t)\right\} + \left\{1 - \prod_{m < k \leq n} F_k(t)\right\}$$

and $\bar{F}_{X_m^*} \leq \bar{F}_{X_n^*}$.

A sequence of i.i.d. random vectors $(X_n, Y_n)_{n \geq 0}$ starting at zero, and with distribution function $F_{X,Y}$ on $\mathbb{R}^2$ defines a bivariate random walk

$$S_n = \sum_{k=1}^{n} X_n, \quad T_n = \sum_{k=1}^{n} Y_n.$$

Thresholds x and $t > 0$ determine the stopping times of the random walks

$$\nu(x) = \min\{n : S_n > x\}, \quad \tau(t) = \min\{n : T_n > t\}$$

such that $S_{\nu(x)} > x$ and $T_{\tau(t)} > t$ if the stopping times are finite, $\nu(x)$ is infinite if $\max_{n \geq 0} S_n \leq x$ and $\tau(t)$ is infinite if $\max_{n \geq 0} T_n \leq t$. With independent sequences $(X_n)_{n \geq 0}$ and $(Y_n)_{n \geq 0}$, the variables $S_{\tau(t)}$ and $T_{\nu(x)}$ have the expectations

$$ES_{\tau(t)} = E(X)\,E\{\tau(t)\}, \quad ET_{\nu(x)} = E(Y)E\{\nu(x)\}.$$

The counting process $N(t) = \sum_{n \geq 1} 1_{\{T_n \leq t\}}$ is such that $T_n > t$ is equivalent to $N(t) < n$ and $\min\{n : N(t) < n\} = k + 1$ if and only if $T_k \leq t < T_{k+1}$, therefore

$$E\{\tau(t)\} = \sum_{k \geq 1} k P(T_{k-1} \leq t < T_k)$$

$$= \int_0^t \left\{ \sum_{k \geq 1} k P(t - s < T_{k-1} \leq t) \right\} dF_Y(s)$$

and the expression of $E\{\nu(x)\}$ is similar with the marginal distribution function F_X. If EY is not zero, T_k tends to infinity and $P(t-s < T_{k-1} \leq t)$ tends to zero as k tends to infinity, the rate of the distribution function F_Y at infinity determines a convergence rate α such that $t^{-\alpha}\tau(t)$ converges to a finite limit as t tends to infinity.

If the sequences $(X_n)_{n \geq 0}$ and $(Y_n)_{n \geq 0}$ are dependent

$$ES_{\tau(t)} = \sum_{n \geq 1} (X_n + S_{n-1}) 1_{\{T_{n-1} \leq t < T_n\}}$$

$$= \int_{-\infty}^t \int_{\mathbb{R}^2} x 1_{\{y > t-s\}} F(dx, dy)\, d\left\{ \sum_{n \geq 1} P(T_{n-1} \leq s) \right\}$$

$$+ \sum_{n \geq 1} E\left\{ S_{n-1} \mid T_{n-1} \right\} \int_{-\infty}^t \{1 - F_Y(t - s)\}\, dP(T_{n-1} \leq s)$$

we have $ES_{\tau(t)} > 0$ if $EY > 0$ and for every $n \geq 1$

$$E\left\{ S_n \mid T_n \right\} = E\left\{ X_n \mid T_{n-1} + Y_n \right\} + E\left\{ S_{n-1} \mid T_{n-1} + Y_n \right\}$$

$$= \frac{1}{ET_n} \left[\int_{\mathbb{R}^3} xy\, F(dx, d(y - y')) dF_Y^{*(n-1)}(y') \right.$$

$$\left. + \int_{\mathbb{R}} E\{ S_{n-1}(T_{n-1} + y) \}\, dF_Y(y) \right].$$

8.7 Shock process

In a shock process, shocks of random magnitude X_n occurring at random times T_n are cumulated. Let $N(t) = \sum_{n \geq 1} 1_{\{T_n \leq t\}}$ be a Poisson process with parameter λ and let $(X_n)_{n \geq 1}$ be an independent sequence of i.i.d. random variables on $\mathbb{R}_+$ with finite expectation and variance. A Poisson shock process is defined as the sum

$$S(t) = \sum_{n=1}^{N(t)} (t - T_n) X_n = \int_0^t (t - s) \, dY_s,$$

with an integral with respect to the weighted point process

$$Y_t = \sum_{n \geq 1} X_n 1_{\{T_n \leq t\}} = \int_0^t X_s \, dN_s,$$

its expectation is $ES(t) = EX \sum_{k \geq 0} P(N(t) = k) \sum_{n=1}^k \int_0^t (t - s) F_{n;k}(ds; t)$, where the conditional distribution function of T_n, $n = 1, \ldots, k$, is determined for all $0 < s < t$ and $1 \leq n \leq k$ by the Poisson distribution

$$
\begin{aligned}
F_{n;k}(s;t) &= P(T_n \leq s \mid N(t) = k) = P(N(s) \geq n \mid N(t) = k) \\
&= \frac{P(N(s) \geq n, N(t - s) \leq k - n)}{P(N(t) = k)} \\
&= \frac{1}{P(N(t) = k)} \sum_{j \geq n, \, l \leq k-n} \frac{\lambda^{j+l} s^j (t - s)^l}{j! \, l!}.
\end{aligned}
$$

The process $S(t)$ is also written according to the sum $S_k = \sum_{n=1}^k X_n$ of the i.i.d. variables X_n as

$$S(t) = (t - T_{N_t}) S_{N_t} + \sum_{n=1}^{N(t)-1} (T_{n+1} - T_n) S_n.$$

The expectation and variance of $S(t)$ depend on the expectation and the variance λt of N_t, they are increasing with t. The expectation of $T_{n+1} - T_n$ is λ^{-1} and its variance is λ^{-2}.

Proposition 8.5. *The expectation of $S(t)$ is*

$$\mu(t) = E(X) \left\{ t - \lambda^{-1}(1 - e^{-\lambda t}) \right\}$$

and its variance is $V(t) = 0(t^4)$, as t tends to infinity.

Proof. For every $n \geq 1$, $ES_n = nEX$ and $\text{Var}(S_n) = n\text{Var}(X)$, the expectation of the variables $(T_{n+1} - T_n)S_n$ is $\mu_n = n\lambda^{-1}E(X)$ and

$$E \sum_{n=1}^{N(t)-1} (T_{n+1} - T_n)S_n = E(X) \sum_{k \geq 0} E\left[1_{\{N_t=k\}}\left\{\sum_{n=1}^{k-1} n(T_{n+1} - T_n)\right\}\right]$$

$$= E(X) \sum_{k \geq 0} E\left[1_{\{N_t=k\}}\left\{\sum_{n=1}^{k-1} n(T_{n+1} - T_n)\right\}\right]$$

$$= \lambda^{-1}E(X) \sum_{k \geq 0} \frac{k(k-1)}{2}P(N_t = k)$$

$$= \frac{\lambda t^2 E(X)}{2}.$$

Their variance is

$$s_n = E\{(T_{n+1} - T_n)^2\}E(S_n^2) - E^2(T_{n+1} - T_n)E^2 S_n$$
$$= n\lambda^{-2}[2\{\sigma^2 + (n-1)E^2 X\} - nE^2 X] = n\lambda^{-2}\{2\sigma^2 + (n-2)E^2 X\}.$$

For all $m < n$, $ES_m S_n = ES_m^2 + ES_m E(S_n - S_m)$ and the covariance of S_m and S_n is the variance of S_m

$$\text{Var}\left\{\sum_{n=1}^{k-1}(T_{n+1} - T_n)S_n\right\} = \sum_{n=1}^{k-1}\left\{s_n + 2\sum_{m=1}^{n-1} s_m\right\}$$

$$= E^2(X)\lambda^{-2}\left\{\sum_{n=1}^{k-1} n(n-2) + 2\sum_{m=1}^{n-1} m(m-2)\right\}$$

$$+ 2\sigma^2\lambda^{-2}\sum_{n=1}^{k-1} n^2$$

$$= \frac{E^2(X)}{3\lambda^2}\sum_{n=1}^{k-1} n(2n^2 - 5) + \frac{2\sigma^2}{\lambda^2}\sum_{n=1}^{k-1} n^2$$

where $\sum_{n=1}^{k-1} n^2 = \frac{1}{6}k(k-1)(2k-1)$ and $\sum_{n=1}^{k-1} n^3 = 0(k^4)$.

The expectation of $S_{N_t}(t - T_{N_t})$ is obtained by independence as

$$E\{S_{N_t}(t - T_{N_t})\} = E(X)\,E\{N_t(t - T_{N_t})\}$$

where $P(t - T_{N_t} \leq x) = P(N_t - N_{t-x} > 0) = P(N_x > 0) = 1 - e^{-\lambda x}$ for all $t \geq x \geq 0$, so the distribution of $t - T_{N_t}$ is a truncated exponential distribution and $E(t - T_{N_t}) = \lambda^{-1}\{1 - e^{-\lambda t}(\lambda t + 1)\}$ is equivalent to λt^2

as t tends to zero. Then

$$E\{N_t(t - T_{N_t})\} = \sum_{k \geq 1} k E\{(t - T_{N_t}) 1_{\{N_t = k\}}\}$$

$$= \sum_{n=1}^{k} k \int_0^t s \, dP(N_t - N_{t-s} > 0, N_{t-s} = k)$$

where $\sum_{n=1}^{k} k P(N_t - N_{t-x} > 0, N_{t-x} = k) = \lambda(t - x)(1 - e^{-\lambda x}) 1_{\{0 \leq x \leq t\}}$, by independence of $N_t - N_{t-x}$ and N_{t-x}, this entails

$$E\{N_t(t - T_{N_t})\} = \lambda \int_0^t s \, d\{(t - s)(1 - e^{-\lambda s})\}$$

$$= -\lambda \int_0^t (t - s)(1 - e^{-\lambda s}) \, ds$$

$$= t - \frac{\lambda t^2}{2} - \lambda^{-1}(1 - e^{-\lambda t}).$$

The variance of $S_{N_t}(t - T_{N_t})$ is

$$E(X^2) \, E\{N_t^2(t - T_{N_t})^2\} - E^2(X) E^2\{N_t(t - T_{N_t})\}$$
$$= E(X^2) \, \mathrm{Var}\{N_t(t - T_{N_t})\} + \mathrm{Var}(X) E^2\{N_t(t - T_{N_t})\},$$

it is calculated like the expectation

$$E\{N_t^2(t - T_{N_t})^2\} = \sum_{k \geq 1} k^2 E\{(t - T_{N_t})^2 1_{\{N(t) = k\}}\}$$

$$= \sum_{n=1}^{k} k^2 \int_0^t s^2 \, dP(N_t - N_{t-s} > 0, N_{t-s} = k).$$

By the independence of the increments of the Poisson process, the probability $P(N_t - N_{t-s} > 0, N_{t-s} = k)$ is the product of the probability of each event $(1 - e^{-\lambda s}) e^{-\lambda(t-s)} \lambda^k (t - s)^k (k!)^{-1}$, then $E\{(t - T_{N_t})^2 1_{\{N(t) = k\}}\}$ is the sum $I_1(t, k) + I_2(t, k)$ where

$$I_1(t, k) = -2 \frac{\lambda^k}{k!} \int_0^t s e^{-\lambda(t-s)} (t - s)^k \, ds,$$

$$I_2(t, k) = 2 e^{-\lambda t} \frac{\lambda^k}{k!} \int_0^t s(t - s)^k \, ds,$$

with the integrals

$$\int_0^t s e^{-\lambda(t-s)} (t - s)^k \, ds = -t \int_0^t e^{-\lambda x} x^k \, dx + \int_0^t e^{-\lambda x} x^{k+1} \, dx,$$

$$\int_0^t s(t - s)^k \, ds = \int_0^t x^{k+1} \, dx - t \int_0^t x^k \, dx$$

$$= -\frac{t^{k+2}}{(k+1)(k+2)}.$$

Since $\sum_{k\geq 1} k(\lambda t)^k (k!)^{-1} = \lambda t e^{\lambda t}$, $\sum_{k\geq 1} k^2(\lambda t)^k (k!)^{-1} = \lambda t(\lambda t + 1)e^{\lambda t}$ and $\sum_{k\geq 1} k^3(\lambda t)^k (k!)^{-1} = \lambda t(\lambda^2 t^2 + 3\lambda t + 1)e^{\lambda t}$, we have

$$\sum_{k\geq 1} k^2 I_2(t, k) = 2[t^2\{1 - e^{-\lambda t}\} - 3\lambda^{-1}t\{1 - e^{-\lambda t}(1 + \lambda t)\}$$

$$+ 4\lambda^{-2}\{1 - e^{-\lambda t}(1 + \lambda t + \lambda^2 t^2)\}],$$

it is equivalent to $2\{t^2 - 3\lambda^{-1}t + 4\lambda^{-2}\}$ as t tends to infinity, and

$$\sum_{k\geq 1} k^2 I_1(t, k) = \frac{\lambda^2 t^4}{6} + \frac{\lambda t^3}{3}.$$

Finally $E\{N_t^2(t - T_{N_t})^2\} \sim \frac{1}{6}\lambda^2 t^4$, the variance of $S_{N_t}(t - T_{N_t}) = O(t^4)$ as t tends to infinity and $\mathrm{Var}\left\{\sum_{n=1}^{N_t - 1}(T_{n+1} - T_n)S_n\right\} = O(N_t^4)$. $\qquad\square$

Like for a random walk, $P(\overline{\lim}_{t\to\pm\infty}|S(t)| = +\infty) = 1$ if $E(X) \neq 0$, with $P(\overline{\lim}_{t\to\infty}N(t) = +\infty) = 1$.

The process $S(t)$ is generalized as a process related to the jump times of the events of a point process N with independent increments. Assuming that the variables $T_{n+1} - T_n$ are i.i.d. with distribution function F, the distribution function of T_n is

$$F_n(t) = P(T_n \leq t) = \int_0^t F_{n-1}(t - x)\, dF(x)$$

and $P(T_{n+1} - T_n > t) = \bar{F}(t)$. The probability

$$P(N_t - N_{t-s} = 0, N_{t-s} = k) = P(T_k \leq t - s, T_{k+1} - T_k > s)$$

is still the product of the probabilities $P(T_{k+1} - T_k > s) = \bar{F}(s)$ and $P(T_k \leq t - s) = F_k(t - s)$, then

$$E\{N_t(t - T_{N_t})\} = \sum_{n=1}^k k \int_0^t s\, d\{\bar{F}(s)F_k(t - s)\}$$

$$= -\sum_{n=1}^k k \int_0^t \bar{F}(s)F_k(t - s)\, ds,$$

$$E\{N_t^2(t - T_{N_t})^2\} = \sum_{n=1}^k k^2 \int_0^t s^2\, d\{\bar{F}(s)F_k(t - s)\}$$

$$= -2\sum_{n=1}^k k^2 \int_0^t s\bar{F}(s)F_k(t - s)\, ds.$$

The expectation and the variance of $S(t)$ are calculated as previously with these expressions.

Another generalization consists of a model where the sequence of sums $(S_n)_n$ is a martingale. Let $(F_n)_{n\geq 1}$ be an increasing sequence of σ-algebras and let $(M_n)_{n\geq 1}$ be a martingale with respect to the filtration $(F_n)_{n\geq 1}$. Defining the variables X_n as the martingale differences $X_n = M_n - M_{n-1}$, they are centered with variances

$$\sigma_n^2 = E(X_n^2) = E(M_n^2) - E(M_{n-1}^2),$$

for every n and with covariances $EX_n X_m = E\{X_m E(X_n \mid F_m)\} = 0$, for every $m < n$, they are uncorrelated.

The Poisson shock process defined with martingale differences variables X_n is centered. The variables $(T_{n+1} - T_n)S_n$ are centered, their variance is $\lambda^{-2}s_n$ where $s_n = \sum_{m=1}^{n} \sigma_m^2$ and

$$\mathrm{Var}\left\{\sum_{n=1}^{k-1}(T_{n+1} - T_n)S_n\right\} = \lambda^{-2}\sum_{n=1}^{k-1}s_n = \lambda^{-2}\sum_{n=1}^{k-1}\{E(M_n^2) - E(M_1^2)\}.$$

The variance of $S_{N_t}(t - T_{N_t})$ is $\sum_{k=1}^{\infty} s_k E\{(t-T_{N_t})^2 1_{\{N(t)=k\}}\} = E\{s_{N_t}(t - T_{N_t})^2\}$ where $E\{(t - T_{N_t})^2 1_{\{N(t)=k\}}\} = O(t^4)$ as t tends to infinity. The variance of $S(t)$ is the sum

$$V(t) = E\{s_{N_t}(t - T_{N_t})^2\} + \lambda^{-2}E\left\{\sum_{n=1}^{N_t-1}(N_t - n)\sigma_n^2\right\}.$$

If there exists an increasing sequence $(b_n)_n$ tending to infinity with n and such that the sum $\sum_{n=1}^{\infty} b_n^{-2}\{E(M_n^2) - E(M_1^2)\}$ is finite, the martingale satisfies the strong law of large numbers, $b_n^{-1}M_n$ converges a.s. to zero as n tends to infinity and $\overline{\lim}_{n\to\infty}|M_n|$ and $\overline{\lim}_{t\to\infty}|S(t)|$ are a.s. infinite.

The variance σ_n^2 is a $o(b_n^2)$ and $s_n = o(\sum_{m=1}^{n} b_m^2)$. Let $\sigma_n^2 = O(n^a)$, $a > 0$, then $s_n = O(n^{2a+1})$ and $Es_{N_t} = O(t^{2a+1})$, the variances of $\sum_{n=1}^{N_t-1}(T_{n+1} - T_n)S_n$ and $S_{N_t}(t - T_{N_t})$ are $O(t^{2(a+1)})$. It follows that there exists a constant M such that for every $A > 0$ and for every t large enough

$$P(t^{-(a+1)}S_t > A) \leq A^{-2}M,$$

the bound does not depend on t and it tends to zero as A tends to infinity, moreover $\overline{\lim}_{t\to\infty}|S(t)|$ is a.s. infinite.

The process $A_t = s_{N_t}(t - T_{N_t})^2\} + \lambda^{-2} \sum_{n=1}^{N_t-1}(N_t - n)\sigma_n^2$ is increasing, positive and the variance of $S(t)$ is EA_t. For all positive increasing real sequences $(B)_t$ and $(\eta)_t$

$$P(S_t > B_t) \leq P(A_t^{-\frac{1}{2}} S_t > \eta_t^{-1} B_t) + P(A_t > \eta_t^2),$$

under the previous conditions, there exist constants M_1 and M_2 such that

$$P(S_t > B_t) \leq M_1 \eta_t^2 B_t^{-2} + M_2 \eta_t^{-2} t^{2(a+1)}.$$

This bound tends to zero as $\eta_t^{-1} B_t$ and $\eta_t t^{-(a+1)}$ tend to infinity with t.

The process A_t being increasing, this inequality generalizes to the variations of the process S_t. Let $(b)_t$ be a positive increasing sequence such that $\sup_{t \geq 0} t^{-\frac{3}{2}} b_t > 0$ and it is finite, then there exists a constant M such that for every $\delta > 0$

$$P\left(\sup_{|t-s| \leq \delta} |S_t - S_s| > b_t \right) \leq b_t^{-2} E\{A_{t+\delta} - A_{t-\delta}\} \leq 8\delta M t^3 b_t^{-2}$$

and the upper bound converges to zero with δ, then the sequence $(A_t^{-\frac{1}{2}} S_t)_{t>0}$ is tight.

8.8 Laplace transform

Let X_t, $t > 0$, be a right-continuous Markov process with left limits, in a measurable state space $(S, \mathcal{S})$. The conditional probability measure P_a of X_t on S, given a starting point a in S, is defined for every measurable set B as

$$P(t, a, B) = P_a(X_t \in B) = P(X_t \in B \mid X_0 = a).$$

The Chapman-Kolmogorov equation for the transition probability of the process X_t starting from a is

$$P(s + t, a, B) = \int_S P(s, b, B) P(t, a, db), \qquad (8.17)$$

where $E_a 1_{\{X_{s+t} \in B\}} = \int_S E_b 1_{\{X_s \in B\}} E_a 1_{\{X_t \in]b, b+db]\}}$.

Consider the exponential process

$$Z_t(\alpha) = \exp\{-\alpha X_t - \log E_a e^{-\alpha X_t}\}$$

for $\alpha > 0$ and a in S.

Proposition 8.6. *For a strong Markov process X_t, the process Z_t satisfies*

$$E_a(Z_s^{-1} Z_t \mid \mathcal{F}_s) = \frac{E_{X_s} e^{-\alpha(X_t - X_s)} E_a e^{-\alpha X_s}}{E_a\{e^{-\alpha X_s} E_{X_s} e^{-\alpha(X_t - X_s)}\}},$$

for every $\alpha > 0$ such that $\log E_a e^{-\alpha X_t}$ and $\log E_a e^{-\alpha X_s}$ are finite.

Proof. For $s < t$, the projection of P_a on $\mathcal{F}_s$ is P_{X_s}

$$E_a(Z_s^{-1} Z_t \mid \mathcal{F}_s) = E_{X_s} \exp\{-\alpha(X_t - X_s)\} \frac{E_a e^{-\alpha X_s}}{E_a e^{-\alpha X_t}}$$

and by (8.17)

$$E_a e^{-\alpha X_t} = E_a \{ e^{-\alpha X_s} E_{X_s} e^{-\alpha(X_t - X_s)} \}.$$

$\square$

Let $\mathcal{D}(S)$ be the space of the bounded right-continuous functions f from S to $\mathbb{R}$ with left limits, then for every decreasing sequence $(t_n)_n$ converging to t_0

$$P\{\lim_n f(X_{t_n}) = f(X_{t_0})\} = 1.$$

It follows that $P(t, a, b)$ is continuous with respect to t. For every $t > 0$ the linear operator $H_t : \mathcal{B}(S) \mapsto \mathcal{B}(S)$ defined as

$$H_t f(a) = E_a f(X_t) = \int_S f(b) P(t, a, db)$$

is continuous, positive with $H_t f \geq 0$ if $f \geq 0$, bounded with $H_t 1 \leq 1$, and

$$\lim_{t \to 0} H_t f(a) = \int_S f(b) P(0, a, db) = f(a).$$

By the Chapman-Kolmogorov equation

$$H_{s+t} f(a) = H_s H_t f(a),$$

and the Laplace transform of the transition probabilities

$$G_\alpha f(a) = \int_0^\infty e^{-\alpha t} H_t f(a)\, dt = E_a \left\{ \int_0^\infty e^{-\alpha t} f(X_t)\, dt \right\}$$

satisfies

$$G_\alpha f(a) = \int_0^\infty \int_S e^{-\alpha t} f(b) P(t, a, db)\, dt$$

$$= \int_S f(b) G_\alpha P(t, a, db).$$

The operator H_t is solution of the differential equation

$$t \Delta H_t f(a) - 2t H_t' f(a) = H_t f(a)$$

and for the Brownian motion starting at a at time zero, with density $\varphi_t(x - a)$, the Laplace transform $L_\alpha \varphi_t(a) = e^{-\frac{\alpha^2 t}{2} - a\alpha}$ and its derivatives are $\Delta L_\alpha \varphi_t(a) = \alpha^2 L_\alpha \varphi_t(a)$ and $\frac{\alpha^2}{2} L_\alpha \varphi_t(a)$, with respect to t

$$\left(\frac{\partial}{\partial t} - \frac{1}{2} \Delta \right) G_\alpha f(a, t) = 0,$$

the Laplace transform of a Gaussian process $B \circ v_t$ with a strictly positive differentiable variance function v and the transform G_α for the distribution of $B \circ v_t$ satisfy

$$\left(\frac{v_t'}{v_t^2}\frac{\partial}{\partial t} - \frac{1}{2}\Delta\right)G_\alpha f(a, t) = 0.$$

Let $\mathcal{C}(S)$ be the space of the bounded continuous functions from S to $\mathbb{R}$.

Theorem 8.2. *For every $\alpha > 0$, G_α is a continuous linear operator on $\mathcal{C}(S)$ such that for every β, $G_\alpha G_\beta = G_\beta G_\alpha$ and*

$$G_\alpha - G_\beta + (\alpha - \beta)G_\beta G_\alpha = 0.$$

Proof. By the Markov property (8.17)

$$\begin{aligned}
G_\beta G_\alpha f(a) &= \int_0^\infty e^{-\beta s} H_s G_\alpha f(a)\, ds \\
&= \int_0^\infty \int_0^\infty e^{-\beta s} e^{-\alpha t} H_{t+s} f(a)\, ds\, dt \\
&= \int_0^\infty \int_0^\infty e^{-(\beta - \alpha)s} e^{-\alpha(s+t)} H_{t+s} f(a)\, ds\, dt \\
&= \int_0^\infty \int_s^\infty e^{-(\beta - \alpha)s} e^{-\alpha x} H_x f(a)\, ds\, dt
\end{aligned}$$

integrating first with respect to $s < x$ implies

$$\begin{aligned}
G_\beta G_\alpha f(a) &= \frac{1}{\beta - \alpha}\int_0^\infty (e^{-\alpha x} - e^{-\beta x}) H_x f(a)\, dx \\
&= \frac{G_\alpha f(a) - G_\beta f(a)}{\beta - \alpha}.
\end{aligned}$$

$\square$

By symmetry, we deduce

$$G_\alpha = G_\beta \circ \{Id + (\alpha - \beta)G_\alpha\},$$

it follows that the image and the kernel of the operator G are constant, $\Im(G_\alpha) = \Im(G_\beta)$ and $\ker(G_\alpha) = \ker(G_\beta) = \{0\}$ by continuity of the transform H_t, for all α and β of $\mathbb{R}_+$.

For every f in $\mathcal{D}(S)$, $H_t f$ is right-continuous with respect to t at zero, if the probability function $P(t, a, B)$ has a derivative with respect to t uniformly in $a > 0$ and B in $\mathcal{S}$, there exists a limit operator

$$\mathcal{A} = \lim_{t \to 0^+} t^{-1}(H_t - Id).$$

This implies

$$\frac{dH_t}{dt} = \lim_{\delta \to 0^+} \frac{H_{t+\delta} - H_t}{\delta} = \mathcal{A}H_t,$$

therefore $H_t = e^{\mathcal{A}t}$ and

$$(\alpha - \mathcal{A})G_\alpha = (\alpha - \mathcal{A}) \int_0^\infty e^{-\alpha t} H_t \, dt = Id.$$

On $Im(G_\alpha)$, the operator G_α does not depend on α and

$$\mathcal{A}u = (\alpha - G_\alpha^{-1})u.$$

Let $u_\alpha(t, a)$ belong to $\Im(G_\alpha)$, for a Brownian motion starting at a at time zero $u_\alpha(t, a) = G_\alpha f(a, t) = L_\alpha H_t f(a)$, f in $\mathcal{D}(S)$, and

$$\left(\frac{\partial}{\partial t} - \frac{1}{2}\Delta \right) \mathcal{A}u_\alpha(t, a) = \frac{1}{2}\Delta f(a).$$

A process X_t is a strong Markov process if for every stopping time τ and for all sets B_1 and B_2 of $\mathcal{S}$

$$P_a(X_t \in B_1, X_{\tau+t} \in B_2) = E_a\{1_{X_t \in B_1} P_{X_\tau}(X_t \in B_2))\}, \ t > 0.$$

The Chapman-Kolmogorov equation for a strong Markov process implies

$$\int_0^\infty e^{-\alpha t} P_a(X_t \in A) \, dt = E_a \int_0^{\tau_a} e^{-\alpha t} P_a(X_t \in A) \, dt$$
$$+ E_a \int_0^\infty e^{-\alpha(t+\tau_a)} P_{X_{\tau_a}}(X_t \in A) \, dt,$$

that property was generalized by Dynkin's theorem.

Theorem 8.3. *Let f be a function of $\mathcal{D}(S)$, let $\alpha > 0$ and let τ be a stopping time of a strong Markov process X_t, the function $u(a) = G_\alpha f(a)$ satisfies*

$$u(a) = E_a \int_0^\tau e^{-\alpha t} f(X_t) \, dt + E_a\{e^{-\alpha\tau} u(X_\tau)\}.$$

Proof. It is sufficient to prove that for every stopping time $E_a \int_\tau^\infty e^{-\alpha t} f(X_t) \, dt = E_a\{e^{-\alpha\tau} u(X_\tau)\}$. By the Markov property of X_t, the mean integral is written as

$$E_a \int_\tau^\infty e^{-\alpha t} f(X_t) \, dt = E_a \int_0^\infty e^{-\alpha(t+\tau)} f(X_{t+\tau}) \, dt$$
$$= \int_0^\infty e^{-\alpha t} E_a[e^{-\alpha\tau} E\{f(X_{t+\tau}) \mid X_\tau\}] \, dt,$$

where the strong Markov property implies

$$E\{f(X_{t+\tau}) \mid X_\tau\} = \int_S f(b)P(X_{t+\tau} \in (b, b+db) \mid X_\tau)$$

$$= \int_S f(b)P(t, X_\tau, db) = E_{X_\tau} f(X_t),$$

it follows

$$E_a \int_\tau^\infty e^{-\alpha t} f(X_t)\, dt = E_a \left\{ e^{-\alpha \tau} E_{X_\tau} \int_0^\infty e^{-\alpha t} f(X_t)\, dt \right\}$$

$$= E_a \{ e^{-\alpha \tau} u(X_\tau) \}.$$

$\square$

Theorem 8.4. *Let $u = G_\alpha f$, f in $\mathcal{D}(S)$, for every stopping time τ of a strong Markov process X_t such that $E_a \tau$ is finite*

$$E_a \int_0^\tau \mathcal{A}u(X_t)\, dt = E_a u(X_\tau) - u(a).$$

Proof. Let $u = G_\alpha f$, then $f = G_\alpha^{-1} u = -(\mathcal{A} - \alpha)u$ and by Theorem 8.3 we have

$$E_a \{ e^{-\alpha \tau} u(X_\tau) - u(a) \} = -E_a \int_0^\tau e^{-\alpha t} f(X_t)\, dt$$

and

$$E_a \{ e^{-\alpha \tau} u(X_\tau) - u(a) \} = -E_a \int_0^\tau e^{-\alpha t} f(X_t)\, dt$$

$$= E_a \int_0^\tau e^{-\alpha t} (\mathcal{A} - \alpha) u(X_t)\, dt$$

the limits as α tends to 0 satisfy

$$E_a \{ u(X_\tau) - u(a) \} = E_a \int_0^\tau \mathcal{A}u(X_t)\, dt.$$

$\square$

For an open set D of S, let $\tau < \infty$ be the first hitting time of a set D by a continuous Markov process X_t starting from a at zero. We have $\tau = 0$ if a belongs to D and $E_a u(X_\tau) - u(a) = 0$ if a does not belong to D. For every function of $\mathcal{D}(S)$, Theorems 8.3 and 8.4 imply

$$u(a) = \{1 - E_a e^{-\alpha \tau}\}^{-1} E_a \int_0^\tau e^{-\alpha t} f(X_t)\, dt,$$

$$E_a \int_0^\tau \mathcal{A}u(X_t)\, dt = 0.$$

These results apply to the random times of a point process starting from T_n, on the set $\{T_n \leq t < Tn + 1\}$, $E_n u(N_{T_n}) - u(n) = 0$, for every integer n.

For every stopping time τ, on the set $\{\tau > t\}$, $\tau = t + \tau_t$ where τ_t is $\mathcal{B}_t$-measurable and for a strong Markov process

$$E[e^{-\{u(X_\tau)-u(X_t)\}}1_{\{\tau>t\}} \mid \mathcal{B}_t] = 1_{\{\tau>t\}}E[e^{-\{u(X_{t+\tau_t})-u(X_t)\}} \mid \mathcal{B}_t]$$
$$= 1_{\{\tau>t\}}E[e^{u(X_t)}E_{X_\tau}e^{-u(X_t)} \mid \mathcal{B}_t]$$

where $E_{X_\tau}\{e^{u(X_t)}E_{X_\tau}e^{-u(X_t)}\} \geq 1$ and $E[e^{-\{u(X_\tau)-u(X_t)\}}1_{\{\tau>t\}} \mid \mathcal{B}_t] \geq 1$. The process $Z_t = \exp\{u(X_t) - u(a) - \int_0^t \mathcal{A}u(X_s)\,ds\}$ is a submartingale, by Theorem 8.4.

8.9 Time-space Markov processes

Consider a time-space borelian state $(S, \mathcal{S})$ and a process in this state space, $X_t = (T_n 1_{T_n \leq t}, \xi_{T_n})_{n \geq 1}$ where $(T_n)_{n \geq 1}$ is a sequence of the random times of an event occurring up to t and $(\xi_t)_{t \geq 0}$ is the Markov location process of X at t. The transition probabilities of the process X_t from a set A at s to a borelian set $B_t = [0, t] \times B$, for $s < t$, are defined as

$$\pi(t, s, A, B_t) = P(T \leq t, \xi_T \in B \mid T \geq s, \xi_s \in A)$$
$$= P(\xi_T \in B \mid T \leq t, T \geq s, \xi_s \in A)P(T \leq t \mid T \geq s, \xi_s \in A)$$
$$= \int_s^t P(u, s, A, B)\,dF_{s,A}(u)$$

where the conditional distribution of T given $\{T \geq s\}$ and $\{\xi_s \in A\}$ is determined for $u \geq s$ by the equations

$$P(T \leq t \mid T \geq s, \xi_s \in A) = \int_s^t P(T \in (u, u + du) \mid T \geq s, \xi_s \in A),$$
$$dF_{s,A}(u) = P(T \in (u, u + du) \mid T \geq u, \xi_s \in A).$$

The Chapman-Kolmogorov equations for the transition probabilities of the process $(T1_{T \leq t}, \xi_T)$, starting in A are determined by the conditional probability $P(T \leq u + v \mid T \geq s, \xi_s \in A) = \int_s^{u+v} dF_{s,A}(z)$ and by the equation for the location

$$P(u + v, s, A, B) = \int_S P(u, v, x, B)P(v, s, A, dx),$$

they imply

$$\pi(s+t,a,A,B) = \int_s^v \int_v^u \int_S P(u,v,x,B)P(v,s,A,dx)\,dF_{s,A}(v)\,dF_{v,x}(u)$$

$$= \int_S \pi(u,v,x,B)\pi(v,s,A,dx).$$

For the sequence $(T_n 1_{T_n \leq t}, \xi_{T_n})_{n \geq 1}$, consider the time-space counting process

$$N(t,B) = \sum_{n \geq 1} 1_{\{T_n \leq t\}} 1_{\{\xi_{T_n} \in B\}},$$

its transition probabilities are defined for $s < t$ as

$$\nu(t,s,A,B) = \sum_{n \geq 1} \sum_{1 \leq k \leq n} P\{N(t,B) = n \mid N(s,A) = k\},$$

$$P\{N(t,B) = n \mid N(s,A) = k\}$$
$$= P(T_n \leq t < T_{n+1}, \xi_{T_n} \in B \mid T_k \leq s < T_{k+1}, \xi_{T_k} \in A)$$
$$= P(\xi_{T_n} \in B \mid \xi_{T_k} \in A, T_n \leq t < T_{n+1}, T_k \leq s < T_{k+1})$$
$$\times P(T_n \leq t < T_{n+1} \mid T_k \leq s < T_{k+1}, \xi_{T_k} \in A)$$

where the conditional transition probabilities of the jump process $(\xi_{T_n})_{n \geq 1}$ are denoted

$$\mu(t,s,A,B) = P(\xi_t \in B \mid \xi_s \in A)$$

and the time transition probabilities are

$$\lambda(t,s,n,k,A) = P(N(t,S) - N(s,S) = n - k \mid N(s,S) = k, \xi_{T_k} \in A).$$

If the process $N(t) = N(t,S)$ has independent increments, the Chapman-Kolmogorov equations for the transition probabilities are

$$P(N(u+v) = n \mid N(s) = k, \xi_s \in A)$$
$$= \sum_{m=0}^{n} P(N(u) = m \mid N(s) = k, \xi_s \in A)$$
$$.P(N(u+v) = n - m \mid N(u) = m, \xi_s \in A)$$

and for the jump Markov process ξ

$$\mu(u+v,s,A,B) = \int_S \mu(u,v,x,B)\mu(v,s,A,dx),$$

they imply the following recurrence equation for the time-space counting process

$$P\{N(u+v,B) = n \mid N(s,A) = k\}$$

$$= \sum_{m=0}^{n} \int_{S} P\{N(u+v,B) = n \mid N(u,(x,x+dx))) = m\} \quad (8.18)$$

$$.P\{N(u,(x,x+dx)) = m \mid N(s,A) = k\}.$$

Let f be a function of $\mathcal{B}(S)$, for $t > s > 0$ the linear operator H_t is defined on $\mathcal{B}(S)$ at (s,x), x in S, as

$$H_t f(s,x) = E_{s,x}\{f(t,\xi_t) \mid \mathcal{B}_s, \xi_s = x\} = \int_{S} f(t,b)\nu(t,s,x,db)$$

$$= \sum_{n \geq 1} \int_{S} f(t,b) P(N(t,(b,b+db)) = n \mid \mathcal{B}_s, \xi_s = x),$$

if the probability measure $\nu(t,s,A,B)$ is right-continuous with respect to t uniformly in $s < t$, A and B in S, we have

$$\lim_{t \to s} H_t f(s,x) = f(x).$$

By the Chapman-Kolmogorov Equation (8.18), we have $H_{s+t} = H_s H_t$ on $\mathcal{B}(S)$ and the Laplace transform of the transition probabilities is such that, for $s < t$

$$G_\alpha f(s,x) = \int_0^\infty e^{-\alpha t} H_t f(t,x)\, dt = E_{s,x}\left\{ \int_0^\infty e^{-\alpha t} f(t,\xi_t)\, dt \right\}.$$

For a strong Markov process ξ_t, Theorems 8.2 and 8.3 are modified to depend on the random times of the process, for every stopping time τ with finite expectation, the function $u(s,x) = G_\alpha f(s,x)$ is

$$u(s,x) = E_{s,x} \int_0^\tau e^{-\alpha t} f(t,\xi_t)\, dt + E_{s,x} a\{e^{-\alpha \tau} u(0,\xi_\tau)\},$$

$$E_{s,x} \int_0^\tau Au(t,X_t)\, dt = E_{s,x} u(0,X_\tau) - u(s,x).$$

For a Poisson process N_t with parameter λ and with a power function $f(x) = z^x$, if $\alpha > \lambda(z-1)$, the transform $G_\alpha f$ is the Laplace transform of the generating function of N_t, $G_\alpha f(z) = \{\alpha - \lambda(z-1)\}^{-1}$.

Let $N(t,x)$ be the number of particles that appear before t from time s at x up to time t at a random times T_n and at random location $\xi(T_n)$. Under the conditions that the time process has independent increments

$(U_n)_n$ and the location of the particles at the random times is a strong Markov process,

$$E_{s,x} \exp\left\{-\int_{[s,t]\times B} f(u,y)\,N(du,dy)\right\}$$

$$= E\left[\prod_{s<T_n\leq t} \exp\{-f(T_n,\xi(T_n))1_{\{\xi(T_n)\in B\}}\} \mid \xi(s)=x\right]$$

$$= E_{s,x}\left[\prod_{n:s\leq T_{n-1},T_n\leq t} E_{\xi_{T_{n-1}}}\exp\{-f(U_n,\xi(U_n))1_{\{\xi(U_n)\in B\}}\}\right].$$

Let f in $\mathcal{D}(S)$ be defined as $f(t,\xi_t) = \sum_n 1_{\{T_n\leq t<T_{n+1}\}}f(T_n,\xi(T_n))$, the function $u(s,x) = G_\alpha f(s,x)$ is

$$u(s,x) = \int_0^\infty e^{-\alpha t}E_{s,x}f(T_{N_t},\xi(T_{N_t}))\,dt$$

$$= \sum_{n=0}^\infty \sum_{k\leq n}\int_0^\infty e^{-\alpha t}P(N_t-N_s=n-k)E_{s,x}f(T_n,\xi(T_n))\,dt$$

$$= \sum_{n=0}^\infty \int_0^\infty \int_y^\infty e^{-\alpha t}f(y,\xi_y)\,dt\,dP_{s,x}(T_n\leq y),$$

where $P(T_n\leq t) = \int_0^t P(T_{n-1}\leq t-s)\,dP(U_n\leq s)$, $n\geq 1$.

For a strong Markov process and denoting the inverse of EU_n by p_n, the sum $T_n = \sum_{k\leq n}U_k$ has an exponential distribution with parameter $\bar{p}_n = \sum_{k=1}^n p_n$ and

$$E_{s,x}\{e^{-\alpha T_n}f(T_n,\xi_{T_n})\} = \frac{\bar{p}_n}{\alpha+\bar{p}_n}e^{-(\alpha+\bar{p}_n)s}E_{s,x}\{f(\widetilde{T}_n,\xi_{\widetilde{T}_n})\}$$

where $\widetilde{T}_n$ has an exponential distribution with parameter $\alpha+\bar{p}_n$.

For every stopping time τ, on the set $\{\tau>t\}$, $\tau=t+\tau_t$ where τ_t is measurable with respect to $\mathcal{B}_t$ and

$$E[e^{-\{u(\tau,\xi_\tau)-u(t,\xi_t)\}}1_{\{\tau>t\}} \mid \mathcal{B}_t] = 1_{\{\tau>t\}}E[e^{-\{u(t+\tau_t,\xi_{t+\tau_t})-u(t,\xi_t)\}} \mid \mathcal{B}_t]$$

$$= 1_{\{\tau>t\}}E[e^{u(t,\xi_t)}E_{\xi_\tau}e^{-u(t,\xi_t)} \mid \mathcal{B}_t].$$

By Theorem 8.4

$$E_{s,x}u(0,\xi_\tau) - u(s,x) = \alpha E_{s,x}\int_0^\tau G_\alpha f(t,\xi_t)\,dt - E_{s,x}\int_0^\tau f(t,\xi_t)\,dt$$

and the process $Z_t = \exp\{u(t,\xi_t)-u(0,a)-\int_0^t \mathcal{A}u(\xi_s)\,ds\}$ is a submartingale.

For a Levy process A_t on $\mathbb{R}_+$, there exists a measure ν such that the process has the Laplace transform

$$L_t(\alpha) = Ee^{-\alpha A_t} = \exp\left\{-t\int_0^\infty (1 - e^{-\alpha u})\, d\nu(u)\right\}$$

where $\int_0^\infty \min(1, u)\, d\nu(u)$ is finite, then

$$Z_t = \exp\left\{-\alpha A_t - \int_0^\infty (e^{-\alpha u} - 1)\, d\nu(u)\right\}$$

is a local martingale.

Chapter 9

Inequalities for Processes

9.1 Introduction

The stationary covariance function of a centered process $(X_t)_{t>0}$ is

$$R_X(t) = E\{X(s)X(s+t)\},$$

for all s and $t > 0$. It is extended to every real t with the covariance $R(-t) = R(t)$ for $t < 0$. Its variance has the property $EX^2(s) = EX^2(0) = R_X(0)$, for every $s > 0$. By the Cauchy-Schwarz inequality

$$E\|X_t X_s\| \leq E\|X_0\|^2, \quad R_X(t) \leq R_X(0).$$

For a Gaussian process, there is equivalence between the stationarity of the distribution and the stationarity of its expectation and covariance functions. For a stationary process

$$E\{X(s+t) - X(s)\}^2 = 2\{R(0) - R_X(t)\},$$

and a necessary and sufficient condition for the continuity of the process is the continuity of the expectation and covariance functions. Thus, the standard Brownian motion is continuous and stationary.

Let X be a centered Gaussian process on $\mathbb{R}_+$ with a covariance function $R(s,t) = E(X_s X_t)$ in $L^2(\mathbb{R}_+^2)$ and let $(e_k(t))_{t\in\mathbb{R}_+, k\geq 0}$ be an orthonormal basis of $L^2(\mathbb{R}_+)$, then the process X and the function $R(s,t)$ have expansion on the basis $(e_k(t))_{t\in\mathbb{R}_+, k\geq 0}$. The variable $Y_k = \int_{\mathbb{R}_+} X_t e_k(t)\, dt$ is a centered Gaussian variable with variance

$$E(Y_k^2) = \int_{\mathbb{R}_+^2} E(X_s X_t) e_k(s) e_k(t)\, ds\, dt$$

$$= \int_{\mathbb{R}_+^2} R(s,t) e_k(s) e_k(t)\, ds\, dt$$

denoted λ_k and $Z_k = \lambda_k^{-\frac{1}{2}} Y_k$ is a normal variable such that X has the expansion

$$X_t = \sum_{k \geq 0} Y_k e_k(t) = \sum_{k \geq 0} \lambda_k^{\frac{1}{2}} Z_k e_k(t)$$

on the orthonormal basis. Its $L^2(\mathbb{R}_+)$ norm is

$$\int_{\mathbb{R}_+} X_t^2 \, dt = \sum_{k \geq 0} \lambda_k Z_k^2.$$

If X belongs to $L^p(\mathbb{R}_+)$, $p \geq 2$, by convexity we have

$$\int |X_t|^p \, dt = \int (|X_t|^2)^{\frac{p}{2}} \, dt \geq \left(\int |X_t|^2, dt \right)^{\frac{p}{2}} = \left(\sum_{k \geq 0} Y_k^2 \right)^{\frac{p}{2}}$$

therefore $\|Y_n\|_{l^2} \leq \|X_t\|_{L^p(\mathbb{R}_+)}$. The covariance function $R(s,t)$ has the expansion

$$R(s,t) = \sum_{k \geq 0} \lambda_k e_k(s) e_k(t)$$

and it has the norm

$$\|R\|_{L^2}^2 = \int_{\mathbb{R}_+^2} R(s,t) \, ds \, dt = \sum_{k \geq 0} \lambda_k^2$$

where $\int_{\mathbb{R}_+} e_k(t) e_j(t) \, dt = 0$ for $k \neq j$.

A stationary process X with expectation function $x = EX$ has the mean process

$$\bar{X}(T) = \frac{1}{T} \int_{[0,T]} X(t) \, dt, \; T > 0,$$

with expectation $\bar{x} = \bar{X}$. For every $T > 0$

$$E \frac{1}{T} \int_{[0,T]} \{X(t) - \bar{x}(t)\}^2 \, dt = R(0),$$

$$E \left[\frac{1}{T} \int_{[0,T]} \{X(t) - \bar{x}(t)\} \, dt \right]^2 = \frac{R(0)}{T} + \frac{1}{T^2} \int_{[0,T]^2} \text{Cov}\{X(s)X(t)\} \, ds \, dt$$

$$= \frac{R(0)}{T} + \frac{1}{T^2} \int_{[0,T]^2} R_{|t-s|} \, ds \, dt,$$

this is the variance of the mean process $\bar{X}_T$ and the first term of its expression tends to zero as T tends to infinity. The Bienaymé-Chebychev implies

$$P\left(\sup_{t \in [0,T]} (\bar{X} - \bar{x})(t) > x \right) \leq \frac{1}{x^2 T^2} \int_{[0,T]^2} R_{|t-s|} \, ds \, dt + \frac{R(0)}{x^2 T^2}.$$

Stochastic inequalities between processes have been established from inequalities between the covariance function of real valued stationary processes (Slepian, 1962). They are generalized in the next section to $\mathbb{R}^n$ and to processes which are not Gaussian. In Section 9.3, sufficient conditions are established for an infinite time in random ruin models, in the expectation models, and the probabilities of ruin in random intervals of the processes are defined. Stochastic orders between several models are deduced.

9.2 Stationary processes

The Gaussian density $f_\sigma(t)$ on $\mathbb{R}$ is decreasing with respect to the variance for t in the interval $I_\sigma = [-\sigma, \sigma]$ and it is increasing outside. The densities of variables X and Y such that the variance of X is larger than the variance of Y are ordered and satisfy the inequalities $f_X(t) \leq f_Y(t)$ in the interval I_σ and $f_X(t) \geq f_Y(t)$ in the tails, then for every $u > \sigma$

$$P(X \geq u) \geq P(Y \geq u).$$

The distribution of a Gaussian vector with expectation zero is determined by its variance matrix $\Sigma_X = (E(X_i - EX_i)(X_j - EX_j))_{i,j=1,\ldots,n}$. In $\mathbb{R}^2$, for Gaussian vectors $X = (X_1, X_2)$ and $Y = (Y_1, Y_2)$ having the same marginal variances σ_1^2 and σ_2^2, and such that $\sigma_{X,12}^2 = E(X_1 X_2) \geq E(Y_1 Y_2) = \sigma_{Y,12}^2$, the quadratic form

$$t^T \Sigma_X^{-1} t - t^T \Sigma_Y^{-1} t = 2t_1 t_2 \frac{\sigma_1 \sigma_2 (\sigma_{X,12}^2 - \sigma_{Y,12}^2)}{\det \Sigma_X \det \Sigma_Y}$$

is positive for $t = (t_1 t_2)$ in $\mathbb{R}_+^2$ or $\mathbb{R}_-^2$, then the densities of X and Y are ordered $f_X < f_Y$ in $\mathbb{R}_+^2$ and for every $u = (u_1, u_2)$ in, $\mathbb{R}_+^2$

$$P(X_1 \geq u_1, X_2 \geq u_2) \geq P(Y_1 \geq u_1, Y_2 \geq u_2).$$

Slepian's lemma (1962) extends this order to Gaussian vectors of $\mathbb{R}^n$ having the same marginal variances and such that $E(X_i X_j) \geq E(Y_i Y_j)$ for all $i \neq j$, then for all real numbers $u_1, \ldots, u_n > 0$

$$P(X_i \geq u_i, i = 1, \ldots, n) \geq P(Y_i \geq u_i, i = 1, \ldots, n).$$

Let $(X_t)_{t \geq 0}$ and $(Y_t)_{t \geq 0}$ be centered Gaussian processes with stationary covariance functions R_X and R_Y. By passage to the limit in Slepian's lemma, the inequality $R_X \geq R_Y$ in an interval $[0, T_0]$ implies

$$P\left(\sup_{t \in [0,T]} X_t \leq c \right) \leq P\left(\sup_{t \in [0,T]} Y_t \leq c \right)$$

for all $c > 0$ and T in $[0, T_0]$.

Theorem 9.1 (Slepian (1962)). *For all times $S \geq 0$ and $T \geq 0$, the inequality $R(t) \geq 0$ in $[0, T+S]$ implies*

$$P\Big(\sup_{t \in [0,T+S]} X_t \leq c \Big) \leq P\Big(\sup_{t \in [0,T]} X_t \leq c \Big) P\Big(\sup_{t \in [0,S]} X_t \leq c \Big).$$

Proof. For all times $S \geq 0$ and $T \geq 0$, the variables $\sup_{t \in [0,T]} X_t$ and $\sup_{t \in [0,S]} X_t$ are independent. By stationarity, $(\max_{t=t_1,\ldots,t_{k+m} \in [0,T+S]} X_t)$ and $\max(\max_{t=t_1,\ldots,t_k \in [0,T]} X_t, \max_{t_{k+1},\ldots,t_{k+m} \in [0,S]} X_t)$ have the same distribution and the covariance matrix of $((X_{t_j})_{j=1,\ldots,k}, (X_{t_j})_{j=k+1,\ldots,k+m})$ is diagonal by blocks, hence the result is an application of Slepian's Lemma (1962), by passage to the limit as the partition of $[0, T+S]$ increases. $\qquad\square$

A centered Gaussian process with values in $\mathbb{R}^n$ also satisfies the inequality of Theorem 9.1 with the Euclidean norm

$$P\Big(\sup_{t \in [0,T+S]} \|X_t\|_{2,n} \leq c \Big) \leq P\Big(\sup_{t \in [0,T]} \|X_t\|_{2,n} \leq c \Big) P\Big(\sup_{t \in [0,S]} \|X_t\|_{2,n} \leq c \Big).$$

It can be extended to balls of $\mathbb{R}^n$ centered at zero and with radius $r > 0$, B_r. If X_t belongs to B_r, its norm is bounded by $\sqrt{n}r$ and reciprocally, therefore we have the following.

Theorem 9.2. *For every $r > 0$, and for all times $S \geq 0$ and $T \geq 0$*

$$P\Big(\sup_{t \in [0,T+S]} X_t \in B_r \Big) \leq P\Big(\sup_{t \in [0,T]} X_t \in B_r \Big) P\Big(\sup_{t \in [0,S]} X_t \in B_r \Big).$$

Let τ_T be the time when a stationary Gaussian process X reaches its supremum on the interval $[0, T]$, by stationarity $\sup_{t \in [0,T+S]} X_t = X_{\tau_{S+T}}$ has the same distribution as the maximum of $\sup_{t \in [0,T]} X_t = X_{\tau_T}$ and $\sup_{t \in [0,S]} X_t = X_{\tau_S}$ therefore τ_{S+T} and $\tau_S \wedge \tau_T$ have the same distribution. Applying Lévy's arcsine law (1.24)

$$P(\tau_T \leq t) = P\Big(\sup_{s \in [0,t]} X_s \geq \sup_{s \in [t,T]} B_s \Big) = \frac{2}{\pi} \arcsin \sqrt{\frac{R(t)}{R(T)}}.$$

The proof is the same as for Proposition 5.27.

A increasing process X of $L^2(P)$ with a stationary covariance function R is such that the centered martingale $M = X - \widetilde{X}$ has the same covariance as X and

$$EM_t^2 = E < M >_t = E(X_t^2 - \widetilde{X}_t^2) = R(t),$$

this implies

$$E(M_{t+s} - M_t)^2 = E(< M >_{t+s} - < M >_t) = R(s).$$

Slepians theorems extend to non Gaussian processes.

Theorem 9.3. *Let X and Y be processes with stationary moment functions $m_{X,k}$ and $m_{Y,k}$ such that for every $k \geq 1$, $m_{X,k} \leq m_{Y,k}$ in an interval $[0, T_0]$. For all $c > 0$ and T in $[0, T_0]$*

$$P\Big(\sup_{t \in [0,T]} X_t \leq c \Big) \geq P\Big(\sup_{t \in [0,T]} Y_t \leq c \Big).$$

Proof. The assumption of ordered moments implies $\varphi_{X_t} \leq \varphi_{Y_t}$ for every t in $[0, T_0]$, and the theorem is a consequence of Chernoff's theorem for the processes X and Y. It applies, for example, to point processes with independent increments. $\qquad\square$

9.3 Ruin models

A marked point process $(T_k, X_k)_{k \geq 1}$ is defined by positive random times T_k, with $T_0 = 0$, and by positive jump size X_k of the process at T_k, for $k \geq 0$, as

$$N_X(t) = \sum_{k \geq 1} X_k 1_{\{T_k \leq t\}}, \, t \geq 0, \tag{9.1}$$

it is also written as $N_X(t) = \sum_{k \leq N(t)} X_k = S_{N(t)}$ where $S_n = \sum_{k \leq n} X_k$, and it generalizes the model (5.9) defined with a Poisson process. Under the conditions of a process $N = (T_k)_{k \geq 1}$ with independent increments and an independent sequence of mutually independent variables $(X_k)_{k \geq 1}$, the process N_X has independent increments and the moment generating function

$$\varphi_{N_X(t)}(\lambda) = e^{\lambda N_X(t)} = \sum_{k \geq 1} e^{\lambda S_k} P(N(t) = k)$$

$$= \sum_{k \geq 1} \prod_{j=1}^{k} \varphi_{X_j}(\lambda) P(N(t) = k).$$

If the inter-arrival times of the point process are identically distributed, with density f, N is a renewal process such that $P(N(t) = k) = F * f^{*(k-1)}(t)$, where the density of T_{k-1} is the convolution $f^{*(k-1)}$. If the variables X_k are identically distributed, denoting $\mu_X = E(X)$ and $\sigma_X^2 = \text{Var}(X)$, the expectation and the variance of $N_X(t)$ are

$$E\{N_X(t)\} = \mu_X E\{N(t)\},$$
$$\text{Var}\{N_X(t)\} = \sigma_X^2 E\{N(t)\} + \mu_X \text{Var}\{N(t)\}$$

and its moment generating function is

$$\varphi_{N_X(t)}(\lambda) = \sum_{k \geq 1} \varphi_X^k(\lambda) P(N(t) = k).$$

In the Sparre Andersen model (Thorin, 1970), a process decreases from an initial model by the random arrival of claims at times T_k and the inter-arrival variables $S_k = T_k - T_{k-1}$ are identically distributed with a distribution function F and a density f in $\mathbb{R}_+$, hence the density of T_k is the convolution f^{*k}. The process is written as

$$Y_t = a + ct - N_X(t), \ t \geq 0,$$

with $N_X(0) = 0$, $a > 0$ and $b > 0$. The time of ruin is the random variable

$$T_a = \inf_{t \geq 0}\{Y_t < 0 | Y_0 = a\},$$

this is a stopping time for the process Y and its probability of occurrence is $P(T_a < \infty) = P(Y_{T_a} < 0 | Y_0 = a)$. Let $\mu = ES_k$, for every $k \geq 1$, the expectation of $T_{N(t)}$ is $\mu^{-1}EN(t)$.

Lemma 9.1. *If* $0 < \mu X_k \leq c$ *a.s., for every* $k \geq 1$, *the time of ruin is a.s. infinite on the set* $\{\mu^{-1}N(t) - T_{N(t)} < c^{-1}a, \forall t > 0\}$.

Proof. The process Y is written as

$$Y_t = a + c(t - T_{N(t)}) + \sum_{k \geq 1}(c\mu^{-1} - X_k)1_{\{T_k \leq t\}} + c\{T_{N(t)} - \mu^{-1}N(t)\}$$

where $t - T_{N(t)} \geq 0$ is the current time since the last occurrence of an event of the process N_X and $U(t) = \sum_{k \geq 1}(c - X_k)1_{\{T_k \leq t\}}$ is strictly positive under the assumption for the variables X_k. Therefore $Y_t > 0$ if the last term is positive. $\qquad\square$

The probability of $\{\mu^{-1}N(t) - T_{N(t)} \geq c^{-1}a, \forall t > 0\}$ is

$$P\left(\sup_{t \geq 0}\{\mu^{-1}N(t) - T_{N(t)}\} \geq c^{-1}a\right) \leq c^2 a^{-2}\{\mu^{-1}N(t) - T_{N(t)}\}^2$$

where $ET_{N(t)} = \mu^{-1}EN(t)$. It follows that a large constant a ensures a large probability for an infinite time of ruin.

The expectation of the current time after the last event of the process N is $t - T_{N(t)} > 0$ is $t - \mu^{-1}EN(t)$, moreover $EN(t) = \sum_{k \geq 1} F * f^{*(k-1)}(t)$, where the convolutions of the distributions are $F^{*k}(t) = \int_0^\infty F(t - s)f^{*(k-1)}(s)\,ds$, $k \leq 2$. Therefore, for every $t > 0$

$$\sum_{k \geq 1} F * f^{*(k-1)}(t) < \mu t. \tag{9.2}$$

The expectation of Y_t is $EY_t = a + ct - EN_X(t)$ and it remains strictly positive up to $t_0 = \arg\min\{t : \sum_{k \geq 1} EX_k \, EF(t - T_{k-1}) \geq a + ct\}$.

Lemma 9.2. *Let N be a renewal process with a mean inter-event time μ, independent of a sequence of independent and identically distributed variables $(X_k)_{k \geq 0}$ with expectation μ_X, and let $c < \mu_X \mu$ be a strictly positive constant. The mean time of ruin is then $t_0 > (\mu_X \mu - c)^{-1} a$.*

Proof. From (9.2), if the variables X_k have the same expectation μ_X
$$EN_X(t) = \mu_X EN(t) < \mu_X \mu t.$$
The bound for t_0 is deduced from the equation of the model
$$EY(t_0) = a + ct_0 - EN_X(t_0) \leq 0.$$

$\square$

Assuming that t_0 is finite, the question is to characterize the smallest values of t such that $Y(t) = a + ct - \sum_{k=1}^{N(t)} X_k 1_{\{T_k \leq t\}}$ is negative. Let
$$\Lambda(t) = \int_0^t \frac{dF}{1 - F}$$
be the cumulative hazard function of the variables S_k. By the Bienaymé-Chebychev inequality, for every $\varepsilon > 0$
$$P(Y(t) < -\varepsilon) = P\left(\sum_{k \geq 1} 1_{\{T_k \leq t\}} = n\right) P\left(\sum_{k=1}^n X_k 1_{\{T_k \leq t\}} > a + ct - \varepsilon\right)$$
$$\leq \frac{\mu_X}{a + ct - \varepsilon} e^{-\Lambda(t)} \sum_{n \geq 1} \frac{\Lambda^n(t)}{n!} \sum_{k=1}^n P(T_k \leq t).$$
Taking the limit as ε tends to zero, we obtain the next result.

Proposition 9.1. *Under the conditions of Lemma 9.2, for every $t > 0$*
$$P(Y(t) < 0) \leq \frac{\mu_X}{a + ct} e^{-\Lambda(t)} \sum_{n \geq 1} \frac{\Lambda^n(t)}{n!} \sum_{k=1}^n F^{*k}(t).$$

The probability that the time of ruin belongs to a random interval $]T_n, T_{n+1}]$ is $P_n = P(Y(T_k) > 0, k = 1, \ldots, n, Y(T_{n+1}) \geq 0)$. Let also $\widetilde{P}_n = P(Y(T_k) > 0, k = 1, \ldots, n+1) = \widetilde{P}_{n-1} - P_n$. For $n = 1$, the independence properties of the model imply
$$P_1 = P(X_1 < a + cT_1, X_1 + X_2 \geq a + cT_2)$$
$$= P(X_1 < a + cT_1) P(X_2 \geq c(T_2 - T_1))$$
$$= \left\{\int_0^\infty F_X(a + cs) \, dF(s)\right\}\left\{1 - \int_0^\infty F_X(cs) \, dF(s)\right\},$$
$$\widetilde{P}_1 = \left\{\int_0^\infty F_X(a + cs) \, dF(s)\right\}\left\{\int_0^\infty F_X(cs) \, dF(s)\right\},$$

they involve additive and multiplicative convolutions. If F has a density f, $\widetilde{P}_1 \leq c^{-2}\|F_X\|_p^2\|f\|_{p'}^2$ and $P_1 \leq c^{-2}\|F_X\|_p\|1-F_X\|_p\|f\|_{p'}^2$, for all conjugates $p \geq 1$ and p'. For $n = 2$, the same arguments entail

$$P_2 = \widetilde{P}_1 P(X_3 > cS_3) = \left\{ \int_0^\infty F_X(a+cs)\, dF(s) \right\}$$
$$\times \left\{ \int_0^\infty F_X(cs)\, dF(s) \right\} \left\{ 1 - \int_0^\infty F_X(cs)\, dF(s) \right\},$$

and

$$\widetilde{P}_2 = \widetilde{P}_1 - P_2 = \left\{ \int_0^\infty F_X(a+cs)\, dF(s) \right\} \left\{ \int_0^\infty F_X(cs)\, dF(s) \right\}^2.$$

Proposition 9.2. *The probability of ruin in the interval* $]T_n, T_{n+1}]$ *is*

$$P_n = \widetilde{P}_{n-1} P(X_{n+1} > cS_{n+1})$$
$$= \left\{ \int_0^\infty F_X(a+cs)\, dF(s) \right\} \left\{ \int_0^\infty F_X(cs)\, dF(s) \right\}^{n-1}$$
$$\times \left\{ 1 - \int_0^\infty F_X(cs)\, dF(s) \right\}.$$

For all conjugates $p \geq 1$ *and* p'

$$P_n \leq c^{-(n+1)}\|F_X\|_p^n\,\|1 - F_X\|_p\,\|f\|_{p'}^{n+1}.$$

The result is proved recursively, from the equality $\widetilde{P}_n = \widetilde{P}_{n-1} - P_n$.

Proposition 9.3. *For every integer* $n \geq 1$, *the time of ruin* T_a *belongs to the interval* $]T_n, T_{n+1}]$ *with probability* P_n *and the interval with the greatest probability is* $]T_1, T_2]$, *except if* $\int_0^\infty F_X(cs)\, dF(s) = \frac{1}{2}$. *In that case, the probabilities that* T_a *belongs to* $]0, T_1]$ *or* $]T_1, T_2]$ *are equal and larger than the probabilities of the other intervals.*

The model has been generalized to allow variations by adding a stochastic diffusion model to a parametric trend $y(t)$ minus $N_X(t)$

$$dY(t) = dy(t) + bY(t)\, dt + \sigma(t)\, dB(t) - dN_X(t),$$

where B is the standard Brownian motion and $Y(0) = y(0)$. Its expectation $m_Y(t) = EY(t)$ is such that $m_Y(t) = y(t) + b\int_0^t m(s)\, ds - EN_X(t)$. The solution of the mean equation is $m_Y(t) = y(t) + (e^{bt} - 1) - EN_X(t)$, denoted $\psi(t) - EN_X(t)$ and Lemma 9.2 applies replacing the linear trend by $\psi(t)$ in the sufficient condition.

Using the solution (5.16) of the stochastic diffusion model

$$Y(t) = \psi(t) + \int_0^t \sigma(s)\, dB(s) - N_X(t). \tag{9.3}$$

Under the condition that the function ψ belongs to $C(\mathbb{R}_+)$, $\psi(T_n) - \psi(T_{n-1}) = (T_n - T_{n-1})\psi'(T_{n-1} + \theta(T_n - T_{n-1}))$, with θ in $]0,1[$. For every $k \geq 1$, the variable $Z_k = \int_{T_{k-1}}^{T_k} \sigma(s)\,dB(s)$ has a Gaussian distribution with variance $v(T_{k-1}, T_k) = \int_{T_{k-1}}^{T_k} \sigma^2(s)\,ds$ conditionally on (T_{k-1}, T_k). The probabilities P_n are therefore convolutions

$$P\Big(X_1 < \psi(T_1) + \int_0^{T_1} \sigma\,dB\Big) = \int_\mathbb{R} \int_0^\infty F(\psi(t) + z)f(t)f_{\mathcal{N}_{v(t)}}(z)\,dt\,dz.$$

The probability that the time of ruin belongs to the interval $]T_1, T_2]$ is

$$P_1 = P\Big(X_1 < \psi(T_1) + \int_0^{T_1} \sigma\,dB, X_2 > \psi(T_2) - \psi(T_1) + \int_{T_1}^{T_2} \sigma\,dB\Big)$$

$$= \int_\mathbb{R} \int_0^\infty \Big\{1 - \int_\mathbb{R} \int_0^\infty F_X(s + \psi'(t + \theta s + z)f(s)f_{\mathcal{N}_{v(t,s+t)}}(z)\,ds\,dz\Big\}$$

$$F_X(\psi(t + z_1)f(t)f_{\mathcal{N}_{v(0,t)}}(z_1)\,dt\,dz_1.$$

The probability that the time of ruin belongs to the interval $]T_n, T_{n+1}]$ is calculated using the notation $s_k = t_k - t_{k-1}$ in the integrals

$$P_n = P\Big(X_1 < \psi(T_1) + \int_0^{T_1} \sigma\,dB, \ldots, X_n < \psi(T_n) - \psi(T_{n-1})$$

$$+ \int_{T_{n-1}}^{T_n} \sigma\,dB, X_{n+1} > \psi(T_{n+1}) - \psi(T_n) + \int_{T_n}^{T_{n+1}} \sigma\,dB\Big)$$

$$= \int_{\mathbb{R}^n} \int_{\mathbb{R}^n_+} \prod_{k=1}^n F_X(cs_k + z_k)f(s_k)f_{\mathcal{N}_{v(t_{k-1}, t_k)}}(z_k)\,ds_k\,dz_k$$

$$\Big\{1 - \int_\mathbb{R} \int_0^\infty F_X(s + \psi'(t + \theta s + z))f(s)f_{\mathcal{N}_{v(0,s)}}(z)\,ds\,dz\Big\}$$

$$ds_1 \ldots ds_n\,dz_1\,dz_n.$$

The process Y_{T_a} is negative and it stops at this value. Adding a stochastic noise to the Sparre Andersen model is therefore necessary to avoid a large probability of an early end of the process as described by Propositions 9.2 and 9.3.

The independence of the variables $Y(T_n) - Y(T_{n-1})$ can also be replaced by an assumption of a positive martingale sequence $(X_k)_{k \geq 0}$ with respect to the filtration $(\mathcal{F}_n)_n$ generated by the process N_X, or by a submartingale (respectively supermartingale) sequence, without an explicit expression for the model also denoted $\psi(t)$. Adding a diffusion noise to the Sparre Andersen model, the process $Y = \psi + \int_0^\cdot \sigma\,dB - N_X$ is a martingale under this assumption.

With a renewal process $(T_k)_{k\geq 1}$ and an independent martingale $(X_k)_{k\geq 1}$, the process N_X has the conditional expectations

$$E\{N_X(t)|\mathcal{F}_n\} = \sum_{k=1}^n X_k 1_{\{T_k\leq t\}} + \sum_{k>n} E\left(X_k 1_{\{\sum_{j=n+1}^k S_j\leq t-T_n\}}|\mathcal{F}_n\right)$$

$$= \sum_{k=1}^n X_k 1_{\{T_k\leq t\}} + \sum_{k>n} X_n \int_0^{t-T_n} f^{*(k-n)}(s)\,ds.$$

If the martingale has the property that X_n is $\mathcal{F}_{n-1}$-measurable, N_X has the predictable compensator

$$\widetilde{N}_X(t) = \int_0^t \left(\sum_{k\geq 1} X_k 1_{\{T_k\geq s\}}\right) d\Lambda(s).$$

The expression of expectation of Y is not modified by the assumption of the dependence between the variables X_k, it is also equal to

$$EY(t) = \psi(t) - EX_0 \sum_{k\geq 1} \int_0^\infty \int_0^t \{1 - F(s-t)\} f^{*(k-1)}(t)\,d\Lambda(s)\,dt$$

if X_n is $\mathcal{F}_{n-1}$-measurable for every n. Lemmas 9.1 and 9.2 are still valid.

9.4   Comparison of models

The weighted processes Y and Z defined (4.5) on a filtered probability space $(\Omega, \mathcal{F}, (\mathcal{F}_n)_{n\geq 0}, P)$, are generalized to an adapted sequence $U = (U_n)_{n\geq 0}$, with adapted weighting variables $(A_n)_n$. Let

$$Y_n = \sum_{k=1}^n A_k(U_k - U_{k-1}), \tag{9.4}$$

$$Z_n = \sum_{k=1}^n A_k^{-1}(U_k - U_{k-1}).$$

By the Cauchy inequality

$$E|Y_n| \leq E\left\{\sum_{k=1}^n E(A_k^2|\mathcal{F}_{k-1})V_n(U)\right\}^{\frac{1}{2}} \tag{9.5}$$

$$\leq \left[E\left\{\sum_{k=1}^n E(A_k^2|\mathcal{F}_{k-1})\right\}\right]^{\frac{1}{2}} \{EV_n(U)\}^{\frac{1}{2}}$$

and Z_n satisfies the same kind of inequality. The bounds are an increasing series and the inequalities hold for non increasing processes Y and Z.

Let U be a local martingale of $L^2(P)$ and A be a predictable process of $L^2(P)$. The sequences $(Y_n)_{n\geq 0}$ and $(Z_n)_{n\geq 0}$ are local martingales of $L^2(P)$ and their expectations are zero. Their L^1-expectations have smaller increasing bounds defined as $u_n = \{E\sum_{k=1}^{n} A_k^2(V_k - V_{k-1})(U)\}^{\frac{1}{2}}$ for $E|Y_n|$ and $v_n = \{E\sum_{k=1}^{n} A_k^{-2}(V_k - V_{k-1})(U)\}^{\frac{1}{2}}$ for $E|Z_n|$. The Kolmogorov inequality (4.7) applies with the random weights A_k and A_k^{-1}. The same inequalities are true with the bounds (9.5) instead of the u_n and v_n, for general adapted processes U and A of $L^2(P)$.

Let us consider ruin models where the random series (9.4) replace the deterministic trend ψ. They are more realistic for applications in other fields where the process N_X is sequentially corrected by inputs. The variables U_k may be defined as the random times of the process N_X or other processes depending on them, they are not necessarily local martingales and their expectations are supposed to be different from zero. Their weights may be predictable or not. The model becomes

$$Y(t) = a + c\sum_{n\geq 1} A_n(U_n - U_{n-1})1_{\{T_n \leq t\}} - N_X(t) \qquad (9.6)$$

and the time of ruin is a.s. infinite if for every $n \geq 1$

$$0 \leq X_n \leq \frac{a}{nc} + A_n(U_n - U_{n-1}), \text{ a.s.}$$

The expectation of the process is everywhere positive if $EN_X(t)$ is sufficiently large with respect to the bound of the inequality (9.5).

Lemma 9.3. *A sufficient condition for a finite expectation time of ruin t_0 is*

$$a + cE\Big(\sum_{k=1}^{n} A_k^2 1_{\{T_k \leq t_0\}}\Big)^{\frac{1}{2}} \{EV_n(U)\}^{\frac{1}{2}} \leq EN_X(t_0).$$

It comes from the Cauchy inequality for the sum $E(\sum_{k=1}^{n} A_k(U_k - U_{k-1})1_{\{T_k \leq t_0\}}) \leq \{E(\sum_{k=1}^{n} A_k^2 1_{\{T_k \leq t_0\}})EV_n(U)\}^{\frac{1}{2}}$ and the inequality $EY(t_0) < 0$ holds if this bound is lower than $c^{-1}\{EN_X(t) - a\}$. The probabilities P_n of ruin in an interval $]T_n, T_{n+1}]$ are written as

$$P(T_a \in]0, T_1]) = P(Y(T_1) \leq 0) = P(X_1 \geq a + cA_1U_1),$$

$$P(T_a \in]T_1, T_2]) = P(X_1 < a + cA_1U_1, X_2 > cA_2(U_2 - U_1),$$

$$P(T_a \in]T_1, T_2]) = P(X_1 < a + cA_1U_1, X_2 > cA_2(U_2 - U_1),$$

$$P(T_a \in]T_n, T_{n+1}]) = P(\{X_1 < a + cA_1U_1\} \cap [\cap_{k=1,\ldots,n}\{X_1 + \cdots + X_k < a$$
$$+ c(A_1U_1 + \cdots + A_n(U_n - U_{n-1}))\}]$$
$$\cap \{X_{n+1} \geq cA_{n+1}(U_{n+1} - U_n)\}).$$

Under the assumption of sequences of independent variables $(X_n)_n$, $(A_n)_n$ and $(U_{n+1} - U_n))_n$, the probabilities are factorized and they are calculated recursively as in Proposition 9.2, by the formula

$$P_n = \widetilde{P}_{n-1} - \widetilde{P}_n = \widetilde{P}_{n-1}\{1 - P(X_{n+1} \geq cA_{n+1}(U_{n+1} - U_n))\}.$$

Under the condition of a nonlinear trend function ψ, the assumption of independent sequences does not hold and the probability P_n is a convolution of $n + 1$ marginal probabilities. The discrete model (9.6) does not include the stochastic diffusion but it can be extended to this model by adding the integral $A_{n+1} \int_{T_n}^t \sigma \, dB$ to the n-th term of the sum in (9.6) where $U_n = \int_0^{T_n} \sigma \, dB$.

The expression of the diffusion in (9.3) does not vary with the past of the process and it can be generalized in the form

$$W(t) = \sum_{n \geq 1} \int_{T_{n-1}}^{t \wedge T_n} \sigma(A_n(s)) \, dB(s)$$

where the integral is the stochastic integral for a predictable sequence $(A_n)_n$. For every sequence of adapted stopping times

$$E\{W(t)|\mathcal{F}_n\} = \sum_{k=1,\ldots,n} \int_{T_{n-1}}^{t \wedge T_n} \sigma(A_n(s)) \, dB(s),$$

it is therefore a local martingale with expectation zero. Under the assumption of a sequence $(A_n)_n$ independent of the Brownian motion, its variance is

$$\mathrm{Var} W(t) = \int_0^t E\Big\{ \sum_{n \geq 1} 1_{[T_{n-1}, T_n]}(t)\sigma^2(A_n(s)) \Big\} \, ds.$$

9.5   Moments of the processes at T_a

From Wald's equations (A.1) for a sequence of independent and identically distributed variables $(X_n)_{n \leq 0}$, such that $EX_i = 0$ and $\mathrm{Var} X_i = \sigma^2$, and a stopping time T with a finite expectation, $ES_T^2 = \sigma^2 T$.

Lemma 9.4. *Let φ be a real function defined on $\mathbb{N}$. For every stopping time T on a filtered probability space $(\Omega, \mathcal{F}, (\mathcal{F}_n)_{n \geq 0}, P)$*

$$\sum_{k \geq 1} \varphi(k) P(T \geq k) = E \sum_{k=1}^T \varphi(k).$$

Proposition 9.4. *On a probability space* $(\Omega, \mathcal{F}, P)$, *let* $(\mathcal{F}_n)_{n \geq 0}$ *be a filtration and let* N_X *be the process defined by* (9.1) *with an* $(\mathcal{F}_n)_{n \geq 0}$-*adapted sequence of independent variables* $(X_n)_{n \geq 0}$ *and an independent renewal process. For every integer-valued stopping time* T *with respect to* $(\mathcal{F}_n)_{n \geq 0}$, $EN_X(T) = EX \, EN(T)$ *and*

$$\mathrm{Var} N_X(T) = E(X^2) \, ET - \{EN_X(T)\}^2].$$

Proof. Since the process Y is only decreasing at the jump times of N_X, T is a stopping time with integer values. For every $n \geq 1$

$$
\begin{aligned}
EN_X(T \wedge n) &= EX \, [EN(T \wedge (n-1)) \\
&\quad + E\{N(n) - N(T \wedge (n-1))\}P(T \geq n)] \\
&= EX \sum_{k=1}^{n} E\{N(k) - N(k-1)\}P(T \geq k)
\end{aligned}
$$

since X_n is independent of $\{T \geq n\}$ and $N_X(T \wedge (n-1))$. Moreover, the expectation of T is $ET = \sum_{k \geq 1} P(T \geq k)$ and by Lemma 9.4, for every process φ on $\mathbb{N}$

$$E \sum_{k \geq 1} \varphi(k) P(T \geq k) = E \sum_{k=1}^{T} \varphi(k),$$

which implies

$$
\begin{aligned}
EN_X(T) &= EX \sum_{k \geq 1} E\{N(k) - N(T \wedge (k-1))\}P(T \geq k) \\
&= EX \, E \sum_{k=1}^{T} \{N(k) - N(T \wedge (k-1))\} = EX \, EN(T).
\end{aligned}
$$

The variance develops as the squared sum

$$
\begin{aligned}
EN_X^2(T \wedge n) &= E[N_X(T \wedge (n-1)) \\
&\quad + \{N_X(n) - N_X(T \wedge (n-1))\}1_{\{T \geq n\}}]^2 \\
&= EN_X^2(T \wedge (n-1)) \\
&\quad + E\{N_X(n) - N_X(T \wedge (n-1))\}^2 P(T \geq n) \\
&= \sum_{k \geq 1} E[\{N_X(n) - N_X(T \wedge (n-1))\}^2 1_{\{T \geq n\}}].
\end{aligned}
$$

Using again Lemma 9.4, we have

$$\sum_{k\geq 1} E[\{N_X(k) - N_X(T \wedge (k-1))\}^2 1_{\{T\geq k\}}]$$

$$= E \sum_{k=1}^{T} \{N_X(k) - N_X(T \wedge (k-1))\}^2$$

$$= E(X^2) E \sum_{k=1}^{T} \{N(k) - N(k-1)\}^2.$$

$\square$

Corollary 9.1. *In the Sparre Andersen model*

$$\frac{ET_a}{EN(T_a)} \geq \frac{a}{cEN(T_a)} + \frac{EX}{c} \geq \frac{EX}{c}.$$

Other consequences of Proposition 9.4 are deduced from the Bienaymé-Chebychev inequality. For every $y < a$

$$P(Y(T_a) < y) = P(N_X(T_a) - cT_a > a - y)$$

$$\leq \frac{EN_X^2(T_a) - 2T_a N_X(T_a) + c^2 T_a^2}{(a-y)^2},$$

where $EN_X^2(T_a)$ is given in Proposition 9.4, $N(0) = 0$ and, by Lemma 9.4, $E\{T_a N_X(T_a)\} = EX \, E \sum_{k=0}^{T_a - 1} \{N(T_a) - N(k)\}$. Since $N_X(T_a) \geq a + cT_a$, for every $t > 0$

$$P(T_a > t) \leq P(N_X(T_a) > a + ct) \leq \frac{EX \, EN(T_a)}{a + ct}.$$

The moments of higher order are calculated in the same way, by an expansion of

$$EN_X^p(T_a \wedge n) = E[N_X(T_a \wedge (n-1))$$

$$+ \{N_X(n) - N_X(T_a \wedge (n-1))\} 1_{\{T_a \geq n\}}]^p.$$

The centered moments of the process $Y(T_a)$ in the Sparre Andersen model are equal to those of $N_X(T_a)$ and to those of $Y(T_a)$ in the model (9.3) with a stochastic diffusion. In model (9.6), $EY(T_a) = a + c \sum_{n\geq 1} \{A_n(U_n - U_{n-1}) - X_n\} 1_{\{T_n \leq T_a\}}$ has the same form and its centered moments are similar if the sequence of variables $(A_n, U_n - U_{n-1}, X_n)_{n\geq 1}$ are independent and identically distributed, writing $E\{A_n(U_n - U_{n-1} - X_n\}$ instead of EX in the expression of EN_X.

If the assumption of independent and identically distributed variables is replaced by an assumption of martingales, conditional expectations are added to the above expressions.

9.6 Empirical process in mixture distributions

Section 3.4 presents the general form of continuous mixture of densities with some parameters lying in an interval I, the density of the observed variable is the convolution $f_X(x) = \int_I f_{X|W}(x - w) f_W(w)\, dw = f_{X|W} * f_W(x)$ and the Fourier transform of f_W is the ratio of the transforms for f_X and $f_{X|W}$.

In a generalized exponential model, the conditional density $f_{X|Y}$ of the variable X, given a dependent real variable Y, is expressed as

$$f_{X|Y}(x;Y) = \exp\{T(x,Y) - b(Y)\},$$

where $b(Y) = \log \int \exp\{T(x,Y)\}\, dx$ is the normalization function of the conditional density and the $T(X,Y)$ only depends on the variables. The model is extended to a semi-parametric conditional density

$$f_{X|Y,\eta}(x;Y) = \exp\{\eta^T T(x,Y) - b(\eta,Y)\}$$

with $b(\eta,Y) = \log \int \exp\{\eta^T T(x,Y)\}\, dx$. When Y is a hidden real variable with distribution function F_Y, the distribution of the variable X has the density $f_\eta(x) = E_Y f_{X|Y,\eta}(x;Y)$ where E_Y is the unknown expectation with respect to F_Y.

Let $E_{X|Y}$ be the conditional expectation with respect to the probability distribution of X conditionally on Y. When the distribution function F_Y is known, the distribution function of X is parametric with parameter η. The derivative with respect to η of $\log f_\eta$ is $f_\eta^{-1} \dot{f}_\eta$ and

$$\dot{f}_\eta(x) = E_Y[\{T(x,Y) - E_{X|Y}T(X,Y)\}f_{X|Y,\eta}(x;Y)] \tag{9.7}$$

with $\dot{b}_\eta(\eta,Y) = E_{X|Y}T(X,Y)$. For the sample $(X_i)_{1 \leq i \leq n}$ of the variable X, the maximum likelihood estimator of η is a solution of the score equation $\dot{l}_n(\eta) = \sum_{i=1}^n f^{-1}(X_i;\eta)\dot{f}_\eta(X_i) = 0$ and its asymptotic behaviour follows the classical theory of the parametric maximum likelihood.

With a nonparametric mixing distribution, the distribution function F_Y is approximated by a countable function

$$F_{Yn}(y) = \sum_{k=1}^{K_n} p_{nk} 1_{\{y_{nk} \leq y\}},$$

where the probabilities

$$p_{nk} = f_{Yn}(y_{nk}) \tag{9.8}$$

sum up to 1, and the distribution function F_X is written as

$$F_X(x) = \sum_{k=1}^{K_n} p_{nk} F_{X|Y,\eta}(x;y_{nk}),$$

with a parametric distribution of X conditionally on Y, $F_{X|Y,\eta}$. The empirical distribution function of the observed variable X is denoted $\widehat{F}_{Xn}$. Let $f_{X|Y,\eta}(x;y)$ be the density of $F_{X|Y,\eta}$, it is supposed to be twice continuously differentiable with respect to the parameter η and with respect to y, with first derivative with respect to η, $\dot{f}_{X|Y,\eta}$, satisfying (9.7), and $f^{(1)}_{X|Y,\eta}(x;y)$, with respect to y.

Proposition 9.5. *Under the constraint $\sum_{k=1}^{K_n} p_{nk} = 1$, the maximum likelihood estimators of the probabilities p_{nk} are*

$$\widehat{p}_{nK_n} = \frac{\int f_{Y|X,\widehat{\eta}_n}(\widehat{y}_{nK_n};x)\, d\widehat{F}_n(x)}{\sum_{k=1}^{K_n} \int \widehat{f}_{X,\widehat{\eta}_n}(\widehat{y}_{nk};x)\, d\widehat{F}_n(x)},$$

$$\widehat{p}_{nk} = \frac{\int f_{Y|X,\widehat{\eta}_n}(\widehat{y}_{nk};x)\, d\widehat{F}_n(x)}{\sum_{k=1}^{K_n} \int f_{Y|X,\widehat{\eta}_n}(\widehat{y}_{nk};x)\, d\widehat{F}_n(x)}, \quad k = 1, \ldots, K_n - 1,$$

where $\widehat{\eta}_n$ and $(\widehat{y}_{nk})_{k \le K_n}$ are the maximum likelihood estimators of the parameters η and $(y_{nk})_{k \le K_n}$.

Proof. The values of $(p_{nk})_{k \le K_n}$ that maximize

$$L_n = \sum_{i=1}^{n} \log \sum_{k=1}^{K_n} p_{nk} f_{X|Y,\eta}(X_i; y_{nk})$$

under the constraint $\sum_{k=1}^{K_n} p_{nk} = 1$ are solutions of the equations

$$0 = \sum_{i=1}^{n} \frac{\sum_{k=1}^{K_n} p_{nk} \dot{f}_{X|Y,\eta}(X_i; y_{K_n,n})}{f_{X,\eta}(X_i)},$$

$$0 = \sum_{i=1}^{n} \frac{\sum_{k=1}^{K_n} p_{nk} f^{(1)}_{X|Y,\eta,k}(X_i; y_{nk})}{f_{X,\eta}(X_i)},$$

$$\sum_{i=1}^{n} \frac{f_{X|Y,\eta}(X_i; y_{K_n})}{f_{X,\eta}(X_i)} = \sum_{i=1}^{n} \frac{f_{X|Y,\eta}(X_i; y_{nk})}{f_{X,\eta}(X_i)}, \quad k = 1, \ldots, K_n - 1,$$

where the last equality is due to the derivatives with respect to p_{nk} under the constraint $\sum_{k=1}^{K_n} p_{nk} = 1$, $f_{X,\eta}(X_i) = \sum_{k=1}^{K_n} p_{nk} f_{X|Y,\eta}(X_i; y_{nk})$. By the definition (9.8) of the probabilities p_{nk} and

$$\frac{f_{X|Y,\eta}(X_i; y_{nk})}{f_{X,\eta}(X_i)} = \frac{f_{Y|X,\eta}(y_{nk}, X_i)}{f_{Y,\eta}(y_{nk})},$$

the last equations are therefore equivalent to

$$\frac{\sum_{i=1}^{n} f_{Y|X,\eta}(X_i; y_{nK_n})}{p_{nK_n}} = \frac{\sum_{i=1}^{n} f_{Y|X,\eta}(X_i; y_{nk})}{p_{nk}}, \quad k = 1, \ldots, K_n - 1,$$

using the constraint yields

$$p_{nK_n} = \frac{\sum_{i=1}^{n} f_{Y|X,\eta}(X_i; y_{nK_n})}{\sum_{k=1}^{K_n} \sum_{i=1}^{n} f_{Y|X,\eta}(X_i; y_{nk})}$$

and the expression of p_{nk} is similar. □

The maximum likelihood estimators of the parameters η and $(y_{nk})_{k \leq K_n}$ are the values of η and $(y_{nk})_{k \leq K_n}$ that maximize $L_n(\widehat{p}_n, y_n, \eta)$, they are solutions of the equations

$$0 = \sum_{i=1}^{n} \frac{\sum_{k=1}^{K_n} \widehat{p}_{nk} \dot{f}_{X|Y,\eta}(X_i; y_{K_n,n})}{f_{X,\eta}(X_i)},$$

$$0 = \sum_{i=1}^{n} \frac{\widehat{p}_{nk} f^{(1)}_{X|Y,\eta,k}(X_i; y_{nk})}{f_{X,\eta}(X_i)}, \ k = 1, \ldots, K_n.$$

By (9.7), the first equation is written

$$\int f_{X|\eta}^{-1}(x) \sum_{k=1}^{K_n} \widehat{p}_{nk} \{T(x, y_{nk}) - E_{X|Y} T(X, y_{nk})\} f_{X|Y,\eta}(x, y_{nk}) \, d\widehat{F}_n(x) = 0$$

and the second equation is equivalent to

$$\int f_{X|\eta}^{-1}(x) \sum_{k=1}^{K_n} \widehat{p}_{nk} \{T_y^{(1)}(x, y_{nk}) - E_{X|Y} T_y^{(1)}(X, y_{nk})\} f_{X|Y,\eta}(x, y_{nk}) \, dx = 0.$$

If η is a vector of dimension d, $K_n + p$ parameters are estimated by these equations and they converge to the true parameter values as n tends to infinity with $K_n = o(n)$, by the classical theory of the maximum likelihood estimation. In an exponential mixture model, the estimator of the density is written as

$$f_n(x) = \sum_{k=1}^{K_n} \theta_{kn} e^{-\theta_{kn} x} p_{kn},$$

where the parameters are solutions of the maximum likelihood equations depending on the empirical distribution $\widehat{F}_n$ of $(X_1, \ldots, X_n)$

$$\frac{1}{\theta_{kn}} = \int_0^\infty x \frac{f_{X|\widehat{\theta}_{kn}}}{\widehat{f}_n(x)} \, d\widehat{F}_n(x),$$

$$\widehat{p}_{kn} = \int_0^\infty \frac{f_{\widehat{\theta}_{kn}|X}}{\sum_{j=1}^{K_n} f_{\widehat{\theta}_{jn}|X}} \, d\widehat{F}_n,$$

they are computed with an iterative algorithm.

9.7   Integral inequalities in the plane

Pitt (1977) proved that for all symmetric convex subsets A and B of $\mathbb{R}^2$, the density of a Gaussian variable X in $\mathbb{R}^2$ satisfies

$$\int_{A \cap B} f(x)\,dx \geq \int_A f(x)\,dx \int_B f(x)\,dx,$$

which is equivalent to

$$P(X \in A \cap B) \geq P(X \in A)P(X \in B),$$

where the intersection of A and B is not empty. Removing the conditions and applying the Hölder inequality to indicator functions 1_A and 1_B implies that for every density function

$$\int_{A \cap B} f(x)\,dx \leq \left\{ \int_A f(x)\,dx \int_B f(x)\,dx \right\}^{\frac{1}{2}}$$

or $P(X \in A \cap B) \leq \{P(X \in A)P(X \in B)\}^{\frac{1}{2}}$.

These inequalities entail that for every positive function h of $L^1(\mu_{\mathcal{N}})$, where $\mu_{\mathcal{N}}$ is a Gaussian distribution, the inequalities apply to the integral of h with respect to the measure $\mu_{\mathcal{N}}$ restricted to symmetric subsets of $\mathbb{R}^2$

$$\int_{A \cap B} h(x)\,d\mu_{\mathcal{N}}(x) \geq \int_A h(x)\,d\mu_{\mathcal{N}}(x) \int_B h(x)\,d\mu_{\mathcal{N}}(x),$$

$$\int_{A \cap B} h(x)\,d\mu_{\mathcal{N}}(x) \leq \left\{ \int_A h(x)\,d\mu_{\mathcal{N}}(x) \int_B h(x)\,d\mu_{\mathcal{N}}(x) \right\}^{\frac{1}{2}}.$$

Moreover, for every positive probability measure μ and every subsets of $\mathbb{R}^2$, positive functions of $L^1(\mu_{\mathcal{N}})$ satisfy

$$\int_{A \cup B} h(x)\,d\mu(x) + \int_{A \cap B} h(x)\,d\mu(x) \geq \int_A h(x)\,d\mu(x) + \int_B h(x)\,d\mu(x),$$

$$\int_{A \cup B} h(x)\,d\mu(x) \leq \int_A h(x)\,d\mu(x) + \int_B h(x)\,d\mu(x),$$

with equality if and only if $A \cap B$ is empty.

The L^p-distance of non intersecting subsets A and B of $\mathbb{R}^n$ is

$$d_p(A, B) = \inf_{x \in A} \inf_{y \in B} \left\{ \sum_{i=}^n |x_i - y_i|^p \right\}^{\frac{1}{p}}, \quad p \geq 1.$$

For independent random subsets A and B of a probability space $(\Omega, \mathcal{A}, P)$, there exist independent variables X and Y defined in $(\Omega, \mathcal{A}, P)$ and with values in $\mathbb{R}^n$ such that

$$P(d_p(A, B) > t) \leq t^{-1} E \inf_{X \in A, Y \in B} \|X - Y\|_p.$$

With a centered Gaussian probability, the bound is

$$t^{-1}(2\pi)^{-\frac{n}{2}}\{\det(\Sigma_1)\det(\Sigma_2)\}^{-\frac{1}{2}} \int_{A\times B} \|x-y\|_p e^{-\frac{1}{2}x^t\Sigma_1^{-1}x} e^{-\frac{1}{2}y^t\Sigma_2^{-1}y}\, dx\, dy.$$

Under the constraint that the distance between sets A and B_α is at least equal to α

$$E \inf_{X\in A, Y\in B_\alpha} \|X-Y\|_p = \alpha P(A\times B_\alpha)$$

and $P(d_p(A,B_\alpha) > t) \leq t^{-1}\alpha P(A\times B_\alpha)$.

Let A and B be subsets of $\mathbb{R}^n$ and let X in A. Applying this inequality conditionally on X gives $P(d_p(X,B) > t|X = x) \leq t^{-1}E\inf_{Y\in B}\|x-Y\|_p$. Under a Gaussian probability P on $\mathbb{R}$, the paths of the process $d(x,B)$ stay a.s. inside frontiers $(2\sigma^2(x)h_x)^{\frac{1}{2}}$ determined by Proposition 6.14, with the variance $Ed^2(x,B) = \int_{\mathbb{R}}(x-s)^2 dP_Y(s) = \sigma^2(x)$ for some Y that achieves the minimum over B and with a function h such that h^{-1} belongs to $L^1(P)$.

9.8 Spatial point processes

A Poisson process N indexed by $\mathbb{R}_+$, with intensity λ, has a covariance function

$$R(s,t) = \mathrm{Cov}(N_s, N_t) = \mathrm{Var}N_{s\wedge t} = \lambda(t\wedge s),$$

the higher moments of the Poisson process are denoted $\mu_k(t) = E\{(N_t - \lambda t)^k\}$, such that $\mu_k = \lambda t$, for $k = 1, 2, 3$.

Lemma 9.5. *Let $t_1, t_2, \ldots, t_k$ be positive real numbers, the crossed-moments of the Poisson process in $\mathbb{R}_+$ are*

$$E\{(N_{t_1} - \lambda t_1)\cdots(N_{t_k} - \lambda t_k)\} = \mu_k(t_{m_k}),$$

where $t_{m_k} = \min(t_1, \ldots, t_k)$ and $k \geq 2$.

Proof. Let $0 < t_1 < t_2 < t_3$, by intersection the moments are written as

$$\begin{aligned}
\nu_3(t_1, t_2, t_3) &= E\{(N_{t_1} - \lambda t_1)(N_{t_2} - \lambda t_2)(N_{t_3} - \lambda t_3)\} \\
&= E\{(N_{t_1} - \lambda t_1)(N_{t_2} - \lambda t_2)^2\} \\
&= \lambda t_1 + E[(N_{t_1} - \lambda t_1)\{(N_{t_2} - \lambda t_2) - (N_{t_1} - \lambda t_1)\} \\
&\qquad\qquad\qquad\quad \{(N_{t_2} - \lambda t_2) + (N_{t_1} - \lambda t_1)\} \\
&= \lambda t_1 + E[(N_{t_1} - \lambda t_1)\{(N_{t_2} - \lambda t_2) - (N_{t_1} - \lambda t_1)\}^2] = \lambda t_1.
\end{aligned}$$

The result for any k is deduced from the independence of the increments of the Poisson process. $\qquad\qquad\square$

Let us consider two Poisson processes with intensities λ_1 and $\lambda_2 = \lambda_1 + x$, $x > 0$, then $P_{\lambda_1}(N_t = k)P_{\lambda_2}^{-1}(N_t = k) = e^{-xt}(1 + \lambda_2^{-1}xt)^k$, it is increasing with respect to x if $k > [xt + \lambda_2]$ and decreasing otherwise, therefore the distributions of Poisson processes cannot be ordered.

A heterogeneous Poisson process with a cumulative intensity Λ has the moments $E[\{N_t - \Lambda(t)\}^k] = \Lambda(t)$, for every integer $k \geq 2$ and

$$E[\{N_{t_1} - \Lambda(t_1)\} \cdots \{N_{t_k} - \Lambda(t_k)\}] = \Lambda(t_{m_k}).$$

Some characteristics of the first two moments of point processes in the plane are introduced in Section 5.9 and they are based on martingales properties. Spatial point processes are characterized by their moments on the balls. In $\mathbb{R}^d$, let $r > 0$ and $B_r(x)$ be the ball of radius r centered at x, its volume is $|B_r(x)| = c_d r^d$ for every center x. The k-th moment of N, $k \geq 2$, is defined by the values of the process in k balls of radius r. For every $x = (x_1, \ldots, x_k)$ in $\mathbb{R}^{kd}$, let k balls $B_r(x_j)$ with a non empty intersection $\cap_{j=1,\ldots,k} B_r(x_j)$,

$$\nu_{k,r}(x) = \frac{1}{(c_d r^d)^{\frac{k}{2}}} E\{N(B_r(x_1)) \cdots N(B_r(x_k))\}.$$

For a spatial stationary process N in $\mathbb{R}^d$, it is invariant by translation and defined for $(k-1)$ location parameters

$$\nu_{k,r}(x) = \frac{1}{(c_d r^d)^{\frac{k}{2}}} E\{N(B_r(x_1 - x_k)) \ldots N(B_r(0))\} = \nu_{k,r}(x_1 - x_k, \ldots, 0).$$

The second moments of a process with independent increments are

$$E\{N(B_r(x_1))N(B_r(x_2))\} = EN(B_r(x_1)) \, EN(B_r(x_2))$$
$$+ \operatorname{Var}N(B_r(x_1) \cap B_r(x_2)).$$

For a stationary process in $\mathbb{R}^d$, this equality becomes

$$E\{N(B_r(0))N(B_r(x_2 - x_1))\} = EN(B_r(0)) \, EN(B_r(x_2 - x_1))$$
$$+ \operatorname{Var}N(B_r(0) \cap B_r(x_2 - x_1)). \quad (9.9)$$

A Poisson process N with a cumulative intensity Λ in $\mathbb{R}^d$ has integer values and its distribution on balls is the distribution of a Poisson variable having as parameter the measure with respect to Λ of the balls

$$P\{N(B_r(x)) = k\} = e^{-\Lambda(B_r(x))} \frac{\Lambda^k(B_r(x))}{k!}$$

and its moments are $EN^k(B_r(x)) = \Lambda(B_r(x))$, $k \geq 1$. Its crossed-moments in balls are given by Lemma 9.5, according to the number of balls intersecting with each ball. Let $\varepsilon > 0$ and consider an ε-net in $\mathbb{R}^d$, balls of radius r and centered at the points of the ε-net have intersections by pairs with their nearest neighbours. Balls having a radius r such that $\varepsilon \geq 2r$ are disjoint, then for every k

$$E[\{N_{B_r(x_1)} - \Lambda(B_r(x_1))\} \cdots \{N_{B_r(x_k)} - \Lambda(B_r(x_k))\}] = 0.$$

Let r be in the interval $I_1(\varepsilon) =]\frac{\varepsilon}{2}, \frac{\varepsilon}{\sqrt{2}}]$ and let $\mathcal{V}_\varepsilon(x)$ be the set of the centers of the nearest balls of $B_r(x_1)$, the values of the counting process on non-overlapping subsets of the balls are independent and they have the same values in the pairwise intersections of the balls, therefore

$$E[\{N_{B_r(x_1)} - \Lambda(B_r(x_1))\} \cdots \{N_{B_r(x_k)} - \Lambda(B_r(x_k))\}]$$
$$= \sum_{i=1}^{k} \sum_{x_j \in \mathcal{V}_\varepsilon(x_i)} \mathrm{Var} N(B_r(x_i) \cap B_r(x_j)).$$

If $\sqrt{2}r \geq \varepsilon$, the number $K_r(x)$ of balls intersecting a ball $B_r(x)$ increases and they are mutually intersecting. In $\mathbb{R}^2$, each ball with a radius r belonging to $I_1(\varepsilon)$ has 4 nearest neighbours (Fig. 9.1). There are $K_r(x) = 8$ balls centered in an ε-net around x, with radius r belonging to the interval $I_2(\varepsilon) =]\frac{\varepsilon}{\sqrt{2}}, \varepsilon]$, (Fig. 9.2). Under the same condition in $\mathbb{R}^3$, a ball $B_r(x)$ has 6 intersecting balls centered in the ε-net if r belongs to $I_1(\varepsilon)$ and $K_r(x) = 24$ intersecting balls centered in the ε-net if r belongs to $I_2(\varepsilon)$.

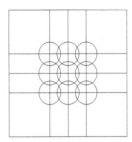

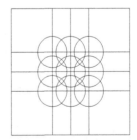

Fig. 9.1 Intersections of balls with radius $r = 3\varepsilon$, in the interval $I_1(\varepsilon)$, and centered in an ε-net.

Fig. 9.2 Intersections of balls with radius $r = .8\varepsilon$, in the interval $I_2(\varepsilon)$, and centered in an ε-net.

Proposition 9.6. *Let N be a heterogeneous Poisson process in $\mathbb{R}^d$, with cumulative intensity Λ and let r be in the interval $I_2(\varepsilon)$. For every $k \geq 2$*

$$
E[\{N_{B_r(x_1)} - \Lambda(B_r(x_1))\} \cdots \{N_{B_r(x_k)} - \Lambda(B_r(x_k))\}]
$$

$$
= \sum_{i=1}^{k} \sum_{j_i=1}^{K_r(x_i)} [\Lambda(B_r(x_i) \cap B_r(x_{j_i}))
$$

$$
- \sum_{k_i \neq j_i, k_i=1}^{K_r(x_i)} \{\Lambda(B_r(x_i) \cap B_r(x_{j_i}) \cap B_r(x_{k_i}))
$$

$$
+ \sum_{l_i \neq j_i, k_i, l_i=1}^{K_r(x_i)} \Lambda(B_r(x_i) \cap B_r(x_{j_i}) \cap B_r(x_{k_i}) \cap B_r(x_{l_i}))\}].
$$

Proof. Let $E_k = E[\{N_{B_r(x_1)} - \Lambda(B_r(x_1))\} \cdots \{N_{B_r(x_k)} - \Lambda(B_r(x_k))\}]$, it is expanded as

$$
E_k = \sum_{i=1}^{k} \sum_{j_i=1}^{K_r(x_i)} \{E[\{N_{B_r(x_i) \cap B_r(x_{j_i})} - \Lambda(B_r(x_i) \cap B_r(x_{j_i}))\}^2]
$$

$$
- \sum_{k_i \neq j_i, k_i=1}^{K_r(x_i)} (E[\{N_{B_r(x_i) \cap B_r(x_{j_i}) \cap B_r(x_{k_i})}
$$

$$
- \Lambda(B_r(x_i) \cap B_r(x_{j_i}) \cap B_r(x_{k_i}))\}^3]
$$

$$
+ \sum_{l_i \neq j_i, k_i, l_i=1}^{K_r(x_i)} E[\{N_{B_r(x_i) \cap B_r(x_{j_i}) \cap B_r(x_{k_i}) \cap B_r(x_{l_i})}
$$

$$
- \Lambda(B_r(x_i) \cap B_r(x_{j_i}) \cap B_r(x_{k_i}) \cap B_r(x_{l_i}))\}^4)\}.
$$

The result is deduced from the expression $EN^k(A) = \Lambda(A)$ for every Borel set of $\mathbb{R}^d$ and for every $k \geq 2$. $\square$

With r in the interval $I_3(\varepsilon) =]\varepsilon, \sqrt{2}\varepsilon]$, the moments calculated in Proposition 9.6 have additional terms including intersections of orders 5 to 8, as in Fig. 9.3.

With a larger radius, the number $K_r(x)$ increases and the k-order moment $E[\{N_{B_r(x_1)} - \Lambda(B_r(x_1))\} \cdots \{N_{B_r(x_k)} - \Lambda(B_r(x_k))\}]$ is a sum of moments of higher order, up to k as ε tends to zero. In the interval $I_4(\varepsilon) =]\sqrt{2}\varepsilon, 2\varepsilon]$, the eight nearest balls of $B_r(x)$ are intersecting by pair. The moments calculated in Proposition 9.6 must include intersections of order larger than 8 (Fig. 9.4).

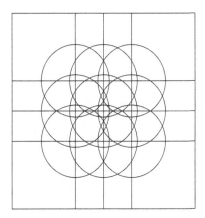

Fig. 9.3   Intersections of balls with radius $r = 1.2\,\varepsilon$, in the interval $I_3(\varepsilon) =]\varepsilon, \sqrt{2}\varepsilon]$, and centered in an ε-net.

A stationary process with independent increments in $\mathbb{R}^d$ has its finite dimensional distributions defined by the distribution of the vector $\{N_{B_r(x_1)} - \Lambda(B_r(x_1)), \ldots, N_{B_r(x_k)} - \Lambda(B_r(x_k))\}$ for every $(x_1, \ldots, x_k)$ and for every $r > 0$. It splits into the variations of $N - \Lambda$ on intersection of balls according to the number of intersecting balls in each subset. Under the assumption of stationarity, it is written as the sum of the values of $N - \Lambda$ in independent subsets having the same pattern, they are therefore identically distributed. It follows that every x in $\mathbb{R}^{kd}$, the variable

$$X_{k,r}(x) = (c_d r^d)^{-\frac{k}{2}} \{N_{B_r(x_1 - x_k)} \cdots N_{B_r(0)} - \nu_{k,r}(x)\}$$

converges weakly as r tends to infinity to a normal variable with variance

$$\sigma^2(x) = \lim_{r \to \infty} (c_d r^d)^{-k} E\{N^2_{B_r(x_1 - x_k)} \cdots N^2_{B_r(0)} - \nu^2_{k,r}(x)\}$$

and for all pairs of distinct points (x_i, x_j) in $\mathbb{R}^{2d}$, the covariance between $N_{B_r(x_i)}$ and $N_{B_r(x_j)}$ is defined by $\mathrm{Var} N_{B_r(0) \cap B_r(x_i - x_j)}$, from (9.9). The covariances differ only by the differences of the centers of the balls and $\mathrm{Cov}(N_{B_r(x_i)}, N_{B_r(x_j)}) = 0$ if $|x_i - x_j| > 2r$.

The variables $X_{k,r}(x)$, where $\varepsilon = |x_i - x_j| > 2r$, satisfy

$$P(\|X_{k,r}(x)\|_k > a) \leq \frac{1}{a^2 c_d r^d} \sum_{i=1}^{k} \mathrm{Var} N_{B_r(x_i - x_k)}$$

$$= \frac{1}{a^2 c_d r^d} \sum_{i=1}^{k} \Lambda(B_r(x_i - x_k)),$$

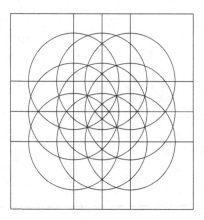

Fig. 9.4 Intersections of balls with radius $r = 1.6\,\varepsilon$, in the interval $I_4(\varepsilon) =]\sqrt{2}\varepsilon, 2\varepsilon]$, and centered in an ε-net.

for all $a > 0$ and $k \geq 2$. The bound is modified according to the domain of r with respect to ε. For all $a > 0$ and r in $\mathbb{R}_+$

$$P\Big\{ \sup_{x \in \mathbb{R}^d} (c_d r^d)^{-\frac{1}{2}} |N_{B_r(x)} - \Lambda(B_r(x))| > a \Big\}$$

$$= P\{ (c_d r^d)^{-\frac{1}{2}} |N_{B_r(0)} - \Lambda(B_r(0))| > a \} \leq \frac{\Lambda(B_r(0))}{a^2 c_d r^d}.$$

The function $\Lambda(B_r(0))$ is increasing with r and under the condition of the convergence of $r^{-d}\Lambda(B_r(0))$ to a limit λ

$$\lim_{r \to \infty} P\Big\{ \sup_{x \in \mathbb{R}^d} (c_d r^d)^{-\frac{1}{2}} |N_{B_r(x)} - \Lambda(B_r(x))| > a \Big\} \leq \frac{\lambda}{a^2 c_d}. \tag{9.10}$$

Proposition 9.7. *The spatial process* $(c_d r^d)^{-\frac{1}{2}} \{N_{B_r(x)} - \Lambda(B_r(x))\}_{x \in \mathbb{R}^d}$ *converges weakly to a centered Gaussian process with variance function* $(c_d r^d)^{-1}\Lambda(B_r(x))$.

Equation (9.10) with a sufficiently large proves the tightness of the process and the weak convergence of its finite dimensional distributions as r tends to infinity proves the result.

Schladitz and Baddeley (2000) gave the expression of second and third-order characteristics of stationary point processes, with explicit formulae for the Poisson with intensity λ and for other parametric processes. For $d = 2$ it is defined as the expected mean number of points of the process contained in $B_r(0)$ and not farther than 1

$$T_2(r) = (2\lambda^2 |B_r(0)|^2)^{-1} E \int_{B_r^2(0)} 1_{\{\|x-y\| \leq 1\}} \, dN(x) \, dN(y)$$

$$= (2\lambda^2 |B_r(0)|^2)^{-1} E \sum_{X_i \in B_r(0)} N_{B_r(0) \cap B_1(X_i)},$$

or with a correction to avoid multiple countings of the centers of the balls

$$\mu_2(r) = (2\lambda^2 |B_r(0)|^2)^{-1} E N_2(r),$$

$$N_2(r) = \sum_{X_i \in B_r(0)} N_{\{B_r(0) \cap B_1(X_i)\} \setminus \{X_i\}}.$$

For $d = 3$, the third-order mean number of points is

$$T_3(r) = \frac{1}{2\lambda^3 |B_r(0)|^3} E \int_{B_r^3(0)} 1_{\{\|x-y\| \leq 1\}} 1_{\{\|y-z\| \leq 1\}}$$

$$1_{\{\|x-z\| \leq 1\}} \, dN(x) \, dN(y) \, dN(z)$$

and it equals to

$$T_3(r) = \frac{1}{2\lambda^3 |B_r(0)|^3} E \int_{B_r^2(0)} N_{B_1(x) \cap B_1(y) \cap B_r(0)} \, dN_{B_1(x)}(y) \, dN(x)$$

$$= \frac{1}{2\lambda^3 |B_r(0)|^3} E \sum_{X_i \in B_r(0)} \sum_{X_j \in B_1(X_i) \cap B_r(0)} N_{B_1(X_i) \cap B_1(X_j) \cap B_r(0)}.$$

With a correction for the centers of the balls, it becomes

$$\mu_3(r) = \frac{1}{2\lambda^3 |B_r(0)|^3} E N_3(r),$$

$$N_3(r) = \sum_i [1_{B_r(0)}(X_i) \sum_{j \neq i} \{1_{\{B_1(X_i) \cap B_r(0)\}\}}(X_j)$$

$$\sum_{k \neq j, i} 1_{\{B_1(X_i) \cap B_1(X_j) \cap B_r(0)\}}(X_k)\}].$$

The functions $T_2(r)$ and $T_3(r)$ are mean convolutions of stochastic measures and they cannot be compared to the mean products $E \prod_{i=1}^{k} N_{B_r(x_i)}$ studied previously. Higher order measures are easily defined in the same way. The processes $N_2(r)$ and $N_3(r)$ cannot be split into sums of the counting processes in disjoint subsets and a condition of mixing, as r increases, is not relevant except for sparse processes.

The same measures can be calculated for a heterogeneous Poisson process with a continuous cumulative intensity measure Λ in $\mathbb{R}^d$. The conditional expectations of the spatial counting processes

$$A_r = \sum_i 1_{B_r(0)}(X_i),$$

$$C_r(X_i) = \sum_{j \neq i} 1_{\{B_1(X_i) \cap B_r(0)\}\}}(X_j),$$

$$D_r(X_i, X_j) = \sum_{k \neq j, i} 1_{\{B_1(X_i) \cap B_1(X_j) \cap B_r(0)\}}(X_k)$$

have the values

$$a_r = \sum_i P\{X_i \in B_r(0)\} = \Lambda(B_r(0)),$$

$$c_r(X_i) = \sum_{j \neq i} P\{X_j \in B_1(X_i) \cap B_r(0)|X_i\}$$
$$= \Lambda(B_1(X_i) \cap B_r(0)),$$

$$d_r(X_i, X_j) = \sum_{k \neq j, i} P\{X_k \in B_1(X_i) \cap B_1(X_j) \cap B_r(0)|X_i, X_j\}$$
$$= \Lambda(B_1(X_i) \cap B_1(X_j) \cap B_r(0)),$$

therefore

$$\mu_2(r) = \frac{1}{2\Lambda^2(B_r(0))} \int_{B_r(0)} \Lambda(B_r(0) \cap B_1(x)) \, d\Lambda(x),$$

$$\mu_3(r) = \frac{1}{2\Lambda^3(B_r(0))} \int_{B_r^2(0)} \Lambda(B_1(x) \cap B_1(y) \cap B_r(0)) \, d\Lambda(y) \, d\Lambda(x).$$

The normalized processes

$$W_{r,A} = \{2\lambda^2(B_r(0))\}^{-1}(A_r - a_r)\{B_r(0)\},$$
$$W_{r,C} = \{2\lambda^2(B_r(0))\}^{-1}(C_r - c_r)\{B_r(0)\},$$
$$W_{r,D} = \{2\lambda^2(B_r(0))\}^{-1}(C_r - c_r)\{B_r(0)\}$$

converge weakly to centered Gaussian processes having variances given by Proposition 9.7. This implies the weak convergence of the normalized and centered processes

$$\{\{2\lambda^2(B_r(0))\}^{-1} \int_{B_r(0)} \{C_r(x, dy) \, A_r(dx) - \mu_2(r)\},$$

$$\{\{2\lambda^3(B_r(0))\}^{-1} \int_{B_r^3(0)} \{D_r(x, y) \, C_r(x, dy) \, A_r(dx) - \mu_3(r)\}$$

to the Gaussian process $\int_{B_r(0)} W_C \, dA + \int_{B_r(0)} C \, dW_A$ and, respectively, $\int_{B_r^2(0)} W_D \, dC \, dA + \int_{B_r(0)} D \, dW_C \, dA + \int_{B_r(0)} D \, dC \, dW_A$.

9.9 Spatial Gaussian processes

Let μ and σ^2 be probability measures on a measurable space $(E, \mathcal{E})$ and let X be a spatial Gaussian process on $(E, \mathcal{E})$ such that for a measurable set A, $X(A)$ has the expectation $\mu(A)$ and the variance $\sigma^2(A)$. For k different measurable sets $(A_1, \ldots, A_k)$ of $E^{\times k}$, the Gaussian variable $X_A = (X(A_1), \ldots, X(A_k))$ has the expectation $\mu_A = (\mu(A_1), \ldots, \mu(A_k))$, the variance matrix Σ_A of X_A has the diagonal $\sigma^2(A_i))_{i=1,\ldots,k})$ and its covariances are

$$\mathrm{Cov}\{X(A_i), X(A_j)\} = \mathrm{Var}\{X(A_i \cap A_j)\} = \sigma^2(A_i \cap A_j),$$

for $i \neq j$ in $\{i = 1, \ldots, k\}$. With distinct sets $(A_1, \ldots, A_k)$, the covariances are zero and the components of X_A are independent. The density of $(X(A_1), \ldots, X(A_k))$ at $x = (x_1, \ldots, x_k)$ is

$$f(x) = \frac{1}{|\Sigma_A|^{\frac{1}{2}} (2\pi)^{\frac{k}{2}}} e^{-\frac{(x-\mu_A)^T \Sigma_A^{-1} (x-\mu_A)}{2}}$$

and for $t = (t_1, \ldots, t_k)$ its Fourier transform is

$$L(t) = E(e^{it^T X_A}) = e^{it^T \mu_A} e^{-\frac{t^T \Sigma_A t}{2}}.$$

For a spatial Brownian motion B on a metric space $(E, \mathcal{E}, d)$, a probability measure α defines the covariance of $B(A_1)$ and $B(A_2)$ as $\alpha(A_1 \cap A_2)$ and

$$E[\{B(A_1) - B(A_2)\}^2] = \alpha(A_1) - 2\alpha(A_1 \cap A_2) + \alpha(A_2).$$

The spatial process B has the properties of a Brownian motion.

Proposition 9.8. *For a spatial Brownian motion B with a probability measure α on $(E, \mathcal{E}, d)$, $B(A_1) - B(A_2)$ has the same distribution as $B(A_1 \setminus A_2)$.*

Proof. For a subset A_2 of A_1, the covariance of $B(A_1)$ and $B(A_2)$ reduces to

$$E[\{B(A_1) - B(A_2)\}^2] = \alpha(A_1) - \alpha(A_2) = \alpha(A_1 \setminus A_2)$$

therefore $E[\{B(A_1) - B(A_2)\}^2] = E\{B^2(A_1 \setminus A_2)\}$. $\qquad\square$

Proposition 9.9. *On $(E, \mathcal{E}, d)$ provided with the filtration generated by the spatial Brownian motion, the process $\{B^2(A) - \alpha(A)\}_{\alpha(A) \geq 0}$ is a martingale.*

Proof. The martingale property of the Brownian motion is expressed for a subset A_2 of A_1 as

$$E\{B(A_1) - B(A_2) \mid B(A_2)\} = E\{B(A_1 \setminus A_2) \mid B(A_2)\} = 0,$$

it implies

$$E[B^2(A_1) - B^2(A_2) \mid B(A_2)] = E[\{B(A_1) - B(A_2)\}^2$$
$$+ 2\{B(A_1) - B(A_2)\}B(A_2) \mid B(A_2)]$$
$$= E[\{B(A_1) - B(A_2)\}^2]$$

and $E[B^2(A_1) - B^2(A_2) \mid B(A_2)] = E[\{B(A_1 \setminus A_2)\}^2] = \alpha(A_1 \setminus A_2)$, and it is equal to $\alpha(A_1) - \alpha(A_2)$. □

Proposition 9.10. *For a spatial Brownian motion B with a probability measure α on $(E, \mathcal{E}, d)$, the process*

$$Y_\theta(A) = \exp\left\{\theta B(A) - \frac{1}{2}\theta^2\alpha(A)\right\}$$

is a martingale with expectation $EY_\theta(A) = 1$.

Proof. For all non empty sets $A_2 \subset A_1$, we have

$$E\left(\frac{Y_\theta(A_1)}{Y_\theta(A_2)} \mid B(A_2)\right) = E\left[\exp\left\{\theta B(A_1 \setminus A_2) - \frac{1}{2}\theta^2\alpha(A_1 \setminus A_2) \mid B(A_2)\right\}\right]$$

$$\geq \exp\left[E\left\{\theta B(A_1 \setminus A_2) - \frac{1}{2}\theta^2\alpha(A_1 \setminus A_2) \mid B(A_2)\right\}\right]$$

$$= 1$$

by convexity and by the martingale property. As the moment generating function of $B(A)$ is $\varphi_A(\theta) = \exp\{\frac{1}{2}\theta^2\alpha(A)\}$, the expectation of $Y_\theta(A)$ is one and the above inequality is an equality. □

For a spatial Brownian motion B with a probability measure α on $(E, \mathcal{E}, d)$, and for $x > 0$ the set $A_x = \inf\{A : B(A) \geq x\}$ is defined as the smallest of increasing measurable sets of $(E, \mathcal{E})$.

Proposition 9.11. *The set A_x is a stopping set with respect to the filtration generated by B on $(E, \mathcal{E})$ and the Laplace transform of $\alpha(A_x)$ is $E\{e^{x\sqrt{2\theta}}\}$.*

Proof. For a measurable set C of E, let $\bar{C}$ be the complementary of C. The event $\{A_x \subset \bar{C}\}$ is equivalent to $\{B(C) < x\}$ and it is measurable with respect to $\mathcal{F}_C$, A_x is therefore a stopping set. The variable B_C has a centered Gaussian distribution with variance $\alpha(C)$, its Laplace transform is

$$E\{e^{-\theta B_C}\} = e^{\frac{\theta^2\alpha(C)}{2}}.$$

At A_x, we have $B_{A_x} = x$ and

$$e^{-\theta\alpha(A_x)} = E\{e^{x\sqrt{2\theta}}\}.$$

□

The Gaussian distribution $\mathcal{N}(0, \alpha(A_x))$ of the set A_x implies it has an inverse Gaussian distribution with density

$$f_{A_x}(A) = \frac{x}{\sqrt{2\pi\alpha^3(A)}} \exp\left\{-\frac{x^2}{2\alpha(A)}\right\}.$$

Let $x < y$ be strictly positive, the independence of the process B on disjoint sets implies that C_x and $C_y \setminus C_x$ are independent and the density function of C_y is the convolution of the densities of $C_y \setminus C_x$ and C_x.

Let $C = \arg\max_{A \in E, \alpha(A) \in [0,1]} B(A)$, it is a stopping time of the Brownian motion B on $(E, \mathcal{E})$ and it has the inverse Gaussian distribution

$$P(C \subset A) = \frac{2}{\pi} \arc\sin(\alpha^{\frac{1}{2}}(A)).$$

Proposition 9.12. *The sets $A_y \setminus A_x$ and A_{y-x} have the same inverse Gaussian distribution.*

Proposition 9.13. *Let P_x be the probability distribution of $x + B$ on $(E, \mathcal{E})$, then for all a and b, $B(A_a) = a$ hence*

$$P_x(A_a \subset A_b) = \frac{b-x}{b-a}, \quad P_x(A_b \subset A_a) = \frac{x-a}{b-a}, \, a < x < b,$$

$$E(T_a \wedge T_b) = \{x - a\}\{b - x\}.$$

This is a consequence of the expansion

$$x = E_x B(A_x \cap A_y) = a P_x(A_x \subset A_y) + b P_x(A_y \subset A_x).$$

For measurable sets $A_1 \subset A_2$ of E, the Laplace transform of $B(A_2) - B(A_1)$ is

$$L_{B(A_2 \setminus A_1)}(x) = \exp\left\{\frac{x^2 \alpha(A_2 \setminus A_1)}{2}\right\}.$$

The Cauchy-Schwarz inequality entails a tail inequality for B^2, for every $x > 2\alpha(A)$

$$P(|B^2(A) - \alpha(A)| > x) \leq \frac{4\alpha^2(A)}{x^2}.$$

Exponential inequalities are consequences of the properties of the Brownian motion. Applying Chernoff's Theorem, we deduce

$$P(B(A) > b) = \exp\left\{-\frac{b^2}{2\alpha(A)}\right\},$$

$$P(B(A_2) - B(A_1) > b) = \exp\left\{-\frac{b^2}{2\alpha(A_2 \setminus A_1)}\right\}.$$

Propositions 5.3 and 5.5 apply to the spatial martingale $\{B^2(A) - \alpha(A)\}_{\alpha(A) \geq 0}$ defined for the Brownian motion. For every measurable set A it satisfies

$$E[\{B^2(A) - \alpha(A)\}^2] = 2\alpha^2(A),$$

it follows that for every real $p \geq 1$, and for every measurable set A

$$2^p \alpha^{2p}(A) \leq E[\{B^2(A) - \alpha(A)\}^{2p}] \leq C_p 2^p \alpha^{2p}(A)$$

and for every p in $]0, 1[$

$$c_p 2^p \alpha^{2p}(A) \leq E[\{B^2(A) - \alpha(A)\}^{2p}] \leq 2^p \alpha^{2p}(A).$$

The inequalities for the supremum of $B^2 - \alpha$ are deduced from Theorem 5.2, for every measurable set A and for every $p > 1$

$$\|B^2(A) - \alpha(A)\|_p \leq \| \sup_{C \subset A} \{B^2(C) - \alpha(C)\}\|_p$$

$$\leq \frac{p}{p-1} \sup_{C \subset A} \|B^2(C) - \alpha(C)\|_p$$

$$\leq \frac{pC_p}{p-1} \sup_{C \subset A} \|\alpha(C)\|_p.$$

Theorem 5.9 provides an exponential bound for the tail probabilities of B^*. For every strictly positive x and for every measurable set A

$$P\left(\sup_{C \subset A} B(C) > x\right) \leq 4C_2 \exp\left(-\frac{x^2}{2\alpha(A)}\right).$$

A spatial diffusion process X on $(E, \mathcal{E})$, with a drift function θ and a variance σ, is defined by the spatial Brownian motion B as

$$X(A) = \int_A \theta(x)\, dx + \int_A \sigma(x)\, dB(x), \quad A \in E,$$

this is a Gaussian process on $(E, \mathcal{E})$ with $E\{X(A)\} = \int_A \theta(x)\, dx$ and the covariance of $X(A)$ and $X(B)$ is

$$\text{Cov}\{X(A), X(B)\} = \int_{A \cap B} \sigma^2(x)\, d\alpha(x),$$

the density of $X(A)$ and its Fourier transform are deduced. For a subset A_2 of A_1, the variable

$$X(A_1) - X(A_2) = \int_{A_1 \setminus A_2} \theta(x)\, dx + \int_{A_1 \setminus A_2} \sigma(x)\, dB(x)$$

has the same distribution as $X(A_1 \setminus A_2)$.

A tail inequality is obtained from the Cauchy-Schwarz inequality for the process $X(A) - \int_A \theta(x)\,dx = \int_A \sigma\,dB$, for every measurable set A in E and for every $x > 2\alpha(A)$

$$P(|X(A) - \theta(A)| > x) \leq \frac{2\int_A \sigma^2(x)\,d\alpha(x)}{x^2}.$$

The transformed Brownian motion $\int_A \sigma\,dB$ defines the spatial martingale $(\{\int_A \sigma\,dB\}^2 - \int_A \sigma^2\,d\alpha)_{\alpha(A)\geq 0}$ such that

$$E\left[\left\{\left(\int_A \sigma\,dB\right)^2 - \int_A \sigma^2\,d\alpha\right\}^2\right] = 2\left\{\int_A \sigma^2(x)\,d\alpha(x)\right\}^2.$$

Propositions 5.3 and 5.5 imply that for every real $p \geq 1$ and for every measurable set A

$$2^p\left\{\int_A \sigma^2(x)\,d\alpha(x)\right\}^{2p} \leq E\left[\left\{X(A) - \int_A \theta(x)\,dx\right\}^{2p}\right]$$

$$\leq C_p 2^p\left\{\int_A \sigma^2(x)\,d\alpha(x)\right\}^{2p}$$

and for every p in $]0,1[$

$$c_p 2^p\left\{\int_A \sigma^2(x)\,d\alpha(x)\right\}^{2p} \leq E\left[\left\{X(A) - \int_A \theta(x)\,dx\right\}^{2p}\right]$$

$$\leq C_p 2^p\left\{\int_A \sigma^2(x)\,d\alpha(x)\right\}^{2p}$$

Exponential inequalities are consequences of the properties of the Brownian motion. For every strictly positive x and for every measurable set A

$$P\left(\sup_{C \subset A}\int_C \sigma\,dB > x\right) \leq 4C_2 \exp\left(-\frac{x^2}{2\int_A \sigma^2\,d\alpha}\right).$$

A general spatial diffusion process X is defined on $(E, \mathcal{E})$ by the equation $dX(z) = \theta(z, X(z))\,dz + \sigma(z, X(z)\,dB(z)$. For every measurable set A, the expectation of $X(A)$ is $E\{X(A)\} = \int_A \theta(z, X(z))\,dz$ and its variance is

$$\mathrm{Var}X(A) = \mathrm{Var}\int_A \theta(z, X(z))\,dz + \int_A \sigma^2(z, X(z))\,d\alpha(z).$$

The transformed Brownian motion $\int_A \sigma(z, X(z))\,dB(z)$ defines the spatial martingale $\left(\{\int_A \sigma(z, X(z))\,dB(z)\}^2 - \int_A \sigma^2(z, X(z))\,d\alpha(z)\right)_{\alpha(A)\geq 0}$ such that for every measurable set A and for every real $p \geq 1$

$$2^p\left\{\int_A \sigma^2(z, X(z))\,d\alpha(z)\right\}^{2p} \leq E\left[\left\{X(A) - \int_A \theta(z, X(z))\,dz\right\}^{2p}\right]$$

$$\leq C_p 2^p\left\{\int_A \sigma^2(z, X(z))\,d\alpha(z)\right\}^{2p}$$

and for every p in $]0, 1[$

$$c_p 2^p \left\{ \int_A \sigma^2(z, X(z)) \, d\alpha(z) \right\}^{2p} \leq E\left[\left\{ X(A) - \int_A \theta(z, X(z)) \, dz \right\}^{2p} \right]$$

$$\leq C_p 2^p \left\{ \int_A \sigma^2(z, X(z)) \, d\alpha(z) \right\}^{2p}.$$

Chapter 10

Inequalities in Complex Spaces

10.1 Introduction

Trigonometric series with real coefficients a_k and b_k are written as

$$S(t) = a_0 + \sum_{k=1}^{\infty} \{a_k \cos(k\omega t) + b_k \sin(n\omega t)\},$$

they are periodic functions with period $T = 2\pi\omega^{-1}$. Let f be a periodic and locally integrable function with period T, then it develops as a trigonometric series with coefficients

$$a_k = \frac{2}{T} \int_{-\frac{T}{2}}^{\frac{T}{2}} f(x) \cos\left(\frac{2\pi}{T} kx\right) dx, \ k \neq 0,$$

$$b_k = \frac{2}{T} \int_{-\frac{T}{2}}^{\frac{T}{2}} f(x) \sin\left(\frac{2\pi}{T} kx\right) dx,$$

$$a_0 = \frac{1}{T} \int_{-\frac{T}{2}}^{\frac{T}{2}} f(x) \, dx.$$

A periodic odd function $f : [-T, T] \to \mathbb{R}$, with period T, develops as a sine series and a periodic odd function develops as a cosine series. The Fourier representation of a periodic function f with period $T = 2\pi\omega^{-1}$ and coefficients λ_k is

$$f(x) = \sum_{k=-\infty}^{\infty} \lambda_k e^{i\omega kx}, \tag{10.1}$$

$$\lambda_k = \frac{1}{T} \int_{-\frac{T}{2}}^{\frac{T}{2}} e^{i\omega kt} f(t) \, dt$$

315

where $\lambda_k = a_k + ib_k$. Parseval's equality for the L^2 norm of the Fourier transform is

$$\|f\|_2 = \left\{ \sum_{k \in \mathbb{Z}} \|\lambda_k\|^2 \right\}^{\frac{1}{2}}. \tag{10.2}$$

This is a consequence of the Fourier representation of the function with a period T and of the integral $T^{-1} \int_{-\frac{T}{2}}^{\frac{T}{2}} e^{i(m-n)wx} \, dx = 1_{\{m=n\}}$.

The expansions of functions as trigonometric series have produced noticeable equalities. For example, for every integer m and for x in $[0, \frac{\pi}{2}]$

$$\int_0^x \frac{\sin(mx)}{\sin x} \, dx = \mathcal{Re}\left\{ \int_0^x e^{i(m-1)x} \frac{1 - e^{-2imx}}{1 - e^{-2ix}} \, dx \right\}$$

$$= \sum_{k=0}^{m} \frac{\sin(m - 2k)x}{m - 2k},$$

with the notation $\frac{\sin 0}{0} = 0$. Therefore

$$\frac{1}{2} \int_0^{\frac{\pi}{2}} \frac{\sin(mx)}{\sin x} \, dx = 0, \quad \text{for every even } m,$$

$$= 1 - \frac{1}{3} + \frac{1}{5} - \frac{1}{7} + \cdots + \frac{1}{m}, \quad \text{for every odd } m.$$

So the integral is finite, with

$$\lim_{m \to \infty} \int_0^{\frac{\pi}{2}} \frac{\sin(2m + 1)x}{\sin x} \, dx = \frac{\pi}{2},$$

using the limit of the series proved in Section 3.1.

Applying functional equations to trigonometric series allows us to solve some of them. For example, the equation $f(x + y) - f^2(x) - f^2(y) + 1 = 0$, where the function $f : \mathbb{R} \mapsto [-1, 1]$ satisfies $f(0) = 1$, $f^{(1)}(0) = 0$ and $f^{(2)}(0) = -2a$, has a solution f defined by $f(x) = \cos(\sqrt{2a}x)$ if and only if $y = x + \frac{k\pi}{\sqrt{2a}}$, k in $\mathbb{Z}$. The functions u that are solutions of the differential equations

$$\frac{d^2 u}{dx^2} \pm \alpha^2 u = 0,$$

$$x \frac{d^2 u}{dx^2} + 2 \frac{du}{dx} - k(k + 1)u = 0,$$

and defined as sums of power functions or as a trigonometric series satisfy polynomial equations depending on the coefficients of the series. They can be solved iteratively, which yield explicit expressions of the solutions. For Fourier's equation

$$\frac{\partial^2 v}{\partial x^2} + \frac{\partial^2 v}{\partial y^2} + \frac{\partial^2 v}{\partial z^2} = 0,$$

expansions of the solution in Fourier series yield explicit solutions (Fourier, 1822). This is not true for all equations and other expansions have been defined such as the development in series defined by the projection of the solution on other functional basis. Legendre's basis of polynomials provides the solution of Legendre's equations for modeling the curve of comets. Several classes of polynomials have been defined for the expansion of solutions of differential equations (Legendre, 1805; Fourier, 1822; Byerly, 1893; Pons, 2015).

In $\mathbb{C}$, the series are defined by two real sequences

$$u_n = v_n + iw_n = \rho_n(\cos\theta_n + i\sin\theta_n),$$

where $(v_n)_{n\geq 0}$ and $(w_n)_{n\geq 0}$ are real series, $(\rho_n)_{n\geq 0}$ and $(\theta_n)_{n\geq 0}$ are defined by the change of variables

$$\rho_n = (u_n^2 + v_n^2)^{\frac{1}{2}}$$

and θ_n is defined modulo $2k\pi$ if $u_n > 0$ and modulo $(2k+1)\pi$ if $u_n < 0$ by

$$\theta_n = \arctan(w_n v_n^{-1}).$$

Reciprocally, the trigonometric functions are expressions of the formulæ

$$e^{i\theta_n} = \cos\theta_n + i\sin\theta_n \text{ and } e^{-i\theta_n} = \cos\theta_n - i\sin\theta_n,$$

the logarithm of a complex number $u_n = v_n + iw_n$ is

$$\log u_n = \log\rho_n + i(\theta_n \pm 2k\pi), \text{ if } v_n > 0,$$
$$= \log\rho_n + i(\theta_n \pm (2k+1)\pi), \text{ if } v_n < 0.$$

For all real x and integer m, the equality $e^{mix} = (e^{ix})^m$ implies

$$\cos(mx) + i\sin(mx) = (\cos x + i\sin x)^m.$$

A necessary and sufficient condition for the convergence of $\sum_{k=0}^{\infty} u_k$ is the convergence of both series $\sum_{k=0}^{\infty} v_k$ and $\sum_{k=0}^{\infty} w_k$ or, equivalently, the convergence of $\sum_{k=0}^{\infty} \rho_k$. Conditions for their convergence is $\rho_{n+1}\rho_n^{-1} < 1$ for every n larger some integer n_0 or $\rho_n^{\frac{1}{n}} < 1$ for every n larger some integer n_0. If one of these ratios remains larger than 1, the series diverges. Inequalities for complex series are inequalities in the two-dimensional space

$\mathbb{C}$, for the real series $(v_n)_{n \geq 0}$ and $(w_n)_{n \geq 0}$ of $u_n = v_n + i w_n$ and they entail inequalities for ρ_n and θ_n.

Cauchy's inequality is written with the scalar product in $\mathbb{C}$, where for all complex numbers z_1 and z_2

$$\|z_1 + z_2\|^2 \geq 4(z_1, \bar{z}_2),$$

with the scalar product (1.1), it is equivalent to

$$0 \leq \|z_1 + z_2\|^2 - 4(z_1, \bar{z}_2) = \|z_1\|^2 + \|z_2\|^2 - 2(z_1, \bar{z}_2) = \|z_1 - z_2\|^2.$$

The geometric equalities (1.2) are deduced from this definition. The Cauchy-Schwarz and the Minkowski inequalities are still valid.

10.2   Polynomials

The algebraic numbers are the roots of polynomials with integer coefficients. Dividing by the coefficient of the higher degree of the variable, this extends to rational coefficients.

The class of polynomials in x with coefficients in $\mathbb{C}$ is denoted by $\mathbb{C}[x]$. Gauss-d'Alembert theorem states that every polynomial of $\mathbb{C}[x]$ has at least a complex root. Let P_k be a polynomial of degree k in $\mathbb{C}[x]$, having a complex root z_0, it is written in the form $P_k(z) = (z - z_0) P_{k-1}(z)$ and a necessary condition ensuring that P_k has at least two complex roots is: P_{k-1} belongs to $\mathbb{C}[x]$. These conditions are not necessarily satisfied.

The roots of the equation $x^n = a + ib$ are the n values $x_{n,k} = \rho^{\frac{1}{n}} e^{\frac{2k\pi}{n}}$, with $k = 1, \ldots, n$ if $a > 0$, and $x_n = \rho^{\frac{1}{n}} e^{\frac{(2k+1)\pi}{n}}$, with $k = 0, \ldots, n-1$ if $a < 0$, and $\rho = (a^2 + b^2)^{\frac{1}{2}}$. Writing $x^n - 1 = (x-1)(x^{n-1} + x^{n-2} + \cdots + x + 1)$, it follows that the $n-1$ roots of the equation $(x^{n-1} + x^{n-2} + \cdots + x + 1) = 0$ are $x_{n,k} = \rho^{\frac{1}{n}} e^{\frac{2k\pi}{n}}$, for $k = 1, \ldots, n-1$ and they belong to $\mathbb{C}$. The cubic root of i is $\sqrt[3]{i} = -i$, its fifth root is $\sqrt[5]{i} = i$, $\sqrt[7]{i} = -i$, etc., $\sqrt[3]{-i} = i$, $\sqrt[5]{-i} = -i$, $\sqrt[7]{-i} = i$, etc.

Proposition 10.1. *The odd roots of i and $-i$ are cyclic in $\mathbb{C}$*

$$
\begin{aligned}
\sqrt[2k+1]{i} &= -i, \quad \text{if } k \text{ is odd}, & \sqrt[4k+1]{i} &= i, \\
\sqrt[2k+1]{-i} &= i, \quad \text{if } k \text{ is odd}, & \sqrt[4k+1]{-i} &= -i.
\end{aligned}
$$

The even roots of i belong to $\mathbb{C}$, in particular

$$\sqrt{i} = \pm\frac{1+i}{\sqrt{2}},$$

$$\sqrt[4]{i} = \pm\frac{\sqrt{\sqrt{2}+1}+i\sqrt{\sqrt{2}-1}}{\sqrt{2\sqrt{2}}},$$

$$\sqrt[6]{i} = \pm\frac{1-i}{\sqrt{2}}.$$

Proof. Writing $\sqrt[4]{i} = a+ib$ implies $a^2-b^2+2iab = \sqrt{i}$, hence $\sqrt{2}(a^2+b^2) = 1$ and $2\sqrt{2}ab = 1$, then b^2 is a zero of $P(x) = 8x^4 + 4\sqrt{2}x^2 - 1 = 0$ in $\mathbb{R}[x^2]$ and P has one positive real root. Let $z = a + ib = \sqrt[6]{i}$, a and b are deduced from $z^2 = -i$ which implies $a = -b$ and $2ab = -2a^2 = -1$. □

As a consequence, all integer roots of -1 belong to $\mathbb{C}$. It follows that $\sqrt[x]{-1}$ belongs to $\mathbb{C}$, for every real x.

A polynomial of $\mathbb{R}[x]$ having a complex root z does not have necessarily the root $\bar{z}$ as proved by the next example, this is true if the roots depends only on the square root of -1. Bernoulli (1742) stated that the polynomial $x^4 - 4x^3 + 2x^2 + 4x + 4$ has four complex roots $1 \pm \sqrt{2 \pm \sqrt{-3}}$ (Bradley, d'Antonio and Sandifer, 2007), so they may depend on roots of degree higher than two of -1 and the product of two roots is still complex. The same method applies for the factorization of 4th degree polynomials of $\mathbb{R}[x]$ of the form

$$\alpha\{x^4 - 4ax^3 + 2x^2(3a^2 - b) - 4ax(a^2 - b) + (a^2 - b)^2 + c\}$$
$$= \alpha\prod(x - a \pm \sqrt{b \pm \sqrt{-c}})$$

with real numbers a, b and c. The coefficients of x^2 and x of the polynomial are not free and this factorization is not general. It can be extended as a product of four complex roots of the form $\pm a \pm \sqrt{\pm b \pm \sqrt{c}}$ with complex constants but 4th degree polynomials of $\mathbb{R}[x]$ are not in this form.

A third-degree polynomial of $\mathbb{R}[x]$ with real roots a, b and c has the form $P(x) = \alpha(x - a)(x - b)(x - c)$ or $P(x) = \alpha(x^3 - Sx^2 + S_2X - P)$ with $S = a + b + c$, $S_2 = ab + bc + ac$, $P = abc$. If $S^2 - 3S_2 \geq 0$, $P(x)$ has three real roots, then a root is between $x_1 = \frac{1}{3}\{S - (S^2 - 3S_2)^{\frac{1}{2}}\}$ and $x_2 = \frac{1}{3}\{S + (S^2 - 3S_2)^{\frac{1}{2}}\}$, another one is smaller than x_1 and the third one is larger than x_2. If $P(x)$ has a double root a, its factorization is easily calculated by solving the equations $S = 2a + b$, $S_2 = a^2 + 2ab$, $P = a^2b$.

If $S^2 - 3S_2 < 0$, $P(x)$ has only one real root and two conjugate complex roots. A polynomial $P(x) = (x - a - \sqrt{b})(x - a + \sqrt{b})(x - c) = x^3 - (2a + c)x^2 + (2ac + a^2 - b)x - (a^2 - b)c$ with real or complex roots can be factorized from the specific expressions of the sums and of the product. There is no general method for the factorization of the polynomials of degree larger than two.

Explicit solutions of third and fourth-degree equations have been established during the 17th century. A third-degree equation

$$x^3 + ax^2 + bx + c = 0 \tag{10.3}$$

is solvable by Cardan's method. It is written in the form $x^3 + px + q = 0$ on $\mathcal{Q}[x]$ by the reparametrization $x \mapsto x - \frac{a}{3}$, where

$$p = b - \frac{a^2}{3}, \quad q = \frac{2a^3}{27} - \frac{ab}{3} + c.$$

Let $\rho = \frac{-1 + i\sqrt{3}}{2}$ be the cubic root of 1.

Theorem 10.1. *The equation $x^3 + ax^2 + bx + c = 0$ has the solutions*

$$x_1 = \left[-\frac{q}{2} + \left\{ \left(\frac{p}{3} \right)^3 + \left(\frac{q}{2} \right)^2 \right\}^{\frac{1}{2}} \right]^{\frac{1}{3}} + \left[-\frac{q}{2} - \left\{ \left(\frac{p}{3} \right)^3 + \left(\frac{q}{2} \right)^2 \right\}^{\frac{1}{2}} \right]^{\frac{1}{3}},$$

$$x_2 = \rho \left[-\frac{q}{2} + \left\{ \left(\frac{p}{3} \right)^3 + \left(\frac{q}{2} \right)^2 \right\}^{\frac{1}{2}} \right]^{\frac{1}{3}} + \rho^2 \left[-\frac{q}{2} - \left\{ \left(\frac{p}{3} \right)^3 + \left(\frac{q}{2} \right)^2 \right\}^{\frac{1}{2}} \right]^{\frac{1}{3}},$$

$$x_3 = \rho^2 \left[-\frac{q}{2} + \left\{ \left(\frac{p}{3} \right)^3 + \left(\frac{q}{2} \right)^2 \right\}^{\frac{1}{2}} \right]^{\frac{1}{3}} + \rho \left[-\frac{q}{2} - \left\{ \left(\frac{p}{3} \right)^3 + \left(\frac{q}{2} \right)^2 \right\}^{\frac{1}{2}} \right]^{\frac{1}{3}}.$$

Proof. According to Cardan's method, let $x = u + v$, the equation is equivalent to

$$u^3 + v^3 + (u + v)(3uv + p) + q = 0$$

and solutions u and v can be found as solutions of the equations

$$u^3 + v^3 + q = 0,$$

$$uv + \frac{p}{3} = 0.$$

Let $U = u^3$ and let $V = v^3$, the equation becomes equivalent to

$$U^2 + qU - \frac{p^3}{27} = 0 = V^2 + qV - \frac{p^3}{27},$$

its solutions are

$$U \text{ and } V = \frac{1}{2} \left\{ -q \pm \left(\frac{4p^3 + 27q^2}{27} \right)^{\frac{1}{2}} \right\},$$

and the solutions of Equation (10.3) are deduced as the cubic roots of U and V. □

A fourth-degree equation

$$x^4 + ax^3 + bx^2 + cx + d = 0 \tag{10.4}$$

is reparametrization as $x^4 + px^2 + qx + r = 0$ by the change of variable $x \mapsto x + \frac{a}{4}$, with

$$p = b - 6a^2, \quad q = c - 4a^3, \quad r = d - a^4.$$

Theorem 10.2. *A fourth-degree equation* (10.4) *has explicit solutions.*

Proof. Equation (10.5) is equivalent to

$$(x^2 + y)^2 = (2y - p)x^2 - qx - r + y^2, \tag{10.5}$$

for every y. A necessary condition for the existence of a solution of (10.4) is therefore that the right-hand term of (10.5) is a square. Considering this expression, it is a square if its discriminant is zero and y is solution of the third-degree equation

$$(2y - p)(y^2 - r) = \frac{q^2}{4}.$$

Then $(x^2 + y)^2 = (Ax + By)^2$ where $A^2 = 2y - p$ and $B^2 = y^2 - r$, equivalently

$$(x^2 + y + Ax + By)(x^2 + y - Ax - By) = 0.$$

Finally the equations $x^2 + Ax + (B+1)y = 0$ and $x^2 - Ax - (B-1)y = 0$ have explicit real or complex solutions which provide four solutions depending on the solutions for y. □

Expansions of $\sin(ma)$ as a polynomial of degree m and $\cos(ma)$ as a polynomial of degree $m - 2$ have the value zero for a in the sequence

$$\left\{ -\frac{(m-1)\pi}{2m}, \ldots, -\frac{3\pi}{2m}, -\frac{\pi}{2m}, \frac{\pi}{2m}, \frac{3\pi}{2m}, \ldots, \frac{(m-1)\pi}{2m} \right\}$$

and they are divided by

$$1 - \frac{\sin^2 a}{\sin^2 \frac{\pi}{2m}}, 1 - \frac{\sin^2 a}{\sin^2 \frac{3\pi}{3m}}, \ldots, 1 - \frac{\sin^2 a}{\sin^2 \frac{(m-1)\pi}{2m}},$$

hence for every even integer m

$$\sin x = m \sin \frac{x}{m} \cos \frac{x}{m} \left(1 - \frac{\sin^2 \frac{x}{m}}{\sin^2 \frac{2\pi}{2m}} \right) \cdots \left(1 - \frac{\sin^2 \frac{x}{m}}{\sin^2 \frac{(m-2)\pi}{2m}} \right),$$

$$\cos x = m \sin \frac{x}{m} \left(1 - \frac{\sin^2 \frac{x}{m}}{\sin^2 \frac{\pi}{2m}} \right) \left(1 - \frac{\sin^2 \frac{x}{m}}{\sin^2 \frac{3\pi}{3m}} \right) \cdots \left(1 - \frac{\sin^2 \frac{x}{m}}{\sin^2 \frac{(m-1)\pi}{2m}} \right)$$

and a similar expansion is written for odd integers m.

They become expansions in terms of $\tan^2$ at the same values using the equality

$$1 - \frac{\sin^2 u}{\sin^2 v} = (\cos^2 u)\left(1 - \frac{\tan^2 u}{\tan^2 v}\right),$$

with m infinite in these expressions we obtain

$$\sin(x\pi) = x\pi(1 - x^2)\left(1 - \frac{x^2}{4}\right)\cdots\left(1 - \frac{x^2}{(2n)^2}\right)\cdots,$$

$$\cos(x\pi) = (1 - 4x^2)\left(1 - \frac{4x^2}{9}\right)\cdots\left(1 - \frac{4x^2}{(2n+1)^2}\right)\cdots.$$

10.3   Fourier and Hermite transforms

A complex function from $\mathbb{R}$ to $\mathbb{C}$ has the form $F(x) = f(x) + ig(x)$ where f and g are real functions defined on the same subset of $\mathbb{R}$. Its complex conjugate is $\bar{F}(x) = f(x) - ig(x)$. Its Euclidean norm is

$$\|F(x)\| = F(x)\bar{F}(x) = (f^2(x) + g^2(x))^{\frac{1}{2}}$$

and its norm $\ell_2(\mu, \mathbb{R})$ is $\|F\|_2 = \{\int_{\mathbb{R}}(f^2 + g^2)\,d\mu\}^{\frac{1}{2}}$.

Let f and g be functions from $\mathbb{C}$ to $\mathbb{C}$ provided with the product Lebesgue, the norm of f is $\|f\|_2 = \int_{\mathbb{C}} f(z)\bar{f}(z)\,d\mu(z)$ and the scalar product of f and g is

$$< f, g > = \int_{\mathbb{C}} f(z)\bar{g}(z)\,d\mu(z) = \int_{\mathbb{C}} g(z)\bar{f}(z)\,d\mu(z),$$

the conjugate of $f\bar{g}$ is $\bar{f}g$. Cauchy's inequality applies and for all functions f and $g : \mathbb{C} \to \mathbb{C}$

$$2\left\{\int_{\mathbb{C}} f(z)\bar{g}(z)\,d\mu(z)\right\}^{\frac{1}{2}} \leq \|f + g\|_2 \leq \|f\|_2 + \|g\|_2.$$

The Fourier transform of a convolution is the product of the Fourier transforms.

A necessary and sufficient condition for the differentiability of a complex function $f(x + iy) = f_1(x + iy) + if_2(x + iy)$ on $\mathbb{C}$ is Cauchy's condition

$$\frac{\partial f_1}{\partial x} = \frac{\partial f_2}{\partial y}, \qquad \frac{\partial f_2}{\partial x} = -\frac{\partial f_1}{\partial y}$$

then the derivative of f at $z = (x + iy)$ is

$$f^{(1)}(x + iy) = \frac{\partial f_1}{\partial x}(z) + i\frac{\partial f_2}{\partial x} = \frac{\partial f_2}{\partial y} - i\frac{\partial f_1}{\partial y}.$$

Lipschitz's condition can be extended to a complex function f on $\mathbb{C}$: for all z and z', the m-th order derivative of f satisfies

$$\|f^{(m)}(z) - f^{(m)}(z')\|_2 \leq k\|z - z'\|_2$$

for a constant k. The condition is equivalent to

$$\left[\left\{\frac{\partial^m f_1(z)}{\partial x^m} - \frac{\partial^m f_1(z')}{\partial x^m}\right\}^2 + \left\{\frac{\partial^m f_2(z)}{\partial x^m} - \frac{\partial^m f_2(z')}{\partial x^m}\right\}^2\right]^{\frac{1}{2}}$$

$$= \left[\left\{\frac{\partial^m f_1(z)}{\partial y^m} - \frac{\partial^m f_1(z')}{\partial y^m}\right\}^2 + \left\{\frac{\partial^m f_2(z)}{\partial y^m} - \frac{\partial^m f_2(z')}{\partial y^m}\right\}^2\right]^{\frac{1}{2}}$$

$$\leq k\|z - z'\|_2.$$

Let X be a real random variable with distribution function F, its Fourier transform is the function with values in $\mathbb{C}$

$$\varphi_X(t) = \int_{-\infty}^{\infty} e^{itx}\, dF(x).$$

Let f be an even function having a finite Fourier transform $\mathcal{F}f$, then all odd moments of f are zero and φ_X develops as

$$\varphi_X(t) = \sum_{k\geq 0} \frac{(-t^2)^k}{(2k)!} E(X^{2k}).$$

Conversely, all even moments of an odd function are zero and, when it is finite, its Fourier transform develops as

$$\varphi_X(t) = i\sum_{k\geq 1} (-1)^k \frac{t^{2k+1}}{(2k+1)!} E(X^{2k+1}).$$

The derivative of a differentiable function f with period $T = 2\pi\omega^{-1}$ is written as $f'(x) = \sum_{k\in\mathbb{Z}} \lambda_k(f') e^{i\omega k x}$ where the coefficients are obtained by integration by parts

$$\lambda_k(f) = \frac{1}{T}\int_{-\frac{T}{2}}^{\frac{T}{2}} e^{i\frac{2\pi}{T}kt} f(t)\, dt = \frac{1}{2ki\pi}\int_{-\frac{T}{2}}^{\frac{T}{2}} e^{i\frac{2\pi}{T}kt} f'(t)\, dt = \frac{1}{iwk}\lambda_k(f')$$

hence $\lambda_k(f') = iwk\lambda_k(f)$ and its norm $L^2([0, T])$ is

$$\|f'\|_2 = 2\pi T^{-1}\left\{\sum_{k\in\mathbb{Z}} k^2\lambda_k^2\right\}^{\frac{1}{2}}.$$

If f is periodic and belongs to $C[0, T]\cap L^2([0, T])$, the coefficients $\lambda_k(f)$ tend to zero as k tends to infinity and the approximation

$$f_n(x) = \sum_{k=1}^{n} \lambda_k e^{iwkt}$$

of the function f by its Fourier series converges in L^2, $\lim_{n \to \infty} \|\widehat{f}_n - f\|_2 = 0$.

If φ_X is $L^1(\mathbb{C})$ and F has a periodic density with period T, it equates to the inverse of the Fourier transform at every continuity point x of f

$$f(x) = \frac{1}{T} \int_{\mathbb{R}} e^{-iwtx} \varphi_X(t)\, dt = \frac{1}{T} \int_{-\frac{T}{2}}^{\frac{T}{2}} \int_{\mathbb{R}} e^{-iw(t-s)x}\, dF(s)\, dt,$$

by Proposition A.4, and the inverse Fourier series has the value $\frac{1}{2}\{f(x^+) + f(x^-)\}$ at every point x where the function f is discontinuous.

Example 10.1. The Fourier transform of the normal density is the function defined on $\mathbb{R}$ by $\varphi(x) = e^{-\frac{x^2}{2}}$. The inverse of the Fourier transform of the normal density f is

$$f(x) = \int_{\mathbb{R}} e^{itx} e^{-\frac{x^2}{2}}\, dt = \int_{\mathbb{R}} \int_{\mathbb{R}} e^{i(t-s)x} f(s)\, ds\, dt.$$

The transform of a normal variable with expectation μ and variance σ^2 is $e^{ix\mu} e^{-\frac{x^2\sigma^2}{2}}$.

Example 10.2. The function $(ax)^{-1} \sin(ax)$, with a constant $a > 0$, is the Fourier transform of the uniform density on $[-a, a]$. Using the inverse transform yields

$$\frac{1}{2\pi} \int_{-\pi}^{\pi} \frac{\sin(ax)}{ax} e^{-itx}\, dx = \frac{1}{2a} \frac{1}{2\pi} \int_{-\pi}^{\pi} \int_{-a}^{a} e^{-itx} e^{its}\, dx\, ds$$

$$= \frac{1}{at^2\pi} \sin(t\pi) \sin(ta)$$

and it differs from $f(t) = a^{-1} 1_{]-a,a[}(t)$.

Let f and g be periodic and integrable functions defined from $[-T, T]$ to $\mathbb{R}$ or $\mathbb{C}$, the expression of their Fourier transform (10.1) provides a scalar product for f and g

$$T^{-1} \int_{-T}^{T} f(x) \bar{g}(x)\, dx = \sum_{n \geq 0} \lambda_n(f) \bar{\lambda}_n(g).$$

Parseval's equality (10.2) for the Fourier transform $\widehat{f}$ of a function f of L^2 is a consequence of this expression

$$\|f\|_2 = \|\widehat{f}\|_2 = \sum_{n \geq 0} \lambda_n^2(f).$$

The operator T_g is defined as $T_g(f) = \int_{-T}^{T} f(x) g(x)\, dx$ for periodic integrable functions f and g on $[-T, T]$. Let g be a periodic function on $[-T, T]$

with Fourier transform $\widehat{g}$, it defines the operator $\widehat{T}_g = T_{\widehat{g}}$. Developing the functions f and g as a Fourier series S_f and S_g yields

$$\int f(x)\bar{S}_g(x)\,dx = \sum_{k\geq 0} a_k(f)\bar{a}_k(g) + \sum_{k\geq 0} b_k(f)\bar{b}_k(g) = \int S_f(x)\bar{g}(x)\,dx,$$

from Fubini's theorem. The scalar product of a function and the Fourier transform of another one is a symmetric operator

$$\widehat{T}_g(f) = T_{\widehat{g}}(f) = T_g(\widehat{f}). \tag{10.6}$$

With the normal density $f_\mathcal{N}$ and a function g on $[-T,T]$, Equation (10.6) becomes

$$\widehat{T}_g(f_\mathcal{N}) = T_{\widehat{g}}(f_\mathcal{N}) = \frac{1}{2T}\int_{-T}^{T} g(x)e^{-\frac{x^2}{2}}\,dx = \frac{1}{2T}\int_{-T}^{T} g(x)\widehat{f}(x)\,dx. \tag{10.7}$$

Corollary 10.1. *Let g be a function of $L^2([-T,T])$, a normal variable X satisfies*

$$E\{g(X)1_{|X|\leq T}\} = \frac{T}{\sqrt{2\pi}}\int_{-\infty}^{\infty} \widehat{g}(x)f_\mathcal{N}(x)\,dx,$$

in particular

$$P(|X| \leq a) = \frac{2}{\sqrt{2\pi}}E\left\{\frac{\sin(aX)}{X}\right\} = 1 - 2e^{-\frac{a^2}{2}}.$$

Proof. This is a consequence of the equality (10.6) with the normal and uniform densities (Examples 7.1 and 7.2)

$$\frac{1}{\sqrt{2\pi}}\int_{\mathbb{R}} \frac{\sin(ax)}{ax}e^{-\frac{x^2}{2}}\,dx = \frac{1}{2a}\int_{-a}^{a} e^{-\frac{x^2}{2}}\,dx = \frac{\sqrt{2\pi}}{2a}P(|X| \leq a)$$

for a variable X with the normal distribution $\mathcal{N}(0,1)$. Chernoff's theorem for the normal variable X implies $1 - 2e^{-\frac{a^2}{2}} = P(|X| \leq a)$. $\qquad\square$

Hermite's polynomials are real functions defined $\mathbb{R}$ from the derivatives of the normal density function $f_\mathcal{N}$ as

$$H_k(t) = (-1)^k \frac{d^k e^{-\frac{t^2}{2}}}{dt^k}e^{\frac{t^2}{2}} = \frac{d^k f_\mathcal{N}(x)}{dx^k}\{f_\mathcal{N}(x)\}^{-1}, \tag{10.8}$$

they are symmetric. Let $\varphi(t) = e^{\frac{-t^2}{2}}$ be Fourier's transform of the normal density, by the inversion formula

$$\frac{d^k f_\mathcal{N}(x)}{dx^k} = H_k(x)f_\mathcal{N}(x) = \frac{1}{2\pi}\int_{-\pi}^{\pi} (it)^k e^{itx}\varphi(t)\,dt, \tag{10.9}$$

or by the real part of this integral. Hermite's polynomials are recursively defined by $H_0 = 1$ and $H_{k+1} = xH_k - H'_k$, for every $k > 2$, hence H_k is a polynomial of degree k, with higher term x^k. With this definition, the functions H_k are orthogonal in $(L^2(\mathbb{R}), \mu_\mathcal{N})$ with the normal distribution $\mu_\mathcal{N}$, and the inequalities of the Hilbert spaces apply to the basis $(H_k)_{k \geq 0}$. From (10.8), their norm $c_k = \|H_k\|_2$ is defined by

$$c_k^2 = \int_\mathbb{R} f_\mathcal{N}(x) H_k^2(x)\, dx = \sqrt{2\pi} \int_\mathbb{R} \{f_\mathcal{N}^{(k)}(x)\}^2 e^{\frac{t^2}{2}}\, dx$$

and it is calculated from the even moments of the normal distribution, $c_k = (k!)^{\frac{1}{2}}$ (Section A.4). The Hermite transform $H_f = \sum_{k \geq 0} a_k(f) H_k$ of a function f has the coefficients

$$a_k(f) = \frac{1}{c_k} \int_\mathbb{R} f(x) H_k(x) f_\mathcal{N}(x)\, dx = \frac{1}{c_k} \int_\mathbb{R} f(x) f_\mathcal{N}^{(k)}(x)\, dx$$

if these integrals are finite. The scalar product of the Hermite expansions of functions f and g in $(L^2(\mathbb{R}), \mu_\mathcal{N})$ is written as $E\{H_f(X) H_g(X)\}$, with a normal variable X, and it develops as $\int_\mathbb{R} H_f H_g\, d\mu_\mathcal{N} = \sum_{k \geq 0} k!\, a_k(f) a_k(g)$. A condition for the convergence of the transform H_f of a function f is the convergence of the series

$$\|H_f\|_2^2 = \sum_{k \geq 1} k!\, a_k^2(f).$$

For every $k \geq 0$

$$\int_\mathbb{R} f(x) c_k^{-1} H_k(x) f_\mathcal{N}(x)\, dx = \int_\mathbb{R} c_k^{-2} H_f(x) H_k(x) f_\mathcal{N}(x)\, dx.$$

A function f is then equal to its normalized Hermite expansion.

The generating functions related to the polynomials H_k provide translations of the normal density and therefore convolutions.

Proposition 10.2. *For every w such that $|w| < 1$*

$$f_\mathcal{N}(x + w) = \sum_{k \geq 0} H_k(x) f_\mathcal{N}(x) \frac{w^k}{k!},$$

$$\sum_{k \neq j \geq 0} H_k(t) H_j(t) e^{-t^2} \frac{w^{j+k}}{(j+k)!} = \sum_{k \neq j \geq 0} \frac{d^{j+k} e^{-\frac{t^2}{2}}}{dt^{j+k}} \frac{w^{j+k}}{(j+k)!}$$

$$= \sqrt{2\pi} f_\mathcal{N}(t + w).$$

For every function g of $(L^2(\mathbb{R}), \mu_\mathcal{N})$

$$\sum_{k \geq 0} \int_\mathbb{R} g(x) H_k(x) f_\mathcal{N}(x) \frac{w^k}{k!}\, dx = \int_\mathbb{R} g(x - w) f_\mathcal{N}(x)\, dx.$$

A normal variable X then satisfies

$$Eg(X - w) = \sum_{k \geq 0} E\{g(X)H_k(X)\}\frac{w^k}{k!},$$

$$= \sum_{k \neq j \geq 0} E\{g(X)H_k(X)H_j(X)e^{\frac{-x^2}{2}}\}\frac{w^{j+k}}{(j+k)!}$$

$$= \sqrt{2\pi} \sum_{k \neq j \geq 0} E\{g(X)f_{\mathcal{N}}^{(k)}(X)f_{\mathcal{N}}^{(j)}(X)\}\frac{w^{j+k}}{(j+k)!}$$

for every function g such that $g(X)$ has a finite variance.

The derivatives of the Hermite transform of a function g of $(L^2(\mathbb{R}), \mu)$ are written as $g^{(j)} = \sum_{k \geq 0} a_k H_k^{(j)}$ where $H_k^{(1)}(x) = xH_k(x) - H_{k+1}(x)$ and the derivative of order j of H_k is a polynomial of degree $k+j$ recursively written in terms of $H_k, \ldots, H_{k+j}$ in the form $H_k^{(j)}(x) = \sum_{i=0}^j P_{ik}(x)H_{k+i}(x)$, where P_{ik} is a polynomial of degree $j - i$.

Proposition 10.3. *The Fourier transform of the Hermite functions H_k has the expansion*

$$\widehat{H}_k(t) = \sum_{j \geq k+1} \frac{(-it)^{j-k}}{(j-k)!} e^{\frac{t^2}{2}} \int_{\mathbb{R}} H_j(x+it)\, dx, \ k \geq 1.$$

Proof. Applying Proposition 10.2 to a function f of $(L^2(\mathbb{R}), \mu_{\mathcal{N}})$, its derivatives are written as

$$f^{(k)}(x+w) = \sum_{j \geq k+1} f^{(j)}(x)\frac{w^{j-k}}{(j-k)!} = \sum_{j \geq k+1} H_j(x)f_{\mathcal{N}}(x)\frac{w^{j-k}}{(j-k)!}$$

and the Fourier transform of H_k defined by (10.8) is

$$\widehat{H}_k(t) = \int_{\mathbb{R}} H_k(x)e^{itx}\, dx = \sqrt{2\pi}e^{\frac{t^2}{2}} \int_{\mathbb{R}} f^{(k)}(x)e^{\frac{(x+it)^2}{2}}\, dx$$

$$= e^{\frac{t^2}{2}} \int f^{(k)}(y-it)f_{\mathcal{N}}^{-1}(y)\, dy$$

where the last integral stands for $y = x + it$ at fixed t. From the expansion of the derivative $f^{(k)}(y+it)$, it is also written as

$$\sum_{j \geq k+1} \frac{(-it)^{j-k}}{(j-k)!} \int_{\mathbb{R}} H_j(x-it)\, dx$$

and it is a complex function of t. $\qquad\square$

The Hermite series of odd functions are zero, by symmetry of the polynomials H_k. The Hermite transform of uniform variables is also zero.

Theorem 10.3. *The Hermite transform of the normal density is the series*
$$H_{f_\mathcal{N}}(x) = \sum_{k \geq 0} a_{2k}(f_\mathcal{N}) H_{2k}(x),$$

its coefficients are strictly positive
$$a_{2k}(f_\mathcal{N}) = \frac{1}{\sqrt{(2k)!}} \int_\mathbb{R} \{f_\mathcal{N}^{(2k)}(x)\}^2 \, dx.$$

Proof. The Hermite transform of the normal density is defined by the integrals $\int_\mathbb{R} f_\mathcal{N}^2(x) H_k(x) \, dx = a_k(f_\mathcal{N}) c_k$, for every $k \geq 0$. Integrating by parts the expansion of the function $H_{f_\mathcal{N}}(x) = \sum_{k \geq 0} a_k(f_\mathcal{N}) H_k(x)$, where $f_\mathcal{N} H_k$ satisfies (10.9), yields

$$a_{2k+1}(f_\mathcal{N}) = \frac{1}{c_{2k+1}} \int_\mathbb{R} f_\mathcal{N}^{(2k+1)}(x) f_\mathcal{N}(x) \, dx$$
$$= \frac{(-1)^k}{c_{2k+1}} \int_\mathbb{R} f_\mathcal{N}^{(k)}(x) f_\mathcal{N}^{(k+1)}(x) \, dx = 0,$$
$$a_{2k}(f_\mathcal{N}) = \frac{1}{c_{2k}} \int_\mathbb{R} H_{2k}(x) f_\mathcal{N}^2(x) \, dx = \frac{1}{c_{2k}} \int_\mathbb{R} f_\mathcal{N}^{(2k)}(x) f_\mathcal{N}(x) \, dx$$
$$= \frac{1}{c_{2k}} \int_\mathbb{R} \{f_\mathcal{N}^{(k)}(x)\}^2 \, dx.$$
$\square$

Corollary 10.2. *The norm of $H_{f_\mathcal{N}}$ is finite.*

The norm of $H_{f_\mathcal{N}}$ is expanded as
$$\|f_\mathcal{N}\|_{L^2, \mu_\mathcal{N}} = \sum_{k \geq 1} c_{2k}^2 a_{2k}^2(f_\mathcal{N}) = \sum_{k \geq 1} \int_\mathbb{R} \{f_\mathcal{N}^{(k)}(x)\}^2 \, dx.$$
The integrals $\sum_{k \geq 1} \int_{|x|>A} \{f_\mathcal{N}^{(k)}(x)\}^2 \, dx$ and $\int_{|x|>A} (1-x^2)^{-1} e^{-x^2} \, dx$ are equivalent as A tends to infinity and they converge to a finite limit.

10.4 Inequalities for the transforms

The Fourier transform is extended to a bounded linear operator on $L^p(\mathbb{R})^n$, $p \geq 1$. From the equivalence between the norms of the spaces L^p, for all conjugates p and p', there exist constants $k_1 \leq p$ and $k_2 \geq (p')^{-1}$ such that for every function f of L^p
$$k_1 \|f\|_{L^p} \leq \|\widehat{f}\|_{p'} \leq k_2 \|f\|_{L^p}$$
and the Fourier transform converges in L^p for every $p \geq 1$.

The Hermite expansion $H_f = \sum_{k \geq 0} a_k(f) H_k$ of a function f of $L^2(\mu_{\mathcal{N}})$ satisfies the following inequality due to the Hölder inequality.

Proposition 10.4. *Let X be a normal variable, functions f and g such that $Ef^2(X)$ and $Eg^2(X)$ are finite satisfy*

$$E\{H_f(X) H_g(X)\} \leq \left(\sum_{k \geq 0} [E\{f(X) H_k(X)\}]^2 \right) \left(\sum_{k \geq 0} [E\{g(X) H_k(X)\}]^2 \right)$$

and $EH_k(X) = 0$ for every $k \geq 0$.

The normalized Hermite polynomials are denoted by

$$h_k = \frac{H_k}{\sqrt{k!}}.$$

The functions h_k are equivalent to $(k!)^{-\frac{1}{2}} x^k$ as x tends to infinity. At zero, the polynomials converge, $H_{2k+1}(x) \sim 0$ and

$$\frac{H_{2k}}{(2k)!} \sim_{x \to 0} \frac{1}{2k}.$$

The expansion of a function f in this basis is denoted by h_f and $H_f \equiv h_f$ for every function f. The order of $\|h_{f_{\mathcal{N}}}\|_{L^2(\mu_{\mathcal{N}})}$ is

$$\lim_{A \to \infty} \int_{|x| > A} \sum_{k \geq 0} \frac{x^{2k}}{(2k)!} e^{-x^2} < \infty.$$

The error $R_n(f) = f - S_n(f)$ in estimating a function f by the sum of the first n terms of its Fourier expansion is $R_n(f) = \sum_{k > n} \lambda_k^2(f)$. In the Hermite expansion, we consider norms $L^p(\mu_{\mathcal{N}})$ of the error of the partial sum $S_n(f; H) = \sum_{k \leq n} a_k(f) H_k(x)$.

Lemma 10.1. *The coefficient a_k of the expansion H_f for a function f belonging to $L^2(\mu_{\mathcal{N}})$ has a norm $\|a_k\|_{L^2(\mu_{\mathcal{N}})} < 1$ if $\|f\|_{L^2(\mu_{\mathcal{N}})} < \|H_k\|_{L^2(\mu_{\mathcal{N}})}$.*

Proposition 10.5. *The partial sums $S_n(f; h)$ of the Hermite expansion h_f in the orthonormal basis $(h_k)_{k \geq 0}$ satisfy*

$$\|n^{-1} S_n(f; h)\|_{L^2(\mu_{\mathcal{N}})} < \|f\|_{L^2(\mu_{\mathcal{N}})}$$

and for all conjugates $p > 1$ and p'

$$\|n^{-1} S_n(f; h)\|_{L^2(\mu_{\mathcal{N}})} < \|f\|_{L^p(\mu_{\mathcal{N}})} n^{-1} \sum_{k > n} \|h_k\|_{L^{p'}(\mu_{\mathcal{N}})}.$$

Proof. The Hermite expansion h_f in the orthonormal basis of polynomials satisfies $\|h_f\|_{L^2(\mu_\mathcal{N})}^2 = \sum_{k \geq 0}\{E(fh_k)(X)\}^2$ and for every $k \geq 1$, the Hölder inequality implies $E|f(X)h_k(X)| \leq \|f\|_{L^2(\mu_\mathcal{N})}$, with equality if and only if $f = h_k$. $\qquad\square$

Proposition 10.6. *For every function f of $L^2(\mu_\mathcal{N})$, the Hermite sum $S_n(f; H)$ has an error*

$$\|R_n(f)\|_{L^2(\mu_\mathcal{N})} \leq \|f\|_2 \|R_n(f_\mathcal{N})\|_{L^2(\mu_\mathcal{N})}$$

and it tends to zero as n tends to infinity.

Proof. The squared norm of $R_n(f)$ is $\|R_n(f)\|_2^2 = \sum_{k>n} c_k^2(f)a_k^2(f)$ with

$$a_k(f) = \frac{1}{c_k(f)} \int_\mathbb{R} f(x)f_\mathcal{N}^{(k)}(x)\, dx,$$

$$c_k^2 a_k^2(f) \leq \left\{\int_\mathbb{R} f^2(x)\, dx\right\}\left\{\int_\mathbb{R} f_\mathcal{N}^{(k)2}(x)\, dx\right\} = c_{2k}a_{2k}(f_\mathcal{N})\|f\|_2^2.$$

From Corollary 10.2, the Hermite transform of the normal density converges therefore $\|R_n(f_\mathcal{N})\|_{L^2(\mu_\mathcal{N})}$ converges to zero as n tends to infinity. $\qquad\square$

A function f having a converging Hermite expansion can then be approximated by finite sums of this expansion, for n large enough.

10.5 Inequalities in $\mathbb{C}$

A map f from a subset $\mathcal{D}$ of $\mathbb{C}$ to $\mathbb{C}$ is *analytic* at z_0 if there exist an open disc $D = D(r, z_0)$ centered at z_0 and with radius r, and a convergent series such that for every z in D

$$f(z) = \sum_{k=0}^\infty a_k(z - z_0)^k.$$

This expansion is unique and all derivatives of f are analytic. Let f_n be the partial sum $f_n(z) = \sum_{k=0}^n a_k(z - z_0)^k$, the coefficients are defined by $a_0 = f(z_0)$ and

$$a_k = \lim_{\|z-z_0\| \to 0} \frac{\|f(z) - f_{k-1}(z)\|}{\|z - z_0\|^k} = \frac{f^{(k)}(z_0)}{k!}.$$

A holomorph function f on a disc $D(r, z_0)$ with frontier the circle $C = C(r, z_0)$ has the derivatives

$$f^{(n)}(x) = \frac{n!}{2\pi} \int_C \frac{f(z)}{(z - z_0)^{n+1}}\, dz.$$

In $\mathbb{C}^2$, a function $f(x, y)$ holomorph on circles $C(r_1, z_1) \times C(r_2, z_2)$ has derivatives

$$\frac{\partial^{n+m} f(x, y)}{\partial x^n \partial y^m} = \frac{n! m!}{(2i\pi)^2} \int_C \frac{f(x, y)}{(x - z_1)^{n+1}(y - z_2)^{m+1}} \, dx \, dy$$

$$= \frac{n! m!}{4\pi^2 r_1^n r_2^m} \int_{[0, 2\pi]^2} f(z_1 + r_1 e^{i\varphi}, z_2 + r_2 e^{i\theta}) e^{-ni\varphi} e^{-mi\theta} \, d\varphi \, d\theta,$$

with integrals on the circles $C_1 = C(r_1, z_1)$ and $C_2 = C(r_2, z_2)$. Then

$$\left| \frac{\partial}{\partial x^n} \frac{\partial}{\partial y^m} f(x, y) \right| < \frac{n! m! \|f\|}{r_1^n r_2^m}.$$

Cauchy's theorem states that the solution of a canonical system of differential equations in $\mathbb{R}^k$

$$\frac{dx_m}{dx_1} = \varphi_m(x_1, \ldots, x_k), \; m = 2, \ldots, k$$

with a set of holomorph functions $(\varphi_1, \ldots, \varphi_{k-1})$ on discs with radius r and centers $a_1, \ldots, a_k$, is a holomorph function of $(x_1, \ldots, a_k)$ in the interior of a disc centered at $a = (a_1, \ldots, a_k)$.

10.6 Complex spaces of higher dimensions

The bijection between the spaces $\mathbb{R}^2$ and $\mathbb{C}$ defined by $h(x, y) = x + iy$ is defined by the square root i of -1 in the Euclidean metric of $\mathbb{R}^2$. In $\mathbb{R}^3$, -1 has two square roots denoted by i and j. The map

$$X = (x, y, z) \mapsto t = x + iy + jz$$

defines a bijection between $\mathbb{R}^3$ and a complex space denoted $\mathbb{C}_2$ (with two roots of -1). The complex conjugate of $t = x + iy + jz$ is $x - iy - jz$ and the space $\mathbb{C}_2$ is a vector space endowed with the Euclidean norm defined as the scalar product of complex conjugates

$$\|t\| = \{(x + iy + jz)(x - iy - jz)\}^{\frac{1}{2}} = \{x^2 + y^2 + z^2\}^{\frac{1}{2}}.$$

The spherical coordinates of a point defined on a sphere with radius r by angles θ in $[0, 2\pi]$ and φ in $[0, \pi]$ are $X = r(\cos\varphi \cos\theta, \cos\varphi \sin\theta, \sin\varphi)$, the norm of X is r while the norm of its projection in the horizontal plane is $r\{(\cos\varphi \cos\theta)^2 + (\cos\varphi \sin\theta)^2\}^{\frac{1}{2}} = r\cos\varphi$.

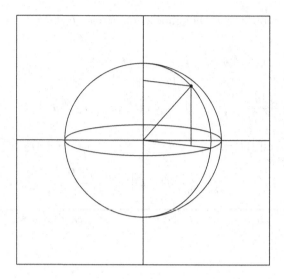

Fig. 10.1   Spherical representation in $\mathbb{R}^3$ and projection of a point in the plane.

Let $\rho_{xy} = \|x + iy\|_2$ and $\rho_{xz} = \|x + jz\|_2$, then $t = x - iy - jz$ is also written as

$$t = \rho_{xy}\rho_{xz}e^{i\theta}e^{j\varphi}, \ \theta \in [0, 2\pi], \varphi \in [0, 2\pi],$$

$$\theta = \arctan\frac{x}{y}, \ y \neq 0,$$

$$\varphi = \arctan\frac{x}{z}, \ z \neq 0,$$

and $\theta = 0$ if $y = 0$, $\varphi = 0$ if $z = 0$.

The product $u = e^{i\theta}e^{j\varphi}$ belongs to $\mathbb{C}^{\otimes 2}$ and its expansion using trigonometric functions is $u = \cos\varphi\cos\theta + ij\sin\varphi\sin\theta + i\cos\varphi\sin\theta + j\sin\varphi\cos\theta$, where $\cos\varphi(\cos\theta + i\sin\theta)$ is the projection of $e^{i\theta}e^{j\varphi}$ in the horizontal plane and $\sin\varphi(\cos\theta + i\sin\theta)$ is its projection in a vertical section of the sphere.

By orthogonality, the squared norm of u can be written as the sum of the squares $\cos^2\varphi\|\cos\theta + i\sin\theta\|^2 + \sin^2\varphi\|\cos\theta + i\sin\theta)\|^2 = 1$. The projections on the orthogonal spaces have a similar form whereas it is not satisfied for the coordinates in the spherical representation.

The equality $t = \bar{t}$ implies that t belongs to $\mathbb{R}$. The real scalar product of x and y in $\mathbb{C}_2$ is defined from the norm by (1.1)

$$(t, t') = \frac{1}{2}\{(t + t')(\overline{t + t'}) - t\bar{t} - t'\bar{t}'\},$$

the inequalities for the norms and the geometric equalities (1.2) are still true. It differs from the complex scalar product

$$t\bar{t}' = xx' + yy' + zz' + i(xy' - x'y) + j(xz' - x'z) - ij(yz' + y'z)$$

where only $xx' + yy' + zz' + i(xy' - x'y) + j(xz' - x'z)$ belongs to $\mathbb{C}_2$ and ij belongs to $\mathbb{C}^2$ and satisfies $(ij)^2 = 1$. It follows that the space $\mathbb{C}_2$ is not a Hilbert space like $\mathbb{C}$.

Let f be a function defined from $\mathbb{R}^3$ to $\mathbb{R}$, its Fourier transform is defined from $\mathbb{R}^3$ to $\mathbb{C}_2$ by

$$\widehat{f}(s,t) = \sum_{k=0}^{\infty} \sum_{l=0}^{\infty} \lambda_{kl} e^{iks} e^{jlt}, \tag{10.10}$$

$$\lambda_{kl} = \frac{1}{(2\pi)^2} \int_{-\pi}^{\pi} \int_{-\pi}^{\pi} e^{-ikx} e^{-jly} f(x,y) \, dx \, dy$$

and the inverse transform of $\widehat{f}$ is

$$f(x,y) = \int_{\mathbb{R}^2} e^{itx} e^{isy} \widehat{f}(s,t) \, ds \, dt.$$

If $f(x,y) = f_1(x)f_2(y)$, $\widehat{f}$ equals the product of the Fourier transforms of f_1 and f_2.

Proposition 10.7. *The Fourier transform of a function f of $L^2(\mathbb{R}^3)$ is*

$$\|f\|_{L^2} = \left\{ \sum_{k=0}^{\infty} \sum_{l=0}^{\infty} |\lambda|_{kl}^2 \right\}^{\frac{1}{2}} = \|\widehat{f}\|_{L^2}.$$

This equality is a consequence of the Fubini theorem and the equality $(2\pi)^{-1} \int_{-\pi}^{\pi} e^{i(m-n)x} \, dx = 1_{\{m=n\}}$ for every real x and for all n and m. Equation (10.6) is also true for the scalar product in $\mathbb{C}_2$.

Let $p \geq 2$, by the same argument as for Proposition 10.7, the norm $L^p(\mathbb{R}^3)$ of f is the sum $\|f\|_{L^p} = \{\sum_{k=0}^{\infty} \sum_{l=0}^{\infty} |\lambda_{kl}|^p\}^{\frac{1}{p}}$. From the expression of the inverse Fourier transform

$$\|f\|_{L^p} = \|\widehat{f}\|_{L^p}.$$

Moreover, each coefficient has the bound

$$|\lambda_{kl}|^p \leq \frac{1}{(2\pi)^2} \int_{-\pi}^{\pi} \int_{-\pi}^{\pi} |f(x,y)|^p \, dx \, dy$$

and there exist constants a and b such that

$$a\|f\|_{L^p}^p \leq \|\widehat{f}\|_{L^{p'}} \leq b\|f\|_{L^p}^p.$$

Let (X, Y) be a random variable with values in $\mathbb{R}^2$, with a joint distribution function F, the function $\psi_{XY}(s, t) = Ee^{isX + jtY}$ has derivatives with respect to (s, t) and

$$\frac{\partial^k}{\partial s^k} \frac{\partial^l}{\partial t^l} \psi_{XY}(s, t) = i^k j^l E\{X^k Y^l e^{isX + jtY}\}$$

whereas in $\mathbb{R}^2$, the function $\varphi_{XY}(s, t) = Ee^{i(sX + tY)}$ has the derivatives

$$\frac{\partial^k}{\partial s^k} \frac{\partial^l}{\partial t^l} \varphi_{XY}(s, t) = i^{k+l} E\{X^k Y^l e^{i(sX + tY)}\}.$$

A real function f of $C(\mathbb{C}_2)$ has a complex derivative $f^{(1)}$ such that

$$f(t + \delta) = f(t) + \delta_x f^{(1)}(t) + i\delta_y f^{(1)}(t) + j\delta_z f^{(1)}(t) + o(\|\delta\|), \ t, \delta \in \mathbb{C}_2.$$

The Cauchy equations for the derivatives in $\mathcal{C}$ are extended to $C(\mathbb{C}_2)$ in the next proposition.

Proposition 10.8. *A function $f(x + iy + jz) = P(x, y, z) + iQ(x, y, z) + jR(x, y, z)$ defined from $\mathbb{C}_2$ to $\mathbb{C}_2$ is continuously differentiable at $t = x + iy + jz$ if and only if the real functions P, Q and R belong to $C(\mathbb{R}^3)$ and*

$$\frac{\partial P(x, y, z)}{\partial x} = \frac{\partial Q(x, y, z)}{\partial y} = \frac{\partial R(x, y, z)}{\partial z},$$

$$\frac{\partial Q(x, y, z)}{\partial x} = -\frac{\partial P(x, y, z)}{\partial y},$$

$$\frac{\partial R(x, y, z)}{\partial x} = -\frac{\partial P(x, y, z)}{\partial z}.$$

Then, its derivative at t is

$$
\begin{aligned}
f^{(1)}(t) &= \frac{\partial P(x, y, z)}{\partial x} + i\frac{\partial Q(x, y, z)}{\partial x} + j\frac{\partial R(x, y, z)}{\partial x} \\
&= \frac{\partial Q(x, y, z)}{\partial y} - i\left\{\frac{\partial P(x, y, z)}{\partial y} - j\frac{\partial R(x, y, z)}{\partial y}\right\} \\
&= \frac{\partial R(x, y, z)}{\partial z} + j\left\{i\frac{\partial Q(x, y, z)}{\partial z} - \frac{\partial P(x, y, z)}{\partial z}\right\}.
\end{aligned}
$$

Proof. Let $t = x + iy + jz$ and $\delta = \delta_x + i\delta_y + j\delta_z$ in $\mathbb{C}_2$, the real functions P, Q, R are defined in $\mathbb{R}^3$ and the derivative of f has the form $f^{(1)} = A + iB + jC$ where the functions A, B and C are defined from $\mathbb{R}^3$ to $\mathbb{R}^3$. There exist real functions ε_k, $k = 1, 2, 3$, defined in $\mathbb{C}_2$ and converging to zero as $\|t\| \to 0$ and such that

$$f(t + \delta) = f(t) + (\delta_x, i\delta_y, j\delta_z)$$
$$\times \{A(x, y, z) + \varepsilon_1, iB(x, y, z) + i\varepsilon_2, jC(x, y, z) + j\varepsilon_3\}^T$$
$$= f(t) + \delta_x\{A(x, y, z) + iB(x, y, z) + jC(x, y, z)\} + \delta_y\{iA(x, y, z)$$
$$- B(x, y, z) + ijC(x, y, z)\} + \delta_z\{jA(x, y, z) + ijB(x, y, z)$$
$$- C(x, y, z)\} + o(\delta_x) + o(\delta_y) + o(\delta_z),$$

it follows that the partial derivatives of f with respect to (x, y, z) satisfy the above conditions, the equalities for $f^{(1)}$ follow. $\qquad\square$

The norm $L^2(\mathbb{C}_2)$ of $f(t)$ is

$$
\begin{aligned}
\|f^{(1)}(t)\|_2 &= \left\{\frac{\partial P(x,y,z)}{\partial x}\right\}^2 + \left\{\frac{\partial Q(x,y,z)}{\partial x}\right\}^2 + \left\{\frac{\partial R(x,y,z)}{\partial x}\right\}^2 \\
&= \left\{\frac{\partial Q(x,y,z)}{\partial y}\right\}^2 + \left\{\frac{\partial P(x,y,z)}{\partial y}\right\}^2 + \left\{\frac{\partial R(x,y,z)}{\partial y}\right\}^2 \\
&= \left\{\frac{\partial R(x,y,z)}{\partial z}\right\}^2 + \left\{\frac{\partial Q(x,y,z)}{\partial z}\right\}^2 + \left\{\frac{\partial P(x,y,z)}{\partial z}\right\}^2.
\end{aligned}
$$

Under the conditions of Proposition 10.8 and as δ tends to zero in $\mathbb{C}_2$

$$
f(t + \delta) = f(t) + \delta f^{(1)}(t) + o(\|\delta\|).
$$

Expansions of a $\mathbb{C}_2$-valued function are similar to the Taylor expansions in $\mathbb{R}^3$ in an orthogonal basis, via the representation of the function as $f(x + iy + jz) = P(x,y,z) + iQ(x,y,z) + jR(x,y,z)$. Let f belong to $C^n(\mathbb{C}_2)$, equivalently the real functions P, Q, R belong to $C^n(\mathbb{R}^3)$ and satisfy equalities similar to those of Proposition 10.8 for all derivatives up to n. As $\|\delta\|$ tends to zero

$$
f(t + \delta) = f(t) + \sum_{k=1}^{n} \frac{\delta^k}{k!} f^{(n)}(t) + o(\|\delta\|^k). \tag{10.11}
$$

The isometry between $\mathbb{R}^3$ and $\mathbb{C}_2$ extends to higher dimensions. Let p be an integer larger or equal to 3 and let $p - 1$ roots $(i_1, \ldots, i_{p-1})$ of -1, they define a complex space $\mathbb{C}_{p-1}$ isometric to $\mathbb{R}^p$ by the bijection $(x_1, \ldots, x_p) \mapsto x_1 + \sum_{k=2}^{p-1} i_{k-1} x_k$. Functions of $C^n(\mathbb{C}_{p-1})$ have expansions like (10.11) under Cauchy conditions of dimension p and order n.

10.7 Stochastic integrals

In Section 5.9, the stochastic integral of a predictable process A with respect to an adapted process M of $\mathcal{M}^2_{0,loc}(\mathbb{R}^2_+)$ is defined in $L^{2,loc}$ from the integral of A^2 with respect to the predictable compensator $< M >$ of M in rectangles $R_z = [0, z]$ and in rectangles $R_{]z,z']}$, with ordered z and z' in $\mathbb{R}^2_+$. If z and z' are not ordered, for example their components satisfy $z_1 < z_2$ and $z_1' > z_2'$, the increment of the martingale between z and z' follows the same rule (5.19) and it is the opposite of its increment between the ordered points (z_1, z_2') and (z_1', z_2). By splitting a surface into rectangles with perimeter tending to zero, the stochastic integral is defined over every Borel set of $\mathbb{R}^2_+$.

The integral satisfies $\int_{R_z} \int_{R_{z_{k-1}}} \cdots \int_{R_{z_2}} dM_{s_k} \cdots dM_{s_1} = (k!)^{-1} M_{R_z}^k$ and, applying this equality, it follows that

$$\int_{R_z} \cdots \int_{R_{z_2}} A_{s_k} \cdots A_{s_1} dM_{s_1} \cdots dM_{s_k} = (k!)^{-1} \left\{ \int_{R_z} A_s \, dM_s \right\}^k.$$

With $k = 2$, if $\int_{R_z} A_s \, dM_s$ belongs to $\mathcal{M}_{0,loc}^2(\mathbb{R}_+^2)$

$$E \int_{R_z} \int_{R_{z_2}} A_{s_2} A_{s_1} dM_{s_1} \, dM_{s_2} = \frac{1}{2} E \left\{ \int_{R_z} A_s \, dM_s \right\}^2$$

$$= \frac{1}{2} E \int_{R_z} A_s^2 \, d < M >_s .$$

Let M_1 and M_2 be local martingales of $\mathcal{M}_{0,loc}^2(\mathbb{R}_+^2)$ and let A_1 and A_2 be predictable processes of $L_{loc}^2(< M >_1)$ and $L_{loc}^2(< M >_2)$, respectively, the process $X_z = \int_{R_z} \int_{R_{z_2}} A_{s_2} B_{s_1} dM_{s_1} \, dM_{s_2}$ is defined from (1.25) as the scalar product of the local martingales $A_1.M_1 = \int_{R_z} A_1 \, dM_1$ and $\int_{R_z} A_2 \, dM_2$.

Let $\mathcal{M}_{0,S,loc}^p(\mathbb{R}_+^2)$ be the space of the L^p local strong martingales with expectation zero. For M belonging to $\mathcal{M}_{0,S,loc}^4(\mathbb{R}_+^2)$, let $M^{(1)}(z)$ be the martingale on $\mathbb{R}_+$ defined at fixed z_2, with respect to the marginal filtration $\mathbb{F}_1$, and let $M^{(2)}(z)$ be the martingale on $\mathbb{R}_+$ defined at fixed z_1, with respect to the marginal filtration $\mathbb{F}_2$. Let A be a $\mathbb{F}_1$ and $\mathbb{F}_2$-predictable process, belonging to $L^2(< M^{(1)} >< M^{(2)} >)$. The integral of A with respect to $M^{(1)}$, then $M^{(2)}$, is denoted $\int_{R_z} A \, dMM$ and the process A belongs to $L^2(MM)$ if it is $L^2(< M^{(1)} >< M^{(2)} >)$-integrable. Cairoli and Walsh (1975) proved that the integral $A.MM_z = \int_{R_z} A \, dMM$ belongs to $\mathcal{M}_{0,S,loc}^2$ and it is continuous if M is, and

$$< A.MM, B.MM >_z = \int_{R_z^2} AB \, d < M^{(1)} > d < M^{(2)} >,$$

$$E(A.MM_z \, M_z) = 0.$$

A Wiener process W in $[0,1]^2$ is a strong martingale with respect to its natural filtration, its expectation is zero and it belongs to $L^p([0,1]^2)$ for every $p \geq 1$. Let $\mathbb{F}_W$ be the filtration generated by a Wiener process W in $\mathbb{R}_+^2$, Wong and Zakaï (1974) established that every M of $\mathcal{M}_{S,loc}^2(\mathbb{R}_+^2, \mathbb{F}_W)$ has an integral representation

$$M_z = M_0 + \varphi.W_z + \psi.WW_z, \quad z \in \mathbb{R}_+^2,$$

where φ is a function of $L^2(W)$ and ψ is a function of $L^2(WW)$, and $E(M_z - M_0)^2 = \int_{R_z} \varphi_z^2 \, d < M >_z + \int_{R_z} \varphi_z^2 \, d < M^{(1)} >_z d < M^{(2)} >_z$.

By a change of variable, for every $z' < z$ in $\mathbb{R}_+^2$, the variations of M in $R_{z',z}$ have the representation

$$M_{R_{z',z}} = M_{z'} + \varphi.W_{R_{z',z}} + \psi.WW_{R_{z',z}}.$$

A Poisson point process with parameter λ has the moment generating function

$$\varphi_{N_{R_z}}(u) = \exp\{\lambda |z_1| \, |z_2|(e^u - 1)\}$$

and a martingale with respect to the natural filtration of the process has a similar representation.

Let us consider integrals in open balls with radius a, $B_a(z)$, z in $\mathbb{R}^2$. By the isometry between $\mathbb{R}^2$ and $\mathbb{C}$, every z' of $B_a(z)$ is written $z' = z + \rho e^{i\theta}$ with $0 < r < a$ and θ in $[-\pi, \pi]$, and the integral of the Brownian motion W in $B_a(z)$ satisfies $E\{\int_{B_a(z)} dW_s\}^2 = \int_{B_a(z)} ds = 0$, z in $\mathbb{C}$, so it is a.s. zero. For every left-continuous function f with right-hand limits in $L^2(B_a(z))$, the integral $\int_{B_a(z)} f(s) \, dW_s$ is defined as a linear map such that

$$E\left\{ \int_{B_a(z)} f(s) \, dW_s \right\}^2 = \frac{1}{2\pi} \int_{-\pi}^{\pi} \int_0^r f^2(z + re^{i\theta})e^{i\theta} \, dr \, d\theta, \quad z \in \mathbb{C},$$

and this defines $\int_{B_a(z)} f(s) \, dW_s$ like in $L^2(\mathbb{R}^2)$. A Wiener process W in $\mathbb{R}^2$ also has an integral $\int_{B_a(z)} dW_s = 0$ a.s., and for every z in $\mathbb{R}^2$ or in $\mathbb{C}$

$$E\left\{ \int_{B_a(z)} f(s) \, dW_s \right\}^2 = \int_{B_a(z)} f^2(s) \, ds - \left\{ \int_{B_a(z)} f(s) \, ds \right\}^2.$$

A Poisson point process with parameter λ has the moment generating function $L_{N_{B_a(z)}}(u) = \exp\{\lambda \pi a^2(e^u - 1)\}$ in balls $B_a(z)$, for every z in $\mathbb{C}$. For every function f of $L^2(\mathbb{R}^2)$, $E\{\int_{B_a(z)} f(s) \, dN_s\}^2 = \lambda \int_{B_a(z)} f^2(s) \, ds$ or it equals $E \int_{B_a(z)} f^2(s) \, d\Lambda(s)$ if N has a cumulative intensity Λ.

Let A be a subset of $\mathbb{R}^2$ and let f be a left-continuous Borel function with right-hand limits in $L^2(A)$, for the Lebesgue measure μ. The function f is the limit as ε tends to zero of a sum $S_\varepsilon = \sum_{i=1}^{n_\varepsilon} \zeta_i 1_{B_r(x_i)}$ where $\zeta_i = f(x_i)$ and n_ε is the number of balls with radius $r = r_\varepsilon$ belonging to the interval $I_2(\varepsilon) =]\frac{\varepsilon}{\sqrt{2}}, \varepsilon]$, centered at points x_i such that $|x_i - x_j| = \varepsilon$ and defining the coverage of A by r_ε-balls in an ε-net in A (Fig. 9.2). Then $n_\varepsilon = O(\varepsilon^{-2}|A|)$, the integral

$$\int_A f(z) \, dz = \lim_{\varepsilon \to 0} \pi r_\varepsilon^2 \sum_{i=1}^{n_\varepsilon} f(x_i)$$

is finite and

$$\int_A f^2(z)\, dz = \lim_{\varepsilon \to 0} \sum_{i=1}^{n_\varepsilon} \sum_{j=1}^{n_\varepsilon} \zeta_i \zeta_j \mu(B_r(x_i) \cap B_r(x_j))$$

$$= \lim_{\varepsilon \to 0} \frac{r_\varepsilon^2}{2}(\theta_\varepsilon - \sin\theta_\varepsilon) \sum_{i=1}^{n_\varepsilon} \sum_{x_j \in \mathcal{V}(x_i)} f(x_i)f(x_j),$$

where $\theta_\varepsilon = 2\arccos\frac{\varepsilon}{2r_\varepsilon}$, hence $\cos\frac{1}{2}\theta_\varepsilon$ belongs to the interval $[\frac{1}{\sqrt{2}}, \frac{1}{2}]$, so $\lim_{\varepsilon \to 0}(\theta_\varepsilon - \sin\theta_\varepsilon)$ is bounded, and this sum converges as ε tends to zero. Its limit also equals $\int_A f^2(re^{i\theta})\, dr\, d\theta$.

The integrals $\int_A f_z\, dW_z$ and $\int_A f_z\, dW_z^{1)}\, dW_z^{2)}$ with respect to the Brownian motion are defined as the linear map with L^2-norms such that

$$E\left(\int_A f_z\, dW_z\right)^2 = \int_A f_z^2\, dz = E\left(\int_A f_z\, dW_z^{1)}\, dW_z^{2)}\right)^2.$$

The Brownian processes have therefore representations in the balls, with the complex and the real parametrizations. Similar representations hold for integrals with respect to the Brownian bridge and with respect to the martingale $N - \Lambda$, for a Poisson processes N with a cumulative intensity Λ.

Appendix A

Probability

A.1 Definitions and convergences in probability spaces

Let $(\Omega, \mathcal{A}, P)$ be a probability space and let $(A_n)_{n \geq 0}$ be a sequence of measurable sets. They are independent if and only if for every finite subset $\mathcal{K}$ of $\mathbb{N}$, $P(\cap_{k \in \mathcal{K}} A_k) = \prod_{k \in \mathcal{K}} P(A_k)$.

Lemma A.1 (Borel-Cantelli's lemma). *A sequence of measurable sets* $(A_n)_{n \geq 0}$ *satisfies* $\limsup_{n \to \infty} A_n = \emptyset$ *a.s. if and only if* $\sum_{n \leq 0} P(A_n)$ *is finite. If the sets* A_n *are independent and* $\sum_{n \leq 0} P(A_n)$ *is infinite, then* $\limsup_{n \to \infty} A_n = \Omega$ *a.s.*

Let $(X_n)_{n \geq 0}$ be a sequence of real random variables on $(\Omega, \mathcal{A}, P)$, the σ-algebra generated by X_n is generated by the sets $\{X_n < x\}$, x in $\bar{\mathbb{R}}$. The variables are independent if and only if for every finite subset $\mathcal{K}$ of $\mathbb{N}$ and every sequence $(x_k)_{k \in \mathcal{K}}$ of $\bar{\mathbb{R}}$, $P(\cap_{k \in \mathcal{K}} \{X_k < x_k\}) = \prod_{k \in \mathcal{K}} P(X_k < x_k)$. An equivalent characterization of the independence of the sequence $(X_n)_{n \geq 0}$ is the equality $E \prod_{k \in \mathcal{K}} f_k(X_k) = E \prod_{k \in \mathcal{K}} f_k(X_k)$, for every sequence of measurable functions $f_k : \mathbb{R} \mapsto \mathbb{R}$ and for every finite subset $\mathcal{K}$ of $\mathbb{N}$. The distribution of a vector of independent random variables is the product of their distributions and their characteristic function (Fourier transform) is the product of the marginal characteristic functions.

A sequence of random variables $(X_n)_{n \geq 0}$ converges a.s. to a variable X if $P(\limsup_n X_n = \liminf_n X_n < \infty) = 1$ and it is equivalent to the existence of a convergent series $(\varepsilon_n)_{n \geq 0}$ such that $\sum_{n \geq 0} P(|X_n - X| > \varepsilon)$ converges. The convergence properties of random variables are not equivalent and their relationships depend on the integrability properties of the variables. For all conjugates p and p', there exist constants $k_1 \leq p$ and $k_2 \geq (p')^{-1}$ such that for every function f and for every random variable X

$$k_1\{E|f(X)|^p\}^{\frac{1}{p}} \le \{E|f(X)|^{p'}\}^{\frac{1}{p'}} \le k_2\{E|f(X)|^p\}^{\frac{1}{p}}$$

and the convergences in expectation are equivalent. The a.s. convergence of a sequence of random variables implies its convergence L^p, for every $p \le 1$, and the convergence in a space L^p implies the convergence in probability. Conversely, every sequence of variables converging in probability to a variable X has a sub-sequence that converges a.s. to X. The weak convergence of a sequence of random variables is equivalent to the convergence of their characteristic functions to the characteristic function of a variable X.

A random variable X belongs to L^1 if and only if $\sum_{n\ge 0} P(|X| \ge n)$ is finite. Let $(X_n)_{n\ge 0}$ be a sequence of random variables of L^p, $p \le 1$. The variable sequence $(X_n)_{n\ge 1}$ is *equi-integrable* if

$$\lim_{a\to\infty} \sup_{n\ge 1} E(|X_n|1_{|X_n|>a}) < \infty.$$

If there exists a variable X of $L^1(P)$ such that $|X_n| \le |X|$ a.s. for every integer n, then $(X_n)_{n\ge 1}$ is equi-integrable. The $L^p(P)$ integrability for $p > 1$ is a sufficient condition for the equi-integrability and the following assertions are equivalent as n tends to infinity (Neveu, 1970)

1. $(X_n)_{n\ge 1}$ is equi-integrable and converges in probability to a variable X,
2. $(X_n)_{n\ge 1}$ converges in $L^1(P)$ to a variable X of $L^1(P)$.

The limiting behaviour of the supremum of weighted normal variables is deduced from the 0-1 law of the Borel-Cantelli Lemma A.1.

Proposition A.1. *Let* $(a_i)_{i=1,\ldots,n}$ *be a decreasing sequence of positive real numbers and let* $(X_n)_{n\ge 1}$ *be a sequence of independent normal variables on* $(\Omega, \mathcal{A}, P)$, *then*

$$P\Big(\sup_{n\ge 1} |a_n X_n| < \infty\Big) = 1 \text{ if } \lim_{n\to\infty} a_n\sqrt{\log n} < \infty,$$

$$= 0 \text{ if } \lim_{n\to\infty} a_n\sqrt{\log n} = \infty.$$

On a probability space $(\Omega, \mathcal{A}, P)$ let $(\mathcal{F}_n)_{n\ge 0}$, with a discrete filtration on $(\Omega, \mathcal{A}, P)$. A sequence of variables $(X_n)_{n\ge 0}$ is *adapted* if X_n is $\mathcal{F}_n$-measurable for every integer n. It follows that for a discrete stopping T, $X_T 1_{\{T\le n\}}$ is $\mathcal{F}_n$-measurable. A *stopping time* T is a random variable such that the event $\{T \le n\}$ is $\mathcal{F}_n$-measurable, then $\{T > n\}$ is also $\mathcal{F}_n$-measurable. It is predictable if $\{T \le n\}$ is $\mathcal{F}_{n-1}$-measurable.

A sequence of adapted random variables $(X_n)_{n\ge 0}$ is a martingale with respect to a filtration $(\mathcal{F}_n)_{n\ge 0}$ if $E(X_{n+1}|\mathcal{F}_n) = X_n$ for every integer n. A well-known example of martingale defined from a stopped sequence is the Snell envelope. It is defined for an adapted sequence $(Z_n)_{n=0,\ldots,N}$ with respect to a filtration $(\mathcal{F}_n)_{n=0,\ldots,N}$ as $U_N = Z_N$ and $U_n = \max\{Z_n, E(U_{n+1}|\mathcal{F}_n)\}$, for every n belonging to $\{0,\ldots,N-1\}$. Then $\nu_0 = \inf\{n : U_n = Z_n\}$ is a stopping time and the sequence $(U_n \wedge \nu_0)_{n=0,\ldots,N}$ is a martingale, with respect to the filtration $(\mathcal{F}_n)_{n=0,\ldots,N}$.

Every positive supermartingale $(X_n)_{n\geq 0}$ converges a.s. to a limit X_∞ and $E(X_\infty|\mathcal{F}_n) = X_n$ for every integer n. A submartingale $(X_n)_{n\geq 0}$ such that $\sup_{n\in\mathbb{N}} EX_n^+$ is finite converges a.s. to a limit in L^1. A martingale of L^1 converges a.s. to a limit in L^1.

Theorem A.1 (Neveu, 1972). *Let $(X_n)_{n\geq 0}$ be a martingale such that $\sup_{n\in\mathbb{N}} E|X_n| (\log|X_n|)^+$ is finite, then $E\sup_{n\in\mathbb{N}} |X_n|$ is finite.*

With a continuous filtration $(\mathcal{F}_t)_{t\geq 0}$, a *stopping time* T is a random variable such that the event $\{T \leq t\}$ is $\mathcal{F}_t$-measurable, then $\{T \geq t\}$ is $\mathcal{F}_{t-}$-measurable, the time variable T is a predictable stopping time if $\{T \leq t\}$ is $\mathcal{F}_{t-}$-measurable. An adapted process X satisfies: X_t is $\mathcal{F}_t$-measurable for every real t and $X_T 1_{\{T\leq t\}}$ is $\mathcal{F}_t$-measurable for every real t and for every stopping time T.

Let $(X_n)_{n\geq 1}$ be a sequence of independent random variables of L^2 and let $S_n = \sum_{i=1}^n X_i$. Wald's equalities for an integrable stopping time T are
$$ES_T = E(X_1)E(T), \text{ if } X_1 \in L^1, T \in L^1,$$
$$ES_T^p = E(X_1^p)E(T), \text{ if } X_1 \in L^p, T \in L^1,\ p > 1. \qquad (A.1)$$
It is proved recursively for integer p and for centered variables writing
$$ES_{T\wedge n}^2 = E(S_{T\wedge(n-1)} + X_n 1_{\{T\geq n\}})^2$$
$$= E(S_{T\wedge(n-1)}^2 + X_n^2 1_{\{T\geq n\}})$$
$$= E(X_1^2)\{\sum_{k=1}^n P(T \geq k)\} = E(X_1^2)\,E(T)$$
since $\{T \geq n\}$ is $\mathcal{F}_{n-1}$-measurable, and the result is deduced for $p = 2$. It is generalized to $p > 1$

From the strong law of large numbers, $n^{-1}S_n$ converges a.s. to EX_1 if the variables belong to $L^2(P)$. It converges in probability to EX_1 if they belong to $L^1(P)$. The criteria of integrability for S_n and for the variables X_i, or their maxima are related, by the Burkholder-Davis-Gundy inequality and by other properties. Let $(a_i)_{i=1,\ldots,n}$ be a sequence of positive real numbers and $A_n = \sum_{i=1}^n a_i$, then
$$P(|S_n| > A_n) \leq \sum_{i=1}^n P(|X_i| > a_i)) \leq \sum_{i=1}^n \frac{E|X_i|}{a_i}.$$

Proposition A.2. *Let $(X_n)_{n\geq 0}$ be a sequence of independent normal variables and let $\alpha = (\alpha_n)_{n\geq 0}$ be a real sequence such that $\|\alpha\|_2 = (\sum_{n\geq 0}\alpha_n^2)^{\frac{1}{2}}$ is finite, then $\sum_{n\geq 0}\alpha_n X_n$ is a normal variable with variance $\|\alpha\|_2^2$.*

The law of the iterated logarithm for sums of independent Gaussian variables can be written as in Section 6.7.

Theorem A.2. *Let $(X_k)_{k\geq 1}$ be a sequence of independent normal variables, let $(\alpha_n)_{n\geq 1}$ be a sequence of real numbers and let $A_n = \sum_{n\geq 1}\alpha_n^2$. For very real function $h > 0$ on $\mathbb{R}$ such that $\sum_{n=1}^{\infty} h_n^{-1}$ is finite, the variable $Y_n = \sum_{n\geq 1}\alpha_n X_n$ satisfies*

$$\limsup_{n\to\infty}\frac{Y_n}{\sqrt{2A_n\log h_n}} \leq 1, \qquad \liminf_{n\to\infty}\frac{Y_n}{\sqrt{2A_n\log h_n}} \geq -1, \text{ a.s.}$$

Proof. The sum of the probabilities of the sets $\{Y_n > \sqrt{2A_n\log n}\}$ is bounded using the Laplace transform of the Gaussian variable Y_n

$$\sum_{n\geq 2} P(Y_n > \sqrt{2A_n\log h_n}) = \sum_{n\geq 2}\exp\left\{-\frac{2A_n\log h_n}{2varY_n}\right\} = \sum_{n\geq 2} h_n^{-1}$$

and it is finite. The result follows from the 0-1 law (Lemma A.1). □

Chow and Lai (1973) related the behaviour of $\|\alpha\|_2$ to other properties of a sum of weighted variables $\sum_{n\geq 0}\alpha_n X_n$, in particular for $\alpha \geq 2$ there exists a constant B_α such that

$$\sup_{n\geq 1} E|\sum_{i=1}^{n}\alpha_i X_i|^\alpha \leq B_\alpha\left(\sum_{n\geq 0}\alpha_n^2\right)^\alpha E|X_1|^\alpha.$$

Let $(X_k)_{k\geq 1}$ be a sequence of i.i.d. centered random variables. For every $\alpha \geq 1$, the next statements are equivalent
(1) the variables X_i belong to $L^\alpha(P)$,
(2) $\lim_{n\to\infty} n^{-\frac{1}{\alpha}}X_n = 0$ a.s.,
(3) there exists a sequence of real numbers $(\alpha_n)_{n\geq 0}$ such that $\sum_{n\geq 0}\alpha_n^2$ is finite and $\lim_{n\to\infty} n^{-\frac{1}{\alpha}}\alpha_{n-i}X_i = 0$ a.s.

Theorem A.3 (Chow and Lai, 1973). *Let $(X_k)_{k\geq 1}$ be a sequence of i.i.d. centered random variables. The following statements are equivalent*
(1) $Ee^{t|X_1|} < \infty$, for every real t,
(2) $\lim_{n\to\infty}\frac{X_n}{\log n} = 0$ a.s.
(3) there exists a sequence of real numbers $(\alpha_n)_{n\geq 0}$ such that $\sum_{n\geq 0}\alpha_n^2$ is finite and $\lim_{n\to\infty}\alpha_{n-i}\frac{X_i}{\log n} = 0$ a.s.

The variable $M_n = \max_{i=1,...,n} X_i$, with independent uniform variables X_i on the interval $[0,1]$ has first moments

$$EM_n = \frac{n}{n+1}, \quad EM_n^p = \frac{n}{n+p}, \quad p \geq 2.$$

For every A in $]0,1[$, $F(A)$ belongs to $]0,1[$ and $P(M_n < A) = F^n(A)$ tends to zero as n tends to infinity therefore $\limsup_{n\to\infty} M_n = 1$ a.s., similarly $\liminf_{n\to\infty} \min_{i=1,...,n} X_i = 0$ a.s. The maximum M_n of n independent variables X_i with a common distribution function F on $\mathbb{R}$ have the expectation $EM_n = n \int_{\mathbb{R}^n} x_n 1_{x_n=\max\{x_1,...,x_n\}} \prod_{i=1,...,n} dF(x_i) = nE\{X_1 F^{n-1}(X_1)\}$ and higher moments $EM_n^p = nE\{X_1^p F^{n-1}(X_1)\}$, its extrema satisfy $\limsup_{n\to\infty} M_n = +\infty$ a.s. and $\liminf_{n\to\infty} \min_{i=1,...,n} X_i = -\infty$ a.s. More generally, $\limsup_{n\to\infty} M_n$ and $\liminf_{n\to\infty} \min_{i=1,...,n} X_i$ are a.s. equal to the maximum and, respectively, the minimum of the support of the distribution of the variables X_i.

Bennett and Hoeffding inequalities. Let $(X_i)_{i=1,...,n}$ be independent variables with respective values in $[a,b]$, then the Bennett inequality is modified as follows. For every $t > 0$ and every integer n

$$P(S_n - ES_n \geq t) \leq \exp\left\{-n\phi\left(\frac{t}{n(b-a)^2}\right)\right\}$$

where $\phi(x) = (1+x)\log(1+x) - x$. In Hoeffding's inequality, the bound is exponential.

Theorem A.4 (Hoeffding's inequality). *Let $(X_i)_{i=1,...,n}$ be independent variables with respective values in $[a_i, b_i]$, then for every $t > 0$*

$$P(n^{-\frac{1}{2}}(S_n - ES_n) \geq t) \leq \exp\left\{-\frac{2t^2}{\sum_{i=1}^n (b_i - a_i)^2}\right\}. \tag{A.2}$$

It is proved along the same arguments as Chernoff's inequality. The following bound is weaker and easily proved. For every $t > 0$

$$P(n^{-\frac{1}{2}}(S_n - ES_n) \geq t) \leq \exp\left\{-\frac{t^2}{2\sum_{i=1}^n (b_i - a_i)^2}\right\}. \tag{A.3}$$

Proof. For every integer n, the moment generating function φ of $S_n - ES_n$ satisfies $\log\varphi(0) = 0$, $\varphi'(0) = 0$ for the centered variables and $\varphi''(0) \leq B_n^2$, with the constants $B_n^2 = \sum_{i=1}^n (b_i - a_i)^2$. Therefore

$$\log P(n^{-\frac{1}{2}}(S_n - ES_n) \geq t) \leq \liminf_{\lambda \to 0}\{\log\varphi(\lambda) - \lambda t\}$$

$$= \inf_\lambda\left\{\frac{1}{2}\lambda^2 B_n^2 - \lambda t\right\} = -\frac{t^2}{2B_n^2}$$

which yields (A.3). $\qquad\square$

The sum S_n has the moments $E(S_n - ES_n)^k \leq \sum_{i=1}^n (b_i - a_i)^k \leq \|a - b\|_\infty^k$ for every $k \geq 2$, from Bennett's inequality this implies

$$P(S_n - ES_n \geq t) \leq \exp\left\{-n\phi\left(\frac{t}{n\|a - b\|_\infty}\right)\right\}.$$

Bernstein inequality for a sum S_n of n independent and centered variables X_i in a metric space such that $\|X_i\|_\infty \leq M$ for every $i \geq 1$ is

$$P(S_n \geq t) \leq \exp\left\{-\frac{t^2}{2\sigma_n + \frac{2}{3}Mt}\right\}, t > 0,$$

where $\operatorname{var} S_n = \sigma_n^2$ (Giné, 1974). It is a slight modification of inequality (A.3) above.

From Freedman (1975), a martingale $S_n = \sum_{i=1}^n X_i$ with quadratic variations $T_n = \sum_{i=1}^n E(X_i^2 | \mathcal{F}_{i-1})$ and such that $|X_n| \leq 1$ satisfies the inequality

$$P(S_n > a, T_n \leq b, n \geq 1) \leq \left(\frac{b}{a+b}\right)^{a+b} \leq \exp\left\{-\frac{a^2}{2(a+b)}\right\}$$

for all $a > 0$ and $b > 0$.

A.2   Boundary-crossing probabilities

Robbins and Siegmund (1970) proved many uniform boundary-crossing probabilities for the Brownian motion W. In particular, with the normal density $f_\mathcal{N}$ and its distribution function $F_\mathcal{N}$, for all real $a > 0$ and b

$$P\left(\sup_{t \geq 1}(W_t - at) \geq b\right) = P\left(\sup_{0 \leq t \leq 1}(W_t - bt) \geq a\right)$$

$$= 1 - F_\mathcal{N}(b + a) + e^{-2ab}F_\mathcal{N}(b - a),$$

$$P\left(\sup_{t \geq 1} t^{-1}W_t \geq a\right) = P\left(\sup_{0 \leq t \leq 1} W_t \geq a\right) = \{1 - F_\mathcal{N}(a)\}$$

and non uniform inequalities such as

$$P(\exists t : |W_t| \geq \sqrt{t}\alpha^{-1}\{\log t^{1-\gamma} + \alpha(a)\}) = 1 - F_\mathcal{N}(a)$$

$$+ f_\mathcal{N}(a)\frac{\int_0^\infty F_\mathcal{N}(a - y)y^{-\gamma}\, dy}{\int_0^\infty f_\mathcal{N}(a - y)y^{-\gamma}\, dy},$$

$$\alpha(x) = x^2 + 2\log\int_0^\infty f_\mathcal{N}(y - x)y^{-\gamma}\, dy,$$

where $\gamma < 1$, $\alpha(x) \sim_{x \to \infty} x^2$ and α^{-1} is the inverse function of α.

Jain, Jogdeo and Stout (1975, Theorem 5.2) proved the a.s. convergence for a martingale sequence $S_n = \sum_{i=1}^n X_i$ such that the martingale differences X_i are stationary and ergodic and $EX_i^2 = 1$: Let $\varphi > 0$ be an increasing function and $I(\varphi) = \int_1^\infty t^{-1}\varphi(t)e^{-\frac{1}{2}\varphi^2(t)}\,dt$, then

$$P\left\{S_n > \frac{\sqrt{n}}{\varphi(n)}\text{a.s.}\right\} = 0 \quad \text{if } I(\varphi) < \infty,$$
$$= 1 \quad \text{if } I(\varphi) = \infty.$$

Other level crossing problems than those presented in Section 6.7 were reviewed by Blake and Lindsey (1973). For a Brownian motion B and with the Lebesgue measure μ, the number of crossings of zero is replaced by the duration over this threshold $\mu\{s \in [0,\tau]; B(s) > 0\}$. From Erdos and Kac (1947)

$$P\{\mu\{s \in [0,\tau] : B(s) > 0\} < t\} = 2\pi^{-1}\arcsin(\tau^{-1}t)^{\frac{1}{2}}$$

and this result has been extended by other authors.

Let $N_c = \min\{k : k^{-\frac{1}{2}}S_k > c\}$, in the Binomial case $EN_c < \infty$ if and only if $c < 1$ (Blakewell and Freeman, 1964) and in the general case of i.i.d. variables X_i with $EX_i = 0$ and $\text{Var}X_i = \sigma^2$, the necessary and sufficient condition is $c^2 < \sigma^2$ (Chow, Robbins and Teicher, 1965). If $\sigma^2 = 1$, $E(N_c^2) < \infty$ if and only if $c < \sqrt{3 - \sqrt{6}}$ (Chow and Teicher, 1966).

For centered Gaussian processes with a stationary covariance function $r(t)$, the number $M(t,u)$ of values of t such that $X_t = u$ has a expectation given by the Rice formula (Dudley, 1973)

$$EM(t,u) = \frac{T}{\pi}\left(-\frac{r''(0)}{r(0)}\right)^{\frac{1}{2}}\exp\left\{-\frac{u^2}{2r(0)}\right\}.$$

A.3 Distances between probabilities

Let P and Q be probabilities on a measurable space $(\mathcal{X}, \mathcal{B})$, with the Borel σ-algebra, and let $\mathcal{F}$ be the set of positive functions f on $(\mathcal{X}, \mathcal{B})$ such that f and f^{-1} are bounded. Kullback's information of P with respect to Q is defined as

$$I_Q(P) = \sup_{f \in \mathcal{F}}\left\{\int_{\mathcal{X}} \log f\,dP - \log\int_{\mathcal{X}} f\,dQ\right\}.$$

Theorem A.5. *Kullback's information $I_Q(P)$ is finite if and only if P is absolutely continuous with respect to Q and $g = \frac{dP}{dQ}$ belongs to $L^1(P)$, then $I_Q(P)$ is a lower semi-continuous real function satisfying*

$$I_Q(P) = \int_{\mathcal{X}} \log\left(\frac{dP}{dQ}\right) dP$$

and there exists a continuous function $\phi : \mathbb{R} \mapsto [0,2]$ such that $\phi(0) = 0$ and

$$\|P - Q\|_1 \leq \sup_{A \in \mathcal{B}} |P(A) - Q(A)| \leq \phi(I_Q(P)).$$

It follows that the convergence of a sequence of probabilities $(P_n)_n$ to a limiting probability P is satisfied if $\lim_{n\to\infty} I_Q(P_n) = I_Q(P)$ for some dominating probability Q, which is equivalent to $\lim_{n\to\infty} \|P_n - P\| = 0$. The distance $\|P - Q\|_1$ defined above is the supremum over the balls and the convergence for this metric implies the convergence in probability, it is equivalent to the convergence in total variation in a probability space endowed with the Borel σ-algebra. Other relationships between probability metrics are reviewed by Gibbs and Su (2002).

The Hausdorff metric on a separable probability space $(\Omega, \mathcal{A}, P)$ is $\rho(A, B) = P(A \Delta B)$ where $A \Delta B = (A \cup B) \setminus (A \cap B)$, where A and B are measurable sets. It is extended to sub-σ fields $\mathcal{B}$ and $\mathcal{C}$ of $\mathcal{F}$

$$\delta(\mathcal{B}, \mathcal{C}) = \max\left\{ \sup_{B \in \mathcal{B}} \inf_{C \in \mathcal{C}} P(B \Delta C), \sup_{C \in \mathcal{C}} \inf_{B \in \mathcal{B}} P(B \Delta C) \right\}.$$

Let (X, δ) be a metric space, with the Hausdorff metric δ.

Lemma A.2 (Rogge's lemma (1974)). *Let $\mathcal{A}$ and $\mathcal{B}$ be sub-σ-algebras of $\mathcal{F}$, then for every $\mathcal{B}$-measurable function $f : X \to [0, 1]$*

$$\|P^{\mathcal{A}} f - f\|_1 \leq 2\delta(\mathcal{A}, \mathcal{B})\{1 - \delta(\mathcal{A}, \mathcal{B})\},$$
$$\|P^{\mathcal{A}} f - f\|_2 \leq [\delta(\mathcal{A}, \mathcal{B})\{1 - \delta(\mathcal{A}, \mathcal{B})\}]^{\frac{1}{2}}.$$

Let Φ be the set of all $\mathcal{F}$-measurables functions $f : X \to [0, 1]$. It follows that for every sub-σ-algebras $\mathcal{A}$ and $\mathcal{B}$ of $\mathcal{F}$

$$\sup_{f \in \Phi} \|P^{\mathcal{A}} f - P^{\mathcal{B}} f\|_2 \leq [2\delta(\mathcal{A}, \mathcal{B})\{1 - \delta(\mathcal{A}, \mathcal{B})\}]^{\frac{1}{2}}.$$

It follows that

$$\delta(\mathcal{A}, \mathcal{B}) \leq \sup_{f \in \Phi} \|P^{\mathcal{A}} f - P^{\mathcal{B}} f\|_1$$

and $\sup_{f\in\Phi} \|P^{\mathcal{F}_n}f - P^{\mathcal{F}_\infty}f\|_1$ tends to zero for every sequence of sub-σ-algebras of $\mathcal{F}$ such that $\delta(\mathcal{F}_n, \mathcal{F}_\infty) \to 0$ as n tends to infinity. This convergence was applied to the equi-convergence of martingales (Boylan, 1971).

Let $(\mathbb{X}, \mathcal{X}, d)$ be a metric space with the Borel σ-algebra and let $\mathcal{P}(\mathbb{X})$ be a space of probability measures on $(\mathbb{X}, \mathcal{X})$. The Prohorov distance between probabilities P and Q of $\mathcal{P}(\mathbb{X})$ is defined by

$$\Pi(P, Q) = \inf\{\varepsilon > 0; P(A) < Q(A^\varepsilon) + \varepsilon, \; Q(A) < P(A^\varepsilon) + \varepsilon, \; \forall A \in \mathcal{X}\}$$

where the ε-neighbourhood of a subset A of $\mathbb{X}$ is

$$A^\varepsilon = \{x \in \mathbb{X}; \inf_{y \in \mathbb{X}} d(x, y) < \varepsilon\}, \; \varepsilon > 0.$$

Equivalently

$$\Pi(P, Q) = \inf\{\varepsilon > 0; |P(A) - Q(B)| < \varepsilon, A, B \in \mathcal{X} \text{ s.t. } d(A, B) < \varepsilon\}.$$

Let $(C[0, 1], \mathcal{C}, \|\cdot\|)$ be the space of continuous real functions defined on $[0, 1]$, provided with the norm uniform $\|x\| = \sup_{t\in[0,1]} |x(t)|$, for every continuous function x of $[0, 1]$, and the Borel σ-algebra. Let $X = (X_t)_{t\in[0,1]}$ and $Y = (Y_t)_{t\in[0,1]}$ be continuous processes indexed by $[0, 1]$, the Prohorov distance between the paths of the processes is defined with respect to sets of the Borel-σ-algebra $\mathcal{B}$ on $\mathbb{R}$

$$\Pi(X, Y) = \inf\{\varepsilon > 0; |P(X_t \in A) - P(Y_t \in B)| < \varepsilon,$$
$$A, B \in \mathcal{B} : d(A, B) < \varepsilon\}.$$

On $C[0, 1]$, it is equivalent to the L^1 distance $\sup_{t\in[0,1]} E|X_t - Y_t|$.

Proposition A.3. *For every function f of $C_b(\mathbb{R})$ and let X and Y be processes with paths in $C_b(\mathbb{R})$. For every $\varepsilon > 0$, there exists $\eta > 0$ such that $\Pi(X, Y) < \eta$ implies $\sup_{t\in[0,1]} E|f(X_t) - f(Y_t)| < \varepsilon$ and $\Pi(f(X), f(Y)) < \varepsilon$.*
Let f be a Lipschitz function of $C_b(\mathbb{R})$, then the Prohorov distances $\Pi(X, Y)$ and $\Pi(f(X), f(Y))$ are equivalent.

A.4 Expansions in $L^2(\mathbb{R})$

Fourier transform

Lemma A.3. *In the expansion* $f(x) = \frac{a_0}{2} + \sum_{k=1}^{\infty} \{a_k \cos(kx) + b_k \sin(kx)\}$, *the coefficients are defined as*

$$a_k = \frac{1}{\pi} \int_{-\pi}^{\pi} f(x) \cos(kx) \, dx,$$

$$b_k = \frac{2}{\pi} \int_{-\pi}^{\pi} f(x) \sin(kx) \, dx,$$

$$a_0 = \frac{1}{2\pi} \int_{-\pi}^{\pi} f(x) \, dx.$$

Proof. The integral $I(f) = \frac{1}{\pi} \int_{-\pi}^{\pi} f(x) \cos(nx) \, dx$, $n \neq 0$ integer, develops as a sum of integrals with coefficients a_k and b_k

$$I_{1n}(f) = \sum_{k=1}^{\infty} \frac{a_k}{2\pi} \int_{-\pi}^{\pi} \cos(kx) \cos(nx) \, dx$$

$$= \sum_{k=1, k \neq n}^{\infty} \frac{a_k}{2\pi} \int_{-\pi}^{\pi} \{\cos(k+n)x + \cos(k-n)x\} \, dx$$

$$+ \frac{a_n}{2\pi} \int_{-\pi}^{\pi} \{1 + \cos(2nx)\} \, dx = a_n,$$

where the other terms of the sum are zero by periodicity of the sine function. Similarly

$$I_{2n}(f) = \sum_{k=1}^{\infty} \frac{b_k}{2\pi} \int_{-\pi}^{\pi} \sin(kx) \sin(nx) \, dx$$

$$= \sum_{k=1, k \neq n}^{\infty} \frac{b_k}{2\pi} \int_{-\pi}^{\pi} \{\cos(k+n)x - \cos(k-n)x\} \, dx$$

$$+ \frac{b_n}{2\pi} \int_{-\pi}^{\pi} \{\cos(2nx) - 1\} \, dx = b_n,$$

the expression of a_0 is a consequence of the periodicity of the trigonometric functions and

$$I_{3n}(f) = \sum_{k=1}^{\infty} \int_{-\pi}^{\pi} \cos(kx) \sin(nx) \, dx$$

$$= \sum_{k=1, k \neq n}^{\infty} \int_{-\pi}^{\pi} \{\sin(k+n)x + \sin(k-n)x\} \, dx + \int_{-\pi}^{\pi} \sin(2nx) \, dx = 0.$$

$\square$

The coefficients of the Fourier transform and the trigonometric series are decreasing as $o(k)$ when k tends to infinity.

Proposition A.4. *The set of functions e_n defined by*

$$e_n(t) = \exp\left\{\frac{2\pi i}{T}nt\right\}, \ t \in [0, T]$$

is an orthogonal basis of $L^2[0, T]$ and every periodic function of $L^2[0, T]$ is the uniform limit of a series of trigonometric polynomials.

Giné (1974) proved a Lipschitz property for the Fourier transform.

Proposition A.5. *Let $f(t) = \sum_{n=-\infty}^{\infty} a_n e^{2\pi i n t}$, t in $[0, 1]$, with coefficients such that $\sum_{n \neq 0} |a_n| |\log n|^\alpha$ is finite, then $|f(s) - f(t)| \leq C_\varepsilon |\log|s - t||^{-\alpha}$ on $[0, 1]$, with the constant*

$$C_\varepsilon = 2\pi \sum_{|n| \leq e^\alpha} \left\{ \left(\frac{n}{\varepsilon}\right)^\alpha |n||a_n| + |a_n|(\log|n|)^\alpha \right\}.$$

Hermite polynomials

Let f_N be the normal density and let $H_0 = 1$ and H_k, $k \geq 1$, be the Hermite polynomials defined by

$$H_k(t) = (-1)^k \frac{d^k e^{-\frac{t^2}{2}}}{dt^k} e^{\frac{t^2}{2}}, \ k \geq 1.$$

The recursive equation $H_{k+1} = xH_k - H_k'$, for $k \geq 2$, provides an algorithm to calculate their expression. The polynomials H_k are normalized by their norm c_k as $h_k = c_k^{-1} H_k$ and we obtain

$$h_1(x) = x,$$
$$h_2(x) = x^2 - 1,$$
$$h_3(x) = x^3 - 3x,$$
$$h_4(x) = x^4 - 6x^2 + 3,$$
$$h_5(x) = x^5 - 10x^3 + 15x,$$
$$h_6(x) = x^6 - 15x^4 + 45x^2 - 15,$$
$$h_7(x) = x^7 - 21x^5 - 105x^3 - 105x, \ \text{etc.}$$

They have the form

$$h_{2k}(x) = \sum_{j=0}^{k-1} b_{2j} x^{2j} + x^{2k},$$

$$h_{2k+1}(x) = \sum_{j=0}^{k-1} b_{2j+1} x^{2j+1} + x^{2k+1}$$

and all coefficients are calculated iteratively: for every $k \geq 2$, the coefficient of x in $H_{2k-1}(x)$ is the even moment $m_{2k} = 3.5 \ldots (2k-1)$ of the normal distribution and this is the constant of $H_{2k}(x)$. The difference $\alpha_{k-2} - \alpha_{k-1}$ of the coefficients α_{k-2} of x^{k-2} in H_k and α_{k-1} of x^{k-1} in H_{k+1} is equal to k, etc. Their norms are calculated from the moments m_{2k} of the normal distribution

$$c_k = \left\{ \int_{\mathbb{R}} H_k^2(x) f_N(x) \, dx \right\}^{\frac{1}{2}} = (k!)^{\frac{1}{2}}.$$

Since $m_{2k} = \{2^k (k)!\}^{-1}(2k)!$, the square of the constant terms of the normalized polynomial $h_{2k}(x)$ is $C_{2k}^k 2^{-2k}$ and it tends to zero as a power of $\frac{1}{2}$ as k increases and their sum converge.

The functions H_k satisfy

$$\sum_{k \geq 0} H_k(t) \frac{w^k}{k!} = e^{-\frac{(t+w)^2}{2}} e^{\frac{t^2}{2}} = e^{-\frac{w^2 + 2wt}{2}}.$$

From the derivatives of

$$e^{-\frac{t^2}{2}} = \frac{1}{\sqrt{\pi}} \int_{\mathbb{R}} e^{-\frac{x^2}{2} + ixt} \, dx,$$

the Hermite polynomials are also expressed as

$$H_k(t) = \frac{1}{\sqrt{\pi}} \int_{\mathbb{R}} (-ix)^k e^{-\frac{(x-it)^2}{2}} \, dx.$$

The polynomials are sometimes defined by similar expressions where $\exp\{-\frac{x^2}{2}\}$ of the normal density is replaced by $\exp\{-x^2\}$, the recursive equation becomes $H_{k+1} = 2xH_k - H_k'$ but these equations generate non orthogonal polynomials. For instance, $H_1(x) = 2x$, $H_2(x) = 2(2x^2 - 1)$, $H_3(x) = x^3 - \frac{3}{2}x$ and $E\{H_1(X)H_3(X)\} = 2$.

Bibliography

Adler, R. J. and Taylor, J. E. (2007). *Random and geometry* (Springer, Berlin).

Alexander, K. (1987). The central limit theorem for empirical processes on Vapnik-Chervonenkis classes, *Ann. Probab.* **15**, pp. 178–203.

Alexander, K. and Pyke, R. (1986). A uniform central limit theorem for set-indexed partial-sum processes with finite variances, *Ann. Probab.* **14**, pp. 582–598.

Alzer, H. (1990a). Inequalities for the arithmetic, geometric and harmonic means, *Bull. London Math. Soc.* **22**, pp. 362–366.

Alzer, H. (1990b). An inequality for the exponential function, *Arch. Math.* **55**, pp. 462–464.

Arcones, M. and Giné, E. (1991). Limit theorems for U-processes, *Ann. Proba.* **21**, pp. 1494–1542.

Assouad, P. (1981). Sur les classes de Vapnik-Chervonenkis, *C. R. Acad. Sci. Paris, I* **292**, pp. 921–924.

Barlow, M. T. and Yor, M. (1981). (Semi-)Martingale inequalities and local times, *Z. Wahrsch. Verw. Gebiete* **55**, pp. 237–254.

Barlow, M. T. and Yor, M. (1982). Semi-martingale inequalities via the Garsia-Rodemich-Rumsey lemma and applications to local times, *J. Funct. Anal.* **49**, pp. 198–229.

Barnett, N. S. and Dragomir, S. S. (2001). A perturbed trapezoid inequality in terms of the third derivative and applications, *Ineq. Theor. Appl.* **5**, pp. 1–11.

Barnett, N. S. and Dragomir, S. S. (2002). A perturbed trapezoid inequality in terms of the fourth derivative, *J. Appl. Math. Comput.* **9**, pp. 45–60.

Beckner, W. (1975). Inequalities in Fourier analysis, *Ann. Math. USA* **102**, pp. 159–182.

Bennet, G. (1962). Probability inequalities for sums of independent random variables, *Amer. Statist. Assoc.* **57**, pp. 33–45.

Berkes, I. and Philipp, W. (1979). Approximation theorems for independent and weakly dependent random vectors, *Ann. Probab.*, pp. 29–54.

Bickel, P. J., Klassen, C. A., Ritov, Y. and Wellner, J. A. (1993). *Efficient and adaptive estimation in semiparametric models* (Johns Hopkins University Press, Baltimore).

Blake, I. F. and Lindsey, W. C. (1973). Level crossing problems for random processes, *IEEE Trans. Inf. Theor.* **19**, pp. 295–315.

Boylan, E. S. (1971). Equi-convergence of martingales, *Ann. Math. Statist.* **42**, pp. 552–559.

Bradley, R. E., d'Antonio, L. A. and Sandifer, C. E. (2007). *Euler at 300: an appreciation* (Math. Ass. Amer., Washington).

Breiman, L. (1968). *Probability* (Addison-Wesley, Reading, Massachusetts).

Burkholder, D. L. (1973). Distribution function inequalities for martingales, *Ann. Prob.* **1**, pp. 19–42.

Burkholder, D. L., Davis, B. J. and Gundy, R. F. (1972). Convex functions of operators on martingales, *Proceedings of the sixth Berkeley Symposium on Mathematical Statistics and Probability, Vol. 2–3*, pp. 789–806.

Byerly, E. B. (1893). *An elementary treatise on Fourier series and spherical, cylindrical, and ellipsoidal harmonics, with applications to problems in mathematical physics* (Ginn and Company, Boston, New York, Chicago London).

Cairoli, R. and Walsh, J. B. (1975). Stochastic integrals in the plane, *Acta. Math.* **134**, pp. 111–183.

Carlson, B. C. (1966). Some inequalities for hypergeometric functions, *Proc. Amer. Math. Soc.* **17**, pp. 32–39.

Cauchy, A. L. (1821). *Cours d'Analyse de l'Ecole Royale Polytechnique, I. Analyse Algébrique* (Editions Jacques Gabay, Sceaux).

Cauchy, A. L. (1833). *Résumés analytiques* (Imprimerie royale, Turin).

Chow, Y. S. and Lai, T. L. (1973). Limit behavior of weighted sums of independent random variables, *Ann. Probab.* **5**, pp. 810–824.

Chow, Y. S., Robbins, H. and Teicher, H. (1965). Moments of randomly stopped sums, *Ann. Math. Statist.* **36**, pp. 789–799.

Chow, Y. S. and Teicher, H. (1966). On second moments of stopping rules, *Ann. Math. Statist.* **37**, pp. 388–392.

Christofides, T. S. and Serfling, R. (1990). Maximal inequalities for multidimensionally indexed submartingale arrays, *Ann. Probab.* **18**, pp. 630–641.

Cox, D. R. (1960). *Point Processes* (Chapman and Hall, Londres).

Csörgö, M., Kolmós, J., Major, P. and Tusnády, G. (1974). On the empirical process when parameters are estimated, *Transactions of the seventh Prague conference on informtion theorey, statistical decision functions, random processes. Academia, Prague* **B**, pp. 87–97.

Cunnigham, F. and Grossman, N. (1971). On Young's inequality, *Amer. Math. Month.* **78**, pp. 781–783.

Dehling, H. (1983). Limit theorems for sums of weakly dependant Banach space valued random variable, *Z. Wahrsch. verw. Geb.* **63**, pp. 393–432.

Dembo, A. and Zeitouni, O. (2009). *Large Deviation Techniques and Applications, 3rd ed. Stochastic Modelling and Applied Probability 38* (Springer, Berlin Heidelberg).

den Hollander, F. (2008). *Large Deviations* (Amer. Math. Soc., London).

Deuschel, J.-D. and Stroock, D. W. (1984). *Large Deviations* (Aademic Press, London).

Doob, J. L. (1953a). *Stochastic Process* (Wiley, New York).

Doob, J. L. (1953b). *Stochastic Processes* (Wiley, New York).

Doob, J. L. (1975). Stochastic process measurability conditions, *Ann. Instit. J. Fourier* **25**, pp. 163–176.

Dragomir, S. S. and Rassias, T. M. (2002). *Ostrowski Type Inequalities and Applications in Numerical Integration* (Kluwer Acaemic Publisher, Dordrecht).

Dragomir, S. S. and Sofo, A. (2000). An integral inequality for twice differentiable mappings and applications, *Tamkang J. Math.* **31**, pp. 257–266.

Dudley, R. M. (1973). Sample functions of the Gaussian process, *Ann. Prob.* **1**, pp. 66–103.

Dudley, R. M. (1974). Metric entropy and the central limit theorem in C(s), *Ann. Inst. Fourier* **24**, pp. 40–60.

Dudley, R. M. (1984). *A course on empirical processes. Ecole d'été de Probabilité de St Flour. Lecture Notes in Math. 1097* (Springer-Verlag, Berlin).

Dunham, W. (1999). *Euler. The Master of Us All* (Math. Ass. Amer., Washington).

Durbin, J. (1971). Boundary-crossing probabilities for the Brownian motion and Poisson processes and techniques for computing the power of the Kolmogorov-Smirnov test, *J. Appl. Prob.* **8**, pp. 431–453.

Durrett, R. (2010). *Probability: Theory and Examples (4th edition)* (Cambridge University Press, New York).

Erdös, P. and Kac, M. (1947). On the number of positive sums of independent random variables, *Bull. Amer. Math. Soc.* **53**, pp. 1011–1020.

Feller, W. (1966). *An Introduction to Probability Theory and its Applications, Vol. 2 (second ed.)* (Wiley, London).

Fernholz, L. T. (1983). *Von Mises calculus for statistical functionals. Lecture Notes in Statistics, 19* (Springer-Verlag, Berlin).

Fourier, J. (1822). *Théorie analytique de la chaleur* (Firmin Didot, Paris).

Fournier, J. J. F. (1977). Sharpness in Young's inequality for convolution, *Pacific. J. Math.* **72**, pp. 383–397.

Freedman, D. A. (1973). Another note on the Borel-Cantelli lemma and the strong law, with the Poisson approximation as a by-product, *Ann. Probab.* **1**, pp. 910–925.

Freedman, D. A. (1975). On tail probabilities for martingales, *Ann. Probab.* **3**, pp. 100–118.

Garsia, A. M. (1973). On a convex unction inequality for martingales, *Ann. Probab.* **1**, pp. 173–174.

Garsia, A. M., Rodemich, E. and Rumsey, H. (1970). A real variable lemma and the continuity of paths of some Gaussian processes, *Indiana Univ. Math. J.* **20**, pp. 565–578.

Geetor, R. K. and Sharpe, M. J. (1979). Excursions of Brownian motion and Bessel processes, *Z. Wahrsch. verw Geb.* **47**, pp. 83–106.

Gibbs, A. L. and Su, F. E. (2002). On choosing and bounding probability metrics, *Instit. Statist. Rev.* **70**, pp. 419–435.

Giné, E., Latala, R. and Zinn, J. (2000). Exponential and moment inequalities for U-statistics, *High dimensional inequalities* **2**, pp. 1–22.

Giné, E. and Zinn, J. (1984). Some limit theorems for empirical prcesses, *Ann. Probab.* **12**, pp. 929–989.

Hardy, G. H. and Littlewood, J. E. (1930). A maximal theorem with function-theoretic applications, *Acta Matematica* **72**, pp. 81–116.

Hardy, G. H., Littlewood, J. E. and Pólya, G. (1952). *Inequalities, 2nd ed.* (Cambridge University Press, Cambridge).

Hutton, C. (1811). *Mathematical tables* (Rivington, Wilkie *et al.*, London).

Itô, K. and McKean, H. P. (1996). *Diffusion processes and their sample paths, 2nd ed.* (Springer, Berlin-Heidelberg-New York).

Jacod, J. and Mémin, J. (1979a). Sur la convergence des semimartingales vers un processus à accroissemnts indépendents, *Sémin. Probab. Springer* **14**, pp. 227–248.

Jacod, J. and Mémin, J. (1979b). Un nouveau critère de compacité relative pour une suite de processus, *Publ. Sémin. Math. Inform. Rennes* **1**, pp. 1–27.

Kac, M. (1947). Random walks and the theory of Brownian motion, *Amer. Math. Month.* **54**, pp. 369–391.

Kaijser, S., Nikolova, L., Perrson, L.-E. and Wedestig, A. (2005). Hardy-type inequalities via convexity, *Math. Inequal. Appl.* **8**, pp. 403–417.

Kaijser, S., Perrson, L.-E. and Öberg, A. (2002). On Carleman and Knopp's inequalities, *J. Approx. Theory* **117**, pp. 140–151.

Karlin, S. and McGregor, J. (1965). Occupation time law for birth and death processes, *Trans. Amer. Math. Soc.* **88**, pp. 249–272.

Keilson, J. and Steutel, F. W. (1974). Mixtures of distributions, moment inequalities and measures of exponentiality and normality, *Ann. Probab.* **2**, pp. 112–130.

Kiefer, J. (1961). On large deviations of the empiric d.f. of vector chance variables and a law of the iterated logarithm, *Pacific J. Math.* **11**, pp. 649–660.

Kolmos, J., Major, P. and Tusnady, G. (1975). An approximation of partial sums of independent rv's, and the sample df. I, *Z. Wahrsch. verw. Geb.* **32**, pp. 111–131.

Kolmos, J., Major, P. and Tusnady, G. (1976). An approximation of partial sums of independent rv's, and the sample df. II, *Z. Wahrsch. verw. Geb.* **34**, pp. 33–58.

Krishnapur, M. (2003). *Probability Theory* (Lecture Notes, Berkely University).

Kufner, A. and Persson, L.-E. (2003). *Weighted inequalities of the Hardy type* (World Scientific Publishing, Singapore).

Lagrange, J.-L. (1826). *Traité de la résolution des équations numériques de tous les degrés* (Bachelier, Paris).

Lamperti, J. (1958). An occupation time theorem for a class of stochastic processes, *Trans. Amer. Math. Soc.* **88**, pp. 380–387.

Legendre, A. M. (1805). *Nouvelles méthodes pour la détermination des orbites des comètes* (Firmin Didot, Paris).

Lenglart, E. (1977). Relation de domination entre deux processus, *Ann. Inst. H. Poincaré* **13**, pp. 171–179.

Lenglart, E., Lépingle, D. and Pratelli, M. (1980). Présentation unifiée de certaines inégalités de la théorie des martingales, *Séminaire Probab. Strasbourg* **14**, pp. 26–48.

Lépingle, D. (1978a). Sur le comportement asymptotique des martingales locales, *Séminaire Probab. Strasbourg* **12**, pp. 148–161.

Lépingle, D. (1978b). Une inégalité de martingales, *Séminaire Probab. Strasbourg* **12**, pp. 134–137.

Lindvall, T. (1977). A probabilistic proof of Blackwell's renewal theorem, *Ann. Probab.* **5**, pp. 482–485.

Lohwater, A. (1982). *Introduction to Inequalities* (Unpublished).

Luor, D.-C. (2009). A variant of a general inequality of the Hardy-Knopp type, *J. Ineq. Pure Appl. Math.* **3**, pp. 1–9.

Massart, P. (1986). Rates of convergence in the central limit theorem for empirical processes, *Ann. Instit. Henri Poincaré* **22**, pp. 381–423.

Maurey, B. (2004). Inégalités de Brunn-Minkowski-Lusternik, et d'autres inégalités géométriques et fonctionnelles, *Séminaire Bourbaki* **928**, pp. 1–19.

Meyer, P.-A. (1969). Les inégalités de Bürkholder en théorie des martingales, d'après Gundy, *Sém. Probab. Strasbourg* **3**, pp. 163–174.

Meyer, P.-A. (1976). Un cours sur les intégrales stochastiques, *Sém. Probab. Strasbourg* **10**, pp. 245–400.

Mona, P. (1994). Remarques sur les inégalités de Bürkholder-Davis-Gundy, *Sém. Probab. Strasbourg* **28**, pp. 92–97.

Mulholland, H. P. (1950). On generalizations of Minkowski's inequality in the form of a triangle inequality, *Proc. London Math. Soc.* **51**, pp. 294–307.

Neveu, J. (1970). *Bases mathématiques du calcul des probabilités* (Masson, Paris).

Neveu, J. (1972). *Martingales à temps discret* (Masson, Paris).

Orey, S. (1973). Conditions for the absolute continuity of two diffusions, *Trans. Am. Math. Soc.* **193**, pp. 413–426.

Pachpatte, B. G. (2005). *Mathematical Inequalities* (North-Holland, Elsevier).

Pechtl, A. (1999). Distributions of occupation times of Brownian motion with drift, *J. Appl. Math. Decis. Sci.* **3**, pp. 41–62.

Pitt, L. D. (1977). A Gaussian correlation inequality for symmetric convex sets, *Ann. Probab.* **5**, pp. 470–474.

Pollard, D. (1981). A central limit theorem for empirical processes, *J. Aust. Math. Soc. A* **33**, pp. 235–248.

Pollard, D. (1982). A central limit theorem for *k*-means clustering, *Ann. Probab.* **10**, pp. 919–926.

Pollard, D. (1984). *Convergence of Stochastic Processes* (Springer, New York).

Pons, O. (1986). A test of independence between two censored survival times, *Scand. J. Statist.* **13**, pp. 173–185.

Pons, O. (2015). *Analysis and differential equations* (World Sci Publ. Co, Singapore).

Pons, O. and Turckheim, E. (1989). Méthode de von-Mises, Hadamard différentiabilité et bootstrap dans un modèle non-paramétrique sur un espace métrique, *C. R. Acad. Sc. Paris, Ser. I* **308**, pp. 369–372.

Pratelli, M. (1975). Deux inégalités concernant les opérateurs de Bürkholder sur les martingales, *Ann. Probab.* **3**, pp. 365–370.

Revuz, D. and Yor, M. (1986). *Continuous martingales and Brownian motion* (Springer, Berlin).

Robbins, H. and Siegmund, D. (1970). Boundary-crossing probabilities for the Wiener process and sample sums, *Ann. Math. Statist.* **41**, pp. 1410–1429.

Rogge, L. (1974). Uniform inequalities for conditional expectations, *Ann. Probab.* **2**, pp. 486–489.

Sahoo, P. K. and Riedel, T. (1998). *Mean value theorems and functional equations* (World Scientific Publishing, Singapore-London).

Sandifer, C. E. (2007). *The Early Mathematics of Leonhard Euler* (Math. Ass. Amer., Washington).

Schladitz, K. and Baddeley, A. J. (2000). Uniform inequalities for conditional expectations, *Scand. J. Statis.* **27**, pp. 657–671.

Serfling, R. J. (1968). Contributions to central limit theory for dependent variables, *Ann. Math. Statist.* **39**, pp. 1158–1175.

Shorack, G. R. and Wellner, J. A. (1986). *Empirical processes and applications to statistics* (Wiley, New York).

Sjölin, P. (1995). A remark on the Hausdorff-Young inequality, *Proc. Am. Math. Soc.* **123**, pp. 3085–3088.

Slepian, D. (1962). The one-sided barrier problem for Gaussian noise, *Bell System Techn. J.* **1**, pp. 463–501.

Stout, W. F. (1970). The Hartman-Wintner law of the iterated logarithm for martingales, *Ann. Math. Statist.* **41**, pp. 2158–2160.

Strassen, V. (1967). Almost sure behaviour of sums of independent random variables and martingales, *Proceed. 5th Berkeley Symp. Mathematical Statistics and Probability, Vol. 2*, pp. 315–343.

Talagrand, M. (1987). Donsker classes and random geometry, *Ann. Probab.* **15**, pp. 1327–1338.

Talagrand, M. (1994). Sharper bounds for Gaussian and empirical processes, *Ann. Probab.* **22**, pp. 28–76.

Talagrand, M. (1995). Concentration of measures and isoperimetric inequalities in product spaces, *Pub. Math. IHES* **81**, pp. 73–205.

Thorin, O. (1970). Some comments on the Sparre Andersen model in the risk theory, *Scand. Actuar. J.* **1**, pp. 29–50.

Van der Vaart, A. and Wellner, J. A. (1995). *Weak Convergence and Empirical Processes* (Springer, New York).

Vapnik, V. and Cervonenkis, A. (1981). Necessary and sufficient conditions for the uniform weak convergence of means to their expectations, *Theor. Probab. Appl.* **26**, pp. 532–553.

Varadhan, S. R. S. (1984). *Large Deviations and Applications* (SIAM, Philadelphia).

Walsh, J. B. (1974). Stochastic integrals in the plane, *Proc. Intern. Congress Math.* **Sect. 11**, pp. 189–194.

Weisz, F. (1994). One-parameter martingale inequalities, *Ann. Univ. Sci. Budapest, Sect. Comp.* **14**, pp. 249–278.

Wong, E. and Zakai, M. (1974). Martingales and stochastic integrals for processes with multidimensional parameter, *Z. Wahrsch. verw. Geb.* **29**, pp. 109–122.

Index

Printed in the United States
by Baker & Taylor Publisher Services